Communications and Control Engineering

Springer
London
Berlin
Heidelberg
New York
Barcelona
Hong Kong
Milan
Paris
Santa Clara
Singapore
Tokyo

Zheng-Hua Luo, Bao-Zhu Guo
and Omer Morgul

Stability and Stabilization of Infinite Dimensional Systems with Applications

Springer

Zheng-Hua Luo
Department of Mechanical Engineering, Nagaoka University of Technology,
Kamitomioka-cho 1603-1, Nagaoka, Niigata 940-2188, Japan

Bao-Zhu Guo
Department of Applied Mathematics, Beijing Institute of Technology, Beijing 100081,
China

Omer Morgul
Department of Electrical and Electronics Engineering, Bilkent University, 06533
Bilkent, Ankara, Turkey

Series Editors

B.W. Dickinson • A. Fettweis • J.L. Massey • J.W. Modestino
E.D. Sontag • M. Thoma

ISBN 1-85233-124-0 Springer-Verlag London Berlin Heidelberg

British Library Cataloguing in Publication Data
Stability and stabilization of infinite dimensional systems
   with applications. - (communication and control engineering
   series)
   1.Engineering mathematics 2.Differentiable dynamical
   systems 3.Control theory
   I.Luo, Zheng-Hua II.Guo, Bao-Zhu III.Morgul, Omer
   629.8'36
ISBN 1852331240

Library of Congress Cataloging-in-Publication Data
A catalog record for this book is available from the Library of Congress

Typesetting: Camera ready by authors
Printed and bound at the Athenæum Press Ltd, Gateshead, Tyne & Wear
69/3830-543210 Printed on acid-free paper

This book is dedicated to our parents

Preface

The time evolution of many physical phenomena in nature can be described by partial differential equations. To analyze and control the dynamic behavior of such systems, infinite dimensional system theory was developed and has been refined over the past several decades. In recent years, stimulated by the applications arising from space exploration, automated manufacturing, and other areas of technological advancement, major progress has been made in both theory and control technology associated with infinite dimensional systems. For example, new conditions in the time domain and frequency domain have been derived which guarantee that a C_0-semigroup is exponentially stable; new feedback control laws have been proposed to exponentially stabilize beam, wave, and thermoelastic equations; and new methods have been developed which allow us to show that the spectrum-determined growth condition holds for a wide class of systems. Therefore, there is a need for a reference book which presents these results in an integrated fashion.

Complementing the existing books, e.g., [1], [41], and [128], this book reports some recent achievements in stability and feedback stabilization of infinite dimensional systems. In particular, emphasis will be placed on the second order partial differential equations, such as Euler-Bernoulli beam equations, which arise from control of numerous mechanical systems such as flexible robot arms and large space structures. We will be focusing on new results, most of which are our own recently obtained research results, until now scattered over a long list of journal articles, conference proceedings, and private communications. Specifically, the book contains a number of new features listed below:

- the integrated semigroup theory in Chapter 2,

- the theorems and characterizations on weak and strong stabilities of C_0-semigroups, and the novel characterization of the growth rate of a C_0-semigroup in Chapter 3,

- the A-dependent operator concept which has proven to be a powerful tool for arguing the well-posedness of some non-standard abstract second order equations in Chapter 4,

- the application of the energy multiplier method to the proof of the closed-loop stability of beam equations with dynamic boundary control in Chapter 5, and

- the exponential stability analysis for a wide class of systems, including wave and thermoelastic systems with boundary stabilizers, based on the verification of the spectrum-determined growth condition in Chapter 6.

So as to keep our book as self-contained as possible, we have tried to include all of the results in semigroup theory which are needed to solve our problems. Most lemmas and theorems, except some well-known theorems whose proofs can easily be found in other books, are given detailed proofs. As a result, readers who are familiar with some basic theorems of functional analysis will have no difficulty in following the book. The book is thus adequate as a textbook for graduate students in applied mathematics or as a reference book for control engineers and applied mathematicians interested in the analysis and control of infinite dimensional systems. It is our hope that the reader will learn not only some new theorems on semigroups and their stabilities, but also some useful techniques for solving practical engineering problems.

This book is a product of international cooperation. In fact, the three authors live in three different countries and it is the Internet that has made it possible for us to communicate with each other promptly and conveniently. We also benefited a great deal from LaTeX, the widely used word processor developed by Professors Donald Knuth and Leslie Lamport.

The authors are indebted to the following Professors who not only guided the authors into the scientific world of infinite dimensional systems, but also offered, and is offering, them supports and various suggestions on research and daily life: Yoshiyuki Sakawa, Kinki University; Toshihiro Kobayashi, Kyushu Institute of Technology; W. L. Chan; and Charles A. Desoer, University of California, Berkeley. Special thanks go to Professor Eduardo D. Sontag, Rutgers University, for his time and efforts towards the publication of this work, and to Ms. Darla Stimer, Ohio State University, for copyediting the manuscript. The authors would like to take this opportunity to thank Prof. H.T.Banks, Prof. Steve Yurkovich, Prof. De-Xing Feng and Ms. Nagano Robin for various helps during the preparation of this work.

Finally, we would like to thank our parents for their support and our wives for their infinite endurance and for sharing the heaviest burden of managing housework and taking care of the children.

Zheng-Hua Luo
Bao-Zhu Guo
Ömer Morgül

September 1998

Contents

Notation and Symbols

$\mathbf{R}$	field of real numbers
$\mathbf{C}$	field of complex numbers
$\mathbf{R}^+$	nonnegative real numbers
$\mathbf{N}$	set of positive integers
$\overline{\lambda}$	complex conjugate of $\lambda \in \mathbf{C}$
$|\lambda|$	absolute value of $\lambda \in \mathbf{C}$
$\mathrm{Re}\lambda$	real part of $\lambda \in \mathbf{C}$
$\mathrm{Im}\lambda$	imaginary part of $\lambda \in \mathbf{C}$
A	(bounded or unbounded) linear operator
$D(A)$	domain of A
$\mathcal{R}(A)$	range of A
$\mathcal{N}(A)$	kernel space of A
$R(\lambda, A)$	$(= (\lambda I - A)^{-1})$ resolvent operator of A
$R^{(k)}(\lambda, A)$	k-th derivative of $R(\lambda, A)$ with respect to λ
$R(\lambda, A)^k$	k multiple of $R(\lambda, A)$
$\sigma(A)$	spectral set of operator A
$\sigma_p(A)$	point spectral set (=eigenvalues) of operator A
$\sigma_c(A)$	continuous spectral set of operator A
$\sigma_r(A)$	residual spectral set of operator A
$\rho(A)$	resolvent set of operator A
$\omega_0(A)$	growth rate of the C_0−semigroup generated by A
$S(A)$	$= \sup\{\mathrm{Re}\lambda | \lambda \in \sigma(A)\}$, the spectral bound of A
$\omega_{ess}(A)$	essential growth rate of the C_0−semigroup generated by A
C_0-	strongly continous
$T(t)$	C_0−group or semigroup
$S(t)$	integrated semigroup
$r(T(t))$	spectral radius of $T(t)$
$\mathcal{L}(X)$	the set of all bounded operators on X
$L^p(0,1), p \geq 1$	Lebesgue integrable space
$L^p(\mathbf{R}; X)$	X-valued Lebesgue integrable space over $\mathbf{R}$
$H^p(0,1)$	Sobolev space of order p
$\in$	belong to
$\subset$	subset
$\cup$	union
$\cap$	intersection

Chapter 1

Introduction

1.1 Overview and examples of infinite dimensional systems

The study of stability originates in mechanics. Early in the 17th century, a principle, called *Torricelli's principle*[154], was already in use; it says that if a system of interconnected heavy bodies is in equilibrium, the center of gravity is at the lowest point. This principle was applied to the study of general motion including, but not limited to, mechanical motion. In fact, any time process in nature can be thought of as motion, and to study stability is actually to study the effect of perturbations to motion.

The fundamental theory of stability was established by the Russian scientist Aleksandr Mikhailovich Lyapunov who published what is now widely known as the *Lyapunov's direct method* for stability analysis in his celebrated memoir "The general problem of the stability of motion" in 1892 [104]. Since then, Lyapunov's direct method has greatly stimulated the research on stability of motion, and further developments have been made possible through the efforts of scientists all over the world during the past 100 years. Nowadays, Lyapunov's stability theory is an indispensable tool for the study of all systems, whether they are finite or infinite, linear or nonlinear, time-invariant or time varying, continuous or discrete. For this reason, it is widely used in system analysis and control for various systems, from electrical systems and mechanical systems to economic systems and solar systems, to name a few.

The purpose of this book is to present recent research results on *stability analysis* for infinite dimensional systems on Banach spaces and its applications to *feedback stabilization* of various control systems described by partial differential equations. Stability analysis is to discuss stability for a given autonomous differential equation (sometimes called a system), whereas feedback stabilization is to design the control input in the feedback form in order for a given system to achieve some desired objectives. We are especially interested in infinite dimensional systems because the stability problems of such systems are much more complicated than in finite dimensional systems, and are far from being completely solved. To demonstrate this, let us start by considering the following time-invariant linear system on the n-dimensional Euclidean space $\mathbf{R}^n$:

$$\left\{ \begin{array}{l} \dfrac{dz(t)}{dt} = Az(t), \\[2mm] z(t_0) = z_0 \end{array} \right. \tag{1.1}$$

where $z(t) \in \mathbf{R}^n$ is the state and A is an $n \times n$ constant matrix. Notice that we could have considered a general nonlinear system with $Az(t)$ on the right-hand side being replaced by a nonlinear function $f(z(t), t)$. However, for simplicity, we only look at the linear case because most of the time we will be considering infinite dimensional linear systems in this book. Clearly, (1.1) has a unique solution which is given by

$$z(t) = e^{At} z_0. \tag{1.2}$$

The solution is said to be *asymptotically stable* or *strongly stable* if $\lim_{t \to \infty} \|z(t)\| = 0$; and *exponentially stable* or *uniformly stable* if there exist positive constants M and ω such that $\|z(t)\| \leq Me^{-\omega t} \|z(0)\|$, where $\|z(t)\|$ denotes the norm of $z(t)$ as a vector in $\mathbf{R}^n$. It is well known that for finite dimensional linear systems such as (1.1), the following statements are equivalent:

(P1) the solution $z(t)$ is asymptotically stable,

(P2) the solution $z(t)$ is exponentially stable,

(P3) all eigenvalues of A locate on the open left-half complex plane.

(P4) for any positive integer p with $p \geq 1$, it holds that $\int_{t_0}^{\infty} \|z(t)\|^p \, dt < \infty$ for any $t_0 \geq 0$.

Furthermore, if we define $T(t) = e^{At}$, the transition matrix, it is easy to see that (P2) also is equivalent to

(P5) $T(t)$ is exponentially stable, i.e., there exist positive constants M and ω such that $\|T(t)\| \leq Me^{-\omega t}$, where $\|T(t)\|$ denotes the operator norm of $T(t) : \mathbf{R}^n \to \mathbf{R}^n$.

$T(t)$ is actually a strongly continuous semigroup of linear operators on $\mathbf{R}^n$. Details are given in Chapter 2.

For infinite dimensional linear systems, however, the equivalence of (P1)-(P5) does not hold in general. To illustrate this, let us look at some typical infinite dimensional linear systems. There are, of course, other types of partial differential equations in practical applications. Even for a specific type of equation, the closed-loop behavior is usually totally different if a different control is applied.

To us, perhaps the most familiar partial differential equation is that known as the heat equation.

• **Heat Equation:**

Suppose we have a thin, narrow, homogeneous rod with unit length. Attach the x-axis to the rod along the longitudinal direction. If $\theta(x,t)$ denotes the temperature of the rod at position x and time t, then a simple model for the temperature distribution is given by

$$
\begin{cases}
\dfrac{\partial \theta(x,t)}{\partial t} = \dfrac{\partial^2 \theta(x,t)}{\partial x^2} + u(x,t), & 0 < x < 1, \\[2mm]
\alpha \theta(i,t) + (1-\alpha)\dfrac{\partial \theta(i,t)}{\partial x} = u_i(t), & i = 0, 1, \\[2mm]
\theta(x,0) = \theta_0(x),
\end{cases}
\tag{1.3}
$$

where $u(x,t)$ and $u_i(t)$ represent the external heat sources and can be thought of as control inputs in the language of control engineering. Specifically, $u(x,t)$ is often called the distributed control input and $u_i(t)$ the boundary control inputs. The second equation of (1.3) describes general boundary conditions including, as special cases, the Dirichlet boundary conditions ($\alpha = 1$) if the temperature at the ends of the rod is controlled, and the Neumann boundary conditions ($\alpha = 0$) if the heat flow is controlled. We note that this equation can be generalized easily to higher dimensions by replacing the spatial derivative with the Laplacian.

We wish now to formulate (1.3) as the compact form, as in (1.1), on some appropriately chosen Hilbert or Banach spaces. For this purpose, let us assume, for simplicity, that all control inputs $u(x,t)$ and $u_i(t)$ are set to zero and $\alpha = 0$. Then (1.3) reduces to

$$
\begin{cases}
\dfrac{\partial \theta(x,t)}{\partial t} = \dfrac{\partial^2 \theta(x,t)}{\partial x^2}, & 0 < x < 1, \\[2mm]
\dfrac{\partial \theta(0,t)}{\partial x} = \dfrac{\partial \theta(1,t)}{\partial x} = 0, \\[2mm]
\theta(x,0) = \theta_0(x).
\end{cases}
\tag{1.4}
$$

Consider this equation on the usual square Lebesgue integrable function space $H = L^2(0,1)$, the most familiar function space. Interpret $\theta(x,t)$ as an H–valued function of time which is denoted by $w(t)$. Formally define the operator A by

$$
Au = \frac{d^2 u}{dx^2}, \quad \forall u \in D(A).
$$

where $D(A)$ denotes the domain of A which is clearly a subspace of H, because not all functions in H are differentiable and nor do they satisfy the associated boundary conditions. In fact,

$$D(A) = \{u(x) \mid u \in L^2(0,1), \frac{d^2u}{dx^2} \in L^2(0,1),$$

$$\frac{du(0)}{dx} = 0 = \frac{du(1)}{dx}\},$$

which is dense in H. With these definitions, (1.4) is now written as

$$\begin{cases} \dfrac{dw(t)}{dt} = Aw(t), \\ w(0) = \theta_0, \end{cases} \tag{1.5}$$

which resembles (1.1). The initial value problem (1.5) is often referred to as the *Cauchy problem*, and the operator A is sometimes called the *system operator*. The solutions of (1.5), however, cannot simply be expressed as $w(t) = \exp(At)$ because A is an unbounded operator in H, hence, $\exp(At)$ does not make sense if it is interpreted as the usual Taylor expansion. Nevertheless, by using the well known "separation of variables" approach, it is easy to show that the solution of (1.5) can be expressed as

$$w(t) = T(t)\theta_0, \tag{1.6}$$

where $T(t)$ is defined by

$$T(t)u(x) = \int_0^1 \left(1 + \sum_{n=1}^{\infty} 2e^{-n^2\pi^2 t} \cos(n\pi x)\cos(n\pi y)\right)u(y)dy, \tag{1.7}$$

and is a bounded operator for $t \geq 0$. Actually $T(t)$ is a C_0-semigroup generated by A. Therefore, (1.6) defines the unique solution of (1.4). It can further be proved that $T(t)$ in (1.7) is analytic in a sector $\sum_\beta = \{z \mid z \neq 0, |\arg(z)| \leq \pi/2 + \beta, \ 0 < \beta < \pi/2\}$ on the complex plane. Because equation (1.5) describes the heat diffusion on the rod under the heat insulation condition, as time passes, the temperature eventually will approach a steady state which is often called the *equilibrium* of system (1.5). Mathematically, the equilibrium is characterized by the eigenvector of A corresponding to its zero eigenvalue. In this special case, the stability of the solution $w(t)$ is determined by the nonzero spectrum of A. Since all the nonzero spectrum of A locate strictly on the left-half complex plane, the solution $w(t)$, or equation (1.5) itself, approaches exponentially to the equilibrium state.

- **Population Equation:**

Let $p(a,t)$ denote the population density at age a and time t. Let $\mu(a)$ be the time independent mortality at age a, $k(a)$ and $h(a)$ the sex ratio and fertility pattern of the female at age a located in the fecundity interval $[a_1, a_2]$, respectively, and a_m the life limit. Let $\beta(t)$ be the total fertility rate (TFR). Then the age-dependent population dynamics can be described by the following partial differential equation

$$
\begin{cases}
\dfrac{\partial p(a,t)}{\partial t} + \dfrac{\partial p(a,t)}{\partial a} = -\mu(a)p(a,t), \quad a \in (0, a_m), \\[2mm]
p(a,0) = p_0(a), \\[2mm]
p(0,t) = \beta(t)\int_{a_1}^{a_2} k(a)h(a)p(a,t)da.
\end{cases}
\tag{1.8}
$$

Many biological processes and birth-death processes also can be modelled using an equation similar to (1.8). Let us now consider (1.8) in the natural state space (in which the norm of the state is the total population) $X = L^1(0, x_m)$ in the case of constant TFR (i.e., $\beta(t) = \beta$) and define

$$
Au(x) = -\frac{du(x)}{dx} - \mu(x)u(x),
$$

$$
D(A) = \{u(x)|\ u, Au \in X,\ u(0) = \beta \int_{a_1}^{a_2} k(a)h(a)u(a)da\}.
$$

Then, (1.8) can be formulated again as the form of (1.5). It is known that if $\beta = \beta_{cr}$, the critical TFR, then there is a unique positive equilibrium state to (1.8) which represents the ideal age structure of population, and mathematically the equilibrium state is characterized by the eigenvector of the operator A corresponding to zero eigenvalue. When $\beta < \beta_{cr}$, there is no nonzero equilibrium for system (1.8). The study of the stability of population dynamics is the study of the behavior of the solutions of (1.8) near its equilibrium. For this problem, it can be shown that the corresponding C_0-semigroup $T(t)$ is compact for $t \geq a_m$; hence, the stability of the system also is determined by the spectrum of the system operator A (see Theorem 3.9).

- **String Equation:**

The one-dimensional equation

$$
\begin{cases}
\rho\dfrac{\partial^2 y(x,t)}{\partial t^2} - T\dfrac{\partial^2 y(x,t)}{\partial x^2} = 0, \quad 0 < x < 1, \\[2mm]
y(0,t) = 0, \\[2mm]
Ty_x(1,t) = -ky_t(1,t), \\[2mm]
y(x,0) = y_0(x), \quad y_t(x,0) = y_1(x)
\end{cases}
\tag{1.9}
$$

governs the vibration of a boundary controlled string for small displacement y from its equilibrium position and is usually called the *wave equation* in the literature. Here, we call it *string equation* to distinguish it from the beam equation to be stated below. The constants ρ and T denote, respectively, the mass of the string and tension in the string, $c = \sqrt{(T/\rho)}$ is usually called the *wave speed*, and $k > 0$ is the feedback gain. Equation (1.9) also describes the longitudinal vibration or torsional vibration of a flexible beam.

Let

$$
\begin{cases}
y_1(x, t) = y(x, t), \\
y_2(x, t) = y_t(x, t).
\end{cases}
$$

Then (1.9) becomes

$$
\begin{cases}
\dfrac{\partial}{\partial t} \begin{bmatrix} y_1 \\ y_2 \end{bmatrix} = \begin{bmatrix} 0 & 1 \\ c^2 \frac{\partial^2}{\partial x^2} & 0 \end{bmatrix} \begin{bmatrix} y_1 \\ y_2 \end{bmatrix}, \\[2mm]
y_1(0, t) = 0, \\
T y_{1x}(1, t) = -k y_2(1, t), \\
y_1(x, 0) = y_0(x), y_2(x, 0) = y_1(x).
\end{cases}
$$

The above equation can be written in the form of (1.5) in the state space (energy space) $X = H_e^2(0, 1) \times L^2(0, 1)$, $H_e^2(0, 1) = \{f \mid f \in H^2(0, 1), f(0) = 0\}$ with

$$
\begin{cases}
A \begin{bmatrix} \phi \\ \psi \end{bmatrix} = \begin{bmatrix} 0 & 1 \\ c^2 \frac{\partial^2}{\partial x^2} & 0 \end{bmatrix} \begin{bmatrix} \phi \\ \psi \end{bmatrix}, \\[2mm]
D(A) = \{(\phi, \psi) \in H^2(0, 1) \times H^1(0, 1) \mid \phi(0) = \psi(0) = 0, \\
\qquad\qquad T\phi'(1) = -k\psi(1)\}.
\end{cases}
$$

It can be shown that equation (1.9) is exponentially stable (the unique equilibrium is 0). The stability can also be determined by the spectrum of the system operator A. However, the same feedback cannot exponentially stabilize a string with a tip mass m. That is, the following equation

$$
\begin{cases}
\rho \dfrac{\partial^2 y(x, t)}{\partial t^2} - T \dfrac{\partial^2 y(x, t)}{\partial x^2} = 0, \quad 0 < x < 1, \\[2mm]
y(0, t) = 0, \\
T y_x(1, t) + m y_{tt}(1, t) = -k y_t(1, t), \\
y(x, 0) = y_0(x), \quad y_t(x, 0) = y_1(x)
\end{cases}
\tag{1.10}
$$

is at most asymptotically stable, but not exponentially stable. This reveals that (P1) is no longer equivalent to (P2) in infinite dimensional systems.

- **Rotating Beam Equation:**

By "beam equation", we mean the following partial differential equation

$$\begin{cases} \dfrac{\partial^2 y(x,t)}{\partial t^2} + \dfrac{\partial^4 y(x,t)}{\partial x^4} = -xu(t), \quad 0 < x < 1, \\ y(0,t) = y_x(0,t) = 0, \\ y_{xx}(1,t) = y_{xxx}(1,t) = 0, \\ y(x,0) = y_0(x), \ y_t(x,0) = y_1(x), \end{cases} \tag{1.11}$$

which is a Euler-Bernoulli dynamic model for the vibration of a rotating beam with free tip end. When there is no control (i.e., $u(t) = 0$ in (1.11)), it is well known that (1.11) is a conservative system, which means that the energy of vibration remains constant as time passes, and hence the solution is never asymptotically stable. Consider now two kinds of feedback laws: (i) $u(t) = ky_t(\cdot,t)$ and (ii) $u(t) = ky_{xx}(0,t)$. For each of these cases, the closed-loop equation can be written as the form (1.5). But as we shall show in Chapter 4, the closed-loop equation corresponding to case (i) is not exponentially stable, whereas the feedback in case (ii) does exponentially stabilize the rotating beam equation. Notice that even in case (ii) the exponential stability does not simply follow from the fact that the spectra of the system operator associated with the closed-loop equation of (1.11) all lie on the open left-half complex plane; the proof of the spectrum-determined growth condition for the latter case was examined recently by using Riesz basis method (see [36]). However, (P3) usually does not imply (P2) in infinite dimensional systems. Some counter-examples are provided in Chapters 3 and 6.

We have seen four types of partial differential equations. A common feature to these equations is that their solutions are strongly related to C_0−semigroups. If there exists a C_0−semigroup generated by the system operator of a specific Cauchy problem, then the system is well posed. Conversely, if we already know that the system is well posed, then we can define a semigroup which maps the initial value to the solution (these will be further explained in Chapter 2). Therefore, the well-posedness and the stability of solutions to abstract Cauchy problems are completely characterized by the semigroups generated by their system operators. For this reason, the study of semigroups is crucial to the understanding of properties of solutions of partial differential equations.

1.2   Organization and brief summary

The rest of this book is arranged as follows. In Chapter 2, basic properties of C_0−semigroups are introduced to explore the well-posedness of abstract Cauchy problems in general Banach spaces. Some special semigroups, such as differentiable semigroups, analytic semigroups, compact semigroups, integrated semigroups, and nonlinear semigroups, are intensively studied. Suppose we are given a densely defined closed linear operator A in a Banach space. The first question that concerns us is under what conditions A generates a C_0−semigroup $T(t)$. This is answered completely in the Hille-Yosida theorem which characterizes the semigroup generation in terms of the resolvent of A. The Hille-Yosida theorem, although theoretically attractive, is hard to use in applications because it requires the estimation of all powers of resolvent, which is usually impossible. For generation of special semigroups, such as contraction semigroups , we need only to verify whether A is m−dissipative, which is given in the Lümer-Phillips theorem (see Theorem 2.27). To test whether a contraction semigroup can be extended to an analytic semigroup, we can use Theorem 2.50 which says $T(t)$ is analytic in a sector $\mathcal{S}_\phi = \{z \in \mathbf{C} \mid |\arg z| < \phi, z \neq 0\}$ if and only if $e^{i\theta}A$ generates a C_0−contraction semigroup for any $\theta \in (-\phi, \phi)$. Theorem 2.48 is also convenient for this purpose. Another special semigroup, compact semigroup, which we shall find useful in Chapter 6, is characterized in Theorem 2.56. A number of examples are contained in this chapter that show the semigroup generations of various operators.

Since the properties of these semigroups are now well documented in the literature, we shall focus on the study of a new kind of semigroup called integrated semigroups. This was proposed in recent years and has applications for solving practical problems, as will be illustrated in Chapter 4. Let A be a closed linear operator (not necessarily densely defined) in a Banach space X. If there exist an integer n, constants M and ω, and a strongly continuous operator family $S(t)$ with $\|S(t)\| \leq Me^{\omega t}$ such that

$$R(\lambda, A)x = \lambda^n \int_0^\infty e^{-\lambda t} S(t)x dt, \quad \forall x \in X, \ \mathrm{Re}\lambda > \omega,$$

then A is called the generator of an *n-times integrated semigroup* $S(t)$. Recall that if a densely defined closed operator A generates a C_0−semigroup $T(t)$, then

$$R(\lambda, A)x = \int_0^\infty e^{-\lambda t} T(t)x dt, \quad \forall x \in X, \ \mathrm{Re}\lambda > \omega.$$

Thus, n-times integrated semigroups are extensions of C_0−semigroups with 0-time integrated semigroup being equal to C_0−semigroup. The attraction of integrated semigroups is that once we know that A generates an n-times integrated semigroup $S(t)$, then the following Cauchy problem

$$\frac{dw(t)}{dt} = Aw(t), \quad w(0) = x$$

has a unique solution which is given by

$$w(t) = S(t)A^n x + \sum_{k=0}^{n-1} \frac{t^k}{k!} A^k x$$

for any $x \in D(A^{n+1})$. Conditions for an operator to generate an n-times integrated semigroup is also derived (see Theorems 2.82, 2.84 and Corollary 2.93).

Chapter 3 is devoted to the stability analysis of C_0-semigroups, which is the central theoretical part of this book. As we have seen, the stability of C_0-semigroups for infinite dimensional systems are very complicated. In addition to the concepts of asymptotic stability and exponential stability, the concept of weak stability can also be useful. The Asymptotic stability of a C_0-semigroup $T(t)$, generated by A, are completely characterized by the spectrum of A. From Theorem 3.26, we know that, for a uniformly bounded C_0-semigroup $T(t)$, if the spectrum of A on the imaginary axis belongs to $\sigma_c(A)$, the continuous spectrum, and the set $\sigma_c(A)$ is countable, then $T(t)$ is asymptotically stable. In particular, if the resolvent of A is compact, then $T(t)$ is asymptotically stable if and only if $\mathrm{Re}\lambda < 0$ for all $\lambda \in \sigma(A)$. For the exponential stability of C_0-semigroups $T(t)$, however, there are cases in which the stability of $T(t)$ is not completely determined by the spectrum of A, as we have mentioned. This is also illustrated by several examples in this chapter. To proceed further, we need the following definition. Define the *growth rate* of $T(t)$ as

$$\omega_0 = \inf_{t>0} \frac{\log \|T(t)\|}{t}.$$

Then there exists a constant $M > 0$ such that $\|T(t)\| \le M e^{(\omega_0+\varepsilon)t}$ for any $\varepsilon > 0$. On the other hand, let

$$S(A) = \sup\{\mathrm{Re}\lambda | \lambda \in \sigma(A)\}.$$

If $\omega_0 = S(A)$, we say that $T(t)$ satisfies the *spectrum-determined growth condition*. In this case, the semigroup $T(t)$ is exponentially stable as long as the spectra of A lie on the open left-half complex plane. We wish to identify a class of C_0-semigroups for which the spectrum-determined growth condition holds. This is made clear in Theorem 3.13 which says that a C_0-semigroup which is continuous in the uniform operator topology on a Banach space satisfies the spectrum-determined growth condition. Since compact semigroups and differentiable semigroups, in particular analytic semigroups, are continuous in the uniform operator topology, they all satisfy the spectrum-determined growth condition.

For feedback stabilization of engineering problems, it is preferable, if possible, to find controllers which exponentially stabilize the given system. This is because the exponential stability provides convergence with a guaranteed convergence rate. For this reason, we will be mainly concerned with exponential stability throughout this book. There are at least four kinds of methods to check the exponential stability of a C_0-semigroup; these are listed below.

1. **Time Domain Criteria:** For a C_0-semigroup $T(t)$ defined on a Ba-

nach space X, if for some $p \geq 1$,

$$\int_0^\infty \|T(t)x\|^p \, dt < \infty, \quad \forall x \in X,$$

then $T(t)$ is exponentially stable.

2. **Frequency Domain Criteria:** Let $T(t)$ be a uniformly bounded C_0-
   semigroup on a Hilbert space with generator A. Then $T(t)$ is exponentially stable if and only if the imaginary axis belongs to the resolvent
   set of A and

   $$\sup_{\tau \in \mathbf{R}} \|R(i\tau, A)\| < \infty.$$

3. **Spectral Analysis Method:** For semigroups which satisfy the
   spectrum-determined growth condition, if $S(A) < 0$, then the semigroup is exponentially stable.

4. **Energy Multiplier Method:** Suppose $E(t)$ is the energy of a system
   which is dissipative, i.e., $\dot{E}(t) \leq 0$ along the solution of the system. If
   we can find an appropriate function $\rho(t)$ such that

   $$|\rho(t)| \leq c_1 E(t) \quad \text{and} \quad \dot{\rho}(t) \leq -c_2 E(t), \quad c_1, c_2 > 0,$$

   then by using an auxiliary function $V(t) = E(t) + \rho(t)$, we can deduce
   easily that $E(t)$ decays exponentially from which we can infer that
   the solution of the system is also exponentially stable if there does
   exist a C_0-semigroup for the given system. Alternatively, the auxiliary
   function $V(t)$ can be chosen as $V(t) = tE(t) + \varepsilon \rho(t)$, where $\varepsilon > 0$ should
   be carefully selected such that $\dot{V}(t) \leq 0$ along the solution of the given
   system.

The time domain criteria can be proved by using the semigroup properties.
The frequency domain criteria appeared in Prüss [133] and Huang [78], but
we give a much simpler proof based on the Fourier analysis for Hilbert space
valued functions. The results of Fourier analysis also enable us to prove the
Paley-Wiener theorem which plays a key role in characterizing the growth
rate of C_0-semigroups on Hilbert spaces. The spectral analysis method applies only to those systems where the spectrum-determined growth condition
holds. The energy multiplier method is actually a combination of the Lyapunov method and the time domain criteria stated above. These methods
will be applied to various practical problems throughout the book.

In the rest of this chapter, we are interested in digging into why the exponential stability of a C_0-semigroup may not be determined by the spectra
of its generator A. It is found that it is those non-isolated spectra and the
isolated eigenvalues with infinite algebraic multiplicities of $T(t)$, in addition

to $S(A)$, that determine the growth rate ω_0 of $T(t)$. This part of the spectrum is called the *essential spectrum*. Denote by ω_{ess} the *essential growth rate* of A (see (3.37) for the precise definition of ω_{ess}). An interesting result is

$$\omega_0 = \max\{S(A), \omega_{ess}\},$$

which completely characterizes the growth rate of a C_0-semigroup in terms of the spectra of A and the essential growth rate. In connection with this, it is further shown that a compact perturbation to A does not affect the essential growth rate $\omega_{ess}(A)$ of A, which is very useful in system analysis and feedback design. A number of other results are also contained in this chapter which are useful in their own rights.

The subsequent chapters will focus on the application of the stability theorems demonstrated in the previous chapters to the stabilization of Euler-Bernoulli beam equations, thermoelastic equations, and wave equations, etc.

In Chapter 4, we consider Euler-Bernoulli beam equations which model the dynamics of vibrations of large space structures and flexible robot arms with or without damping. We are especially interested in the sensor output feedback control, such as strain or shear force feedback control of the Euler-Bernoulli beam equations. For a rotating beam, the strain feedback results in a closed-loop equation:

$$\begin{cases} \dfrac{\partial^2 y(x,t)}{\partial t^2} + kx\dfrac{\partial^3 y(0,t)}{\partial t \partial x^2} + \dfrac{\partial^4 y(x,t)}{\partial x^4} = 0, & 0 < x < 1, \\ y(0,t) = y_x(0,t) = 0, \\ y_{xx}(1,t) = y_{xxx}(1,t) = 0, \\ y(x,0) = y_0(x), \ \dot{y}(x,0) = y_1(x), \end{cases} \tag{1.12}$$

and for a translating beam, shear force feedback results in the following closed-loop equation:

$$\begin{cases} \dfrac{\partial^2 y(x,t)}{\partial t^2} - k\dfrac{\partial^4 y(0,t)}{\partial t \partial x^3} + \dfrac{\partial^4 y(x,t)}{\partial x^4} = 0, & 0 < x < 1, \\ y(0,t) = y_x(0,t) = 0, \\ y_{xx}(1,t) = y_{xxx}(1,t) = 0, \\ y(x,0) = y_0(x), \ \dot{y}(x,0) = y_1(x). \end{cases} \tag{1.13}$$

The well-posedness and stability of (1.12) and (1.13) are proved by using the concept of A-dependent operators, which are special unbounded operators (see Definition 4.3). With the aid of the A-dependent operators, we also are able to analyze the semigroup generation of the following "non-standard" partial differential equation:

$$\begin{cases} \dfrac{\partial^2 y(x,t)}{\partial t^2} + \dfrac{\partial^4 y(x,t)}{\partial x^4} - \dfrac{\partial^3 y(0,t)}{\partial x^3} = 0, \quad 0 < x < 1, \\[2mm] y(0,t) = y_x(0,t) = y_{xx}(1,t) = y_{xxx}(1,t) = 0, \\[2mm] y(x,0) = y_0(x), \quad \dot{y}(x,0) = y_1(x). \end{cases} \tag{1.14}$$

For the shear force feedback control of a rotating beam, we arrive at the following closed-loop equation:

$$\begin{cases} \dfrac{\partial^2 y(x,t)}{\partial t^2} - kx\dfrac{\partial^4 y(0,t)}{\partial t \partial x^3} + \dfrac{\partial^4 y(x,t)}{\partial x^4} = 0, \quad 0 < x < 1, \\[2mm] y(0,t) = y_x(0,t) = 0, \\[2mm] y_{xx}(1,t) = y_{xxx}(1,t) = 0, \\[2mm] y(x,0) = y_0(x), \quad \dot{y}(x,0) = y_1(x). \end{cases} \tag{1.15}$$

Notice that, although the second term of the first equation in (1.15) is only slightly different from that in (1.12) and (1.13), (1.15) requires a totally new method to argue its well-posedness and stability because the A-dependent operator method is no longer applicable, and it is also not easy to find a Lyapunov function for (1.15). It is the theory about integrated semigroups developed in Chapter 2 that enables us to successfully solve (1.15). It is shown that there exists a unique solution to (1.15) for sufficiently smooth initial conditions, and that the solution is exponentially stable.

Also included in this chapter are

- the spectral analysis for strain or shear force feedback controlled Euler-Bernoulli beam equations with or without damping,

- stability analysis for a sensor output feedback controlled hybrid system composed of a partial differential equation and an ordinary differential equation based on the compact perturbation theory developed in Chapter 3, and

- stability analysis of a nonlinear equation arising from gain adaptive strain feedback control of Euler-Bernoulli beams.

Chapter 5 covers topics on dynamic boundary control of passive vibration systems. We first build a general framework for system passivity which encompasses equations such as wave equation, Euler-Bernoulli beam equations and Timoshenko beam equations with adequate system outputs and boundary control inputs. For this general class of passive systems, we then design *positive real controllers* to achieve the closed-loop asymptotic or exponential stability. The concept of passivity originates from electric circuit systems and is now generalized to linear or nonlinear abstract control systems. In finite dimensional system theory, it is well known that the closed-loop system, consisting of a passive open-loop system and a strictly positive real (hence passive) controller in feedback connection, is asymptotically stable. The arguments in this chapter demonstrate that this result also holds for passive infinite dimensional systems.

The issue of closed-loop stability robustness, with respect to small time delay in dynamic feedback law, also is discussed. The main machinery used to prove the exponential stability of various feedback loops is the Energy Multiplier Method mentioned above.

Chapter 6 concentrates on stability analysis and feedback stabilization of string equations and thermoelastic equations which are two types of equations where the satisfaction of the spectrum-determined growth condition is rigorously proved. We first study the following general hyperbolic system

$$
\begin{cases}
\dfrac{\partial}{\partial t}\begin{bmatrix} u(x,t) \\ v(x,t) \end{bmatrix} + K(x)\dfrac{\partial}{\partial x}\begin{bmatrix} u(x,t) \\ v(x,t) \end{bmatrix} + C(x)\begin{bmatrix} u(x,t) \\ v(x,t) \end{bmatrix} = 0, 0 < x < 1, \\[2mm]
\dfrac{d}{dt}[v(1,t) - Du(1,t)] = Fu(1,t) + Gv(1,t), \\[2mm]
u(0,t) = Ev(0,t)
\end{cases}
\tag{1.16}
$$

which includes string equations as special cases. In order to exploit the solution properties that this general system possesses, we first consider three reduced systems, which are simplified equations, either by ignoring the couplings between state variables, or by replacing the dynamic boundary with a static boundary. For the most simple reduced system among the three, we show that the spectrum-determined growth condition holds. This, together with a number of results on the relationship among semigroups of each reduced system, demonstrates that the growth rate of the general system is determined completely by the spectral bound of itself and that of the most simple reduced system. Using this key result, we are able to show, in Sections 6.2-6.4, that the spectrum-determined growth condition is satisfied for wave equations describing dynamics of serially connected vibrating strings with point stabilizers as well as a vibrating cable with a tip mass. In Section 6.5 and 6.6, we consider a thermoelastic system with both Dirichlet-Dirichlet and Neumann-Dirichlet boundary conditions. It is shown that the spectrum-determined growth condition also holds for these systems with the aid of a result due to Renardy [136] on spectrum-determined growth condition for a class of normal operators with bounded perturbations in Hilbert spaces. The satisfaction of the spectrum-determined growth condition allows us to prove the exponential stability of these systems by utilizing the spectral analysis method. Finally, in Section 6.7, we present a counter-example given by Renardy [136] which shows that the spectrum alone cannot determine the growth rate for a two-dimensional hyperbolic system even with lower order perturbations.

1.3 Remarks on notation

We use the basic notations listed in "Notations and Symbols". Various notations are used for differentiation, namely $\frac{\partial u(x,t)}{\partial x}$, $u_x(x,t)$, and $u'(x,t)$, whichever is convenient, all denote the partial derivative of $u(x,t)$ with respect to x, as has already appeared in this chapter. Similarly, for a time function $u(t)$, we use $\frac{du(t)}{dt}$, $\dot{u}(t)$, or $u'(t)$ to denote its time derivative. Unless otherwise stated, X always denotes a Banach space and H always denotes a Hilbert space. For a bounded operator, we say that it is defined "on" a space; for an unbounded operator, we say that it is defined "in" a space. The notations $O(\cdot)$ and $o(\cdot)$ are defined as follows: for $t > 0$ and a real number p

$$f(t) = O(t^p) \quad \text{as } t \to 0 \iff t^{-p}|f(t)| \text{ is bounded as } t \to 0$$
$$f(t) = o(t^p) \quad \text{as } t \to 0 \iff t^{-p}|f(t)| \to 0 \text{ as } t \to 0.$$

1.4 Notes and references

The literature on stability and stabilization of infinite dimensional systems is now extensive; see the survey papers by Pritchard and Zabczyk [132] and by Russell [141]. For stabilizability of general boundary control we refer to Bensoussan et al.[18] and for optimal control of infinite dimensional systems we refer to Lions [92]. Zabczyk's book [174] also contains a number of interesting results.

Chapter 2

Semigroups of Linear Operators

2.1 Motivation and definitions

Semigroups of linear operators are closely related to the solution of the following linear differential equation:

$$\dot{u}(t) = Au(t), \quad u(0) = x \in X, \tag{2.1}$$

where X is a Banach space and $A : D(A)(\subset X) \to X$ is a linear operator. To motivate the concept, we consider (2.1) with linear bounded A. Equation (2.1) is said to be *well-posed* (for bounded A) if

(i) for each $x \in D(A) = X$, there is a solution $u(t; x)$ to (2.1) which is differentiable for $t > 0$, continuous at $t = 0$, and $u(t, x)$ satisfies (2.1) for $t > 0$;

(ii) $u(t; x)$ depends continuously on the initial condition x, that is,

$$x \to 0 \ \text{ implies } u(t; x) \to 0 \ \text{ for each } t > 0;$$

(iii) $u(t; x)$ is unique for each $x \in D(A) = X$.

Now, assume that (2.1) is well-posed. We can then define an operator $T(t)$ by $T(t)x = u(t; x)$ for each $t \geq 0$. From the existence and uniqueness of the solution $u(t; x)$, we know that $T(t), t \geq 0$ is well defined on X. Moreover, the following facts hold:

(a) $T(t)$ is linear for every $t \geq 0$ and $T(0)x = x$ for all $x \in X$, that is, $T(0) = I$;

(b) for any $t, s \geq 0$, note that $u(t + s; x)$ is the unique solution of $\dot{u}(t) = Au(t)$ with $u(0) = u(s; x)$, that is, $T(t)u(s; x) = u(t + s; x)$. On the other hand, $u(s; x) = T(s)x$ and $u(t + s; x) = T(t + s)x$. Therefore, $T(t + s)x = T(t)T(s)x$, which means $T(t + s) = T(t)T(s)$;

(c) since $u(t; x)$ is continuous with respect to t, we have

$$T(t)x \to T(s)x \text{ as } t \to s$$

for all $t, s \geq 0$ and hence $T(t)$ is strongly continuous with respect to $t \geq 0$. Moreover, the continuous dependence on the initial condition (ii) implies that

$$T(t)x \to 0 \text{ as } x \to 0,$$

that is, $T(t)$ is continuous.

We claim that for linear bounded A, equation (2.1) is well-posed. Indeed, since A is bounded, solving (2.1) is equivalent to finding a continuous solution of the following integral equation:

$$u(t; x) = x + \int_0^t Au(s; x)ds = x + A \int_0^t u(s; x)ds. \qquad (2.2)$$

This is a Volterra type integral equation in X. It is well known that there is a unique continuous solution to (2.2) which can be found by iterative process and is given by

$$u(t; x) = \sum_{n=0}^{\infty} \frac{t^n}{n!} A^n x. \qquad (2.3)$$

From (2.3), it follows that $\|u(t; x)\| \leq e^{\|A\|t} \|x\|$. Hence the solution depends continuously on the initial condition. All conditions in (i)-(iii) are thus satisfied and a family of strongly continuous operators $\{T(t), t \geq 0\}$ can be defined as

$$T(t) = \sum_{n=0}^{\infty} \frac{t^n}{n!} A^n = e^{At}. \qquad (2.4)$$

For A being a matrix, (2.4) is just the well known fundamental matrix associated with A. Once $T(t)$ is defined, the solution of (2.1) can be expressed as

$$u(t; x) = T(t)x, \qquad (2.5)$$

and clearly we can recover A from $T(t)$ through the following equation

$$\lim_{t \downarrow 0} \frac{T(t)x - x}{t} = Ax. \qquad (2.6)$$

For a large number of problems encountered in partial differential equations, the resulting operator A is unbounded. For such cases, the exponential expression (2.4) cannot be defined, but the basic requirements of well-posedness for equation (2.1) are still necessary. It is the concept of *semigroups* that successfully generalizes the exponential expression for unbounded operator A on abstract spaces, and establishes a relationship, resembling (2.5), between the solution of (2.1) and the semigroups associated with A.

Definition 2.1 *Let X be a Banach space and let $T(t) : X \to X$ be a family of bounded linear operators for $0 \le t < \infty$. $T(t)$ is called a semigroup of bounded linear operators, or simply a semigroup, on X if*

(i)

$$T(0) = I, \qquad (2.7)$$

(ii)

$$T(t + s) = T(t)T(s), \quad t \ge 0, \quad s \ge 0. \qquad (2.8)$$

A semigroup $T(t)$ is called uniformly continuous *if*

$$\lim_{t \downarrow 0} \|T(t) - I\| = 0, \qquad (2.9)$$

and is called strongly continuous *(or C_0-semigroup for short) if*

$$\lim_{t \downarrow 0} T(t)x = x, \qquad \forall x \in X. \qquad (2.10)$$

We note that the conditions (2.9) and (2.10) are the continuity requirements for $T(t)$ at $t = 0$ in uniform and strong operator topologies, respectively.

Actually, the above defined concept of semigroups can be extended to one-parameter groups as follows.

Definition 2.2 *A one-parameter family $\{T(t), t \in \mathbf{R}\}$ of bounded linear operators on a Banach X is called a strongly continuous group of linear operators, or a C_0-group for short, if it satisfies*

(i) $T(0) = I$.

(ii) $T(t + s) = T(t)T(s)$ for $t, s \in (-\infty, \infty)$.

(iii) $\lim_{t \to 0} T(t)x = x$, $\forall x \in X$.

Definition 2.3 *Let $T(t)$ be a C_0-semigroup on X. The operator A defined by*

$$Ax = \lim_{t \downarrow 0} \frac{T(t)x - x}{t}, \qquad \forall x \in D(A) \qquad (2.11)$$

with domain

$$D(A) = \{x \in X \mid \lim_{t \downarrow 0} \frac{T(t)x - x}{t} \quad exists\} \qquad (2.12)$$

is called the infinitesimal generator of the semigroup $T(t)$.

Similarly, we can define the infinitesimal generators of C_0−groups.

Definition 2.4 *Let $T(t)$ be a C_0−group on X. The operator A defined by*

$$Ax = \lim_{t \downarrow 0} \frac{T(t)x - x}{t}, \quad \forall x \in D(A),$$

$$D(A) = \{x \in X \mid \lim_{t \downarrow 0} \frac{T(t)x - x}{t} \quad exists \}$$

is called the infinitesimal generator of the group $T(t)$.

It is obvious that if $T(t)$ is a C_0-group with generator A, then both $\{T(t), t \geq 0\}$ and $\{T(-t), t \geq 0\}$ are C_0-semigroups with generators A and $-A$, respectively. Conversely, if both A and $-A$ generate C_0-semigroups $T_+(t)$ and $T_-(t)$, respectively, then A is the infinitesimal generator of the C_0-group $T(t)$ given by

$$T(t) = \begin{cases} T_+(t), t \geq 0, \\ T_-(-t), t < 0. \end{cases}$$

We shall find that sometimes it is convenient to use e^{At} to denote the C_0−semigroup generated by A, even if A is unbounded.

We now give some examples of C_0−semigroups and their generators.

Example 2.5 *Let $X = L^p[0, \infty)$, $p \geq 1$, be a Banach space which is specified by*

$$L^p[0, \infty) = \{f(x) : [0, \infty) \to \mathbf{C} \mid \int_0^\infty |f(x)|^p dx < \infty \}$$

with the norm

$$\|f\| = (\int_0^\infty |f(x)|^p dx)^{1/p}.$$

Consider the following shift operator

$$T(t)f(x) = f(t + x), \quad f(x) \in X, \quad x \in [0, \infty), \quad t \geq 0.$$

Obviously, $T(0) = I$ and $T(t + s) = T(t)T(s)$ for $t, s \geq 0$. For strong continuity, let $h \in X$ be any continuous function with compact support on $[0, \infty)$. Then

$$\lim_{t \downarrow 0} \|T(t)h - h\|^p = \lim_{t \downarrow 0} \int_0^\infty |h(t + x) - h(x)|^p dx = 0.$$

*Since such functions are dense in X, it then follows that $\lim_{t\downarrow 0} T(t)f = f$
$\forall f \in X$, hence $T(t)$ is a C_0-semigroup on X. The infinitesimal generator
of $T(t)$ is found to be*

$$(Af)(x) = \lim_{t\downarrow 0} \frac{f(t+x) - f(x)}{t} = \frac{d^+ f(x)}{dx}.$$

i.e., $Af = \frac{d^+ f}{dx}$ with the domain

$$D(A) = \{ f \in L^p[0,\infty) \mid \frac{d^+ f}{dx} \in L^p[0,\infty) \}.$$

Example 2.6 *Let $\{\phi_n, n \geq 1\}$ be a set of orthonormal basis in a Hilbert
space H, and let $\{\lambda_n, n \geq 1\}$ be a set of real numbers. For any $x \in H$ and
$t \geq 0$, let us define the family of linear operators $T(t)$ as follows.*

$$T(t)x = \sum_{n=1}^{\infty} e^{\lambda_n t} \langle x, \phi_n \rangle \phi_n.$$

*Obviously, $T(t)$ is a bounded linear operator if and only if $\sup_{n \geq 1} \lambda_n < \infty$,
which will be assumed in the sequel. Obviously, $T(0) = I$. and by using
orthonormality, it can easily be shown that (2.8) holds. For strong conti-
nuity, let $\epsilon > 0$ be given. Note that since $\|x\|^2 = \sum_{n=1}^{\infty} |\langle x, \phi_n \rangle|^2$. there
exists an $N > 0$ such that $\sup_{t \in [0,1]} \sum_{N+1}^{\infty} (e^{\lambda_n t} - 1)^2 |\langle x, \phi_n \rangle|^2 < \epsilon$. Since
$\lim_{t\downarrow 0} \sum_{n=1}^{N} (e^{\lambda_n t} - 1)^2 |\langle x, \phi_n \rangle|^2 = 0$ for any $N > 0$, it follows that
$\lim_{t\downarrow 0} \|T(t)x - x\|^2 < \epsilon$. Since $\epsilon > 0$ is arbitrary, it follows that $T(t)$ is a
C_0-semigroup. The infinitesimal generator of $T(t)$ is given by*

$$Ax = \lim_{t\downarrow 0} \sum_{n=1}^{\infty} \frac{e^{\lambda_n t} - 1}{t} \langle x, \phi_n \rangle \phi_n = \sum_{n=1}^{\infty} \lambda_n \langle x, \phi_n \rangle \phi_n$$

with the domain

$$D(A) = \{ x \in H \mid \sum_{n=1}^{\infty} \lambda_n^2 |\langle x, \phi_n \rangle|^2 < \infty \}.$$

2.2 Properties of semigroups

We first state the following theorem which characterizes the uniformly continuous semigroups.

Theorem 2.7 *Let X be a Banach space. For any bounded linear operator
A on X, $T(t) = e^{At}$ given by (2.4) is a uniformly continuous semigroup and
A is the infinitesimal generator of $T(t)$ with $D(A) = X$. Conversely, for any*

uniformly continuous semigroup $T(t)$ on X, there exists a unique bounded linear operator A on X such that $T(t) = e^{At}$ and A is the infinitesimal generator of $T(t)$ with $D(A) = X$. Moreover, $T(t)$ is differentiable in norm and

$$\frac{d^n T(t)}{dt^n} = A^n T(t) = T(t) A^n, \quad n = 1, 2, \cdots. \tag{2.13}$$

Proof. For any bounded linear operator A, (2.4) clearly defines a uniformly continuous semigroup. It also follows easily that

$$\lim_{t \downarrow 0} \frac{1}{t} \| T(t)x - x - Atx \| = \lim_{t \downarrow 0} \frac{1}{t} \left\| \sum_{i=2}^{\infty} \frac{(At)^i}{i!} x \right\| = 0, \quad \forall x \in X.$$

Hence, A is the infinitesimal generator of A with $D(A) = X$.

The converse statement is proved as follows. Since $T(t)$ is uniformly continuous on $[0, \infty)$, it follows that $T(t)$ is bounded on an interval $[0, T_1]$ for some $T_1 > 0$, i.e., there is a constant M such that $\sup_{t \in [0, T_1]} \| T(t) \| \leq M$. Let $T > 0$ be given and let N be a sufficiently large integer satisfying $T < T_1 N$. Then by using the semigroup property, we obtain $T(t) = [T(\frac{t}{N})]^N$. For every $t \in [0, T]$, since $\frac{t}{N} \leq \frac{T}{N} \leq T_1$, we have $\sup_{t \in [0, T]} \| T(t) \| \leq M^N$. Hence $\| T(t) \|$ is uniformly bounded on compact intervals.

Let $s > 0$ be fixed. For $t > s$, we set $r = t - s > 0$, and the following holds.

$$\lim_{t \downarrow s} \| T(t) - T(s) \| \leq \lim_{r \downarrow 0} \| T(r) - I \| \; \| T(s) \| = 0.$$

For $s > t$, set $r = s - t > 0$. Then for sufficiently small $\varepsilon > 0$, $s - \varepsilon > 0$, and

$$\sup_{t \in [s-\varepsilon, s]} \| T(t) - T(s) \| \leq \| T(r) - I \| \sup_{t \in [s-\varepsilon, s]} \| T(t) \|.$$

It follows that

$$\lim_{t \uparrow s} \| T(t) - T(s) \| = 0.$$

Therefore, $\lim_{t \to s} \| T(t) - T(s) \| = 0$, i.e., $T(t)$ is continuous for $t \geq 0$.

Let us define $F(t) = \int_0^t T(\sigma) d\sigma$. Since $F(0) = 0$ and $\dot{F}(t) = T(t)$, we have $\lim_{h \to 0} \frac{F(h)}{h} = I$. Therefore, for any $\gamma > 0$ such that $\left\| \frac{F(\gamma)}{\gamma} - I \right\| < 1$, $F(\gamma)$ is bounded and invertible. Now consider the following.

$$\frac{T(h) - I}{h} F(\gamma) = \frac{1}{h} \int_0^{\gamma} [T(\sigma + h) - T(\sigma)] d\sigma = \frac{1}{h} [F(\gamma + h) - F(\gamma) - F(h)].$$

Letting $h \to 0$, we obtain

$$\lim_{h \to 0} \frac{T(h) - I}{h} = [\dot{F}(\gamma) - I][F(\gamma)]^{-1} = [T(\gamma) - I][F(\gamma)]^{-1}. \tag{2.14}$$

Therefore, $A = [T(\gamma) - I][F(\gamma)]^{-1}$ is the generator for which we are seeking. Note that A is a bounded linear operator because $F(\gamma)$ has bounded inverse as stated.

The remaining arguments easily follow from the mathematical induction.

$\square$

For the rest of this chapter, we will consider C_0−semigroups, unless stated otherwise.

Theorem 2.8 *Let $T(t)$ be a C_0−semigroup on a Banach space X. Then the following hold.*

(i) There exist constants $M \geq 1$ and $\omega \geq 0$ such that

$$\|T(t)\| \leq M e^{\omega t}, \quad t \geq 0. \tag{2.15}$$

(ii) $T(t)$ is strongly continuous on X, i.e., for any $x \in X$, the map $t \to T(t)x$ is continuous.

Proof. (i): There exists a $\delta > 0$ such that $\|T(t)\|$ is uniformly bounded on $[0, \delta]$. Otherwise there exists a sequence $\{t_n\}$ such that $t_n > 0$, $\lim_{n \to \infty} t_n = 0$ and $\|T(t_n)\| \geq n$. Hence, from the uniform boundedness theorem it follows that for some $x \in X$, $\|T(t_n)x\|$ also is unbounded, which contradicts (2.10). Therefore, there exist constants $\delta > 0$ and $M \geq 1$ such that $\|T(t)\| \leq M$ for $t \in [0, \delta]$. Given any $t \geq 0$, let n be the integer such that $t = n\delta + r$ where $r \in [0, \delta)$. From (2.8) it follows that

$$\|T(t)\| = \|T(r)T(\delta)^n\| \leq M^{n+1} \leq M e^{\omega t}.$$

where $\omega = \frac{\log M}{\delta}$.

(ii): Let $x \in X$, $t, h \geq 0$. Then we have

$$\|T(t + h)x - T(t)x\| \leq M e^{\omega t} \|T(h)x - x\|,$$

and for $t \geq h \geq 0$ we have

$$\|T(t - h)x - T(t)x\| \leq \|T(t - h)\| \|x - T(h)x\| \leq M e^{\omega t} \|x - T(h)x\|.$$

The strong continuity of $T(t)$ then follows from (2.10). $\square$

We now introduce the *growth rate* of a C_0−semigroup.

Theorem 2.9 *Let $T(t)$ be a C_0−semigroup, and let ω_0 be defined as*

$$\omega_0 = \inf_{t > 0} \frac{\log \|T(t)\|}{t}. \tag{2.16}$$

Then the following hold.

(i)

$$-\infty \le \omega_0 = \lim_{t \to \infty} \frac{\log \|T(t)\|}{t} < \infty. \qquad (2.17)$$

(ii) For any $\omega > \omega_0$, there exists an $M \ge 1$, which may depend on ω, such that (2.15) holds. (Note that in (2.15) we require $\omega \ge 0$, however, here $\omega < 0$ also is possible).

(iii) The spectral radius $r(T(t))$ of $T(t)$ is given by $r(T(t)) = e^{\omega_0 t}$.

Proof. (i): Let $w(t) = \log \|T(t)\|$, $t > 0$. It follows from (2.8) that $w(t)$ is subadditive, i.e., $w(t + s) \le w(t) + w(s)$ for $t, s > 0$. We note that $\omega_0 = -\infty$ also is possible. First, assume that ω_0 is finite. Let $\epsilon > 0$ be given. Then there exists a $\delta > 0$ such that $\omega_0 \le \frac{w(\delta)}{\delta} \le \omega_0 + \epsilon$. Let $t > 0$ be given and n be the integer satisfying $t = n\delta + r$, $r \in [0, \delta)$. Then we have

$$\omega_0 \le \frac{w(t)}{t} \le \frac{nw(\delta)}{n\delta + r} + \frac{w(r)}{t}.$$

Hence, for t sufficiently large, or equivalently n sufficiently large, we have

$$\omega_0 \le \frac{w(t)}{t} \le \omega_0 + \epsilon + \frac{w(r)}{t}.$$

Since $\|T(t)\|$ is bounded on compact intervals, so is $w(r)$, and since $\epsilon > 0$ is arbitrary, it follows that (2.17) holds when ω_0 is finite. If $\omega_0 = -\infty$, then for any positive integer n we can find $\delta > 0$ such that $\frac{w(\delta)}{\delta} < -n$, and by repeating the arguments given above we obtain $\lim_{t \to \infty} \frac{w(t)}{t} < -n$. Since the integer n is arbitrary, (2.17) follows.

(ii): Let $\omega > \omega_0$ be given. From (2.17) it follows that there is a $t_0 > 0$ such that $\frac{w(t)}{t} < \omega$ for $t \ge t_0$, and hence $\|T(t)\| \le e^{\omega t}$ for $t \ge t_0$. Since $\|T(t)\|$ is bounded on compact intervals, there exists an $M_0 \ge 1$ such that $\|T(t)\| \le M_0$ for $t \in [0, t_0]$. It then follows that (2.15) holds for some $M \ge 1$.

(iii): Let $t > 0$ be fixed. Note that $r(T(t))$ is defined by

$$r(T(t)) = \lim_{n \to \infty} \|T(t)^n\|^{1/n} = \lim_{n \to \infty} \|T(nt)\|^{1/n},$$

see, e.g., [84]. By taking the logarithms we obtain

$$\log r(T(t)) = t \lim_{n \to \infty} \frac{\log \|T(nt)\|}{nt} = \omega_0 t.$$

Hence, we have $r(T(t)) = e^{\omega_0 t}$. $\square$

Remark 2.10 *We note that in general we cannot take $\omega = \omega_0$ in (2.15). To see this, let $A = \begin{bmatrix} 1 & 1 \\ 0 & 1 \end{bmatrix}$ and the semigroup $T(t) = e^{At} = \begin{bmatrix} e^t & te^t \\ 0 & e^t \end{bmatrix}$. If we choose the operator norm induced by ℓ^1 norm, see e.g. [84], then we have $\|T(t)\| = e^t(t+1)$. It then follows from (2.17) that $\omega_0 = 1$ and (2.15) cannot be satisfied for $\omega = 1$.*

Corollary 2.11 *Let $T(t)$ be a C_0−semigroup. Then (2.15) is satisfied for some $\omega < 0$ if and only if $\|T(t_0)\| < 1$, or $r(T(t_0)) < 1$, for some $t_0 > 0$.*

Proof. This result follows directly from Theorem 2.9. If (2.15) is satisfied for some $\omega < 0$, then necessarily we have $\omega_0 < 0$, and it follows from (2.16) that we must have $\|T(t_0)\| < 1$ for some $t_0 > 0$. Conversely, if $\|T(t_0)\| < 1$ for some $t_0 > 0$, then from (2.16) we obtain $\omega_0 < 0$, hence by Theorem 2.9 it follows that (2.15) holds for any $\omega_0 < \omega < 0$. $\qquad\qquad\square$

Theorem 2.12 *Let $T(t)$ be a C_0−semigroup on a Banach space X and let A be its infinitesimal generator. Then the following hold.*

(i) For any $x \in X$ and $t \geq 0$,

$$\lim_{h \downarrow 0} \frac{1}{h} \int_t^{t+h} T(s)x\,ds = T(t)x. \tag{2.18}$$

(ii) For any $x \in X$ and $t \geq 0$,

$$\int_0^t T(s)x\,ds \in D(A), \quad A \int_0^t T(s)x\,ds = T(t)x - x. \tag{2.19}$$

(iii) For any positive integer n and $x \in D(A^n)$, where $D(A^n)$ is the domain of A^n, and for any $1 \leq k \leq n$, $t \geq 0$ we have $T(t)x \in D(A^k)$ and

$$\frac{d^k}{dt^k}(T(t)x) = A^k T(t)x = T(t)A^k x. \tag{2.20}$$

where for $t = 0$ the derivatives in (2.20) mean right derivatives.

Proof. (i): Since $t \to T(t)x$ is continuous, $F(t) = \int_0^t T(s)x\,ds$ is differentiable and $\dot{F}(t) = T(t)x$. Since

$$\dot{F}(t) = \lim_{h \downarrow 0} \frac{F(t+h) - F(t)}{h} = \lim_{h \downarrow 0} \frac{1}{h} \int_t^{t+h} T(s)x\,ds,$$

we obtain (2.18).

(ii): Since for $h > 0$. we have

$$\lim_{h\downarrow 0}\frac{T(h)-I}{h}\int_0^t T(s)xds = \lim_{h\downarrow 0}\frac{1}{h}\int_0^t [T(h+s)x - T(s)x]ds$$

$$= \lim_{h\downarrow 0}\frac{1}{h}\int_t^{t+h} T(s)xds - \lim_{h\downarrow 0}\frac{1}{h}\int_0^h T(s)xds,$$

it then follows that (2.19) holds.

(iii): Let $n = 1$, and let $x \in D(A)$. For $h > 0$ we have

$$T(t)Ax = \lim_{h\downarrow 0}T(t)\frac{T(h)-I}{h}x = \lim_{h\downarrow 0}\frac{T(h)-I}{h}T(t)x = AT(t)x,$$

hence $T(t)x \in D(A)$ and $AT(t)x = T(t)Ax$ for $t \geq 0$. For the derivative, we have

$$\lim_{h\downarrow 0}\frac{T(t+h)x - T(t)x}{h} = \lim_{h\downarrow 0}\frac{T(h)-I}{h}T(t)x = AT(t)x = T(t)Ax,$$

i.e., (2.20) holds for the right derivative. For the left derivative, let $t \geq h > 0$. We have

$$\lim_{h\downarrow 0}\frac{T(t)x - T(t-h)x}{h} = \lim_{h\downarrow 0}T(t-h)[\frac{T(h)-I}{h}x - Ax]$$

$$+ \lim_{h\downarrow 0}T(t-h)Ax = T(t)Ax,$$

where we used the fact that $\|T(t-h)\|$ is bounded for $0 < h \leq t$ and the strong continuity of $T(t)$. This proves that (2.20) holds for $n = 1$.

For arbitrary $n > 1$, since $A^n x = AA^{n-1}x$, it follows that if $x \in D(A^n)$, then $x \in D(A^k)$ for $1 \leq k \leq n$. By using this fact, and that $AT(t)x = T(t)Ax$ for $x \in D(A)$, it follows that if $x \in D(A^n)$ then we have $T(t)A^k x = A^k T(t)x$; hence, $T(t)x \in D(A^k)$ for $1 \leq k \leq n$. Note that (2.20) holds for $n = 1$. By using mathematical induction, assume that (2.20) holds for $k - 1 < n$. Then we have

$$\frac{d^k}{dt^k}[T(t)x] = \frac{d}{dt}[T(t)A^{k-1}x] = T(t)A^k x = A^k T(t)x,$$

hence, (2.20) follows. □

Theorem 2.13 *Let $T(t)$ be a C_0-semigroup on X and let A be its infinitesimal generator. We have the following.*

(i) $D(A)$ is dense in X.

(ii) A is a closed operator.

(iii) For any $n \geq 1$. $D(A^n)$ is dense in X. The set $D = \cap_{n=1}^{\infty} D(A^n)$ also is dense in X and is invariant under $T(t)$, i.e.. for $x \in D$. $T(t)x \in D$ for $t \geq 0$. Moreover, if we define $D^{\infty} = \{x \in X \mid t \to T(t)x \in C^{\infty}\}$. then we have $D = D^{\infty}$.

Proof. (i): Let $x \in X$ be given. Set $h_n = 1/n$ and $x_n = \frac{1}{h_n}\int_0^{h_n} T(s)x ds$. From Theorem 2.12 we have $x_n \in D(A)$ and $\lim_{n\to\infty} x_n = x$. Hence $D(A)$ is dense in X.

(ii): Assume that $x_n \in D(A)$, $\lim_{n\to\infty} x_n = x$, and that $\lim_{n\to\infty} Ax_n = y$. By simple integration of (2.20), we obtain $T(h)x_n - x_n = \int_0^h T(s)Ax_n ds$ for $h > 0$. Since $\|T(s)\|$ is uniformly bounded for $s \in [0,h]$, we obtain $T(h)x - x = \int_0^h T(s)y ds$, as $n \to \infty$. Dividing both sides of this equation by h and then letting $h \downarrow 0$. we obtain $x \in D(A)$ and $Ax = y$. Hence, A is a closed operator.

(iii): Let $\rho_n : \mathbf{R} \to \mathbf{R}$ be a C^{∞} function with support in $(0, 1/n)$, $\rho_n \geq 0$ and $\int_0^{\infty} \rho_n(s) ds = 1$. Such functions are called *regularizing* functions, see e.g. [33]. Let $x \in X$ be given and set $x_n = \int_0^{\infty} \rho_n(s)T(s)x ds$. Then by $\|x_n - x\| \leq \sup_{s\in[0,\frac{1}{n}]} \|T(s)x - x\|$, we have $\lim_{n\to\infty} x_n = x$. It remains to show that $x_n \in D(A^n)$ for all n. Observe that for any $h > 0$ we have

$$\frac{T(h) - I}{h}x_n = \int_0^{\infty} \rho_n(s)\frac{T(s+h)x - T(s)x}{h}ds$$
$$= \int_0^{\infty} \frac{\rho_n(s-h) - \rho_n(s)}{h}T(s)x ds.$$

Letting $h \downarrow 0$, we obtain $Ax_n = -\int_0^{\infty} \rho_n'(s)T(s)x ds$, hence $x_n \in D(A)$. By repeating the same argument we obtain $x_n \in D(A^k)$ for any $k \geq 1$, hence $D(A^k)$ is dense in X; moreover $A^k x_n = (-1)^k \int_0^{\infty} \rho_n^{(k)}(s)T(s)x ds$. Therefore, $x_n \in D$, and hence D is dense in X. If $x \in D$, then $x \in D(A^k)$ for any k. It then follows from Theorem 2.12 that $T(t)x \in D(A^k)$, hence $T(t)x \in D$. For the last part, if $x \in D$, then $x \in D(A^k)$ for any k and by (2.20) $T(t)x \in C^{\infty}$, hence $x \in D^{\infty}$. Conversely, if $x \in D^{\infty}$ then $T(t)x$ is differentiable at $t = 0$, hence $x \in D(A)$. Therefore, $\frac{d}{dt}[T(t)x] = T(t)Ax$, hence $Ax \in D^{\infty}$. By differentiating once more, we obtain $\frac{d^2}{dt^2}[T(t)x] = T(t)A^2 x$, hence $x \in D(A^2)$. By repeating this argument inductively we obtain $x \in D(A^k)$ for any k, hence $x \in D$. □

The following theorem proves that C_0−semigroups are uniquely determined by their generators.

Theorem 2.14 *Let $T(t)$ and $S(t)$ be C_0−semigroups, and let A and B be their infinitesimal generators, respectively. If $A = B$. then $T(t) = S(t)$ for $t \geq 0$.*

Proof. Since $A = B$ we have $D(A) = D(B)$. Let $x \in D(A)$ and $s > 0$ be fixed. Define $z(t) = T(s - t)S(t)x$ for $0 \leq t \leq s$. From Theorem 2.12 it follows that $z(t)$ is differentiable and we have

$$\frac{dz(t)}{dt} = -T(s - t)S(t)Ax + T(s - t)S(t)Bx = 0,$$

where we used (2.20). Hence, $z(s) = z(0)$, and we have $T(t)x = S(t)x$ for any $x \in D(A)$ and $t \geq 0$. Since $D(A)$ is dense in X and since $T(t)$ and $S(t)$ are bounded, it follows that $T(t)x = S(t)x$ for all $x \in X$. $\qquad \square$

Theorem 2.14 indicates that a C_0−semigroup $T(t)$ is uniquely determined by its generator A.

2.3 Generation theorems for semigroups

In this section, we will give some classical results, namely the Hille-Yosida and the Lümer-Phillips theorems, which give necessary and/or sufficient conditions for an operator A to be the infinitesimal generator of a C_0−semigroup. These conditions depend on the behaviour of the resolvent of A in the resolvent set. Let us first recall some important properties of the resolvent.

Let A be a closed linear operator on a Banach space X. If for some complex number λ, $(\lambda I - A)^{-1}$ is a bounded linear operator, then we say that $\lambda \in \rho(A)$, the resolvent set of A, and $R(\lambda, A) = (\lambda I - A)^{-1}$ is called the resolvent operator, or simply resolvent, of A. Recall that $\rho(A)$ is an open set in the complex plane, and that $R(\lambda, A)$ is holomorphic on the resolvent set and

$$\frac{d^n}{d\lambda^n}R(\lambda, A) = (-1)^n n! R(\lambda, A)^{n+1}, \quad n = 1, 2, 3, \cdots \qquad (2.21)$$

For further details, see e.g. [84].

Theorem 2.15 *Let $T(t)$ be a C_0−semigroup on a Banach space X, and let A be its infinitesimal generator. Let ω_0 be the growth rate defined by (2.16).*

(i) *If $\lambda \in \mathbf{C}$ satisfies $Re\lambda > \omega_0$, then $\lambda \in \rho(A)$ and for any $x \in X$, the following holds.*

$$R(\lambda, A)x = \int_0^\infty e^{-\lambda\tau}T(\tau)x d\tau. \qquad (2.22)$$

Moreover, we have

$$\lim_{Re\lambda \to \infty} \|R(\lambda, A)\| = 0. \qquad (2.23)$$

(ii) *For any* $x \in X$, *the following holds.*

$$\lim_{\substack{Re\lambda \to \infty}} \lambda R(\lambda, A)x = x. \tag{2.24}$$

(iii) *For any* $x \in D(A)$, *the following holds.*

$$\lim_{\substack{Re\lambda \to \infty}} \lambda[\lambda R(\lambda, A) - I]x = Ax. \tag{2.25}$$

(iv) *For any* $Re\lambda > \omega > \omega_0$, *there exists a constant* $M(\omega)$ *such that the following holds.*

$$\|R(\lambda, A)^n\| \leq \frac{M(\omega)}{(Re\lambda - \omega)^n}, \qquad n = 1, 2, 3, \cdots \tag{2.26}$$

Proof. (i): Let $\omega \in \mathbf{R}$ be such that $\omega > \omega_0$ and let $x \in X$. By Theorem 2.9, there exists a constant $M \geq 1$ such that (2.15) is satisfied. Let us define $R(\lambda)$ as

$$R(\lambda)x = \int_0^\infty e^{-\lambda\tau} T(\tau)x d\tau. \tag{2.27}$$

By using (2.15), we obtain

$$\|R(\lambda)x\| \leq \int_0^\infty e^{-(\sigma-\omega)\tau} M\|x\| d\tau \leq \frac{M}{\sigma - \omega}\|x\|, \tag{2.28}$$

where $\sigma = Re\lambda$. Hence, (2.27) defines a bounded linear operator on X for $Re\lambda > \omega_0$. Furthermore, for $h > 0$,

$$\begin{aligned}
\frac{T(h) - I}{h}R(\lambda)x &= \frac{1}{h}\int_0^\infty e^{-\lambda\tau}[T(\tau + h)x - T(\tau)x]d\tau \\
&= \frac{e^{\lambda h} - 1}{h}\int_0^\infty e^{-\lambda\tau} T(\tau)x d\tau - \frac{e^{\lambda h}}{h}\int_0^h e^{-\lambda\tau} T(\tau)x d\tau.
\end{aligned} \tag{2.29}$$

In a similar way as we proved (2.18), (actually $e^{-\lambda t}T(t)$ is obviously a C_0-semigroup), we know that $\lim_{h\downarrow 0} \frac{1}{h}\int_0^h e^{-\lambda\tau} T(\tau)x d\tau = x$. By letting $h \downarrow 0$ in (2.29), we obtain $R(\lambda)x \in D(A)$ and $AR(\lambda)x = \lambda R(\lambda)x - x$, hence

$$(\lambda I - A)R(\lambda)x = x, \qquad x \in X. \tag{2.30}$$

Let $x \in D(A)$. By using (2.20) and the closedness of A, we obtain

$$R(\lambda)Ax = \int_0^\infty e^{-\lambda\tau} T(\tau) Ax d\tau = A\int_0^\infty e^{-\lambda\tau} T(\tau)x d\tau = AR(\lambda)x. \tag{2.31}$$

Combining (2.31) and (2.30) gives

$$R(\lambda)(\lambda I - A)x = x, \qquad x \in D(A). \tag{2.32}$$

(2.30) and (2.32) indicate that $R(\lambda) = R(\lambda, A)$. Letting $\sigma \to \infty$ in (2.28) we obtain (2.23).

(ii): Let $x \in D(A)$. It follows from (2.23) that $\|R(\lambda, A)Ax\| \leq \|R(\lambda, A)\|$ $\|Ax\| \to 0$ as $\text{Re}\lambda \to \infty$, which, when incorporated into (2.32), shows that (2.24) holds for $x \in D(A)$. By using (2.28), we have $\|\lambda R(\lambda, A)\| \leq M$ for $\text{Re}\lambda$ sufficiently large. Since $D(A)$ is dense in X, it then follows that (2.24) holds for $x \in X$.

(iii): Let $x \in D(A)$. It follows from (2.32) that $\lambda^2 R(\lambda, A)x - \lambda x = \lambda R(\lambda, A)Ax$. By letting $\sigma \to \infty$ and using (2.24), we obtain (2.25).

(iv): By using (2.22) and Fubini's theorem we obtain

$$R(\lambda, A)^n x = \int_0^\infty \cdots \int_0^\infty e^{-\lambda(\tau_1 + \ldots + \tau_n)} T(\tau_1 + \ldots + \tau_n) x \, d\tau_1 \ldots d\tau_n, \tag{2.33}$$

for any $x \in X$. Using (2.15) and (2.28) leads to (2.26).

$\square$

Remark 2.16 *For any $\lambda \in \rho(A)$. define the bounded linear operator A_λ by*

$$A_\lambda = \lambda^2 R(\lambda, A) - \lambda I, \tag{2.34}$$

which is called the Yosida approximation of A. Note that (2.25) can be re-stated as

$$\lim_{\text{Re}\lambda \to \infty} A_\lambda x = Ax, \qquad x \in D(A). \tag{2.35}$$

Because of (2.20), when A generates a C_0−semigroup $T(t)$. then (2.1) is well-posed in the sense that there exists a unique solution for every initial value in $D(A)$ which depends continuously on the initial value. But under what conditions does A generate a C_0−semigroup? Yosida provided a method to attack this problem. The idea is to replace an unbounded operator A by the Yosida approximation A_λ, defined in (2.34), and to consider the abstract Cauchy problem $z'_\lambda(t) = A_\lambda z_\lambda(t), z_\lambda(0) = x$, which possesses a unique solution since A_λ is bounded. Equation (2.35) shows that $A_\lambda x$ converges to Ax for $x \in D(A)$. It is therefore reasonable to expect that the above approximated solution z_λ converges to the solution of (2.1). The following theorem gives an affirmative answer to this conjecture.

Theorem 2.17 (Hille-Yosida). *Let X be a Banach space, and let A be a linear (not necessarily bounded) operator in X. A is the infinitesimal generator of a C_0−semigroup $T(t)$ if and only if*

(i) *A is closed and $D(A)$ is dense in X.*

(ii) *There exist positive constants M and ω such that $(\omega, \infty) \subset \rho(A)$ and for any $\sigma > \omega$ the following holds.*

$$\|R(\sigma, A)^n\| \leq \frac{M}{(\sigma - \omega)^n}, n = 1, 2, \ldots . \tag{2.36}$$

Proof. The necessity follows from Theorem 2.13 and Theorem 2.15.

For the sufficiency, we let the Yosida approximation A_σ of A for $\sigma > \omega$ be defined as in (2.34), i.e., $A_\sigma = \sigma^2 R(\sigma, A) - \sigma I$. Then A_σ is a bounded operator, and by using (2.3) and (2.5), we can find the uniformly continuous semigroup $T_\sigma(t)$ generated by A_σ as follows.

$$T_\sigma(t) = e^{A_\sigma t} = e^{-\sigma t} \sum_{n=0}^{\infty} \frac{(\sigma^2 t)^n}{n!} R(\sigma, A)^n. \tag{2.37}$$

We will show that the strong limit of $T_\sigma(t)$ as $\sigma \to \infty$ exists and is the required semigroup, as implied by (2.34) and (2.35). Using (2.36) in (2.37) yields

$$\|T_\sigma(t)\| \leq e^{-\sigma t} \sum_{n=0}^{\infty} \frac{(\sigma^2 t)^n}{n!} \frac{M}{(\sigma - \omega)^n} \leq M e^{\frac{\omega \sigma}{\sigma - \omega} t}, \quad t \geq 0. \tag{2.38}$$

For any $\sigma, \mu > \omega$ we have $\sigma, \mu \in \rho(A)$, and since resolvents commute, i.e., $R(\sigma, A)R(\mu, A) = R(\mu, A)R(\sigma, A)$, we have $A_\sigma A_\mu = A_\mu A_\sigma$ as well. Hence, we also have $T_\sigma(t)A_\mu = A_\mu T_\sigma(t)$. Therefore, for $x \in D(A)$,

$$[T_\sigma(t) - T_\mu(t)]x = \int_0^t \frac{d}{ds}[T_\mu(t - s)T_\sigma(s)]x \, ds$$

$$= \int_0^t T_\mu(t - s)T_\sigma(s)(A_\sigma - A_\mu)x \, ds. \tag{2.39}$$

If $\omega < 0$ in (2.38), then for $\sigma > 0$ we have $\frac{\omega \sigma}{\sigma - \omega} < 0$. On the other hand, if $\omega > 0$, then for any $k > 1$ and $\sigma > \frac{k\omega}{k-1}$ we have $\frac{\omega \sigma}{\sigma - \omega} < k\omega$. Hence, we may conclude that for any $k > 1$, if $\sigma > \frac{k|\omega|}{k-1}$, then $\frac{\omega \sigma}{\sigma - \omega} < k|\omega|$, from which we obtain $\|T_\sigma(t)\| \leq M e^{k|\omega| t}$. Therefore for $k > 1$ and $\sigma, \mu > \frac{k|\omega|}{k-1}$, it follows from (2.39) that

$$\|(T_\sigma(t) - T_\mu(t))x\| \leq M^2 t e^{k|\omega| t} \|(A_\sigma - A_\mu)x\|, \quad x \in D(A). \tag{2.40}$$

However, by (2.35), we have $\lim_{\sigma \to \infty} A_\sigma x = Ax$, hence $\|(A_\sigma - A_\mu)x\| \to 0$ as $\sigma, \mu \to \infty$. Therefore, for each $t \geq 0$, $T_\sigma(t)x$ is a Cauchy sequence. From the Banach-Steinhaus theorem and (2.38), it follows that there exists a linear operator $T(t)$ on X satisfying

$$T(t)x = \lim_{\sigma \to \infty} T_\sigma(t)x, \quad x \in X. \tag{2.41}$$

Moreover, by the uniform boundedness theorem, $T(t)$ is a bounded linear operator. For any $x \in X$ and $t, s \geq 0$ we have

$$T(t+s)x = \lim_{\sigma \to \infty} T_\sigma(t+s)x = \lim_{\sigma \to \infty} T_\sigma(t)T_\sigma(s)x = T(t)T(s)x.$$

Obviously we have $T(0) = I$, and the strong continuity is a consequence of uniform convergence in (2.41) on compact intervals. Hence $T(t)$ is a C_0-semigroup. It then remains to show that A is the infinitesimal generator of $T(t)$. From (2.35) and (2.41) it follows that $\lim_{\sigma \to \infty} T_\sigma(t)A_\sigma x = T(t)Ax$ for $x \in D(A)$, where the convergence is uniform on compact intervals. Hence, by using (2.41) and Theorem 2.12 (i.e., integrating (2.20) from 0 to t for $k = 1$), we obtain

$$T(t)x - x = \lim_{\sigma \to \infty} T_\sigma(t)x - x = \lim_{\sigma \to \infty} \int_0^t T_\sigma(\tau)A_\sigma x d\tau = \int_0^t T(\tau)Ax d\tau \tag{2.42}$$

for all $x \in D(A)$. Let B be the infinitesimal generator of $T(t)$. From (2.42) we have

$$Bx = \lim_{t \downarrow 0} \frac{T(t)x - x}{t} = \lim_{t \downarrow 0} \frac{1}{t} \int_0^t T(\tau)Ax d\tau = Ax \quad x \in D(A). \tag{2.43}$$

Hence, $D(A) \subseteq D(B)$. From (2.38) and (2.41) it follows that

$$\|T(t)\| \leq \lim_{\sigma \to \infty} \inf \|T_\sigma(t)\| \leq Me^{\omega t}, \tag{2.44}$$

hence for any $\sigma > \omega$ we have $\sigma \in \rho(B)$ and $(\sigma I - B)D(B) = X$. Since $\sigma \in \rho(A)$, we have $(\sigma I - B)D(A) = (\sigma I - A)D(A) = X$. But then $D(B) = R(\sigma, B)X = D(A)$, and hence from (2.43) we have $B = A$. $\qquad \square$

In applications, the main difficulty lies in checking the condition (2.36) for $n = 1, 2, \dots$. There are, however, two cases where this condition is easily checked by estimating $\|R(\lambda, A)\|$ only. One case is when A generates a contraction semigroup to be discussed below, and another case is when A generates an analytic semigroup which will be discussed in details in Section 2.5.

Definition 2.18 *Let $T(t)$ be a C_0-semigroup on a Banach space X and let $M \geq 1$ and $\omega \in \mathbf{R}$ be the constants in (2.15). If $\omega \leq 0$, then we have $\|T(t)\| \leq M$ for $t \geq 0$ and $T(t)$ is called* uniformly bounded. *Moreover, if we have $M = 1$, (i.e. $\|T(t)\| \leq 1$), then $T(t)$ is called a* contraction.

The following version of the Hille-Yosida Theorem characterizes the generators of C_0−semigroups of contractions.

Corollary 2.19 *Let X be a Banach space and let A be a linear (not necessarily bounded) operator in X. Then, A is the infinitesimal generator of a C_0−semigroup of contractions $T(t)$ on X if and only if the following hold.*

(i) A is closed and $D(A)$ is dense in X.

(ii) For any $\sigma > 0$, $\sigma \in \rho(A)$ and

$$\|R(\sigma, A)\| \leq \frac{1}{\sigma}. \tag{2.45}$$

Proof. Since $T(t)$ is a contraction, we can take $M = 1$ and $\omega \leq 0$. Since $\|R(\sigma, A)^n\| \leq \|R(\sigma, A)\|^n$, the result then follows from the Hille-Yosida theorem. $\qquad\qquad\qquad\qquad\qquad\qquad\qquad\qquad\qquad\qquad\qquad\qquad\qquad\qquad$ □

Example 2.20 *Consider the C_0- semigroup $T(t)$ introduced in Example 2.5. Since $\|T(t)f\|^p = \int_0^\infty |f(t+x)|^p dx \leq \int_0^\infty |f(s)|^p ds = \|f\|^p$, we have $\|T(t)\| \leq 1$; hence, $T(t)$ is a contraction.*

Example 2.21 *Let $H = L^2(0,1)$ and let us define an operator A as $A\varphi = -\varphi'$ for $\varphi \in D(A)$ where*

$$D(A) = \{\varphi \in H \mid \varphi' \in H, \ \varphi(0) = 0 \}.$$

Since $(A^{-1}f)(x) = -\int_0^x f(s)ds$ for any $f \in H$, A^{-1} is a bounded linear operator, hence both A^{-1} and A are closed, see [84]. It is clear that $D(A)$ is dense in H. For any $\lambda > 0$ and $f \in H$, the equation $(\lambda I - A)u = f$, i.e. $\lambda u + u' = f$, has a solution $u \in D(A)$ given by

$$u(x) = \int_0^x e^{-\lambda(x-s)} f(s)ds \quad .$$

Hence, for $\lambda > 0$, $R(\lambda, A)$ exists, and therefore, $(0, \infty) \in \rho(A)$. Moreover,

$$\int_0^1 |f(s)|^2 ds = \int_0^1 |\lambda u(s) + u'(s)|^2 ds$$

$$= \lambda^2 \int_0^1 |u(s)|^2 ds + 2\lambda \int_0^1 u(s)\overline{u'(s)}ds + \int_0^1 |u'(s)|^2 ds.$$

Since $2\int_0^1 u(s)\overline{u'(s)}ds = |u(1)|^2 \geq 0$, it follows that

$$\int_0^1 |f(s)|^2 ds \geq \lambda^2 \int_0^1 |u(s)|^2 ds,$$

i.e. $\|u\| \leq \frac{1}{\lambda}\|f\|$, where $f = (\lambda I - A)u$. Hence (2.45) holds and by Corollary 2.19, A generates a C_0−semigroup of contractions on H.

On some special occasions, we can derive much simpler conditions in order for a specific operator to generate semigroups. To show this, let us first consider a C_0−semigroup of contractions $T(t)$ on a Hilbert space H. In this case, $\|T(t)\| \leq 1$, so that

$$|\langle T(t)x,\ x \rangle| \leq \|T(t)\|\,\|x\|^2 \leq \|x\|^2$$

for every $x \in H$. Thus,

$$\mathrm{Re}\langle T(t)x - x,\ x \rangle = \mathrm{Re}\langle T(t)x,\ x \rangle - \|x\|^2 \leq 0.$$

Restricting x to $x \in D(A)$, dividing t on both sides of the above equation, and letting $t \to 0$, we obtain

$$\mathrm{Re}\langle Ax,\ x \rangle \leq 0,\ \forall x \in D(A). \tag{2.46}$$

Our purpose in this section is to see whether we can derive a similar condition to the above when $T(t)$ is a semigroup of contractions on a Banach space. And more importantly, we are interested in investigating whether a condition such as (2.46) guarantees the generation of semigroups of contractions on Banach spaces, or, in particular, Hilbert spaces. For this purpose, we need the concept of *duality set* $F(x) \in X^*$, the duality space of X, for any $x \in X$, which is defined by

$$F(x) = \{x^* \in X^* | \langle x,\ x^* \rangle = \|x\|^2 = \|x^*\|^2 \}, \tag{2.47}$$

where $\langle \cdot,\ \cdot \rangle$ denotes the duality paring between X and X^*. It should be noted that $F(x) \neq \emptyset$ for any $x \in X$ by the Hahn-Banach theorem.

Definition 2.22 *Let X be a Banach space and let $F(x)$ be the duality set. A linear operator A in X is said to be* dissipative *if for every $x \in D(A)$ there is an $x^* \in F(x)$ such that*

$$\mathrm{Re}\langle Ax,\ x^* \rangle \leq 0. \tag{2.48}$$

Theorem 2.23 *Let X be a Banach space. Let $x, y \in X$. Then*

$$\|x\| \leq \|x - hy\|,\ \forall h > 0$$

if and only if there exists an $x^* \in F(x)$ such that $\mathrm{Re}\langle y,\ x^* \rangle \leq 0$.

Proof. Suppose that $\mathrm{Re}\langle y,\ x^* \rangle \leq 0$ for some $x^* \in F(x)$. Then for any $h > 0$,

$$\|x - hy\|\,\|x\| \geq |\langle x - hy,\ x^* \rangle| \geq \mathrm{Re}\langle x - hy,\ x^* \rangle \geq \|x\|^2,$$

hence, $\|x\| \leq \|x - hy\|$. Conversely, suppose $x \neq 0$ (the case of $x = 0$ is trivial) and $\|x\| \leq \|x - hy\|$ for all $h > 0$. Let $x_h^* \in F(x - hy)$ and let $y_h^* = \|x_h^*\|^{-1} x_h^*$. Then

$$\|x\| \leq \|x - hy\| = \langle x - hy, \, y_h^* \rangle = \mathrm{Re}\langle x, \, y_h^* \rangle - h\,\mathrm{Re}\langle y, \, y_h^* \rangle \leq \|x\| - h\,\mathrm{Re}\langle y, \, y_h^* \rangle.$$

Therefore,

$$\mathrm{Re}\langle y, \, y_h^* \rangle \leq 0, \quad \text{and} \quad \langle x, \, y_h^* \rangle \geq \|x\| + h\langle y, \, y_h^* \rangle.$$

Since the closed unit ball of X^* is weak* compact, we may assume, without loss of generality, that $y_h^* \to y_0^*$ weakly as $h \to 0$. Letting $h \to 0$, we arrive at

$$\mathrm{Re}\langle y, \, y_0^* \rangle \leq 0, \quad \text{and} \quad \|x\| \leq \langle x, \, y_0^* \rangle \leq \|x\|.$$

Let $x^* = \|x\| y_0^*$. Then the above relations show that $x^* \in F(x)$ and $\mathrm{Re}\langle y, \, x^* \rangle \leq 0$. This completes the proof. $\qquad\square$

Corollary 2.24 *Let A be a linear operator in a Banach space X. Then A is dissipative if and only if $\|x\| \leq \|x - hAx\|$ for each $h > 0$ and all $x \in D(A)$.*

Proof. The proof easily follows from Theorem 2.23 by taking $y = Ax$. $\qquad\square$

Proposition 2.25 *Let A be a linear operator in a Banach space X. Then A generates a C_0-semigroup of contractions on X if and only if*

(i) $\overline{D(A)} = X$.

(ii) A is dissipative and $\mathcal{R}(\lambda - A) = X$, $\forall \lambda > 0$.

Proof. Obviously, condition (ii) is equivalent to (2.45) by Corollary 2.24. The result then follows from Corollary 2.19. $\qquad\square$

Definition 2.26 *A linear operator A in a Banach space X is called m-dissipative if A is dissipative and $\mathcal{R}(\lambda_0 - A) = X$ for some $\lambda_0 > 0$.*

Theorem 2.27 (Lümer-Phillips) *Let A be a linear operator in a Banach space X. Then A generates a C_0- semigroup of contractions if and only if*

(i) $\overline{D(A)} = X$.

(ii) A is m-dissipative.

Proof. The necessity is obvious. We show the sufficiency. By Proposition 2.25, we need only to show that A is dissipative and $\mathcal{R}(\lambda_0 - A) = X$ imply $\mathcal{R}(\lambda - A) = X$ for all $\lambda > 0$. To this end, denote

$$\Omega = \{\lambda > 0 | \mathcal{R}(\lambda - A) = X\}. \tag{2.49}$$

Then $\Omega \neq \emptyset$ because $\lambda_0 \in \Omega$. Let $\lambda \in \Omega$. By Corollary 2.24, $\lambda \in \rho(A)$ and $\|R(\lambda, A)\| \leq \frac{1}{\lambda}$. Since $\rho(A)$ is open, so is Ω. If Ω is closed then $\Omega = (0, \infty)$.

Let $\lambda_n \in \Omega$ and $\lambda_n \to \lambda > 0$. We want to show that $\mathcal{R}(\lambda - A) = X$. Given $y \in X$, there exists an $x_n \in D(A)$ such that

$$\lambda_n x_n - A x_n = y.$$

Then $\|x_n\| \leq \|(\lambda_n - A)^{-1} y\| \leq 1/\lambda_n \|y\| \leq C$ for some $C > 0$. Observing that

$$\begin{aligned} \lambda_m \|x_n - x_m\| &\leq \|\lambda_m(x_n - x_m) - A(x_n - x_m)\| = |\lambda_n - \lambda_m| \|x_n\| \\ &\leq C|\lambda_n - \lambda_m|, \end{aligned}$$

we see that $\{x_n\}$ is a Cauchy sequence. Assume that $x_n \to x$. Then

$$A x_n \to \lambda x - y.$$

Since $\lambda_0 \in \rho(A)$, A is closed. Therefore, $x \in D(A)$ and $\lambda x - Ax = y$, which means $\mathcal{R}(\lambda - A) = X$ as required. $\qquad\qquad\square$

The following consequence of the Lümer-Phillips theorem, which is frequently used in applications, provides an alternative characterization of contraction semigroups in terms of both the operator itself and its adjoint operator.

Corollary 2.28 *Let A be a linear operator in a Banach space X. Then A generates a C_0−semigroup of contractions on X if and only if*

(i) A is densely defined and closed,

(ii) both A and A^* are dissipative.

Proof. The necessity is obvious since if A generates a C_0−semigroup of contractions, then A is densely defined, dissipative, and satisfies (2.45). Because $\lambda \in \rho(A)$ implies that $\lambda \in \rho(A^*)$ and $R(\lambda, A^*) = R^*(\lambda, A)$, we have

$$\|R(\lambda, A^*)\| = \|R^*(\lambda, A)\| = \|R(\lambda, A)\| \leq \frac{1}{\lambda}, \ \forall \lambda > 0,$$

which shows that A^* is dissipative by Corollary 2.24. To prove sufficiency, it suffices to prove that $\mathcal{R}(I - A) = X$ by the Lümer-Phillips theorem. Since

A is dissipative, $\mathcal{R}(I - A)$ is a closed subspace of X by Corollary 2.24. If $\mathcal{R}(I-A) \neq X$ then there exists an $x^* \in X^*, x^* \neq 0$ such that $\langle x - Ax, \; x^* \rangle = 0$ for all $x \in D(A)$ and so $\langle Ax, \; x^* \rangle = \langle x, \; x^* \rangle$ which implies that $x^* \in D(A^*)$ and $x^* = A^* x^*$. But since A^* is also dissipative, it follows from Corollary 2.24 that $x^* = 0$, a contradiction. Thus, $\mathcal{R}(I - A) = X$. $\qquad\square$

When X is reflexive, the condition $\overline{D(A)} = X$ can be removed in the Lümer-Phillips theorem.

Theorem 2.29 *Let A be a linear operator in a Banach space X. If X is reflexive, then A generates a C_0-semigroup of contractions on X if and only if A is dissipative with $\mathcal{R}(\lambda_0 - A) = X$ for some $\lambda_0 > 0$.*

The proof of Theorem 2.29 is based on the following Lemma

Lemma 2.30 *Let A be a linear operator in a reflexive Banach space X such that there exist $\omega, M > 0$ verifying the property: for all $\lambda > \omega, \lambda \in \rho(A)$ and*

$$\|R(\lambda, A)\| \leq \frac{M}{\lambda - \omega},$$

then $\overline{D(A)} = X$.

Proof. If $\overline{D(A)} \neq X$, then by the Hahn-Banach theorem, there exists an $x_0^* \in X^*$ such that

$$\langle x_0, \; x_0^* \rangle = 1, \; \langle x, \; x_0^* \rangle = 0, \; \forall x \in \overline{D(A)}$$

for any given $x_0 \in X \backslash \overline{D(A)}$. Since the sequence $\|nR(n, A)x_0\|, n \in N, n > \omega$ is bounded and X is reflexive, there exists a subsequence $\{n_k R(n_k, A)x_0\}$ converging weakly to some $y \in X$. Hence, $AR(n_k, A)x_0$ converges weakly to $y - x_0$. As A is weakly closed and $R(n_k, A)x_0 \to 0$, we have $y = x_0$. But $\{n_k R(n_k, A)x_0\} \subset D(A)$ and $0 = \lim_{k \to \infty} \langle n_k R(n_k, A)x_0, \; x^* \rangle = \langle x_0, \; x_0^* \rangle = 1$. This is a contradiction. Therefore, we obtain the conclusion. $\qquad\square$

Proof of Theorem 2.29. Only the proof of the sufficiency is needed. Since A is dissipative and $\mathcal{R}(\lambda_0 - A) = X$ we have shown in the proof of Theorem 2.27 that $(0, \infty) \subset \rho(A)$ and $\|R(\lambda, A)\| \leq \dfrac{1}{\lambda}$ for a $\lambda > 0$. It follows from Lemma 2.30 that $\overline{D(A)} = X$. By the Lümer-Phillips theorem, we conclude that A generates a C_0-semigroup of contractions. $\qquad\square$

Theorem 2.31 *Let A be a dissipative operator in a Banach space X.*

(i) If $D(A)$ is dense, then A is closable.

(ii) If A is closable, then $\bar{A}$, the closure of A, is also dissipative.

Proof. (i): Note that A is closable if and only if for a sequence $u_n \in D(A)$, $u_n \to 0$ and $Au_n \to y$ implies $y = 0$. By contradiction, assume that A is not closable. Then there exists a sequence $u_n \in D(A)$ such that $u_n \to 0$ and $Au_n \to y \neq 0$. Then for any $u \in D(A)$ and $\lambda > 0$ it follows from Corollary 2.24 that

$$\|\lambda(u + \lambda u_n) - A(u + \lambda u_n)\| \geq \lambda \|u + \lambda u_n\|.$$

By letting $n \to \infty$, and then dividing by λ, we obtain

$$\left\| u - y - \frac{1}{\lambda} Au \right\| \geq \|u\|, \quad u \in D(A), \ \lambda > 0.$$

However, since $\|u - y - \frac{1}{\lambda} Au\| \leq \|u - y\| + \frac{1}{\lambda}\|Au\|$ and $D(A)$ is dense in H, this is impossible if $y \neq 0$. Hence A is closable.

(ii): Let $u \in D(\bar{A})$. Then there exists a sequence $u_n \in D(A)$ such that $u_n \to u$ and $Au_n \to \bar{A}u$. Since A is dissipative, for any $\lambda > 0$ we have $\|(\lambda I - \bar{A})u\| \geq \lambda \|u\|$. Hence by Corollary 2.24, $\bar{A}$ is dissipative. $\quad\square$

Finally, we state a very important theorem which is called Stone's theorem.

Theorem 2.32 (Stone) *Let A be a linear operator in a Hilbert space H. Then A generates a C_0-group of unitary operators on H if and and only if iA is self-adjoint, or A is skew-adjoint: $A^* = -A$.*

Proof. Recall that an isometric operator U on a Hilbert space H is called unitary if U is an isometry and $\mathcal{R}(U) = H$, or, in other words, U is bounded and $U^* = U^{-1}$.

If A generates a C_0-group of unitary operators $T(t)$, then A is densely defined, so is A^*. It can easily be shown that A^* is the generator of the C_0-semigroup $\{T^*(t), t \geq 0\}$ (see also Proposition 2.75). Since $T(t)$ is unitary, we have

$$-Ax = \lim_{t \downarrow 0} \frac{T(-t)x - x}{t} = \lim_{t \downarrow 0} \frac{T^*(t)x - x}{t} = A^* x$$

for all $x \in D(A)$ or $x \in D(A^*)$, which implies that $A = -A^*$ and so iA is self-adjoint.

Conversely, if iA is self-adjoint then A is densely defined and $A^* = -A$. Thus for every $x \in D(A)$, since

$$\langle Ax, \ x \rangle = \langle x, \ A^* x \rangle = \langle x, \ -Ax \rangle = -\overline{\langle Ax, \ x \rangle},$$

it follows that $\mathrm{Re}\langle Ax, \ x \rangle = 0$ for all $x \in D(A)$, from which we see that both A and A^* are dissipative. We show that $\mathcal{R}(I - A) = H$ and $\mathcal{R}(I - A^*) = H$. Indeed, if there is a $y \in H$ such that

$$\langle x - Ax, \ y \rangle = 0$$

for all $x \in D(A)$, then $\langle Ax, y \rangle = \langle x, y \rangle$ for all $x \in D(A)$. This shows that $y \in D(A^*)$ and $A^* y = y$. By virtue of Corollary 2.24, we arrive at $\|y\| \leq \|y - A^* y\| = 0$. Therefore, $\mathcal{R}(I - A) = H$. Similarly it can be shown that $\mathcal{R}(I - A^*) = H$. It follows from the Lümer-Phillips theorem that A and $A^* = -A$ generates C_0-semigroups of contractions $T_+(t)$ and $T_-(t)$, respectively. Therefore, A generates a C_0-group $T(t)$ given by

$$T(t) = \begin{cases} T_+(t), t \geq 0, \\ \\ T_-(-t), t < 0. \end{cases}$$

Since $T^{-1}(t) = T(-t)$, $T(t)$ is an isometry and $\mathcal{R}(T(t)) = H$. Therefore, $T(t)$ is a unitary C_0-group on H. $\qquad\square$

Example 2.33 *Let $H = L^2(0,1)$ and let us define an operator A as $A\varphi = -\varphi'$ for $\varphi \in D(A)$ where*

$$D(A) = \{ \varphi \in H \mid \varphi' \in H, \ \varphi(0) = 0 \ \}.$$

In Example 2.21, by using the Hille-Yosida theorem it was shown that A generates a C_0-semigroup of contractions on H. Here we will obtain the same result by using the Lümer-Phillips theorem. The standard inner product on H is given as

$$\langle u, v \rangle = \int_0^1 u(s)\overline{v(s)}ds \quad u, v, \in H.$$

For $u \in D(A)$ we have

$$Re\langle Au, u \rangle = -\int_0^1 u'(s)\overline{u(s)}ds = -\frac{1}{2}|u(1)|^2 \leq 0,$$

hence A is dissipative. To apply the Lümer-Phillips theorem it remains to show that the range of $\lambda I - A$ is H for some $\lambda > 0$. Let $\lambda > 0$ and $f \in H$ be given. Then $(\lambda I - A)u = \lambda u + u' = f$ has a unique solution $u(x) = \int_0^x e^{-\lambda(x-s)} f(s)ds \in D(A)$. Hence, the range condition is satisfied, and by the Lümer-Phillips theorem A generates a C_0-semigroup on H.

Example 2.34 *Let H be a Hilbert space and consider the following second order system*

$$\ddot{y}(t) + B\dot{y}(t) + Ay(t) = 0, \quad y(0) = y_0, \quad \dot{y}(0) = y_1, \qquad (2.50)$$

where A and B are linear operators on H. We assume the following.

(i) $D(A)$ is dense in H. and A is self-adjoint and coercive. i.e., for some $\alpha > 0$,

$$\langle Ax, \ x\rangle \geq \alpha\|x\|^2, \quad x \in D(A). \tag{2.51}$$

(ii) $D(B)$ is dense in H. and B is self-adjoint and is nonnegative. i.e. $\langle Bx, \ x\rangle \geq 0$ for $x \in D(B)$.

We note that since A is self-adjoint and coercive, it has a unique square root $A^{1/2}$ which is self-adjoint and nonnegative, with $D(A^{1/2}) \supset D(A)$, and moreover since A is positive, so is $A^{1/2}$, see [41] and [84].

For the system (2.50), we consider the following energy form

$$E(y, \dot{y}) = \frac{1}{2}\langle \dot{y}, \ \dot{y}\rangle + \frac{1}{2}\langle A^{1/2}y, \ A^{1/2}y\rangle. \tag{2.52}$$

Note that since $A^{1/2}$ is positive, $E(y, \dot{y}) \geq 0$ and $E(y, \dot{y}) = 0$ if and only if $y = 0$ and $\dot{y} = 0$. If $y(t)$ is a twice strongly and continuously differentiable solution of (2.50) with $y(t) \in D(A)$ and $\dot{y}(t) \in D(B)$, then by differentiating (2.52) along the solutions of (2.50) we obtain

$$\dot{E}(y, \dot{y}) = \langle \dot{y}, \ \ddot{y} + Ay\rangle = -\langle \dot{y}, \ B\dot{y}\rangle \leq 0. \tag{2.53}$$

By using (2.53), we may show the existence of solutions for (2.50) in an appropriate space. By using $z = (y, \ \dot{y})^T$, where the superscript T denotes transpose, we can write (2.50) as

$$\dot{z} = \mathcal{A}z, \quad z(0) \in \mathcal{H}, \tag{2.54}$$

where

$$\mathcal{A} = \begin{bmatrix} 0 & I \\ -A & -B \end{bmatrix}, \tag{2.55}$$

and $\mathcal{H}$ will be defined in the sequel. It is clear that if we choose $\mathcal{H}$ such that $\mathcal{A}$ is dissipative, then we may use the Lümer-Phillips theorem to prove the existence of a C_0−semigroup. From (2.52) and (2.53) it follows that a natural choice is $\mathcal{H} = D(A^{1/2}) \times H$. For $z_1 = (u_1, \ v_1)^T, z_2 = (u_2, \ v_2)^T \in \mathcal{H}$, we can define, as usual, the inner product in $\mathcal{H}$ as

$$\langle z_1, \ z_2\rangle = \langle A^{1/2}u_1, \ A^{1/2}u_2\rangle + \langle v_1, \ v_2\rangle, \tag{2.56}$$

and the norm induced by (2.56). We note that if a linear operator Q is self-adjoint and nonnegative, then Q is coercive if and only if Q^{-1} is bounded, see [41] and [84]. It then follows that $A^{-1/2}$ is bounded and hence $\mathcal{H}$ with the norm induced by (2.56) is complete. In the sequel, we will assume that $D(A^{1/2}) \cap D(B)$ is dense in $\mathcal{H}$. Then $D(\mathcal{A}) = D(A) \times (D(A^{1/2}) \cap D(B))$. For $z = (u, \ v)^T \in D(\mathcal{A})$ we have

$$\langle z, \ \mathcal{A}z\rangle = \langle A^{1/2}u, \ A^{1/2}v\rangle + \langle v, \ -Au - Bv\rangle = -\langle v, \ Bv\rangle \leq 0,$$

hence $\mathcal{A}$ is dissipative on $\mathcal{H}$. The dual operator $\mathcal{A}^*$ is easily found to be

$$\mathcal{A}^* = \begin{bmatrix} 0 & -I \\ A & -B \end{bmatrix},$$

with $D(\mathcal{A}^*) = D(\mathcal{A})$. It follows that for $z = (u, v)^T \in D(\mathcal{A}^*)$, we have $\langle z, \mathcal{A}^* z \rangle = -\langle v, Bv \rangle \leq 0$, which indicates that $\mathcal{A}^*$ is also dissipative. Hence, if $\mathcal{A}$ is closed, then by Corollary 2.28, $\mathcal{A}$ generates a C_0-semigroup of contractions on $\mathcal{H}$. Obviously, the closedness of $\mathcal{A}$ depends on B, and in general it is not easy to characterize such operators. In the sequel, we will show that if $A^{-1/2}B$ is bounded, then $\mathcal{A}$ is closed. To see this, assume that there exists a sequence $z_n = (u_n, v_n)^T \in D(\mathcal{A})$ and $z = (u, v)^T \in \mathcal{H}$, $y = (w, x)^T \in \mathcal{H}$ such that $z_n \to z$ and $\mathcal{A}z_n \to y$ as $n \to \infty$. Then $u_n \in D(A) \to u \in D(A^{1/2})$, $v_n \in D(A^{1/2}) \to v \in H$, $v_n \in D(A^{1/2}) \to w \in D(A^{1/2})$, and $-Au_n - Bv_n \in H \to x \in H$. Obviously, $v = w \in D(A^{1/2})$, and since $A^{-1/2}$ is bounded, we have $A^{-1/2}(-Au_n - Bv_n) = -A^{1/2}u_n - A^{-1/2}Bv_n \to A^{-1/2}x$. Since $A^{-1/2}B$ is also bounded, we have $A^{-1/2}Bv_n \to A^{-1/2}Bv$; hence, $-A^{1/2}u_n \to A^{-1/2}x + A^{-1/2}Bv$. Since $A^{1/2}$ is closed, the latter implies $-A^{1/2}u_n \to -A^{1/2}u = A^{-1/2}x + A^{-1/2}Bv$. But $A^{-1/2}(x + Bv) \in D(A^{1/2})$ hence, $A^{1/2}u \in D(A^{1/2})$, $u \in D(A)$, and $-Au - Bv = x$. Therefore, $\mathcal{A}$ is closed.

The class of operators such that $A^{-1/2}B$ is bounded may not be characterized easily. If B is of the form $B = \alpha I + \beta A^{1/2}$, $\alpha \geq 0$, $\beta \geq 0$, this property is satisfied and hence $\mathcal{A}$ generates a C_0-semigroup of contractions for such cases. If B is of the form $B = \rho A$, $\rho > 0$, which may occur in some applications, $\mathcal{A}$ is not closed, see [31]. However, in such cases since $\mathcal{A}$ is dissipative and $D(\mathcal{A})$ is dense, by Theorem 2.31, $\mathcal{A}$ is closable and $\bar{\mathcal{A}}$ is dissipative. Hence, we may use $\bar{\mathcal{A}}$ for the existence of a C_0-semigroup.

Example 2.35 *Consider the undamped wave equation given below .*

$$\begin{cases} y_{tt}(x, t) - y_{xx}(x, t) = 0, & 0 < x < 1, \quad t \geq 0, \\ y(0, t) = y(1, t) = 0, \\ y(x, 0) = f(x), \quad y_t(x, 0) = g(x), \end{cases}$$

where $f(x)$ and $g(x)$ are appropriate functions. Following Example 2.34, we define the operator A on $H = L^2(0, 1)$ as $Az = -z''$, and $D(A)$ as

$$D(A) = \{ z \in H \mid z, z', z'' \in H, \quad z(0) = z(1) = 0 \}.$$

It is well known that A is self-adjoint and nonnegative. Moreover, for any $z \in D(A)$ we have

$$Re\langle Az, z \rangle = -\int_0^1 z''(s)\overline{z(s)}ds = \int_0^1 |z'(s)|^2 ds.$$

But we also have $z(x) = \int_0^x z'(s)ds$ for $z \in D(A)$, hence by Schwartz inequality we have $\int_0^1 |z(s)|^2 ds \leq \int_0^1 |z'(s)|^2 ds$. Hence, we obtain $Re\langle Az, z\rangle \geq \langle z, z\rangle$ for $z \in D(A)$, i.e. A is coercive. Therefore $A^{1/2}$ exists and we can transform the wave equation given above into the following abstract Cauchy problem

$$\dot{z} = \mathcal{A}z, \quad z(0) \in \mathcal{H}, \tag{2.57}$$

where

$$\mathcal{A} = \begin{bmatrix} 0 & I \\ -A & 0 \end{bmatrix},$$

with $\mathcal{H} = D(A^{1/2}) \times H$ and $D(\mathcal{A}) = D(A) \times D(A^{1/2})$. It follows from Example 2.34 that $\mathcal{A}$ generates a C_0-semigroup of contractions on $\mathcal{H}$.

Usually it is not easy to obtain an explicit form of $A^{1/2}$, but for this specific example, it is shown in [170] that

$$A^{1/2}z = Q(iz'), \quad z \in D(A^{1/2})$$
$$D(A^{1/2}) = \{z \in H \mid z, z' \in H, z(0) = 0 \}$$

where Q is a bounded operator on H.

Example 2.36 Consider the following Euler-Bernoulli beam equation.

$$\begin{cases} y_{tt}(x,t) + y_{xxxx}(x,t) = 0, & 0 < x < 1, \ t \geq 0 \quad, \\ y(0,t) = y_x(0,t) = y_{xx}(1,t) = y_{xxx}(1,t) = 0, \\ y(x,0) = f(x), \quad y_t(x,0) = g(x), \end{cases}$$

where $f(x)$ and $g(x)$ are appropriate functions. Following Example 2.34 we define the operator A in $H = L^2(0,1)$ as $Az = z''''$ with $D(A)$ defined as

$$D(A) = \{z \in H \mid z, z', z'', z''', z'''' \in H, \ z(0) = z'(0) = z''(1) = z'''(1) = 0 \}.$$

It can easily be shown that A is self-adjoint and nonnegative. Using integration by parts, we obtain $Re\langle z, Az\rangle = \int_0^1 |z''(s)|^2 ds$ for $z \in D(A)$. Also for $z \in D(A)$, we obtain $\int_0^1 |z(s)|^2 ds \leq \int_0^1 |z'(s)|^2 ds \leq \int_0^1 |z''(s)|^2 ds$, see Example 2.35. Hence, we have $\langle z, Az\rangle \geq \langle z, z\rangle$ for $z \in D(A)$, i.e., A is coercive. Therefore, $A^{1/2}$ exists. Following Examples 2.34 and 2.35, we can transform the Euler-Bernoulli beam equation into the abstract form given by (2.57) and following Example 2.34, $\mathcal{A}$ generates a C_0-semigroup of contractions in $\mathcal{H} = D(A^{1/2}) \times H$ where

$$D(A^{1/2}) = \{z \in H \mid z, z', z'' \in H, z(0) = z'(0) = 0\}.$$

We now state two perturbation results on C_0-semigroup generation.

Theorem 2.37 (perturbations by bounded linear operators) Let X be a Banach space and let A be the infinitesimal generator of a C_0-semigroup $T(t)$ on X, satisfying $\|T(t)\| \leq Me^{\omega t}$. If B is a bounded linear operator on X, then $A+B$ generates a C_0-semigroup $S(t)$ on X and $\|S(t)\| \leq Me^{(\omega + M\|B\|)t}$.

Proof. Define a norm

$$|x| = \sup_{t \geq 0} \left\| e^{-\omega t} T(t)x \right\|$$

on X. It is obvious that

$$\|x\| \leq |x| \leq M \|x\|,$$

i.e., the two norms $|\cdot|$ and $\|\cdot\|$ on X are equivalent. Moreover,

$$
\begin{aligned}
|T(t)x| &= \sup_{s \geq 0} \left\| e^{-\omega s} T(s)T(t)x \right\| \\
&= \sup_{s \geq 0} \left\| e^{\omega t} e^{-\omega(s+t)} T(s+t)x \right\| \leq e^{\omega t} |x|.
\end{aligned}
$$

Therefore, we need only to check the Hille-Yosida condition under this new norm. For this purpose, let us note that

$$|BR(\lambda, A)| \leq \frac{|B|}{\lambda - \omega} < 1$$

when $\lambda > \omega + |B|$. Thus,

$$R(\lambda, A + B) = R(\lambda, A)[I - BR(\lambda, A)]^{-1} = R(\lambda, A) \sum_{n=0}^{\infty} [BR(\lambda, A)]^n$$

and

$$|R(\lambda, A + B)| \leq \frac{1}{\lambda - \omega} [1 - |BR(\lambda, A)|]^{-1} \leq \frac{1}{\lambda - \omega - |B|}$$

which says that $A + B$ generates a C_0-semigroup. Let $S(t) = e^{(A+B)t}$. Then $|S(t)| \leq e^{(\omega+|B|)t}$. Returning to the original norm $\|\cdot\|$ we have $\|S(t)\| \leq Me^{(\omega+M\|B\|)t}$. $\qquad \square$

Theorem 2.38 (perturbations by unbounded linear operators) *Let $T(t)$ be a C_0-semigroup with generator A in a Banach space X. Assume that $B : D(A) \to D(A)$ is linear and continuous in the graph norm on $D(A)$. Then $A + B$ with domain $D(A + B) = D(A)$ is the generator of a C_0-semigroup.*

Proof. We first show that $(I - BR(\lambda, A))$ is invertible for some $\lambda \in \mathbf{C}$. Choose $\lambda_0 \in \rho(A)$. Then $S = (\lambda_0 - A)BR(\lambda_0, A)$ is bounded on X by the closed graph theorem since S is closed with domain $D(S) = X$. Let $\lambda > 0$ be sufficiently large such that $\|SR(\lambda, A)\| < 1$. Since $T = (\lambda_0 - A)BR(\lambda, A)$ is also bounded, $TR(\lambda_0, A)$ and $R(\lambda_0, A)T$ have the same non-zero spectrum. But

$$R(\lambda_0, A)T = BR(\lambda, A)$$
$$TR(\lambda_0, A) = (\lambda_0 - A)BR(\lambda, A)R(\lambda_0, A) = SR(\lambda, A),$$

hence $1 \in \rho(BR(\lambda, A))$ because $1 \in \rho(SR(\lambda, A))$, which means $(I - BR(\lambda, A))^{-1}$ is bounded on X. Let $C = (A - \lambda)B(A - \lambda)^{-1}$ which is bounded. Then $A + C$ is the generator of a C_0−semigroup by Theorem 2.37. Let $U = I - BR(\lambda, A)$. Then U is an isomorphism on X with $UD(A) = D(A)$. Moreover,

$$
\begin{aligned}
U(A + C)U^{-1} &= U(A - \lambda + C)U^{-1} + \lambda \\
&= U[A - \lambda - (A - \lambda)BR(\lambda, A)]U^{-1} + \lambda \\
&= U(A - \lambda)[I - BR(\lambda, A)]U^{-1} + \lambda \\
&= U(A - \lambda) + \lambda = A + B.
\end{aligned}
$$

Therefore, $A + B$ generates a C_0−semigroup $Ue^{(A+C)t}U^{-1}$.   □

2.4   Relation with the Laplace transform

Let $T(t)$ be a C_0−semigroup on a Banach space X satisfying $\|T(t)\| \leq Me^{\omega t}$, and let A be its infinitesimal generator. By Theorem 2.15 we have

$$R(\lambda, A)x = \int_0^\infty e^{-\lambda t}T(t)x\,dt, \quad \mathrm{Re}\lambda > \omega, \quad x \in X, \qquad (2.58)$$

i.e., $R(\lambda, A)$ is the Laplace transform of $T(t)$. Actually, the converse statement also holds.

Theorem 2.39 *Let A be a densely defined and closed linear operator in a Banach space X and let $T(t)$ be a strongly continuous family of bounded linear operators for $t \geq 0$ satisfying $\|T(t)\| \leq Me^{\omega t}$ for some $M > 0$ and $\omega \in \mathbf{R}$. If (2.58) holds, then $T(t)$ is the C_0−semigroup generated by A.*

Proof.   Note that the following resolvent equation holds for all λ, $\mu \in \rho(A)$.

$$R(\lambda, A) - R(\mu, A) = (\mu - \lambda)R(\lambda, A)R(\mu, A). \qquad (2.59)$$

For simplicity, assume that $\mathrm{Re}\mu > \mathrm{Re}\lambda$. By using (2.58) and Fubini's theorem we obtain

$$\frac{1}{\mu - \lambda}[R(\lambda, A) - R(\mu, A)] = \int_0^\infty e^{-(\mu - \lambda)t}dt[\int_0^\infty e^{-\lambda \tau}T(\tau)d\tau$$

$$- \int_0^\infty e^{-\mu\tau} T(\tau) d\tau \ ]$$

$$= \int_0^\infty \int_0^\infty e^{-(\mu-\lambda)t} e^{-\lambda\tau} T(\tau) d\tau dt$$

$$- \int_0^\infty \int_0^\infty e^{-(\mu-\lambda)(t+\tau)} e^{-\lambda\tau} T(\tau) d\tau dt$$

$$= \int_0^\infty (\int_0^\tau e^{-(\mu-\lambda)s} ds \ ) e^{-\lambda\tau} T(\tau) d\tau$$

$$= \int_0^\infty (\int_s^\infty e^{-\lambda\tau} T(\tau) d\tau) e^{-(\mu-\lambda)s} ds$$

$$= \int_0^\infty \int_0^\infty e^{-\lambda t} e^{-\mu s} T(t+s) ds dt. \qquad (2.60)$$

We also have

$$R(\lambda, A) R(\mu, A) = \int_0^\infty \int_0^\infty e^{-\lambda t} e^{-\mu s} T(t) T(s) dt ds. \qquad (2.61)$$

By using (2.59)-(2.61), the strong continuity of $T(t)$ and the uniqueness of the Laplace transform, we obtain $T(t+s) = T(t)T(s)$ for $t, s \geq 0$. Hence, we have $T(t) = T(t)T(0)$ for $t \geq 0$. Substituting this into (2.58) we obtain $R(\lambda, A)x = R(\lambda, A)T(0)x$, hence $T(0)x = x$, $\forall x \in X$, i.e., $T(0) = I$. Consequently, we have shown that $T(t)$ is a C_0−semigroup. Let B be its infinitesimal generator. By Theorem 2.15 we have

$$R(\lambda, B)x = \int_0^\infty e^{-\lambda t} T(t)x dt, \quad \text{Re}\lambda > \omega, \quad x \in X.$$

This, together with (2.58), implies that $R(\lambda, A) = R(\lambda, B)$ for Reλ sufficiently large, hence $A = B$. $\qquad \square$

The next result relates the semigroup $T(t)$ to the inverse Laplace transform of $R(\lambda, A)$.

Theorem 2.40 *Let $T(t)$ be a C_0−semigroup satisfying $\|T(t)\| \leq Me^{\omega t}$ for some $M > 0$ and $\omega \in \mathbf{R}$ on a Banach space X and let A be its infinitesimal generator.*

(i) For any $\gamma > \omega$ and $x \in D(A)$

$$T(t)x = \lim_{\tau \to \infty} \frac{1}{2\pi i} \int_{\gamma-i\tau}^{\gamma+i\tau} e^{\lambda t} R(\lambda, A)x d\lambda, \qquad (2.62)$$

where the limit is a strong limit in X.

(ii) For any $\gamma > \omega$ and $x \in D(A^2)$

$$T(t)x = \frac{1}{2\pi i} \int_{\gamma - i\infty}^{\gamma + i\infty} e^{\lambda t} R(\lambda, A)x d\lambda, \qquad (2.63)$$

where the integral converges uniformly in $t > 0$.

Remark 2.41 *Note that for an arbitrary $x \in X$, the existence of the limit in (2.62) does not guarantee the existence of the improper integral given in (2.63). On the other hand, if the improper integral in (2.63) exists, then the limit in (2.62) also exists and they are both equal, see [34, p. 193], [51, p. 160]. When X is a Hilbert space, it is shown that formula (2.62) holds for all $x \in X$. See [169] for details.*

Proof. If $x \in D(A)$, then by Theorem 2.12, $T(t)x$ is differentiable. Since $R(\lambda, A)x$ is the Laplace transform of $T(t)x$, (2.62) follows from classical results on the inversion of Laplace transform, see [34, p. 199], [51, p. 157].

If $x \in D(A^2)$, then by Theorem 2.12, $T(t)x$ is twice differentiable. Let us define $f(t) = e^{-\gamma t} T(t)x$. Then $f(t)$ is also twice differentiable with $f, \dot{f}, \ddot{f} \in L^1$, and $f(t), \dot{f}(t) \to 0$ as $t \to \infty$. By using classical results on Fourier analysis and the relation between the Fourier and Laplace transforms, we obtain (2.63), see e.g. [87, p. 298, p. 373].   □.

The following theorem gives sufficient conditions for an operator to generate a C_0–semigroup.

Theorem 2.42 *Let A be a densely defined and closed linear operator on a Banach space X. Assume that A satisfies the following conditions.*

(i) For some $\theta \in (0, \pi/2)$, $\Sigma_\theta = \{\lambda \in \mathbb{C} \mid |\arg \lambda| < \pi/2 + \theta\} \cup \{0\} \subset \rho(A)$.

(ii) There exists a constant $M > 0$ such that

$$\|R(\lambda, A)\| \leq \frac{M}{|\lambda|}, \qquad \forall \lambda \in \Sigma_\theta, \quad \lambda \neq 0. \qquad (2.64)$$

Then A generates a uniformly bounded C_0–semigroup $T(t)$. Moreover, $T(t)$ can be expressed as

$$T(t) = \frac{1}{2\pi i} \int_\Gamma e^{\lambda t} R(\lambda, A) d\lambda, \qquad (2.65)$$

where Γ is a smooth curve in Σ_θ running from $\infty e^{-i\nu}$ to $\infty e^{i\nu}$ for $\pi/2 < \nu < \pi/2 + \theta$, and the integral converges for $t > 0$ in the uniform operator topology.

Proof. Let $T(t)$ be defined by (2.65). It follows from (2.64) that the integral converges for $t > 0$ in the uniform operator topology. Since $R(\lambda, A)$ is analytic in $\sum_\theta$, by Cauchy's principle we can shift the path Γ to an appropriate path Γ_n in $\sum_\theta$ without changing the value of the integral. Let us choose $\Gamma_n = \Gamma_1 \cup \Gamma_2 \cup \Gamma_3$ where $\Gamma_1 = \{re^{-i\nu} \mid t^{-1} \le r < \infty\}$, $\Gamma_2 = \{t^{-1}e^{i\psi} \mid -\nu \le \psi < \nu\}$, $\Gamma_3 = \{re^{i\nu} \mid t^{-1} \le r < \infty\}$. By evaluating the integral in (2.65) on Γ_n and by using (2.64) we see that $T(t)$ is uniformly bounded, i.e., for some $C > 0$ we have $\|T(t)\| \le C$, $\forall t \ge 0$. Next, we will show that $R(\lambda, A)$ is the Laplace transform of $T(t)$. Let $\mathrm{Re}\lambda > 0$ and consider the following.

$$\int_0^T e^{-\lambda t}T(t)dt = \frac{1}{2\pi i}\int_0^T e^{-\lambda t}\int_\Gamma e^{\mu t}R(\mu, A)d\mu dt$$

$$= \frac{1}{2\pi i}\int_\Gamma \frac{1}{\mu - \lambda}(e^{(\mu-\lambda)T} - 1)R(\mu, A)d\mu, \quad (2.66)$$

where we used Fubini's theorem. By using the extended Cauchy integral formula, see e.g. [34, p. 181], we obtain

$$\frac{1}{2\pi i}\int_\Gamma \frac{1}{\lambda - \mu}R(\mu, A)d\mu = R(\lambda, A).$$

Also, since $\mathrm{Re}\mu \le 0$ for $\mu \in \Gamma$ and $\mathrm{Re}\lambda > 0$, by using (2.64) we obtain

$$\left\|\int_\Gamma e^{(\mu-\lambda)T}\frac{1}{\lambda - \mu}R(\mu, A)d\mu\right\| \le Me^{-\mathrm{Re}\lambda T}\int_\Gamma \frac{|d\mu|}{|\mu|\,|\mu - \lambda|},$$

hence, as $T \to \infty$, this integral vanishes. Therefore, by letting $T \to \infty$ in (2.66) we obtain

$$R(\lambda, A) = \int_0^\infty e^{-\lambda t}T(t)dt.$$

By Theorem 2.39, $T(t)$ is the semigroup generated by A. $\qquad\square$

The importance of this theorem is that $T(t)$ can be determined uniquely by $R(\lambda, A)$, which can be calculated in most of practical problems when A is available. Another generation theorem due to Crandall-Liggett is motivated from solving the abstract Cauchy problem (2.1) by Euler implicit scheme obtained by replacing $z'(t)$ in (2.1) with the differential quotient $(z(t+h) - z(t))/h$. Namely, given $t > 0$ and a partition $0 < \frac{t}{n} < \cdots < \frac{nt}{n} = t$, solving

$$\frac{z_n(t + k/n) - z_n(t + (k-1)/n)}{h} = Az_n(t + k/n), \quad h = \frac{t}{n}, \, 1 \le k \le n$$

with $z(0) = x$, we have

$$z_n(t) = \left(I - \frac{t}{n}A\right)^{-n}x.$$

Again, we expect that $z_n(t)$ converges to $z(t)$ as $n \to \infty$.

We close this section with the proof of this idea.

Theorem 2.43 (The exponential formula) *Let* $T(t)$ *be a* C_0-*semigroup on a Banach space* X, *and let* A *be its infinitesimal generator. Then the following holds.*

$$T(t)x = \lim_{n\to\infty} \left(I - \frac{tA}{n}\right)^{-n} x = \lim_{n\to\infty} \left[\frac{n}{t}R(\frac{n}{t}, A)\right]^n x, \quad x \in X, \qquad (2.67)$$

where the limit is uniform in t *on bounded intervals.*

Proof. Since $R(\lambda, A)$ is analytic for $\mathrm{Re}\lambda > \omega$ and (2.22) holds, for a given $t > 0$, by differentiating (2.22) n times with respect to λ, then setting $\lambda = \frac{n}{t}$, defining a new variable as $s = \tau/t$ and then by using (2.21) we obtain

$$\left[\frac{n}{t}R(\frac{n}{t}, A)\right]^{n+1} x = \frac{n^{n+1}}{n!} \int_0^\infty (se^{-s})^n T(st)x\,ds. \qquad (2.68)$$

Using integration by parts successively, it can easily be shown that

$$\frac{n^{n+1}}{n!} \int_0^\infty (se^{-s})^n ds = 1. \qquad (2.69)$$

Using (2.68) and (2.69) we obtain

$$\left[\frac{n}{t}R(\frac{n}{t}, A)\right]^{n+1} x - T(t)x = \frac{n^{n+1}}{n!} \int_0^\infty (se^{-s})^n [T(st)x - T(t)x]ds. \qquad (2.70)$$

Since $T(t)$ is strongly continuous, for any $\epsilon > 0$ we can find constants $0 < s_1 < 1 < s_2 < \infty$ such that

$$\|T(st)x - T(t)x\| < \epsilon \quad \forall s \in [s_1, s_2]. \qquad (2.71)$$

Moreover, if t belongs to a compact interval Ω, then s_1, s_2 could be chosen accordingly such that (2.71) holds $\forall t \in \Omega$. By breaking the integral in (2.70) into three parts we obtain the following estimates.

$$\lim_{n\to\infty} \frac{n^{n+1}}{n!} \left\| \int_0^{s_1} (se^{-s})^n [T(st)x - T(t)x]ds \right\| = 0, \qquad (2.72)$$

$$\frac{n^{n+1}}{n!} \left\| \int_{s_1}^{s_2} (se^{-s})^n [T(st)x - T(t)x]ds \right\| < \epsilon, \qquad (2.73)$$

$$\lim_{n\to\infty} \frac{n^{n+1}}{n!} \left\| \int_{s_2}^\infty (se^{-s})^n [T(st)x - T(t)x]ds \right\| = 0, \qquad (2.74)$$

where (2.72) follows from $se^{-s} \leq s_1 e^{-s_1} < e^{-1}$ for $0 \leq s \leq s_1 < 1$; (2.73) follows from (2.69) and (2.71). To show (2.74), we used (2.15), chose $n > \omega t$ and integrated by parts. Note that the limits in (2.72) and (2.74) are uniform in t on compact intervals. Since $\epsilon > 0$ is arbitrary, by using (2.72)-(2.74) and (2.24) in (2.70) we obtain (2.67). $\qquad \square$

2.5   Differentiability and analytic semigroups

Definition 2.44 *Let $T(t)$ be a C_0-semigroup on a Banach space X. $T(t)$ is called differentiable if for every $x \in X$, $T(t)x$ is differentiable for $t > 0$.*

Recall that by Theorem 2.12, if $T(t)$ is a C_0-semigroup generated by A and if $x \in D(A)$, then $T(t)x$ is differentiable for $t \geq 0$, and $\frac{d}{dt}T(t)x = AT(t)x$. If $T(t)$ is differentiable, then this property holds for all $x \in X$ and $t > 0$. Also note that we do not require differentiability at $t = 0$, otherwise for any $x \in X$ we would have

$$\dot{T}(0)x = \lim_{t \downarrow 0} \frac{T(t) - I}{t} x = Ax,$$

hence $x \in D(A)$. Therefore, $D(A) = X$, and A must be a bounded operator and $T(t)$ must be a uniformly continuous semigroup.

Theorem 2.45 *Let $T(t)$ be a differentiable C_0-semigroup on a Banach space X and let A be its infinitesimal generator; then the following hold.*

(i) For every $x \in X$ and $t > 0$, $T(t)x \in D = \cap_{n=1}^{\infty} D(A^n)$.

(ii) For $t > 0$ and $n = 1, 2, \ldots$, we have $T^{(n)}(t) = A^n T(t)$, $T^{(n)}(t)$ is a bounded operator and is continuous in the uniform operator topology and the following holds.

$$T^{(n)}(t) = [AT(\frac{t}{n})]^n = [\frac{d}{dt}T(\frac{t}{n})]^n . \tag{2.75}$$

Proof.   We will prove these statements by induction. For $x \in X$, by using differentiability of $T(t)$ we obtain

$$\frac{d}{dt}T(t)x = \lim_{\Delta \downarrow 0} \frac{T(t+\Delta) - T(t)}{\Delta} x = \lim_{\Delta \downarrow 0} \frac{T(\Delta) - I}{\Delta}T(t)x = AT(t)x,$$

hence $T(t)x \in D(A)$, and (2.75) holds for $n = 1$. Since A is closed and $T(t)$ is bounded, $AT(t)$ is closed and is defined for all $x \in X$. Therefore, by the closed graph theorem, $AT(t)$ is bounded. Let $\|T(t)\| \leq C_1$ for $0 \leq t \leq 1$ for some $C_1 > 0$ and let $0 < s \leq t \leq s + 1$. Then we have

$$
\begin{aligned}
T(t)x - T(s)x &= \int_s^t \frac{d}{d\tau}T(\tau)x d\tau \\
&= \int_s^t AT(\tau)x d\tau = \int_s^t T(\tau - s)AT(s)x d\tau;
\end{aligned}
$$

hence,

$$\|T(t)x - T(s)x\| \leq (t - s)C_1\|AT(s)\|\|x\|,$$

which implies the continuity of $T(t)$ for $t > 0$ in the uniform operator topology.

We will complete the proof by induction. Assume that $T(t)x \in D(A^n)$ and the statement (ii) holds for n. For any fixed $\tau > 0$ such that $0 < \tau < t$ and $x \in X$ we have

$$T^{(n)}(t)x = T(t - \tau)A^n T(\tau)x. \tag{2.76}$$

The right-hand side of (2.76), hence also $T^{(n)}x$, is differentiable. By differentiating (2.76) we obtain $T^{(n+1)}(t)x = A^{n+1}T(t)x$. This implies that $T(t)x \in D(A^{n+1})$ and that $T^{(n+1)} = A^{n+1}T(t)$. This also proves, exactly as given above, that $A^{n+1}T(t)$ is bounded and that $T^{(n+1)}(t)$ is continuous in the uniform operator topology. Since $T(t)x \in D$ for any $x \in X$ and $t > 0$, from Theorem 2.12 it follows that $A^n T(t)x = A^n T(\frac{t}{n})^n x = AT(\frac{t}{n})A^{n-1}T(\frac{t}{n})^{n-1}x$. By repeatedly applying Theorem 2.12, we obtain $A^n T(t)x = [AT(\frac{t}{n})]^n x$ for every $x \in X$. Since $T^{(n)}(t) = A^n T(t)$, (2.75) follows immediately. $\square$

In the previous sections, we considered the semigroups defined on the real nonnegative axis. In some cases, it may be possible to extend the domain of the semigroup to a region in the complex plane containing the real nonnegative axis. In the following, we will consider only some special regions which consist of angles around the nonnegative real axis.

Definition 2.46 *Let X be a Banach space, let $\phi_1, \phi_2 \in [-\pi/2, \pi/2]$ be given, let the sector S be defined as $S = \{ z \in \mathbf{C} \mid \phi_1 < \arg z < \phi_2 \}$ and for $z \in S$, let $T(z)$ be a bounded linear operator. The family of operators $T(z)$, $z \in S$, is called an* analytic semigroup *in S if the following conditions hold.*

(i) $T(z)$ is analytic in S.

(ii) $T(0) = I$ and $\lim_{S \ni z \to 0} T(z)x = x$ for all $x \in X$.

(iii) $T(z_1 + z_2) = T(z_1)T(z_2)$ for all $z_1, z_2 \in S$.

A semigroup $T(t)$ on X is called analytic *if it is analytic in some sector S as defined above.*

Recall that if A is the infinitesimal generator of a C_0-semigroup $T(t)$, then $A - \epsilon I$ is the infinitesimal generator of the C_0-semigroup $e^{-\epsilon t}T(t)$. Since this operation does not change the analyticity properties of $T(t)$, without loss of generality, we may assume that $T(t)$ is uniformly bounded and that $0 \in \rho(A)$.

Theorem 2.47 *Let $T(t)$ be a uniformly bounded C_0-semigroup. Let A be its infinitesimal generator and let $0 \in \rho(A)$. Then the following statements are equivalent.*

(i) $T(t)$ can be extended to an analytic semigroup in a sector $\mathcal{S}_\phi = \{z \in \mathbf{C} \mid |\arg z| < \phi \}$, and $T(z)$ is uniformly bounded in every closed sub-sectors $\bar{\mathcal{S}}_{\phi_*}$ where $\phi_* < \phi$.

(ii) There exists a constant $M_1 > 0$ such that for every $z = \sigma + i\tau$, $\sigma > 0$, $\tau \neq 0$, the following holds.

$$\|R(z, A)\| \leq \frac{M_1}{|\tau|}. \tag{2.77}$$

(iii) There exist a $\phi \in (0, \pi/2)$ and an $M_2 > 0$ such that

$$\rho(A) \supset \sum = \{z \in \mathbf{C} \mid |\arg z| < \phi + \pi/2 \} \cup \{0\}, \tag{2.78}$$

and for every $z \in \sum$, $z \neq 0$, the following holds.

$$\|R(z, A)\| \leq \frac{M_2}{|z|}. \tag{2.79}$$

(iv) $T(t)$ is differentiable for $t > 0$, and for a constant $M_3 > 0$, the following holds for $t > 0$.

$$\|AT(t)\| \leq \frac{M_3}{t}. \tag{2.80}$$

Proof. (i) $\Rightarrow$ (ii): From the uniform boundedness of $T(t)$ and (2.22), we obtain the following.

$$R(z, A)x = \int_0^\infty e^{-zt} T(t) x \, dt, \tag{2.81}$$

where Re $z > 0$. By using the analyticity and the uniform boundedness of $T(t)$ in $\bar{\mathcal{S}}_{\phi_*}$, we can shift the path of integration to a ray of the form $\rho e^{i\phi}$ for $0 < \rho < \infty$, $|\phi| < \phi_*$. Let $z = \sigma + i\tau$, $\sigma > 0$ and $\tau \neq 0$. For $\tau > 0$, we shift the path of integration to $\rho e^{i\phi_*}$. Then from (2.81) it follows that

$$\|R(z, A)x\| \leq C_1 \|x\| \int_0^\infty e^{-\rho(\sigma \cos \phi_* + \tau \sin \phi_*)} d\rho \leq \frac{M_1}{\tau} \|x\|,$$

for some constants $C_1 > 0$ and $M_1 > 0$. Similarly, for $\tau < 0$, we shift the path of integration to $\rho e^{-i\phi_*}$ and obtain $\|R(z, A)x\| \leq -\frac{M_1}{\tau} \|x\|$, hence (2.77) holds.

(ii) $\Rightarrow$ (iii): Since $R(z, A)$ is an analytic function of z, by using the Taylor expansion around $z = \sigma + i\tau$, $\sigma > 0$, $\tau \neq 0$ and (2.21) we obtain

$$R(z, A) = \sum_{n=0}^\infty R(\sigma + i\tau, A)^{n+1} (\sigma + i\tau - z)^n, \tag{2.82}$$

where the series is convergent in uniform operator topology for $\|R(\sigma + i\tau, A)\|\|\sigma + i\tau - z\| \leq C < 1$ for any $C < 1$. Let us choose z such that $\mathrm{Im}z = \tau$. Then from (2.77) and (2.82) it follows that the series converges in the uniform operator topology for $|\sigma - \mathrm{Re}z| < \frac{C|\tau|}{M_1}$. Since $\sigma > 0$ is arbitrary and $\mathrm{Im}z = \tau$, dividing this inequality by $|\tau|$ we see that the resolvent set contains all z satisfying $\mathrm{Re}z \leq 0$ and $|\mathrm{Re}z|/|\mathrm{Im}z| < C/M_1$. Since by assumption $0 \in \rho(A)$, it follows that (2.78) holds for $\phi = \tan^{-1} C/M_1$. By using (2.77) and (2.82), it follows that in the region $\sum \backslash \{0\}$, the following holds

$$\|R(z, A)\| \leq \frac{M_1}{|\tau|(1 - C)}.$$

For $\mathrm{Re}z \leq 0$, $z \in \sum \backslash \{0\}$, it follows that $|z|/|\tau| \leq 1/\cos \phi = \sqrt{1 + \tan^2 \phi} = \sqrt{C^2 + M_1^2}/M_1$, hence $1/|\tau| \leq \sqrt{M_1^2 + C^2}/(M_1|z|)$. Therefore, we have

$$\|R(z, A)\| \leq \frac{\sqrt{M_1^2 + C^2}}{|z|(1 - C)}. \tag{2.83}$$

For $\mathrm{Re}z > 0$, the Hille-Yosida theorem implies that $\|R(z, A)\| \leq M/\mathrm{Re}z$ for some $M > 0$. By combining this result with (2.77) and (2.83) we obtain (2.79).

(iii) $\Rightarrow$ (iv): From Theorem 2.42 it follows that

$$T(t) = \frac{1}{2\pi i} \int_\Gamma e^{\lambda t} R(\lambda, A) d\lambda, \tag{2.84}$$

here the path Γ consists of the rays $\rho e^{i\theta}$, $\rho e^{-i\theta}$ where $0 < \rho < \infty$ and $\theta \in (\pi/2, \pi/2 + \phi)$ and oriented along the increasing direction of $\mathrm{Im}\lambda$. The integral in (2.84) converges in the uniform operator topology. By formally differentiating (2.84) we obtain

$$\frac{d}{dt}T(t) = \frac{1}{2\pi i} \int_\Gamma \lambda e^{\lambda t} R(\lambda, A) d\lambda. \tag{2.85}$$

To justify this formal differentiation, first note that the integral in (2.85) converges in the uniform operator topology for $t > 0$. In fact, by using (2.77) we obtain

$$\left\|\frac{d}{dt}T(t)\right\| \leq \frac{M_1}{\pi} \int_0^\infty e^{-\rho|\cos\theta|t} d\rho = \frac{M_1}{\pi|\cos\theta|t}. \tag{2.86}$$

Hence, (2.80) follows from (2.75) and (2.86).

(iv) $\Rightarrow$ (i): It follows from Theorem 2.45 that $T^{(n)}(t) = [\frac{d}{dt}T(\frac{t}{n})]^n$, hence $\|T^{(n)}(t)\| \leq \|[\frac{d}{dt}T(\frac{t}{n})]\|^n$. By using this result, (2.80) and $n!e^n \geq n^n$, we obtain

$$\|T^{(n)}(t)\| \leq n!\left(\frac{M_3 e}{t}\right)^n. \tag{2.87}$$

We now consider the following power series.

$$T(z) = T(t) + \sum_{n=1}^{\infty} \frac{T^{(n)}(t)}{n!}(z - t)^n, \tag{2.88}$$

where the series is convergent in the uniform operator topology for $| z - t | < Ct/M_3e$ for any $C \in (0, 1)$. It then follows easily that $T(z)$ is analytic in the sector $\mathcal{S} = \{z \in \mathbf{C} \mid |\arg z| < \tan^{-1}(1/M_3e) \}$. It can be shown, by analyticity and semigroup property of $T(z)$ on the real line, that $T(z)$ satisfies the semigroup property and that $T(z)$ is uniformly bounded on any closed subsector of $\mathcal{S}$.   □

The following theorem is more easily verified and thus useful in applications.

Theorem 2.48 *Let A be a densely defined linear operator in a Banach space X. A is the generator of an analytic semigroup $T(z)$ on X if and only if there exists an $\omega_0 \in \mathbf{R}$ such that*

(i) $\rho(A) \supset \{\lambda \mid Re\lambda \geq \omega_0\}$.

(ii) $\|R(\lambda, A)\| \leq \dfrac{M}{1 + |\lambda|}$ *for all* $Re\lambda \geq \omega_0$.

Proof.   (Necessity). Let $T(t)$ be an analytic semigroup in a sector

$$\mathcal{S}_\phi = \{z \in \mathbf{C} \mid |\arg z| < \phi, \, \phi < \frac{\pi}{2}\}$$

and strongly continuous on $\overline{\mathcal{S}_\phi}$. It can be proved that there exist some constants C and ω such that

$$\|T(z)\| \leq Ce^{\omega|z|} \text{ for all } z \in \overline{\mathcal{S}_\phi}. \tag{2.89}$$

From (2.89) it follows that the analytic operator function $f(z) = e^{-\lambda z}T(z)$, for all $z \in \mathcal{S}_\phi$, satisfies

$$\|f(z)\| \leq Ce^{-\rho(\sigma \cos \vartheta - \tau \sin \vartheta - \omega)}, \quad \text{for } \lambda = \sigma + i\tau, z = \rho e^{i\vartheta}. \tag{2.90}$$

Let Γ_1 be a ray $z = \rho e^{i\vartheta_0} (0 \leq \rho < \infty)$, where ϑ_0 is some fixed number in $(0, \phi)$. It follows from (2.90) that

$$\max_{0 \leq \arg z \leq \vartheta_0} \|zf(z)\| \to 0 \text{ as } z \to \infty \text{ when } \sigma > \omega/\cos \vartheta_0 \text{ and } \tau < 0.$$

Consequently, for these λ the following relation is satisfied:

$$\int_0^\infty e^{-\lambda t}T(t)dt = \int_{\Gamma_1} e^{-\lambda z}T(z)dz.$$

On the other hand, it follows from (2.58) that

$$R(\lambda, A)x = \int_0^\infty e^{-\lambda t}T(t)x\,dt, \quad \text{for } Re\lambda > \omega \text{ and } x \in X.$$

Hence, for any $\lambda = \sigma + i\tau$ with $\sigma > \omega/\cos\vartheta_0$ and $\tau < 0$

$$R(\lambda, A) = \int_{\Gamma_1} e^{-\lambda z}T(z)dz. \tag{2.91}$$

Similarly, for any $\lambda = \sigma + i\tau$ with $\sigma > \omega/\cos\vartheta_0$ and $\tau > 0$

$$R(\lambda, A) = \int_{\Gamma_2} e^{-\lambda z}T(z)dz \tag{2.92}$$

where Γ_2 is the ray $z = \rho e^{-i\vartheta_0}, 0 \le \rho < \infty$.

From these two formulas for $R(\lambda, A)$ and inequality (2.90), it follows that for $\lambda = \sigma + i\tau$ with $\sigma \ge \omega_0 > \omega/\cos\vartheta_0$ the following inequality holds:

$$\|R(\lambda, A)\| \le \frac{C}{\sigma\cos\vartheta_0 + |\tau|\sin\vartheta_0 - \omega} \le \frac{M}{1+|\lambda|} \quad \text{for some } M > 0.$$

This completes the proof of the necessity.

(Sufficiency): It follows from (ii) that

$$\|R(\lambda, A)\| \le \frac{M}{1+|\lambda|} \le \frac{M}{|\tau|} \quad \text{for all } Re\lambda \ge \omega_0, \lambda = Re\lambda + \tau i. \tag{2.93}$$

Let $B = A - \omega_0$. Then (2.93) implies

$$\|R(\sigma + i\tau, B)\| \le \frac{M}{|\tau|}, \quad \text{for all } Re\lambda \ge 0, \lambda = Re\lambda + \tau i. \tag{2.94}$$

It follows from the proof of $(ii) \Rightarrow (iii)$ of Theorem 2.47 that there exist $0 < \alpha < \pi/2$ and $C \ge 1$ such that

$$\rho(B) \supset \sum = \{\lambda \big| |\arg\lambda| < \frac{\pi}{2} + \alpha\} \cup \{0\} \tag{2.95}$$

and

$$\|R(\lambda, B)\| \le \frac{C}{|\lambda|} \quad \text{for } \lambda \in \sum, \lambda \ne 0. \tag{2.96}$$

By Theorem 2.47, B generates a bounded analytic semigroup and so A generates an analytic semigroup. $\qquad\square$

Remark 2.49 *It is seen from the proof of Theorem 2.48 that if A generates an analytic semigroup, then there exist $0 < \alpha < \pi/2$ and ω_0 such that*

$$\rho(A) \supset \sum = \{\lambda + \omega_0 \,\big|\, |\arg(\lambda)| < \frac{\pi}{2} + \alpha\} \cup \{0\}. \tag{2.97}$$

In the following, we will give an alternative characterization of analytic semigroup of contractions without finding the resolvent of generators. For further details, see [83].

Theorem 2.50 *Let $T(t)$ be a C_0-semigroup of contractions on a Banach space X and let A be its infinitesimal generator and let the sector $\mathcal{S}_\phi$ be defined as $\mathcal{S}_\phi = \{z \in \mathbf{C} \mid |\arg z| < \phi, \ z \neq 0\}$. Then, $T(t)$ can be extended to an analytic semigroup of contractions in the sector $\mathcal{S}_\phi$ if and only if $e^{i\theta}A$ generates a C_0- semigroup of contractions for any $\theta \in (-\phi, \phi)$.*

Proof. Assume that $T(t)$ can be extended to an analytic sector on $\mathcal{S}_\phi$ and define $T_\theta(t) = T(e^{i\theta}t)$, $\theta \in (-\phi, \phi)$. Obviously, $T_\theta(t)$ is a C_0-semigroup of contractions. By using differentiability and the analyticity of $T_\theta(t)$, we obtain

$$A_\theta = \lim_{h \downarrow 0} \frac{T_\theta(h) - I}{h} = \lim_{h \downarrow 0} \frac{T(e^{i\theta}h) - I}{h} = e^{i\theta}A.$$

Conversely, assume that $e^{i\theta}A$ generates a C_0-semigroup of contractions $T_\theta(t)$ for any $\theta \in (-\phi, \phi)$. By Theorem 2.43 we have

$$T_\theta(t)x = \lim_{n \to \infty} \left(I - \frac{t}{n}e^{i\theta}A\right)^{-n} x = \lim_{n \to \infty} \left[\frac{n}{z}R(\frac{n}{z}, A)\right]^n x, \quad \forall x \in X, \quad (2.98)$$

where we set $z = te^{i\theta}$. Since for $z = te^{i\theta} \in \mathcal{S}_\phi$ we have $\text{Re}\frac{n}{z} = \frac{n}{t}\cos\theta > 0$, and since A generates a C_0-semigroup of contractions, (e.g. choose $\theta = 0$), it follows that $[\frac{n}{z}R(\frac{n}{z}, A)]^n$ is analytic in $\mathcal{S}_\phi$. It then follows from (2.98) that $T_\theta(t)$ can be extended analytically on $\mathcal{S}_\phi$. Obviously, $T_\theta(0) = I$ and $T_\theta(z_1+z_2) = T_\theta(z_1)T_\theta(z_2)$ for any $z_1, z_2 \in \mathcal{S}_\phi$. To prove the strong continuity, let $x \in D(A) = D(e^{i\theta}A)$. We set $z = te^{i\theta}$, $T(z) = T_\theta(t)$. Integrating (2.20), we obtain

$$\|T(z)x - x\| \leq \int_0^t \|T_\theta(s)e^{i\theta}Axds\| \leq t\|Ax\| = |z|\|Ax\|.$$

Hence $\lim_{z \to 0} \|T(z)x - x\| = 0$ for $x \in D(A)$. Since $T_\theta(t)$ is a contraction and $D(A)$ is dense in X, the strong continuity follows. $\square$

The following result is a simple consequence of Theorem 2.50.

Corollary 2.51 *Let H be a Hilbert space, let $T(t)$ be a C_0-semigroup of contractions on H, and let A be the infinitesimal generator of $T(t)$. Then $T(t)$ can be extended to an analytic semigroup of contractions on a sector $\mathcal{S}_\phi$ if and only if $zI - A$ is onto for all $z \in \mathcal{S}_\phi$ and*

$$Re\{e^{i\theta}\langle Ax, x\rangle\} \leq 0, \quad \forall x \in D(A), \quad \forall \theta \in (-\phi, \phi). \quad (2.99)$$

Proof. The proof follows from Lümer-Phillips theorem (Theorem 2.27) and Corollary 2.28. □

An interesting property of analytic semigroups is given in the following corollary.

Corollary 2.52 *Let $T(t)$ be an analytic C_0-semigroup and let A be its infinitesimal generator. on a Banach space X. Then $T(t)$ is uniformly continuous in the uniform operator topology for $t > 0$.*

Proof. For any $x \in X$, and for any $0 < s \le t < \infty$, integrating (2.20) we obtain

$$T(t)x - T(s)x = \int_s^t AT(\tau)x d\tau. \qquad (2.100)$$

Substituting (2.80) into (2.100) gives

$$\|T(t)x - T(s)x\| \le M_3(\log t - \log s)\|x\|,$$

hence $T(t)$ is uniformly continuous in the uniform operator topology for $t > 0$. □

We state two theorems on perturbation of analytic semigroups which will be used in the subsequent chapters.

Theorem 2.53 *Let A be the infinitesimal generator of an analytic semigroup on a Banach space such that*

$$\|R(\lambda, A)\| \le \frac{M}{1 + |\lambda|}, \qquad \mathrm{Re}\lambda > \omega$$

for some $M > 0$ and $\omega \in \mathbf{R}$. Let B be a linear operator satisfying $D(A) \subset D(B)$ and

$$\|Bx\| \le a\|Ax\| + b\|x\| \qquad \text{for } x \in D(A)$$

where $a < 1/(1 + M)$. Then $A + B$ generates an analytic semigroup on X.

Proof. Let $0 < \varepsilon < \frac{1}{a(1+M)} - 1$ and $\hat{\omega} = \max\{\omega, \frac{bM}{\varepsilon(1+M)}\}$. Then for $\mathrm{Re}\lambda > \hat{\omega}$,

$$\|BR(\lambda, A)\| \le a\|AR(\lambda, A)\| + b\|R(\lambda, A)\|$$

$$\le a(1 + M) + \frac{bM}{1 + |\lambda|} < (1 + \varepsilon)a(1 + M) < 1.$$

Hence, for all λ satisfying $\mathrm{Re}\lambda > \hat{\omega}$, we have $\lambda \in \rho(A + B)$ and

$$\|R(\lambda, A + B)\| \le \|R(\lambda, A)\| \|I - BR(\lambda, A)\|$$

$$\le \frac{M}{1 + |\lambda|} \frac{1}{1 - (1 + \varepsilon)a(1 + M)}.$$

By Theorem 2.48, $A + B$ generates an analytic semigroup on X. $\quad\square$

Theorem 2.54 *Let A be the infinitesimal generator of an analytic semigroup on a Banach space X. Suppose that T is a linear bounded operator of finite rank, i.e.,*

$$Tx = \sum_{i=1}^{n} f_i(x)x_i$$

where $f_i \in X^*$, $x_i \in X$, $i = 1, 2, \cdots, n$. If $D(A^*)$ is dense in X^*, then $A + TA$ generates an analytic semigroup.

Proof. By Theorem 2.48, there are constants M and ω such that

$$\|R(\lambda, A)\| \leq \frac{M}{1 + |\lambda|} \quad \text{for some } \operatorname{Re}\lambda > \omega.$$

Since $D(A^*)$ is dense in X^*, we can find g_i, $i = 1, 2, \cdots, n$ such that

$$\|g_i - f_i\| < \frac{1}{n\,\|x_i\|}\frac{1}{(1 + M)}, \quad i = 1, 2, \cdots, n.$$

Define $Qx = \sum_{i=1}^{n} g_i(x)x_i$ for $x \in X$. Then

$$
\begin{aligned}
\|TAx\| &\leq \|QAx\| + \|(T - Q)Ax\| \\
&\leq \left\| \sum_{i=1}^{n}(A^*g_i)(x)x_i \right\| + \sum_{i=1}^{n} \|g_i - f_i\|\,\|x_i\|\,\|Ax\| \\
&\leq \sum_{i=1}^{n} \|A^*g_i\|\,\|x_i\|\,\|x\| + \sum_{i=1}^{n} \|g_i - f_i\|\,\|x_i\|\,\|Ax\|.
\end{aligned}
$$

Let $a = \sum_{i=1}^{n} \|g_i - f_i\|\,\|x_i\|$ and $b = \sum_{i=1}^{n} \|A^*g_i\|\,\|x_i\|$. Then $B = TA$ satisfies all the conditions of Theorem 2.53. Hence, $A + TA$ generates an analytic semigroup. $\quad\square$

As an example of differentiable and analytic semigroups, we consider the semigroup $T(t)$ defined in Example 2.6, i.e.,

$$T(t)x = \sum_{n=1}^{\infty} e^{\lambda_n t}\langle x, \phi_n\rangle\phi_n,$$

with $\lambda_{n+1} < \lambda_n < 0$, $\forall n = 1, 2, \ldots$. Note that here we restrict $\lambda_n < 0$ for simplicity of computations, while in Example 2.6, λ_n is only required to satisfy $\sup_{n \geq 1} \lambda_n < \infty$. It has been shown that the infinitesimal generator of $T(t)$ is given by

$$Ax = \sum_{n=1}^{\infty} \lambda_n \langle x, \ \phi_n \rangle \phi_n$$

with the domain

$$D(A) = \{x \in H \mid \sum_{n=1}^{\infty} \lambda_n^2 |\langle x, \ \phi_n \rangle|^2 < \infty\}.$$

We show that the semigroup $T(t)$ defined above is differentiable for $t > 0$ and is also analytic by showing that $T(t)x \in D(A)$ for $x \in H$ and that $\|AT(t)x\| \leq \frac{M}{t} \|x\|$ for a positive constant M. Since

$$\lim_{h \downarrow 0} \frac{T(t+h)x - T(t)x}{h}$$

$$= \lim_{h \downarrow 0} \frac{1}{h} \left(\sum_{n=1}^{\infty} e^{\lambda_n(t+h)} \langle x, \ \phi_n \rangle \phi_n - \sum_{n=1}^{\infty} e^{\lambda_n t} \langle x, \ \phi_n \rangle \phi_n \right)$$

$$= \sum_{n=1}^{\infty} \lambda_n e^{\lambda_n t} \langle x, \ \phi_n \rangle \phi_n,$$

and

$$\sum_{n=1}^{\infty} \lambda_n^2 e^{2\lambda_n t} |\langle x, \ \phi_n \rangle|^2 \leq \frac{1}{t^2} e^{-2} \sum_{n=1}^{\infty} |\langle x, \ \phi_n \rangle|^2 = \frac{1}{t^2} e^{-2} \|x\|^2,$$

we have that $T(t)x \in D(A)$ which means $T(t)$ is differentiable for $t > 0$ and that

$$\|AT(t)x\|^2 = \sum_{n=1}^{\infty} (\lambda_n e^{\lambda_n t})^2 |\langle x, \ \phi_n \rangle|^2$$

$$\leq \frac{1}{t^2} \frac{1}{e^2} \|x\|^2.$$

In deriving the above inequalities, we have used the fact that the real function $f(s) = se^s$ defined on $s \in (-\infty, 0)$ attains its minimum value $-e^{-1}$ at $s = -1$. From the last inequality, we see that $T(t)$ is also analytic by virtue of Theorem 2.47.

2.6 Compact semigroups

Definition 2.55 *A C_0−semigroup $T(t)$ on a Banach space X is called compact if $T(t)$ is a compact operator for each $t > 0$.*

Note that if $T(t)$ is compact for $t \geq 0$, then $T(0) = I$ must be compact, and hence X must be finite dimensional. Also, if $T(t_0)$ is compact for some $t_0 > 0$, since $T(t) = T(t - t_0)T(t_0)$ and $T(t - t_0)$ is bounded, it follows that $T(t)$ is compact for $t \geq t_0$.

Theorem 2.56 *Let $T(t)$ be a C_0−semigroup and let A be its infinitesimal generator, on a Banach space X. Then, $T(t)$ is compact if and only if $T(t)$ is continuous in the uniform operator topology for $t > 0$ and $R(\lambda, A)$ is compact for some $\lambda \in \rho(A)$.*

Remark 2.57 *Note that for any $\lambda, \mu \in \rho(A)$, the following resolvent equation holds*

$$R(\lambda, A) = [I + (\mu - \lambda)R(\lambda, A)]R(\mu, A). \tag{2.101}$$

Since the resolvent is a bounded operator, it follows from (2.101) that if $R(\mu, A)$ is compact for some $\mu \in \rho(A)$, then $R(\lambda, A)$ is compact for all $\lambda \in \rho(A)$.

Proof. Let $T(t)$ be a compact semigroup. Then for a fixed $t > 0$, the set $\mathcal{R} = \{T(t)x \mid \|x\| \leq 1 \}$ is compact. Hence, for any given $\epsilon > 0$, there exist finitely many points $x_1, \ldots, x_n$ and spheres $S(x_k, \epsilon)$ centered at x_k with radius ϵ, $k = 1, \ldots, n$, which cover $\mathcal{R}$. Because of strong continuity, we can find $\delta > 0$ such that $\|(T(h) - I)x_k\| < \epsilon$ for $k = 1, \ldots, n$ and $0 \leq h < \delta$. For any $\|x\| \leq 1$, $T(t)x \in S(x_l, \epsilon)$ for some $1 \leq l \leq n$. Hence, we have

$$(T(t + h) - T(t))x = (T(h) - I)x_l + (T(h) - I)(T(t)x - x_l).$$

Since we have $\|T(h) - I\| \leq C$ for some $C > 0$ and $0 \leq h < \delta$, we obtain $\|(T(t + h) - T(t))x\| \leq (C + 1)\epsilon$. Since $\epsilon > 0$ is arbitrary, it follows that $T(t)$ is right continuous in the uniform operator topology. For the left continuity, when $t > 0$, we can find a $\delta > 0$ such that $t > 2\delta$. Then for all $0 < h < \delta$, we have

$$T(t - h) - T(t) = [T(t - \delta - h) - T(t - \delta)]T(\delta) = T(t - \delta - h)[I - T(h)]T(\delta).$$

Note that $T(\delta)$ is compact. We can thus use similar arguments as above to show that $T(t)$ is left continuous in the uniform operator topology.

Since $\|T(t)\| \leq Me^{\omega t}$, and since $T(t)$ is continuous in the uniform operator topology, $R(\lambda, A)$ satisfies

$$R(\lambda, A) = \int_0^\infty e^{-\lambda t}T(t)dt, \quad \operatorname{Re}\lambda > \omega, \tag{2.102}$$

where the integral converges in the uniform operator topology. For any $\epsilon > 0$, $L > 0$, $\epsilon < L < \infty$, we define

$$R_{\epsilon, L}(\lambda) = \int_\epsilon^L e^{-\lambda t}T(t)dt, \quad \operatorname{Re}\lambda > \omega. \tag{2.103}$$

Since $T(t)$ is compact, it follows that $R_{\epsilon,L}$ also is compact. By using (2.102) and (2.103) we obtain

$$\|R(\lambda, A) - R_{\epsilon,L}(\lambda)\| \leq \| \int_0^\epsilon e^{-\lambda t} T(t) dt \| + \| \int_L^\infty e^{-\lambda t} T(t) dt \|$$

$$\leq \epsilon M e^{\omega \epsilon} + \frac{M}{\mathrm{Re}\lambda - \omega} e^{-(\mathrm{Re}\lambda - \omega)L},$$

and since $\mathrm{Re}\lambda > \omega$, the right-hand side of this inequality approaches 0 as $\epsilon \to 0$ and $L \to \infty$. Hence $R(\lambda, A)$ also is compact since it is the uniform limit of compact operators.

Conversely, assume that $T(t)$ is continuous in the uniform operator topology and that $R(\lambda, A)$ is compact for some $\lambda \in \rho(A)$, hence for all $\lambda \in \rho(A)$. It then follows that (2.102) and the following holds.

$$\lambda R(\lambda, A) T(t) - T(t) = \lambda \int_0^\infty e^{-\lambda \tau} (T(t + \tau) - T(t)) d\tau. \qquad (2.104)$$

Let us choose λ as real and $\lambda > \omega$. For any $\epsilon > 0$, splitting the integral in (2.104) into two parts on the intervals $[0, \epsilon]$ and $[\epsilon, \infty)$, and using $\|T(t)\| \leq M e^{\omega t}$ we obtain

$$\lim_{\lambda \to \infty} \|\lambda R(\lambda, A) T(t) - T(t)\| \leq \sup_{\sigma \in [0,\epsilon]} \|T(t + \sigma) - T(t)\|. \qquad (2.105)$$

Since $\epsilon > 0$ is arbitrary and $T(t)$ is continuous in the uniform operator topology, it follows that the limit in (2.105) is 0. Since $\lambda R(\lambda, A) T(t)$ is compact for $\lambda > \omega$, it follows that $T(t)$ also is compact. $\qquad \square$

Theorem 2.58 *Let $T(t)$ be a compact semigroup on a Banach space X and let A be its infinitesimal generator. Then the following statements hold.*

(i)

$$\lim_{|\tau| \to \infty} \|R(\sigma + i\tau, A)\| = 0, \quad \forall \sigma \in \mathbf{R}. \qquad (2.106)$$

(ii) *The spectrum of A contains countably many isolated eigenvalues with finite algebraic multiplicities.*

Proof. Without loss of generality we assume that $T(t)$ is uniformly bounded, since otherwise we may consider $S(t) = e^{-\mu t} T(t)$ for $\mu > \omega$. Since $T(t)$ is uniformly continuous in the uniform operator topology for $t > 0$, (2.102) holds for $\mathrm{Re}\lambda > 0$. Letting $\lambda = \sigma + i\tau$, $\sigma > 0$ in (2.102) we obtain

$$R(\sigma + i\tau, A) = \int_0^\infty e^{-i\tau t} e^{-\sigma t} T(t) dt. \qquad (2.107)$$

Since $e^{-\sigma t}\|T(t)\|$ is integrable for $\sigma > 0$, by using Riemann-Lebesgue lemma we obtain (2.106). For $\sigma \leq 0$, by using $\lambda = \sigma + i\tau$, Taylor expansion and (2.21) we obtain

$$R(\sigma + i\tau, A) = \sum_{n=0}^{\infty} R(1 + i\tau, A)^{n+1}(1 - \sigma)^n. \tag{2.108}$$

Since $\lim_{|\tau| \to \infty} \|R(1 + i\tau, A)\| = 0$, for each $\sigma \leq 0$ there exists a $\tau_\sigma > 0$ such that for $|\tau| > \tau_\sigma$, the series in (2.108) converges, and moreover (2.106) holds.

Since $R(\lambda, A)$ is compact for $\lambda \in \rho(A)$, the spectrum of A is necessarily discrete, i.e., contains countably many isolated eigenvalues with finite algebraic multiplicities, see [53] and [84]. In the next chapter, we show a stronger result which says that for any $-\infty < m \leq M < \infty$, the strip $m \leq \mathrm{Re}\lambda \leq M$ contains at most a finite number of eigenvalues of A. □

Since determining the continuity of $T(t)$ in the uniform operator topology is not easy in general (a characterization by resolvent in Hilbert space is developed in [169]), the characterization of compact semigroups by using Theorem 2.56 is not completely satisfactory. For analytic semigroups, the following result gives a simpler characterization.

Corollary 2.59 *Let $T(t)$ be an analytic semigroup on a Banach space X and let A be its infinitesimal generator. Then $T(t)$ is compact if and only if $R(\lambda, A)$ is compact for some $\lambda \in \rho(A)$.*

Proof. The proof easily follows from Corollary 2.52 and Theorem 2.56. □

Note that Corollary 2.59 could be quite useful in applications since most of the operators encountered in classical boundary problems have compact resolvent, see [84, p. 187].

As an example, let us now examine the conditions under which the semigroup given in Example 2.6 is compact. First, we need the following lemma.

Lemma 2.60 *Let H be a Hilbert space with an orthonormal basis $\{\phi_n\}_{n \geq 1}$. For any $x \in H$, let $x_n = \langle x, \phi_n \rangle, \forall n$. Given $\delta_n > 0$, define a set S by*

$$S = \{x \in H \mid |x_n| \leq \delta_n, \forall n\}.$$

Then the set S is compact if and only if $\sum_{n=1}^{\infty} \delta_n^2 < \infty$.

Proof. Sufficiency: Since $\sum_{n=1}^{\infty} \delta_n^2 < \infty$, for any $\epsilon > 0$, there exists an N, such that $\sum_{n=N+1}^{\infty} \delta_n^2 < \epsilon^2/2$. Denote by

$$S_1 = \{x \mid x = \sum_{n=1}^{N} x_n \phi_n, |x_n| = |\langle x, \phi_n \rangle| \leq \delta_n\}$$

a finite dimensional subspace of H which is bounded and hence compact. Thus, there exist finite $\epsilon^2/2$-nets which cover S_1. That is, there exist $x^i \in S_1$, $i = 1, \cdots, M$, such that for any $x \in S_1$, there is an x^k, $1 \leq k \leq M$ which renders

$$\sum_{n=1}^{N} |x_n - x_n^k|^2 < \frac{\epsilon^2}{2},$$

where $x_n = \langle x, \phi_n \rangle$ and $x_n^k = \langle x^k, \phi_n \rangle$.

We now show that $x^i, i = 1, \cdots, M$, are ϵ-nets of S. In fact, for any $x \in S$,

$$
\begin{aligned}
\|x - x^k\| &= \left(\sum_{n=1}^{\infty} |x_n - x_n^k|^2 \right)^{1/2} \\
&= \left(\sum_{n=1}^{N} |x_n - x_n^k|^2 + \sum_{n=N+1}^{\infty} |x_n - x_n^k|^2 \right)^{1/2} \\
&\leq \left(\sum_{n=1}^{N} |x_n - x_n^k|^2 + \sum_{n=N+1}^{\infty} \delta_n^2 \right)^{1/2} \\
&\leq \left(\frac{\epsilon^2}{2} + \frac{\epsilon^2}{2} \right)^{1/2} \\
&= \epsilon,
\end{aligned}
$$

which shows that S is compact.

Necessity: For any positive integer N, set $x^N = \sum_{n=1}^{N} \delta_n \phi_n \in S$. Then, $\|x^N\|^2 = \sum_{n=1}^{N} \delta_n^2$ which is uniformly bounded for all N because S is compact. Letting $N \to \infty$ yields

$$\sum_{n=1}^{\infty} \delta_n^2 < \infty,$$

which is the desired result.   □

Example 2.61 *Consider again the C_0-semigroup $T(t)$ defined in Example 2.6, i.e.,*

$$T(t)x = \sum_{n=1}^{\infty} e^{\lambda_n t} \langle x, \phi_n \rangle \phi_n, \quad \forall x \in H,$$

where $\{\lambda_n\}_{n \geq 1}$ are real numbers, and $\{\phi_n\}_{n \geq 1}$ form an orthonormal basis in H. Our question here is under what conditions is the semigroup $T(t)$ compact. To answer this question, let us define a set

$$S = \{ T(t)x = \sum_{n=1}^{\infty} e^{\lambda_n t} \langle x, \phi_n \rangle \phi_n \mid \|x\|^2 = \sum_{n=1}^{\infty} |\langle x, \phi_n \rangle|^2 \leq 1 \}.$$

If S is compact, then we know that $T(t)$ is compact because S is the image of $T(t)$ mapping on the unit ball $\{x \in H \mid \|x\| \leq 1\}$. Since

$$|e^{\lambda_n t} \langle x, \phi_n \rangle| \leq e^{\lambda_n t} := \delta_n.$$

from the previous lemma, we know that if $\sum_{n=1}^{\infty} \delta_n^2 < \infty$ then S is compact. Since

$$\sum_{n=1}^{\infty} \delta_n^2 \leq \sum_{n=1}^{\infty} e^{2\lambda_n t},$$

obviously,

$$\lim_{n \to \infty} \frac{e^{2\lambda_{n+1} t}}{e^{2\lambda_n t}} = \lim_{n \to \infty} e^{2(\lambda_{n+1} - \lambda_n)t} < 1$$

guarantees the convergence of $\sum_{n=1}^{\infty} e^{2\lambda_n t}$. Therefore, we conclude that if for any $\varepsilon > 0$, $\{\lambda_n\}_{n \geq 1}$ satisfy the following condition of separation of spectrum

$$\lambda_{n+1} - \lambda_n < -\varepsilon,$$

then $T(t)$ is compact for $t > 0$.

The results obtained in this example are very useful, because the semigroups associated with many practical problems can be expressed as the form in this example.

It also should be mentioned that many systems, such as delay equations, generate compact semigroups.

2.7 Abstract Cauchy problem

2.7.1 Homogeneous initial value problems

Let X be a Banach space and let A be a linear operator with domain $D(A) \subset X$. Consider the following differential equation on X.

$$\frac{du(t)}{dt} = Au(t), \quad u(0) = x \in X, \quad t > 0. \tag{2.109}$$

The abstract Cauchy problem is to find a solution $u(t)$ of (2.109). The meaning of a solution is clarified in the next definition.

Definition 2.62 *A function $u(\cdot) : \mathbf{R}_+ \to X$ is said to be a solution (or classical solution) of (2.109) if $u(t)$ is continuous for $t \geq 0$, continuously differentiable for $t > 0$ and satisfies (2.109).*

We first state the following well-known uniqueness result for (2.109).

Theorem 2.63 (Ljubic) *Let A be a linear operator in a Banach space X with $(\omega, \infty) \subset \rho(A)$ for some $\omega \in \mathbf{R}$. If*

$$\overline{\lim_{\lambda \to \infty}} \lambda^{-1} \log \|R(\lambda, A)\| = 0,$$

then (2.109) has at most one solution for every $x \in X$.

Proof. We assume that $R(\lambda, A)$ exists for all real $\lambda, \lambda \geq 0$ since otherwise, consider $\alpha - A$ instead of A for some appropriate real α. Let $u(t)$ be a solution of (2.109) satisfying $u(0) = 0$. We show that $u(t) \equiv 0$. To this end, consider the function $t \to R(\lambda, A)u(t)$ for $\lambda > 0$. Since $u(t)$ is a solution of (2.109), we have

$$\frac{d}{dt} R(\lambda, A)u(t) = R(\lambda, A)Au(t) = \lambda R(\lambda, A)u(t) - u(t)$$

which implies

$$R(\lambda, A)u(t) = - \int_0^t e^{\lambda(t-\tau)} u(\tau) d\tau.$$

From the assumption, it follows that for every $\sigma > 0$

$$\lim_{\lambda \to \infty} e^{-\sigma \lambda} \|R(\lambda, A)\| = 0$$

and, therefore,

$$\lim_{\lambda \to \infty} \int_0^{t-\sigma} e^{\lambda(t-\sigma-\tau)} u(\tau) d\tau = 0.$$

Thus, for fixed t and $\sigma < t$, there exists an $N > 0$ such that

$$\left\| \int_0^{t-\sigma} e^{\lambda(t-\sigma-\tau)} u(\tau) d\tau \right\| \leq M, \quad \text{as } \lambda > N$$

for some $M > 0$.

Let $x^* \in X^*$ and set $\phi(\tau) = \langle u(\tau), x^* \rangle$. Then ϕ is clearly continuous on $[0, t]$ and

$$\left| \int_0^{t-\sigma} e^{\lambda(t-\sigma-\tau)} \phi(\tau) d\tau \right| \leq M \|x^*\|, \quad \text{as } \lambda > N.$$

We will show that the above implies $\phi(\tau) \equiv 0$ on $[0, t-\sigma]$, and since $x^* \in X^*$ and t, σ are arbitrary, it follows that $u(t) \equiv 0$.

Consider the series

$$\sum_{k=1}^{\infty} \frac{(-1)^{k-1}}{k!} e^{kn\tau} = 1 - e^{-n\tau}.$$

This series converges uniformly in τ on bounded intervals. Therefore, for $n > N$,

$$\left| \int_0^{t-\sigma} \sum_{k=1}^{\infty} \frac{(-1)^{k-1}}{k!} e^{kn(\tau-s)} \phi(s)ds \right|$$

$$\leq \sum_{k=1}^{\infty} \frac{1}{k!} e^{kn(\tau-t+\sigma)} \left| \int_0^{t-\sigma} e^{kn(t-\sigma-s)} \phi(s)ds \right|$$

$$\leq M \|x^*\| [\exp(e^{n(\tau-t+\sigma)}) - 1].$$

For $\tau < t - \sigma$ the RHS above tends to zero as $n \to \infty$. On the other hand, we have

$$\int_0^{t-\sigma} \sum_{k=1}^{\infty} \frac{(-1)^{k-1}}{k!} e^{kn(\tau-s)} \phi(s)ds = \int_0^{t-\sigma} [1 - \exp(-e^{n(\tau-s)})]\phi(s)ds.$$

Using Lebesgue's dominated convergence theorem we see that the right-hand side of the above equation converges to $\int_0^{\tau} \phi(s)ds$ as $n \to \infty$. So $\int_0^{\tau} \phi(s)ds = 0$ which implies that $\phi(\tau) \equiv 0$ on $[0, t - \sigma]$. □

The next result establishes the link between the solution of the abstract Cauchy problem and the semigroup theory.

Theorem 2.64 *Let A be a densely defined linear operator in a Banach space X with $\rho(A) \neq \emptyset$. The Cauchy problem has unique solution for $x \in D(A)$ which is continuously differentiable for $t \geq 0$ if and only if A generates a C_0−semigroup $T(t)$ on X.*

Proof. If A generates a C_0 semigroup $T(t)$, then from Theorem 2.12 it follows that for $x \in D(A)$, $u(t) = T(t)x$ is a solution of (2.109), which is continuously differentiable for $t \geq 0$. Now assume that there is another solution $y(t)$. Then the difference $e(t) = u(t) - y(t)$ satisfies (2.109) with $e(0) = 0$. For a given $t > 0$, let us define $z(s) = T(t - s)e(s)$ for $0 \leq s \leq t$. Then $z(s)$ is strongly differentiable for $0 \leq s \leq t$, and for any $0 < \delta < t$ we have $z(t) - z(\delta) = \int_\delta^t \frac{dz(s)}{ds} ds$. However, $\frac{dz(s)}{ds} = -T(t - s)Ae(s) + T(t - s)Ae(s) = 0$, hence we have $z(t) = e(t) = z(\delta) = T(t - \delta)e(\delta)$. Since $T(t)$ is strongly continuous and $e(0) = 0$, by taking the limit as $\delta \to 0$ we obtain $e(t) = 0$, hence $u(t) = y(t)$ for $t > 0$.

Conversely, assume that for any $x_0 \in D(A)$, (2.109) has a unique solution denoted by $u(t, x_0)$. For any $x \in D(A)$, let us denote the graph norm $\|\|\cdot\|\|$ on $D(A)$ by $\|\|x\|\| = \|x\| + \|Ax\|$. Since $\rho(A) \neq \emptyset$, A is closed, hence $D(A)$ with the graph norm given above becomes a Banach space, denoted by $[D(A)]$, and A becomes a bounded operator on $[D(A)]$, see e.g. [84]. Let, for any $t_0 > 0$, $C(I, D(A))$ denote the set of continuous functions from $I = [0, t_0]$

to $[D(A)]$. This set is also a Banach space with the usual supremum norm. Consider the mapping $S : [D(A)] \rightarrow C(I, D(A))$ defined by $Sx = u(t, x)$ for $t \in [0, t_0]$. Since (2.109) is linear and its solution is unique, it follows easily that S is a linear operator. To show that S is closed, let $x_n \rightarrow x$ in $[D(A)]$ and $Sx_n \rightarrow v$ in $C(I, D(A))$. Since $Sx_n = u(t, x_n)$ is the solution of (2.109) with $u(0) = x_n$, we have

$$u(t, x_n) = x_n + \int_0^t Au(\tau, x_n)d\tau,$$

and since A is closed, it follows that

$$v(t) = x + \int_0^t Av(\tau)d\tau,$$

which shows that $v(t) = u(t, x) = Sx$, hence S is closed. By the closed graph theorem, S is a bounded linear operator. Let us define a mapping $T(t) : [D(A)] \rightarrow [D(A)]$ by $T(t)x = u(t, x) = Sx$. By using uniqueness it follows that $T(t)$ satisfies the semigroup properties. Moreover, since $u(\cdot, x)$ is continuous it follows that $T(t)$ is uniformly bounded on $[0, t_0]$. Hence, by following the proof of Theorem 2.8 it can be shown that $T(t)$ can be extended to a semigroup on $[D(A)]$ satisfying $|||T(t)x||| \leq Me^{\omega t}|||x|||$ for $t > 0$. Next, we will show that $T(t)$ can be extended to X. Let $x \in D(A^2)$ and set

$$z(t) = x + \int_0^t T(s)Ax\,ds. \qquad (2.110)$$

Differentiating (2.110) we obtain

$$\frac{dz(t)}{dt} = T(t)Ax = Ax + A\int_0^t T(s)Ax\,ds = Az(t), \qquad (2.111)$$

where we used (2.19) and the fact that A is closed. Since $z(0) = x$, it follows from the uniqueness assumption that $z(t) = u(t, x) = T(t)x$. Hence, from (2.111) we have

$$T(t)Ax = AT(t)x, \quad \forall x \in D(A^2). \qquad (2.112)$$

Let $z \in D(A)$ and $\lambda \in \rho(A)$. The equation $z = (\lambda I - A)x$ has a unique solution $x = R(\lambda, A)z \in D(A^2)$. Hence, we have

$$\|T(t)z\| \leq |\lambda|\|T(t)x\| + \|T(t)Ax\| \leq (1 + |\lambda|)|||T(t)x||| \leq M_1 e^{\omega t}|||x|||,$$
$$\qquad (2.113)$$

where $M_1 = (1 + |\lambda|)M$, and we used (2.112). Since $R(\lambda, A)$ and $AR(\lambda, A)$ are bounded, it follows that $|||x||| \leq C\|z\|$ for some $C > 0$. Hence we have

$$\|T(t)z\| \leq M_2 e^{\omega t}\|z\| \quad \forall z \in D(A). \qquad (2.114)$$

Therefore, $T(t)$ can be extended to a C_0−semigroup on X. To complete the proof, we have to show that A is the infinitesimal generator of $T(t)$. Let A_1 be the infinitesimal generator of $T(t)$ and let $z \in D(A)$. Since $u(t) = T(t)z$ is the solution of (2.109) by evaluating (2.109) at $t = 0$ we obtain $Az = A_1z$, hence $A \subset A_1$, i.e. A is the restriction of A_1 to $D(A)$. By using (2.112) we obtain

$$e^{-\lambda t}AT(t)x = e^{-\lambda t}T(t)Ax = e^{-\lambda t}T(t)A_1x, \qquad x \in D(A^2), \qquad (2.115)$$

where $\lambda > \omega$. By integrating (2.115) on $[0, \infty)$, using (2.22) and the fact that A is closed, we obtain

$$AR(\lambda, A_1)x = R(\lambda, A_1)A_1x = A_1R(\lambda, A_1)x. \qquad (2.116)$$

Since $A_1R(\lambda, A_1)$ is uniformly bounded, A is closed and $D(A^2)$ is dense in X, it follows that (2.116) holds for $x \in X$. This implies that $D(A_1) = \text{Range}R(\lambda, A_1) \subset D(A)$, hence $A_1z = Az$ for $z \in D(A_1)$. This proves that $A_1 \subset A$, and hence $A = A_1$. □

Note that the condition $\rho(A) \neq \emptyset$ in Theorem 2.64 cannot be removed. In fact, the following example demonstrates that there exists a densely defined closed operator A with $\rho(A) = \emptyset$ such that for every $x \in D(A)$ the abstract Cauchy problem associated with A has a unique solution. However, A is not the generator of a C_0−semigroup.

Example 2.65 *Let B be a densely defined, unbounded, and closed operator in a Banach space X. Let $Y = X \times X$ and define $A : D(A) \to Y$ as follows:*

$$A = \begin{bmatrix} 0 & B \\ 0 & 0 \end{bmatrix}, \qquad D(A) = X \times D(B).$$

Then A is a densely defined closed operator in Y. For any $(x, y) \in X \times D(B)$, the abstract Cauchy problem

$$\begin{cases} \frac{d}{dt}u(t) = Au(t), \\ u(0) = (x, y) \end{cases}$$

has a unique solution $u(t) = (x + tBy, y)$. But for any $\lambda \in \mathbf{C}$,

$$(\lambda - A)D(A) = \{(\lambda x - By, \lambda y)\big| x \in X, y \in D(B)\} \subset X \times D(B) \neq Y,$$

which means $\rho(A) = \emptyset$. By the Hille-Yosida theorem, A cannot be the generator of a C_0−semigroup.

In general, if x is not in $D(A)$, then (2.109) may not have a solution at all. In this case, we may define a *mild solution* of (2.109), which also is called the *generalized solution*, as follows.

Definition 2.66 *Let $T(t)$ be a C_0-semigroup on a Banach space X and let A be its infinitesimal generator. For every $x_0 \in X$, $x(t) = T(t)x_0$ is called a mild solution of (2.109).*

If the semigroup has some additional properties, we may prove the equivalence of mild and classical solutions as follows.

Theorem 2.67 *If A generates a differentiable C_0-semigroup $T(t)$ on a Banach space X, then (2.109) has unique solution $u(t)$ for every $x \in X$, which is equal to the mild solution, i.e., $u(t) = T(t)x$.*

Proof. From the assumptions, it follows that $T(t)x$ is differentiable for $x \in X$ and that $\frac{d}{dt}T(t)x = AT(t)x$. By Theorem 2.45, $AT(t)$ is bounded and hence is Lipschitz for $t > 0$. Therefore, (2.109) has unique solution, and since $u(t) = T(t)x$ is also a solution, it is the unique solution. □

Corollary 2.68 *If A generates an analytic semigroup $T(t)$ on a Banach space X, then for every $x \in X$, (2.109) has unique solution $u(t)$, which is also the mild solution, i.e., $u(t) = T(t)x$.*

Proof. The proof follows from Theorems 2.47 and 2.67. □

2.7.2   Inhomogeneous initial value problems

In this section, we consider the following abstract Cauchy problem.

$$\frac{du(t)}{dt} = Au(t) + f(t), \quad u(0) = x \in X, \quad t > 0, \qquad (2.117)$$

where X is a Banach space, $f : [0, T) \to X$ is a given function. First we clarify the meaning of a solution of (2.117).

Definition 2.69 *A function $u(\cdot) : [0, T) \to X$ is called a solution (or classical solution) of (2.117) if $u(\cdot)$ is continuous on $[0, T)$, is continuously differentiable and $u(t) \in D(A)$ on $(0, T)$ and (2.117) is satisfied on $[0, T)$.*

Definition 2.70 *Let A be the infinitesimal generator of a C_0-semigroup $T(t)$ on X, let $x \in X$ and let $f \in L^1([0, T); X)$. Then the function $u \in C([0, T); X)$ given by*

$$u(t) = T(t)x + \int_0^t T(t - s)f(s)ds, \qquad (2.118)$$

is called the mild solution of (2.117) on $[0, T]$.

Theorem 2.71 *Let A be the infinitesimal generator of a C_0-semigroup $T(t)$ on X. If $f \in L^1([0, T); X)$, then for every $x \in X$, (2.117) has at most one solution. If (2.117) has a solution, then it is also a mild solution.*

Proof. Let $u(t)$ be a solution of (2.117) and define $v(s) = T(t - s)u(s)$. Obviously, $v(s)$ is differentiable for $s \in (0, t)$ and

$$\frac{dv(s)}{ds} = T(t - s)f(s), \qquad (2.119)$$

where we used (2.19). By integrating (2.119) on $[0, t)$, and noting that $v(0) = T(t)x$, we obtain

$$u(t) = T(t)x + \int_0^t T(t - s)f(s)ds. \qquad (2.120)$$

Since any solution of (2.117) satisfies (2.120), it follows easily that there can be at most one solution, which is also a mild solution. $\square$

Next we will give a sufficient condition which guarantees the existence of a solution.

Theorem 2.72 *Let A be the infinitesimal generator of a C_0-semigroup $T(t)$ on X. If $f(\cdot)$ is continuously differentiable on $[0, T]$, then (2.117) has a unique solution on $[0, T)$ for every $x \in D(A)$.*

Proof. Note that the mild solution $u(t)$ is given by (2.118). Let us denote the integral part of (2.118) by $z(t)$, i.e.,

$$z(t) = \int_0^t T(t - s)f(s)ds. \qquad (2.121)$$

Hence, we have $u(t) = T(t)x_0 + z(t)$. Obviously, $z(t)$ is continuously differentiable for $t \in [0, T]$. Moreover, for any $t \in (0, T)$ and any $h > 0$ such that $t + h \in (0, T)$ we have

$$\frac{z(t + h) - z(t)}{h} = \frac{T(h) - I}{h} \int_0^t T(t - s)f(s)ds + \frac{1}{h} \int_t^{t+h} T(t + h - s)f(s)ds.$$

By letting $h \to 0$, we have $\dot{z}(t) = Az(t) + f(t)$. Since $f(t)$ is continuous, it follows that $z(t) \in D(A)$ for $t \in (0, T)$. By differentiating (2.118), we obtain $u'(t) = AT(t)x + Az(t) + f(t) = Au(t) + f(t)$ for $t \in (0, T)$. Since $z(0) = 0$, we have $u(0) = x$, hence the mild solution given by (2.118) is a solution of (2.117). Uniqueness follows from Theorem 2.71. $\square$

2.7.3   Lipschitz perturbations

Let X be a Banach space, $A : D(A) \subset X \to X$ a linear operator and $f : [0, \infty) \times X \to X$ a nonlinear function. In this section we consider the following problem :

$$\begin{cases} \dot{u}(t) = Au(t) + f(t, u(t)), & t \geq t_0 \geq 0 \\ u(t_0) = x \in X. \end{cases} \tag{2.122}$$

We assume that A generates a C_0-semigroup $T(t)$ on X. In general, (2.122) may not have a classical or mild solution, see Definitions 2.69, 2.70. If a (classical) solution $u(\cdot)$ to (2.122) exists, then similar to (2.118), it can be shown that $u(\cdot)$ satisfies the following integral equation

$$u(t) = T(t - t_0)x + \int_{t_0}^t T(t - \tau)f(\tau, u(\tau))d\tau. \tag{2.123}$$

As in Definition 2.70, a continuous solution $u(\cdot)$ of (2.123) will be called a mild solution of (2.122).

Theorem 2.73 *Consider the system (2.122). Let A be the generator of a C_0-semigroup $T(t)$ on X, and let $f : [0, \infty) \times X \to X$ be continuous in t, and locally Lipschitz continuous in u, uniformly in t on bounded intervals. Then, for any $t_0 \geq 0$ and $x \in X$, there exists a $t_{max} > t_0$ such that (2.122) has unique mild solution on $[t_0, t_{max})$. Moreover, if $t_{max} < \infty$, then $\lim_{t \uparrow t_{max}} \|u(t)\| = \infty$.*

Proof.   Let $x \in X$ and $t_0 \geq 0$ be given. For $t_1 > t_0$, let $C([t_0, t_1]; X)$ denote the space of continuous functions $g : [t_0, t_1] \to X$ endowed with the usual supremum norm. An appropriate value for t_1 will be determined later. Let $T \in (0, \infty)$ be fixed and define $M(t_0) = \sup\{\|T(t)\| \mid t \in [0, t_0 + T]\}$, $r(t_0) = 2M(t_0)\|z_0\|$, $h(t_0) = \sup\{\|f(t, 0)\| \mid t \in [0, t_0 + T]\}$. Let us define the closed balls $\mathcal{S} \subset C([t_0, t_1]; X)$ and $B \subset X$ as follows

$$\begin{cases} \mathcal{S} = \{z(\cdot) \in C([t_0, t_1]; X) \mid \|z(\cdot)\|_\infty \leq r(t_0)\} \\ B = \{y \in X \mid \|y\| \leq r(t_0)\}, \end{cases}$$

where the subscript ∞ denotes the supremum norm in $C([t_0, t_1]; X)$. By assumptions on f, there exists a constant $k(t_0)$ such that the following holds

$$\|f(t, u) - f(t, v)\| \leq k(t_0)\|u - v\|,$$

for $u, v \in B$ and $t \in [0, t_0 + T]$. Let us define an operator $F : C([t_0, t_1]; X) \to C([t_0, t_1]; X)$ as

$$(Fu)(t) = T(t - t_0)x + \int_{t_0}^t T(t - \tau)f(\tau, u(\tau))d\tau. \tag{2.124}$$

Note that (2.123) can be written as $Fu = u$, i.e. any solution of (2.123) is a fixed point of F.

First note that for $u(\cdot) \in \mathcal{S}$,

$$\|(Fu)(t)\|_\infty \leq M(t_0)\|x\| + \int_{t_0}^t M(t_0)\|f(\tau, u(\tau)) - f(\tau, 0) + f(\tau, 0)\|\, d\tau$$

$$\leq M(t_0)\|x\| + M(t_0)[k(t_0)r(t_0) + h(t_0)](t_1 - t_0). \qquad (2.125)$$

Now fix a $\rho \in (0, 1)$ and choose t_1 as

$$t_1 - t_0 = \delta = \min\{T, \frac{\|x\|}{k(t_0)r(t_0) + h(t_0)}, \frac{\rho}{M(t_0)k(t_0)}\}. \qquad (2.126)$$

It follows from (2.125) and (2.126) that $F : \mathcal{S} \to \mathcal{S}$. Next, we will show that F is a contraction on $\mathcal{S}$. Let $u(\cdot), v(\cdot) \in \mathcal{S}$. By using

$$\|(Fu)(t) - (Fv)(t)\|_\infty \leq M(t_0)k(t_0)(t_1 - t_0)\|u - v\|_\infty$$

$$\leq \rho\|u - v\|_\infty, \qquad (2.127)$$

we see that F is a contraction on $\mathcal{S}$. It follows from the contraction mapping theorem that F has unique fixed point, hence (2.123) has unique solution $u(\cdot)$ in $\mathcal{S}$. Next, we will show that any solution $v(\cdot) \in C([t_0, t_1]; X)$ of (2.123) is necessarily in $\mathcal{S}$. Assume that $\|v(t_2)\| > r(t_0)$ holds for some $t_2 \in [t_0, t_1]$. Then there exists a $t_3 \in [t_0, t_2]$ such that $\|v(t_3)\| = r(t_0)$ and $\|v(t)\| < r(t_0)$ for $t \in [t_0, t_3]$. Then by repeating the calculations of (2.125) for $t = t_3$, we obtain $\|v(t_3)\| < r(t_0)$, which is a contradiction. Hence $v(\cdot) \in \mathcal{S}$, and therefore (2.123) has unique solution in $C([t_0, t_1]; X)$.

The above procedure shows that if $u(\cdot)$ is a mild solution on $[t_0, t_1]$, then this solution can be extended to $[t_0, t_1 + \delta]$ by using t_1 as initial time and $x = u(t_1)$ in (2.123), (2.124), for the interval $[t_1, t_1 + \delta]$, where $\delta > 0$ can be calculated similar to (2.126). By using this procedure the solution can be extended to a maximal interval of existence $[t_0, t_{max})$. If $t_{max} < \infty$, then necessarily $\lim_{t \uparrow t_{max}} \|u(t)\| = \infty$. For otherwise, there exists a sequence $t_n \uparrow t_{max}$ and $C > 0$ such that $\|u(t_n)\| < C$, $\forall n$. Hence, from (2.126) it follows that there exists a $\delta_* > 0$, independent of n, such that the solution on $[t_0, t_n]$ can be extended to $[t_0, t_n + \delta_*]$. By choosing t_n such that $t_{max} - t_n < \delta_*$, it follows that the solution can be extended beyond t_{max}, contradicting the definition of t_{max}.

Finally, if $u(\cdot)$ and $v(\cdot)$ are two mild solutions, then on any closed interval $[t_0, t_1]$, we have $u = v$. Hence, both solutions have the same t_{max}, and consequently the mild solution is unique on $[t_0, t_{max})$. $\qquad \square$

Theorem 2.74 *Consider the system (2.122). Let A be the generator of a C_0-semigroup $T(t)$ on X, and let $f : [0, \infty) \times X \to X$ be continuously differentiable both in t and u. Then, for every $t_0 \geq 0$ and $x \in D(A)$, there exists a*

$t_{max} > t_0$ such that (2.122) has unique mild solution $u(\cdot)$ on $[t_0, t_{max})$, which is also the unique (classical) solution of (2.122). Moreover, if $t_{max} < \infty$, then $\lim_{t \uparrow t_{max}} \|u(t)\| = \infty$.

Proof. First note that the continuous differentiability of f implies that f is continuous in t and is locally Lipschitz continuous in u, uniformly in t on bounded intervals. Hence, by Theorem 2.73, there exists unique mild solution $u(t)$ to (2.123) on $[t_0, t_1]$ for some $t_1 > t_0$. We will show that this solution is differentiable and satisfies (2.122).

Since $x \in D(A)$, it follows that $T(t-t_0)x$ is differentiable and the following holds

$$\frac{d}{dt}T(t - t_0)x = AT(t - t_0)x, \tag{2.128}$$

see Theorem 2.12. Now consider the following

$$\int_{t_0}^{t+h} T(t + h - \tau)f(\tau, u(\tau))d\tau - \int_{t_0}^{t} T(t - \tau)f(\tau, u(\tau))d\tau =$$

$$[T(h) - I] \int_{t_0}^{t} T(t - \tau)f(\tau, u(\tau))d\tau + \int_{t}^{t+h} T(t + h - \tau)f(\tau, u(\tau))d\tau. \tag{2.129}$$

By using the identity

$$f(\tau, u(\tau)) = f(t, u(t)) + [f(\tau, u(\tau)) - f(\tau, u(t))] + [f(\tau, u(t)) - f(t, u(t))], \tag{2.130}$$

and the differentiability of f we obtain

$$f(\tau, u(\tau)) - f(\tau, u(t)) = Df_u(\tau)[u(\tau) - u(t)] + r_1(\tau, t), \tag{2.131}$$
$$f(\tau, u(t)) - f(t, u(t)) = Df_\tau(\tau)[\tau - t] + r_2(\tau, t), \tag{2.132}$$

where $Df_u(\tau) = \frac{\partial f}{\partial u}(\tau)$, $Df_\tau(\tau) = \frac{\partial f}{\partial \tau}(\tau)$, and for $i = 1, 2$, we have $\lim_{h \to 0} h^{-1}\|r_i(\tau, t)\| = 0$ for $\tau \in [t, t+h]$, uniformly in t on $[t_0, t_1]$. Using (2.130)-(2.132), we obtain

$$\lim_{h \to 0} h^{-1} \int_{t}^{t+h} T(t + h - \tau)f(\tau, u(\tau))d\tau = f(t, u(t)), \tag{2.133}$$

where we used Theorem 2.12,

$$h^{-1} \left\| \int_{t}^{t+h} T(t + h - \tau)[f(\tau, u(\tau)) - f(\tau, u(t))]d\tau \right\| \leq h^{-1}Ma$$

$$\times \int_{t}^{t+h} \|u(\tau) - u(t)\| \, d\tau + h^{-1} \int_{t}^{t+h} \|r_1(\tau, t)\| \, d\tau, \tag{2.134}$$

$$h^{-1}\left\|\int_t^{t+h} T(t+h-\tau)[f(\tau,u(t))-f(t,u(t))]d\tau\right\| \le Mbh$$

$$+h^{-1}\int_t^{t+h}\|r_2(\tau,t)\|\,d\tau, \qquad (2.135)$$

where $M = \sup\{\|T(t)\| \mid t \in [0,t_1]\}$, $a = \sup\{\|Df_u(\tau)\| \mid \tau \in [t_0,t_1]\}$, $b = \sup\{\|Df_\tau(\tau)\| \mid \tau \in [t_0,t_1]\}$. Since $u(\cdot)$ is continuous, we have $\lim_{\tau\to t}\|u(\tau)-u(t)\| = 0$, hence we have $\lim_{h\to 0} h^{-1}\int_t^{t+h}\|u(\tau)-u(t)\|\,d\tau = 0$. Similarly, the remaining terms in (2.134) and (2.135) tend to 0 as $h \to 0$. By using (2.128)-(2.135) and (2.123), we obtain

$$\lim_{h\to 0}\frac{u(t+h)-u(t)}{h} = AT(t-t_0)x + \lim_{h\to 0}\frac{T(h)-I}{h}$$

$$\times \int_{t_0}^t T(t-\tau)f(\tau,u(\tau))d\tau + f(t,u(t))$$

$$= Au(t) + f(t,u(t)). \qquad (2.136)$$

Hence $u(\cdot)$ is a classical solution of (2.122). Since a classical solution is also a mild solution, and the mild solution is unique, it follows that $u(\cdot)$ is the unique classical solution. The remaining part of the theorem is similar to Theorem 2.73. $\qquad\square$

2.8 Integrated semigroups

This section is devoted to the study of integrated semigroups which are developed in recent years to cover nondensely defined operators as well as operators whose resolvents are not necessarily Laplace transforms of C_0-semigroups. Let us first give some examples to indicate the importance of relaxing the denseness and the Hille-Yosida condition in C_0-semigroups.

Proposition 2.75 *Let A generate a C_0-semigroup $T(t)$ on a Banach space X. Denote by X^* the dual space of X, by A^* the adjoint of A, and by $T^*(t)$ the adjoint of $T(t)$, respectively. Then $T^*(t)$ is a C_0-semigroup on X^* if X is reflexive.*

Proof. Since A generates a C_0-semigroup, A is densely defined and closed in X. Suppose that $D(A^*)$ is not dense in X^*. Then there exists an $x_0 \in X$ such that $x_0 \ne 0$ and $\langle x_0, x^*\rangle = 0$ for every $x^* \in D(A^*)$. Since A is closed, its graph in $X \times X$ is closed and does not contain $(0,x_0)$. From the Hahn-Banach theorem it follows that there are $x_1^*, x_2^* \in X^*$ such that $\langle x, x_1^*\rangle - \langle Ax, x_2^*\rangle = 0$ for every $x \in D(A)$ and $\langle 0, x_1^*\rangle - \langle x_0, x_2^*\rangle \ne 0$. From the second equation it follows that $x_2^* \neq 0$ and that $\langle x_0, \ x_2^* \rangle \neq 0$. But, from the first equation it follows that $x_2^* \in D(A^*)$ which implies $\langle x_0, \ x_2^* \rangle = 0$, a contradiction. Thus, $D(A^*)$ is dense in X^*. Next, from standard functional analysis, $\lambda \in \rho(A)$ implies $\lambda \in \rho(A^*)$ and

$$R(\lambda, A^*) = R^*(\lambda, A).$$

Since A satisfies the Hille-Yosida condition, there are constants ω and M such that for all real $\lambda, \lambda > \omega, \lambda \in \rho(A)$ and

$$\| \ R(\lambda, A^*)^n \ \| = \| \ R^*(\lambda, A)^n \ \| = \| \ R(\lambda, A)^n \ \| \leq \frac{M}{(\lambda - \omega)^n}, \quad n = 1, 2, \ldots.$$

Therefore, A^* generates a C_0-semigroup $T_0(t)$ on X^*. For $x \in X$ and $x^* \in X^*$ we have, by definition,

$$\left\langle (I - \frac{t}{n}A)^{-n}x, \ x^* \right\rangle = \left\langle x, \ (I - \frac{t}{n}A^*)^{-n}x^* \right\rangle, \quad n = 1, 2, \ldots.$$

Letting $n \to \infty$ above, and using the exponential formula (Theorem 2.43), we obtain

$$\langle T(t)x, \ x^* \rangle = \langle x, \ T_0(t)x^* \rangle.$$

Since $\langle T(t)x, \ x^* \rangle = \langle x, \ T^*(t)x^* \rangle$, and X is reflexive, we have $T^*(t) = T_0(t)$. Thus, $T^*(t)$ is a C_0-semigroup with generator A^*.   □

If X is nonreflexive, $T^*(t)$ is usually not a C_0-semigroup since the mapping $T(t) \to T^*(t)$ does not necessarily reserve the strong continuity of $T(t)$. A typical example is the shift semigroup on $L^1(\mathbf{R})$.

Example 2.76 Let $X = L^1(\mathbf{R})$ and define the shift operator $T(t)$ as follows

$$T(t)x(s) = x(s + t), \quad \forall x(s) \in L^1(\mathbf{R}).$$

Then $X^* = L^\infty(\mathbf{R})$ and

$$T^*(t)\phi(s) = \phi(s - t), \quad \forall \phi(s) \in L^\infty(\mathbf{R}).$$

For any non-trivial characteristic function $\phi(s)$, we have $\|T^*(t)\phi - \phi\| = 1$ for $t > 0$ and so $T^*(t)$ is not strongly continuous.

Example 2.77 Let $X = C[0, 1]$ denote the continuous function space over $[0, 1]$. Define the operator A by

$$Au = -u', \quad D(A) = C_0^1[0, 1],$$

where $C_0^1[0,1]$ denotes the space consisting of continuously differentiable functions with compact support. We have $\overline{D(A)} = C_0^1[0,1] \neq X$. Moreover, for each $\lambda > 0$ and $u \in X$,

$$(R(\lambda, A)u)(s) = \int_0^s e^{-\lambda t}u(s-t)dt, \quad s \in [0,1],$$

hence

$$\|(R(\lambda, A)u)(s)\| \leq \|u\| \int_0^\infty e^{-\lambda t}dt = \frac{1}{\lambda}\|u\|.$$

Here, the Hille-Yosida condition holds, but A is not a generator because A is not densely defined.

We have seen in the previous sections that for a linear operator A to generate a C_0−semigroup, it is necessary that A is densely defined and furthermore the resolvent estimate of A in the Hille-Yosida theorem should hold for any integer $n = 1, 2, \cdots$. These can be very restrictive in practical applications, and it is desired that these requirements be relaxed. In the following, we shall see that it is the integrated semigroup, first introduced by W.Arendt [3],[4], that successfully relaxes these two requirements in C_0-semigroups.

Definition 2.78 *Let A be a closed linear operator in a Banach space X. If there exists an integer n, constants M, ω and a strongly continuous family $S(t)$ in $\mathcal{L}(X)$ with $\|S(t)\| \leq Me^{\omega t}$ for all $t \geq 0$ such that*

$$R(\lambda, A)x = \lambda^n \int_0^\infty e^{-\lambda t}S(t)xdt, \text{ for } x \in X \text{ and } \lambda > \omega, \qquad (2.137)$$

then A is called the generator of an n-times integrated semigroup $S(t)$.

It is seen from Definition 2.78 and Theorem 2.39 that the 0-time integrated semigroup is just the C_0-semigroup if A is densely defined. Besides, if A generates a C_0-semigroup $T(t)$, when integrating by parts one sees that $S(t)$ defined by $S(t)x = \int_0^t T(s)xds$ for all $x \in X$ is a 1-time integrated semigroup with generator A. This is the motivation of the terminology "integrated semigroup". Generally, if A generates an n-times integrated semigroup $(n \geq 0)$, then A generates also an m-times integrated semigroup for all $m > n$. We do not assume the denseness of the generator in the definition of integrated semigroups.

Paralleling to C_0-semigroups, we have the following basic properties of integrated semigroups.

Theorem 2.79 *Suppose that $S(t)$ is an n-times integrated semigroup in a Banach space X with generator A and $\|S(t)\| \leq Me^{\omega t}$. If $n \geq 1$, then*

(i) $S(t)x = 0$ for all $t \geq 0$ implies $x = 0$.

(ii) For every $x \in D(A)$, $S(t)x \in D(A)$, $S(t)Ax = AS(t)x$, and

$$S(t)x \in C^1([0, \infty); X) \text{ and } S(t)x = \int_0^t S(\tau)Ax d\tau + \frac{t^n}{n!}x.$$

In particular, $S(0) = 0$ and if $x \in D(A^{n+1})$, then

$$u(t) = S(t)A^n x + \sum_{k=0}^{n-1} \frac{t^k}{k!} A^k x \qquad (2.138)$$

is the unique solution of the following abstract Cauchy problem

$$\frac{du(t)}{dt} = Au(t), \quad u(0) = x.$$

(iii) For every $x \in X$ we have $\int_0^t S(s)x ds \in D(A)$ and

$$A \int_0^t S(s)x ds = S(t)x - \frac{t^n}{n!}x.$$

In particular, $S(t)x \in \overline{D(A)}$ for every $x \in X$.

(iv) $S(t)$ is uniquely determined by A.

(v) $\{\lambda \mid Re\lambda > \omega\} \subset \rho(A)$ and (2.137) holds for all $Re\lambda > \omega$.

Proof.   The statement (i) is obvious from Definition 2.78. Let $\lambda, \mu > \omega$ and $y \in X$. Then

$$\lambda^n \int_0^\infty e^{-\lambda t} S(t) R(\mu, A) y dt = R(\lambda, A) R(\mu, A) y$$

$$= R(\mu, A) R(\lambda, A) y$$

$$= R(\mu, A) \lambda^n \int_0^\infty e^{-\lambda t} S(t) y dt.$$

By the uniqueness of Laplace transforms, we have

$$S(t) R(\mu, A) y = R(\mu, A) S(t) y \in D(A) \qquad (2.139)$$

for all $t \geq 0$. Hence, $S(t)D(A) \subset D(A)$. For any $x \in D(A)$, let $(\mu - A)x = y$. From (2.139)

$$
\begin{aligned}
AS(t)x &= AS(t)R(\mu, A)y \\
&= AR(\mu, A)S(t)y \\
&= \mu R(\mu, A)S(t)y - S(t)y \\
&= \mu S(t)R(\mu, A)y - S(t)y \\
&= \mu S(t)x - S(t)(\mu - A)x = S(t)Ax.
\end{aligned}
$$

Since $\int_0^\infty \lambda^{n+1} \frac{t^n}{n!} e^{-\lambda t} dt = 1$, we have

$$
\begin{aligned}
\int_0^\infty \lambda^{n+1} \frac{t^n}{n!} e^{-\lambda t} x dt &= (\lambda - A) R(\lambda, A) x \\
&= \lambda R(\lambda, A) x - A R(\lambda, A) x \\
&= \int_0^\infty \lambda^{n+1} e^{-\lambda t} S(t) x dt - \int_0^\infty \lambda^n e^{-\lambda t} S(t) A x dt \\
&= \int_0^\infty \lambda^{n+1} e^{-\lambda t} S(t) x dt \\
&\quad - \int_0^\infty \lambda^{n+1} e^{-\lambda t} \left(\int_0^t S(\tau) A x d\tau \right) dt,
\end{aligned}
$$

or

$$
\int_0^\infty \lambda^{n+1} e^{-\lambda t} \left[S(t) x - \int_0^t S(\tau) A x d\tau - \frac{t^n}{n!} x \right] dt = 0, \text{ for all } \lambda > \omega.
$$

Again, due to the uniqueness of Laplace transforms,

$$
S(t) x = \int_0^t S(\tau) A x d\tau + \frac{t^n}{n!} x \tag{2.140}
$$

for any $x \in D(A)$. This verifies the first part of (ii). Keeping (2.140) in mind, a direct computation shows that $u(t)$ defined by (2.138) is a solution of the abstract Cauchy problem. Since A generates an integrated semigroup, the conditions of Theorem 2.63 are satisfied; hence, (2.138) is the unique solution of the abstract Cauchy problem.

Next, for $x \in D(A)$, by noting (2.139), we have

$$
\begin{aligned}
\int_0^t S(\tau) x d\tau &= \int_0^t S(\tau) (\lambda - A) R(\lambda, A) x d\tau \\
&= \lambda R(\lambda, A) \int_0^t S(\tau) x d\tau - \int_0^t S(\tau) A R(\lambda, A) x d\tau \\
&= \lambda R(\lambda, A) \int_0^t S(\tau) x d\tau - [S(t) R(\lambda, A) x - \frac{t^n}{n!} R(\lambda, A) x]
\end{aligned}
$$

where we have used (2.140) to obtain the last equality. It is evident that each term on the right-hand side belongs to $D(A)$; hence, the first part of (iii) follows. Operating on both sides by $\lambda - A$, and using the commuting property (2.139), yields the second part of (iii). Finally, for every $x \in X$,

$$
S(t) x = \lim_{h \downarrow 0} \frac{1}{h} \int_t^{t+h} S(s) x ds \in \overline{D(A)}.
$$

(iv) is a direct consequence of the uniqueness of Laplace transforms. Finally, since it is well known that if the Laplace transform of a continuous function exists for some λ_0, then it exists for all complex numbers λ with $\text{Re}\lambda > \text{Re}\lambda_0$. Hence, (v) holds.   □

Similar to C_0-semigroups, we shall find some characterization conditions for a linear operator to generate an n-times integrated semigroup. The first result paralleling to the Hille-Yosida condition for C_0-semigroups is developed by Arendt. To state this result, we need the following representation theorem in Laplace transforms.

Theorem 2.80 (Widder) *The following statements are equivalent:*

(i) $r(\lambda) \in C^\infty(0,\infty)$ *and* $\left|\frac{\lambda^{k+1}}{k!}r^{(k)}(\lambda)\right| \leq M$ *for some* $M \geq 0$ *and* $\lambda \in (0,\infty)$ *for all* $k = 0,1,\cdots$, *where* $r^{(k)}(\lambda) = \frac{d^k}{d\lambda^k}r(\lambda)$.

(ii) *There exists a function* $f \in L^\infty(0,\infty)$ *with* $|f(t)| \leq M$ *for all* $t \in (0,\infty)$ *such that* $r(\lambda) = \int_0^\infty e^{-\lambda t}f(t)dt$, *for* $\lambda > 0$.

Proof. See Widder [163].   □

Recalling the Hille-Yosida condition, one observes the striking similarity between the statement (i) of Widder's theorem and the Hille-Yosida condition. Therefore, it is quite reasonable to expect a close connection between Widder's theorem generalized to Banach spaces and the Hille-Yosida theorem. Unfortunately, Widder's theorem does not hold for functions taking values in a general Banach space. The following result is a generalization of Widder's theorem and was developed by Arendt [4].

Theorem 2.81 *Let* $R(\cdot) : (\omega,\infty) \to X$ *be a function where* X *is a Banach space . The following statements are equivalent:*

(i) $R(\cdot) \in C^\infty((\omega,\infty);X)$ *and there exists constant* $M \geq 0$ *such that for* $k = 0,1,2,\cdots$ *and all* $\lambda > \omega$,

$$\left\| \frac{R^{(k)}(\lambda)}{k!} \right\| \leq \frac{M}{(\lambda-\omega)^{k+1}}.$$

(ii) *There exists a function* $F(\cdot) : [0,\infty) \to X$ *satisfying*

$$F(0) = 0 \text{ and } \| F(t+h) - F(t) \| \leq M e^{\omega(t+h)}h, \quad \forall t,h \geq 0$$

such that

$$R(\lambda) = \lambda \int_0^\infty e^{-\lambda t}F(t)dt, \quad \lambda > \omega.$$

Proof. Suppose that (i) holds. Let $x^* \in X^*$ and consider the scalar valued function $r(\lambda) = x^*(R(\lambda+\omega))$, $\lambda > 0$. Clearly, r satisfies the Widder's theorem since we have

$$\left\| \frac{r^{(k)}(\lambda)}{k!} \right\| = \left\| \frac{x^*(R^{(k)}(\lambda + \omega))}{k!} \right\| \leq \frac{M}{\lambda^{k+1}} \|x^*\|, \quad \text{for all } \lambda > 0, k = 0, 1, \cdots.$$

Hence, there exists a function $f(t, x^*) \in L^\infty(0, \infty)$ depending on x^* such that $|f(t, x^*)| \leq M \|x^*\|$ for all $t > 0$ and

$$r(\lambda) = x^*(R(\lambda + \omega)) = \int_0^\infty e^{-\lambda t} f(t, x^*) dt$$

or

$$x^*(R(\lambda)) = \int_0^\infty e^{-\lambda t} e^{\omega t} f(t, x^*) dt, \quad \lambda > \omega.$$

Define

$$g(t, x^*) = \int_0^t e^{\omega \tau} f(\tau, x^*) d\tau. \tag{2.141}$$

Then $x^*(R(\lambda))$ can be written as

$$x^*(R(\lambda)) = \lambda \int_0^\infty e^{-\lambda t} g(t, x^*) dt, \quad \lambda > \omega.$$

The function $t \to g(t, x^*)$ is continuous and is also linear bounded on X^*, and hence for each $t \in [0, \infty)$ there exists an $F(t) \in X^{**}$, the bidual of X, such that $g(t, x^*) = \langle F(t), x^* \rangle$ for all $x^* \in X^*$. In other words,

$$R(\lambda) = \lambda \int_0^\infty e^{-\lambda t} F(t) dt, \quad \lambda > \omega$$

with $F(t) \in X^{**}$ for $t \geq 0$. We prove that $F(t) \in X$. Identify X as a closed subspace of X^{**} under the canonical embedding $X \to X^{**}$, and let Φ denote the quotient map $X^{**} \to X^{**}/X$. Since $R(\lambda) \in X$, we have

$$0 = \Phi(R(\lambda)/\lambda) = \int_0^\infty e^{-\lambda t} \Phi(F(t)) dt, \; \lambda > \omega.$$

It follows from the uniqueness of Laplace transforms that $\Phi(F(t)) = 0$ for all $t \geq 0$. This means that $F(0) \in X$ for all $t \geq 0$. Since $g(0, x^*) = 0$ we have $F(0) = 0$; and further, it follows from (2.141) that

$$\begin{aligned}
|\langle F(t + h) - F(t), x^* \rangle| &= |g(t + h, x^*) - g(t, x^*)| \\
&= \left| \int_t^{t+h} e^{\omega \tau} f(\tau, x^*) d\tau \right| \\
&\leq M \|x^*\| \int_t^{t+h} e^{\omega \tau} d\tau \\
&\leq M e^{\omega(t+h)} h \|x^*\|.
\end{aligned}$$

Hence, $\|F(t+h) - F(t)\| \leq Me^{\omega(t+h)}h$. In particular, letting $t = 0$, we have
$$\|F(t)\| \leq Me^{\omega t}t, \text{ for all } t \geq 0.$$
The reverse implication is trivial. This completes the proof. $\square$

As a consequence of Theorem 2.81, we have the following result.

Theorem 2.82 *Let A be a linear operator in a Banach space X with $(\omega, \infty) \subset \rho(A), \omega > 0$. Then the following statements are equivalent:*

(i) A generates an $(n+1)$- times integrated semigroup $S(t)$ with
$$\|S(t+h) - S(t)\| \leq Me^{\omega(t+h)}h, \quad \forall t, h > 0.$$

(ii) $\left\|[\lambda^{-n}R(\lambda, A)]^{(k)}/k!\right\| \leq \frac{M}{(\lambda-\omega)^{k+1}}, \quad \forall \lambda > \omega, k = 0, 1, \cdots.$

Proof. Assume (ii). Take any $x \in X$ and define $R(\lambda) = \lambda^{-n}R(\lambda, A)x$. It follows from Theorem 2.81 that there exists a family of linear operators $\{S(t), t \geq 0\}$ in X such that
$$R(\lambda) = \lambda^{-n}R(\lambda, A)x = \lambda \int_0^\infty e^{-\lambda t}S(t)x dt \text{ and } S(0) = 0.$$
Further, it follows from (ii) of Theorem 2.81 that
$$\|S(t+h) - S(t)\| \leq Me^{\omega(t+h)}h, \forall t, h > 0.$$
This shows that (ii) is the necessary and sufficient condition for A to be the generator of an $(n+1)$-times integrated semigroup with the above Lipschitz continuity.

Assume (i). (ii) follows from Theorem 2.81 by setting $F(t) = S(t)$. $\square$

Remark 2.83 *When $n = 0$, the condition (ii) of Theorem 2.82 is just the Hille-Yosida condition. So, A generates a locally Lipschitz continuous 1-time integrated semigroup if and only if the Hille-Yosida condition holds. Consequently, if A generates a C_0-semigroup, then A generates a locally Lipschitz continuous 1-time integrated semigroup.*

Next, we see what happens to Theorem 2.82 if A is densely defined .

Theorem 2.84 *Let A be a densely defined linear operator in a Banach space X with $(\omega, \infty) \subset \rho(A), \omega > 0$. Then the following statements are equivalent:*

(i) A generates an n-times integrated semigroup $S(t)$.

(ii) A generates an $(n+1)$- times integrated semigroup $T(t)$ with
$$\|T(t+h) - T(t)\| \leq Me^{\omega(t+h)}h, \quad \forall t, h > 0.$$

(iii) $\left\|[\lambda^{-n}R(\lambda, A)]^{(k)}/k!\right\| \leq \frac{M}{(\lambda - \omega)^{k+1}}, \quad \forall \lambda > \omega, k = 0, 1, \cdots.$

Proof. The equivalence of (ii) and (iii) is just Theorem 2.81. We show equivalence of (i) and (ii). Suppose (ii). Let $Y = \{x \in X | T(t)x \in C^1[0, \infty)\}$. By (ii) Y is a closed subspace of X. For $x \in D(A)$, by (ii) of Theorem 2.79, it follows that $T(t)x \in D(A)$ and $T(t)Ax = AT(t)x$, and $T'(t)x = T(t)Ax + \frac{t^n}{n!}x$. Hence $T'(t)x$ is continuous and $D(A) \subset Y$. Since Y is closed and $\overline{D(A)} = X$, we have $Y = X$. So, $T'(t)$ is well defined and strongly continuous on whole X.

Let $S(t) = T'(t)$. By (ii), $\|S(t)\| \le Me^{\omega t}$. Since $T(t)$ is an $(n+1)$-times integrated semigroup, it follows that

$$R(\lambda, A)x = \lambda^{n+1} \int_0^\infty e^{-\lambda t} T(t)x dt = \lambda^n \int_0^\infty e^{-\lambda t} S(t)x dt$$

for all $x \in X$. Therefore, A generates an n-times integrated semigroup. The converse part follows from a simple fact that if A generates an n-times integrated semigroup $S(t)$, then A generates an $(n+1)$-times integrated semigroup $T(t) = \int_0^t S(\tau)d\tau$, with $T(t)$ being differentiable in the sense of strong topology. This completes the proof. □

It should be remarked that the Lipschitz type continuous condition in (ii) of Theorem 2.84 is not removable. Actually, for every $n \ge 1$, there exists a densely defined linear operator A in a Banach (Hilbert) space X such that A is the generator of an $(n+1)$-times integrated semigroup on X, but not an n-times integrated semigroup (Proposition 2.4 in [123]). However, the equivalence of (i) and (iii) of Theorem 2.84 is certainly surprising, since, in general, Widder's theorem does not hold in general Banach spaces [4]. In fact, it is the additional property $\overline{D(A)} = X$ that makes up the gap between them.

The adjoint semigroup discussed at the beginning of this section could be explained now by integrated semigroup as follows

Corollary 2.85 *Let A be a densely defined linear operator on a Banach space X. If A generates an n-times Lipschitz continuous integrated semigroup, then its adjoint A^* generates an $(n+1)$-times integrated semigroup. When X is reflexive, then A generates an n-times integrated semigroup if and only if its adjoint A^* generates an n-times integrated semigroup.*

Proof. The first part follows immediately from (ii),(iii) of Theorem 2.84 since $R(\lambda, A)^* = R(\lambda, A^*)$ for all $\lambda \in \rho(A)$. For the second part, it should be noted that the denseness of A in X implies the denseness of A^* in X^* when X is reflexive. □

Another interesting consequence of Theorem 2.82 arises when we consider resolvent operators on which an algebraic condition, but no norm condition, is imposed.

Corollary 2.86 *Let X be an ordered Banach space with normal and generating cone. Let A be an operator on X such that $(\omega, \infty) \subset \rho(A)$ for some $\omega \in \mathbf{R}$ and $R(\lambda, A) \geq 0$ for all $\lambda > \omega$. Then A is the generator of a 2-times Lipschitz continuous integrated semigroup. If $D(A)$ is dense, then A generates a 1-time integrated semigroup.*

The proof is based on the following lemma.

Lemma 2.87 *Let A be an operator and $\lambda \in \rho(A)$. Then for every integer $m \geq 1$*

$$(-1)^m \lambda^{m+1} [\lambda^{-1} R(\lambda, A)]^{(m)} / m! = \sum_{k=0}^{m} \lambda^k R(\lambda, A)^{k+1}.$$

Proof. This is immediate by developing $[\lambda^{-1} R(\lambda, A)]^{(m)}$ and using

$$(-1)^k R(\lambda, A)^{(k)} / k! = R(\lambda, A)^{k+1}.$$

$\square$

Proof of Corollary 2.86. Considering $A - \omega$ instead of A if necessary, we may assume that $[0, \infty) \subset \rho(A)$ and $R(\lambda, A) \geq 0$ for all $\lambda \geq 0$. Then for all integers $m \geq 1$,

$$\sum_{k=0}^{m-1} \lambda^k R(\lambda, A)^{k+1} = R(0, A) - \lambda^m R(\lambda, A)^m R(0, A).$$

In fact, $m = 1$ follows from the resolvent equation $R(\lambda, A) = R(0, A) - \lambda R(\lambda, A) R(0, A)$. Suppose it is true for m, consider the case of $m+1$. Again, by the resolvent equation

$$\begin{aligned}
\sum_{k=0}^{m} \lambda^k R(\lambda, A)^{k+1} &= R(0, A) - \lambda^m R(\lambda, A)^m R(0, A) + \lambda^m R(\lambda, A)^{m+1} \\
&= R(0, A) - \lambda^m R(\lambda, A)^m R(0, A) \\
&\quad + \lambda^m [R(\lambda, A)^m R(0, A) - \lambda R(\lambda, A)^{m+1} R(0, A)] \\
&= R(0, A) - \lambda^{m+1} R(\lambda, A)^{m+1} R(0, A).
\end{aligned}$$

Therefore,

$$0 \leq \sum_{k=0}^{m-1} \lambda^k R(\lambda, A)^{k+1} \leq R(0, A).$$

Since X is an ordered Banach space with normal and generating cone, there exists a constant $M > 0$ such that for all linear operators S and T on X one has

$$0 \leq S \leq T \text{ implies } \|S\| \leq M \|T\|.$$

This, together with Lemma 2.87, gives

$$\left\| [\lambda^{-1} R(\lambda, A)]^{(m)}/m! \right\| = \left\| \lambda^{-m-1} \sum_{k=0}^{m} \lambda^k R(\lambda, A)^{k+1} \right\|$$

$$\leq M\lambda^{-m-1} \left\| R(0, A) \right\|, \forall \lambda > 0, m \geq 1.$$

So, the claims follow from (ii) and (iii) of Theorems 2.82 and 2.84.   □

Similar to C_0-semigroups, the Hille-Yosida type condition for integrated semigroups developed in Theorem 2.82 is only theoretically important because it is not usually easily verifiable in applications. We now wish to simplify the characterization conditions for generation of integrated semigroups by using the well-posedness results of abstract Cauchy problems (ACP).

It is seen from (ii) of Theorem 2.79 that if A generates an n-times integrated semigroup, then for any $x \in D(A^{n+1})$ there exists a unique solution to (ACP) which is given by (2.138). Conversely, we have the following interesting counterpart of this conclusion: the integrated semigroup is determined by the solution of (ACP).

Theorem 2.88 *Suppose that $S(t)$ is an n-times integrated semigroup on a Banach space X with generator A and $\|S(t)\| \leq Me^{\omega t}$. If $u(\cdot)$ is the unique solution of the following abstract Cauchy problem (ACP):*

$$\frac{du(t)}{dt} = Au(t), \quad u(0) = x \in D(A^{n+1}),$$

then

$$S(t)x = \int_0^t \frac{(t-s)^{n-1}}{(n-1)!} u(s)ds. \qquad (2.142)$$

Proof. By the uniqueness of solution, $u(t) = S(t)A^n x + \sum_{k=0}^{n-1} \frac{t^k}{k!} A^k x$. Since $\int_0^t \frac{(t-s)^{n-1}}{(n-1)!} \frac{s^k}{k!} ds = \frac{t^{n+k}}{(n+k)!}$, by (ii) of Theorem 2.79, we have

$$S(t)x = \int_0^t S(\tau) Ax d\tau + \frac{t^n}{n!} x \qquad (x \in D(A))$$

$$= \int_0^t \left[\int_0^\tau S(r) A^2 x dr + \frac{\tau^n}{n!} Ax \right] d\tau + \frac{t^n}{n!} x$$

$$= \int_0^t (t-r) S(r) A^2 x dr + \frac{t^{n+1}}{(n+1)!} Ax + \frac{t^n}{n!} x \quad (x \in D(A^2))$$

$$= \int_0^t (t-r) \left[\int_0^r S(\tau) A^3 x d\tau + \frac{r^n}{n!} A^2 x \right] dr + \frac{t^{n+1}}{(n+1)!} Ax + \frac{t^n}{n!} x$$

$$= \int_0^t \frac{(t-\tau)^2}{2!} S(\tau) A^3 x d\tau + \int_0^t r \frac{(t-r)^n}{n!} A^2 x dr + \frac{t^{n+1}}{(n+1)!} Ax + \frac{t^n}{n!} x$$

$$= \int_0^t \frac{(t-\tau)^2}{2!} S(\tau) A^3 x \, d\tau + \frac{t^{n+2}}{(n+2)!} A^2 x + \frac{t^{n+1}}{(n+1)!} Ax + \frac{t^n}{n!} x$$
$$(x \in D(A^3)).$$

Iterating the above process, we obtain, for $x \in D(A^{n+1})$,

$$
\begin{aligned}
S(t)x &= \int_0^t \frac{(t-\tau)^{n-1}}{(n-1)!} S(\tau) A^n x \, d\tau + \sum_{k=0}^{n-1} \frac{t^{n+k}}{(n+k)!} A^k x \\
&= \int_0^t \frac{(t-\tau)^{n-1}}{(n-1)!} S(\tau) A^n x \, d\tau + \sum_{k=0}^{n-1} \int_0^t \frac{(t-\tau)^{n-1}}{(n-1)!} \frac{\tau^k}{k!} A^k x \, d\tau \\
&= \int_0^t \frac{(t-s)^{n-1}}{(n-1)!} u(s) \, ds.
\end{aligned}
$$

This is (2.142).   □

Because of (2.142), we sometimes call $S(t)$ the n-times integrated solution of (ACP).

Furthermore, if A generates an n-times integrated semigroup, then the unique solution $u(t) = S(t) A^n x + \sum_{k=0}^{n-1} \frac{t^k}{k!} A^k x$ of (ACP) depends continuously on its initial condition $x \in D(A^{n+1})$ in the following way

$$\|u(t)\| \le M_0 e^{\omega t} |x|_n, \quad |x|_n = \sum_{k=0}^n \|A^k x\|. \tag{2.143}$$

Definition 2.89 *Let A be a closed linear operator on a Banach space X. The (ACP) is called exponentially n-well-posed if there exist constants $M_0 \ge 0, \omega \in \mathbf{R}$ such that for every $x \in D(A^{n+1})$ there exists a unique solution $u(\cdot)$ of (ACP) with (2.143) holding for all $t \ge 0$.*

The connection between generators of integrated semigroups and exponentially n-well-posedness of (ACP) is given by the following theorem.

Theorem 2.90 *Let A be a linear operator on a Banach space X with nonempty resolvent set. Then*

(i) *If A generates an n-times integrated semigroup, then (ACP) is exponentially n-well-posed;*

(ii) *If A is densely defined and if (ACP) is exponentially n-well-posed, then A generates an n-times integrated semigroup.*

Proof. (i) follows from (2.143). For (ii), let $x \in D(A^{n+1})$ and assume that $u(t)$ is the unique solution of (ACP) with $\|u(t)\| \leq Me^{\omega t}|x|_n$. Motivated by (2.142), we define $S(t) : D(A^{n+1}) \to X$ by

$$S(t)x = \int_0^t \frac{(t-s)^{n-1}}{(n-1)!} u(s)ds.$$

For λ in the resolvent set of A, the function defined by $w(t) = R(\lambda, A)u(t)$ is a solution of (ACP) with $\|w(t)\| \leq M_1 e^{\omega t}|x|_{n-1}$. Let $v(t) = \int_0^t u(s)ds$ be the 1-time integrated solution. Then

$$v(t) = \int_0^t (\lambda - A)w(s)ds = \lambda \int_0^t w(s)ds - w(t) + R(\lambda, A)x.$$

Hence, $\|v(t)\| \leq M_2 e^{\omega_0 t}|x|_{n-1}$ for suitable constants M_2 and ω_0. By induction, we obtain that such defined n-times integrated solutions $S(t)x$ are exponentially bounded.

Since $D(A)$ is dense in X, it is well known that $D(A^n)$ also is dense in X. Hence, the linear operator $S(t)$ has a unique extension on X, which is denoted by the same symbol. We shall show that $S(t)$ is the required n-times integrated semigroup. From $\|S(t)\| \leq \bar{M} e^{\bar{\omega} t}$ for all $t \geq 0$ for suitable $\bar{M}$ and $\bar{\omega}$, we conclude that $S(t)$ is strongly continuous. For $\mathrm{Re}\lambda > \bar{\omega}$, define a bounded linear operator $R(\lambda)x = \lambda^n \int_0^\infty e^{-\lambda t} S(t)x dt$. Let $x \in D(A^{n+1})$. By the closedness of A, one obtains that $S(t)x \in D(A)$ for every $t \geq 0$ and

$$AS(t)x = \frac{d}{dt}S(t)x - \frac{t^{n-1}}{(n-1)!}x.$$

Multiplying both sides by $\lambda^{n-1}e^{-\lambda t}$ and integrating with respect to t from 0 to T, we obtain, by integration by parts

$$A\left(\lambda^{n-1}\int_0^T e^{-\lambda t}S(t)x dt\right) = \lambda^{n-1}e^{-\lambda T}S(T)x + \lambda^n \int_0^T e^{-\lambda t}S(t)x dt$$

$$-\lambda^{n-1}\int_0^T e^{-\lambda t}\frac{t^{n-1}}{(n-1)!}x dt.$$

Letting $T \to \infty$, and noting that $e^{-\lambda T}S(T) \to 0$, $\int_0^\infty e^{-\lambda t}t^{n-1}dt = (n-1)!/\lambda^n$, we see that for every $x \in D(A^{n+1})$, $R(\lambda)x \in D(A)$ and $(\lambda - A)R(\lambda)x = x$. Now let $x \in X$ and $x_k \in D(A^{n+1})$ with $x_k \to x$ as $k \to \infty$. Then $AR(\lambda)x_k = \lambda R(\lambda)x_k - x_k \to \lambda R(\lambda)x - x$ and $R(\lambda)x_k \to R(\lambda)x$. Therefore, $R(\lambda)x \in D(A)$ and $AR(\lambda)x = \lambda R(\lambda)x - x$ or $(\lambda - A)R(\lambda)x = x$. Hence, $(\lambda - A)$ maps $D(A)$ onto X for every $\lambda \in \mathbf{C}$ with $\mathrm{Re}\lambda > \bar{\omega}$. Suppose there is a $y \in D(A)$ with $Ay = \lambda y$ where $\mathrm{Re}\lambda > \bar{\omega}$. Then $u(t) = e^{\lambda t}y$ is a solution of

(ACP) and $y \in D(A^{n+1})$. But

$$
\begin{aligned}
S(t)y &= \int_0^t \frac{(t-s)^{n-1}}{(n-1)!} u(s)ds = \int_0^t \frac{(t-s)^{n-1}}{(n-1)!} e^{\lambda s} y \, ds \\
&= (\lambda^{1-n} e^{\lambda t} - \sum_{k=0}^{n-1} \lambda^{1-n+k} \frac{t^k}{k!}) y,
\end{aligned}
$$

and hence

$$
\bar{M} e^{\bar{\omega} t} \geq \|S(t)y\| \geq |\lambda^{1-n}| e^{\operatorname{Re}\lambda t} \|y\| - \sum_{k=0}^{n-1} |\lambda^{1-n+k}| \frac{t^k}{k!} \|y\|
$$

which is impossible for large t. Therefore, λ is in the resolvent set of A and $R(\lambda, A) = R(\lambda)$. From the definition of $R(\lambda)$ given above and Definition 2.78, it follows that A generates an n-times integrated semigroup, which is $S(t)$. □

Now, we state a result which is very convenient in applications.

Theorem 2.91 *Let A be a linear operator in a Banach space X. If there are constants M, ω such that $R(\lambda, A)$ exists and satisfies*

$$
\|R(\lambda, A)\| \leq M(1 + |\lambda|^k) \quad \text{for some } k \geq -1 \tag{2.144}
$$

for all $\lambda \in \mathbf{C}$ with $\operatorname{Re}\lambda > \omega$, then (ACP) has a unique solution $u(\cdot)$ for every $x \in D(A^{[k]+3})$ such that $\|u(t)\| \leq M_\alpha e^{\alpha t} |x|_{[k]+3}$ for $\alpha > \omega$. Hence, (ACP) is at least exponentially $[k]+3$ well-posed. Here, $[k]$ is the largest integer not exceeding k.

Before proving this theorem, we state a lemma

Lemma 2.92 *Suppose that the function $u(t)$ is continuous on $[0,\infty)$ and continuously differentiable on $(0,\infty)$, and that its derivative $u'(t)$ has a limit as $t \to 0$. If the linear operator A is closed and the function $u(t)$ satisfies $u'(t) = Au(t)$ on $(0,\infty)$, then it is a solution of $u'(t) = Au(t)$ on $[0,\infty)$.*

Proof. We need only to verify that the function $u(t)$ satisfies the equation at $t = 0$. It is right differentiable at $t = 0$. Indeed, passing to the limit in the equation

$$
u(t) - u(\epsilon) = \int_\epsilon^t u'(s)ds,
$$

as $\epsilon \to 0$, we find that

$$
u(t) - u(0) = \int_0^t u'(s)ds,
$$

from which it follows that

$$u'(0) = \lim_{t \downarrow 0} u'(t).$$

Using the fact that A is closed, taking the limit $t \downarrow 0$ in the equation $u'(t) = Au(t)$ and we arrive at the equation $u'(0) = Au(0)$.   □

Proof of Theorem 2.91. Uniqueness of the solution follows from Theorem 2.63. Let $\lambda_0 > \alpha > \omega$ and let $m = [k] + 3$. For $x \in D(A^m)$, letting $y = (\lambda_0 - A)^m x$, we have

$$x = R(\lambda_0, A)^m y.$$

Then

$$\begin{aligned}
R(\lambda, A)x &= R(\lambda, A)R(\lambda_0, A)^m y = \frac{R(\lambda_0, A) - R(\lambda, A)}{\lambda - \lambda_0} R(\lambda_0, A)^{m-1} y \\
&= \frac{R(\lambda_0, A)^m y}{\lambda - \lambda_0} - \frac{R(\lambda_0, A)^{m-1} y}{(\lambda - \lambda_0)^2} + \cdots \\
&\quad + (-1)^{m-1} \frac{R(\lambda_0, A)y}{(\lambda - \lambda_0)^m} + (-1)^m \frac{R(\lambda, A)y}{(\lambda - \lambda_0)^m}.
\end{aligned}$$

If we multiply both sides above by $e^{\lambda t}$ and integrate along the line $Re\lambda = \alpha$, then the integrals of the functions of the form

$$e^{\lambda t} \frac{R(\lambda_0, A)^s y}{(\lambda - \lambda_0)^{m+1-s}}, \quad 1 \le s \le m$$

will vanish in the principal value sense for $t > 0$. In fact, for any $N > 0$ and $n \ge 1$,

$$\int_{\alpha-iN}^{\alpha+iN} \frac{e^{\lambda t}}{(\lambda - \lambda_0)^n} d\lambda = e^{\alpha t} \int_{-iN}^{iN} \frac{e^{\mu t}}{(\alpha - \lambda_0 + \mu)^n} d\mu.$$

Integrating the analytic function $e^{\lambda t}(\lambda - \lambda_0)^{-n}$ over the square with corners $\pm iN$, $-N \pm iN$, we obtain

$$\begin{aligned}
\int_{\alpha-iN}^{\alpha+iN} \frac{e^{\lambda t}}{(\lambda - \lambda_0)^n} d\lambda &= -e^{\alpha t} \int_0^{-N} \frac{e^{\sigma t} e^{iNt}}{(\alpha - \lambda_0 - \sigma - iN)^n} d\sigma \\
&\quad + e^{\alpha t} \int_{-N}^{N} \frac{e^{-Nt} e^{i\tau t}}{(\alpha - \lambda_0 + N - i\tau)^n} d\tau \\
&\quad + e^{\alpha t} \int_0^{-N} \frac{e^{\sigma t} e^{-iNt}}{(\alpha - \lambda_0 - \sigma + iN)^n} d\sigma = I_1 + I_2 + I_3.
\end{aligned}$$

It can be verified that $|I_2| \to 0$ as $N \to \infty$ and

$$|I_3| \le e^{\alpha t} \int_0^{-N} \left| \frac{e^{\sigma t} e^{-iNt}}{(\alpha - \lambda_0 - \sigma + iN)^n} \right| d\sigma \le e^{\alpha t} \int_0^{-N} \frac{e^{\sigma t}}{|\alpha - \lambda_0 - \sigma + iN|^n} d\sigma$$

$$\le e^{\alpha t} \frac{1}{N^n} \int_0^{-N} e^{\sigma t} d\sigma = e^{\alpha t} \frac{1}{N^n} \frac{1}{t} [1 - e^{-Nt}] \to 0 \text{ as } N \to \infty.$$

Similarly, it can be shown that $|I_1| \to 0$. Hence

$$\int_{\alpha - i\infty}^{\alpha + i\infty} \frac{e^{\lambda t}}{(\lambda - \lambda_0)^n} d\lambda = 0$$

in the principal value sense for $t > 0$. In addition, for $\lambda = \sigma + i\tau$, since

$$\left\| e^{\lambda t} \frac{R(\lambda, A)y}{(\lambda - \lambda_0)^m} \right\| \le M \, \| y \| \, e^{\alpha t} \frac{1 + (\sigma^2 + \tau^2)^{k/2}}{[(\sigma - \lambda_0)^2 + \tau^2]^{m/2}}$$

$$\le M \, \|y\| \, e^{\alpha t} \frac{1 + (\sigma^2 + \tau^2)^{k/2}}{|\tau|^m}, \qquad (2.145)$$

the integral of

$$e^{\lambda t} \frac{R(\lambda, A)y}{(\lambda - \lambda_0)^m}$$

converges absolutely for $t \ge 0$, hence the integral

$$\frac{1}{2\pi i} \int_{\alpha - i\infty}^{\alpha + i\infty} e^{\lambda t} R(\lambda, A)x d\lambda$$

exists in the principal value sense for $t > 0$. Motivated by these observations, we define

$$u(t) = \frac{1}{2\pi i} \int_{\alpha - i\infty}^{\alpha + i\infty} e^{\lambda t} R(\lambda, A)x d\lambda$$

$$= (-1)^m \frac{1}{2\pi i} \int_{\alpha - i\infty}^{\alpha + i\infty} e^{\lambda t} \frac{R(\lambda, A)y}{(\lambda - \lambda_0)^m} d\lambda, \quad t \ge 0.$$

By (2.145), we know that $u(t)$ is continuous for $t \ge 0$. Furthermore, since

$$\frac{\|R(\lambda, A)\|}{|\lambda - \lambda_0|^{m-1}} \le M \frac{1 + (\sigma^2 + \tau^2)^{k/2}}{[(\sigma - \lambda_0)^2 + \tau^2]^{(m-1)/2}} \le M \frac{1 + (\sigma^2 + \tau^2)^{k/2}}{|\tau|^{m-1}}$$

we have

$$\int_{\alpha - i\infty}^{\alpha + i\infty} \frac{\| R(\lambda, A) \|}{|\lambda - \lambda_0|^{m-1}} |d\lambda| < \infty.$$

Therefore, $u(t)$ is continuously differentiable. We claim that $u(t)$ is a solution of the equation $u'(t) = Au(t)$. Indeed, for $t > 0$

$$u'(t) = (-1)^m \frac{1}{2\pi i} \int_{\alpha-i\infty}^{\alpha+i\infty} Ae^{\lambda t} \frac{R(\lambda, A)y}{(\lambda - \lambda_0)^m} d\lambda$$

$$+ (-1)^m \frac{1}{2\pi i} \int_{\alpha-i\infty}^{\alpha+i\infty} e^{\lambda t} \frac{y}{(\lambda - \lambda_0)^m} d\lambda$$

$$= Au(t).$$

It follows from Lemma 2.92 that $u(t)$ is a solution of $u'(t) = Au(t)$ for $t \geq 0$. The solution $u(t)$ satisfies the initial condition

$$u(0) = (-1)^m \frac{1}{2\pi i} \int_{\alpha-i\infty}^{\alpha+i\infty} \frac{R(\lambda, A)y}{(\lambda - \lambda_0)^m} d\lambda.$$

For any $N > 0$, integrating $\frac{R(\lambda,A)y}{(\lambda-\lambda_0)^m}$ over the square with corners $\alpha \pm iN$, $N \pm iN$, and using the residue theorem, we obtain

$$u(0) = (-1)^m \frac{1}{2\pi i} \int_{\alpha-i\infty}^{\alpha+i\infty} \frac{R(\lambda, A)y}{(\lambda - \lambda_0)^m} d\lambda$$

$$= (-1)^m \frac{1}{(m-1)!} \frac{d^{m-1} R(\lambda_0, A)y}{d\lambda^{m-1}} = R(\lambda_0, A)^m y = x.$$

Therefore, $u(t)$ solves (ACP). Moreover,

$$\|u(t)\| = \left\| \frac{1}{2\pi i} \int_{\alpha-i\infty}^{\alpha+i\infty} e^{\lambda t} \frac{R(\lambda, A)y}{(\lambda - \lambda_0)^m} d\lambda \right\|$$

$$\leq \frac{1}{2\pi} e^{\alpha t} \int_{\alpha-i\infty}^{\alpha+i\infty} \frac{\|R(\lambda, A)\|}{|\lambda - \lambda_0|^m} |d\lambda| \|y\|$$

$$\leq M_\alpha e^{\alpha t} |x|_m,$$

for some $M_\alpha > 0$. This completes the proof.

By Theorem 2.90, we have immediately

Corollary 2.93 *Let A be a densely defined linear operator in a Banach space X. If there are constants M, ω such that $R(\lambda, A)$ exists and satisfies*

$$\|R(\lambda, A)\| \leq M(1 + |\lambda|^k) \text{ for some } k \geq -1,$$

for all $\lambda \in \mathbf{C}$ with $\operatorname{Re}\lambda > \omega$, then A generates at least $[k] + 3$-times integrated semigroup.

Remark 2.94 *It is easily verified that for* $-1 \le k < 0$ *condition* (2.144) *may be weakened somewhat and replaced by*

$$\|R(\lambda, A)\| \le M(1 + |\tau|)^k, \lambda = \sigma + i\tau, \text{ for some } k \ge -1, \qquad (2.146)$$

for all $\sigma \ge \omega$. *In this case,* (ACP) *has a unique solution* $u(\cdot)$ *for every* $x \in D(A^2)$ *and, for* $\alpha > \omega$, $\|u(t)\| \le M_\alpha e^{\alpha t}|x|_2$. *and hence* (ACP) *is at least exponentially 2-well-posed.*

Theorem 2.95 *Let* A *be a densely defined linear operator in a Banach space* X. *Then the following statements are equivalent:*

(i) A *generates an integrated semigroup.*

(ii) *There exist real constants* M, ω *and* $k \ge -1$ *such that* $R(\lambda, A)$ *exists and satisfies* $\| R(\lambda, A) \| \le M(1 + |\lambda|^k)$ *for all* $\lambda \in \mathbf{C}$ *with* $\mathrm{Re}\lambda > \omega$.

Proof. $(i) \Rightarrow (ii)$. Suppose that A generates an n-times integrated semigroup $S(t)$ with $\|S(t)\| \le Me^{\omega t}$. Then from (v) of Theorem 2.79, for $\mathrm{Re}\lambda > \omega$, and all $x \in X$,

$$\|R(\lambda, A)x\| = \left\| \lambda^n \int_0^\infty e^{-\lambda t} S(t)x dt \right\|$$

$$\le |\lambda^n| \int_0^\infty e^{-\mathrm{Re}\lambda t} \|S(t)x\| dt$$

$$\le \frac{M}{\mathrm{Re}\lambda - \omega} |\lambda|^n \|x\| \le M|\lambda|^n \|x\|, \text{ if } \mathrm{Re}\lambda \ge \omega + 1.$$

$(ii) \Rightarrow (i)$. This is a consequence of Theorems 2.90 and 2.91.   □

It is observed from Theorem 2.95 that, unlike C_0−semigroups, the resolvents of the generators of integrated semigroups are allowed to grow in polynomial order.

Finally, we give a perturbation result on the integrated semigroups [175].

Theorem 2.96 *Suppose that* A *generates an* n-*times integrated semigroup* $S(t)$ *on a Banach space* X *with* $\|S(t)\| \le Me^{\omega t}, \omega \ge 0$. *Let* $B \in \mathcal{L}(X)$ *satisfy*

$$BR(\lambda, A) = R(\lambda, A)B \text{ for all large } |\lambda|.$$

Then $A + B$ *generates an* n-*times integrated semigroup* $S_B(t)$. *Moreover,*

$$\| S_B(t) \| \le M_\omega (1 + \frac{\| B \|}{\omega + \| B \|}) e^{(\omega + \|B\|)t}, \text{ for some } M_\omega \text{ and } t \ge 0.$$

Proof. From (ii) of Theorem 2.79, for any $x \in D(A)$

$$S'(t)x = S(t)Ax + \frac{t^{n-1}}{(n-1)!}x.$$

Generally, one can easily show by induction that for each $x \in D(A^k), k \le n$,

$$\begin{aligned}
S^{(k)}(t)x &= S(t)A^k x + \frac{t^{n-1}}{(n-1)!}A^{k-1}x + \frac{t^{n-2}}{(n-2)!}A^{k-2}x + \cdots + \frac{t^{n-k}}{(n-k)!}x \\
&= S(t)A^k x + \sum_{m=1}^{k} \frac{t^{n-m}}{(n-m)!}A^{k-m}x \\
&= S(t)A^k x + \sum_{m=n-k}^{n-1} \frac{t^m}{m!}A^{k-n+m}x.
\end{aligned} \tag{2.147}$$

Similarly, if $k > n$, then for each $x \in D(A^k)$, one has

$$\begin{aligned}
S^{(k)}(t)x &= S(t)A^k x + \frac{t^{n-1}}{(n-1)!}A^{k-1}x + \frac{t^{n-2}}{(n-2)!}A^{k-2}x + \cdots \\
&\quad + \frac{t}{1!}A^{k-n+1}x + A^{k-n}x \\
&= S(t)A^k x + \sum_{m=0}^{n-1} \frac{t^m}{m!}A^{k-n+m}x.
\end{aligned} \tag{2.148}$$

We write (2.147) and (2.148) in the following compact form

$$S^{(k)}(t)x = S(t)A^k x + \sum_{m=\max\{0,n-k\}}^{n-1} \frac{t^m}{m!}A^{k-n+m}x, \forall x \in D(A^k). \tag{2.149}$$

For each $x \in X$, differentiating n times both sides of the following equality with respect to $\lambda > \omega$

$$R(\lambda, A)x = \lambda^n \int_0^\infty e^{-\lambda t}S(t)x dt,$$

we have

$$\begin{aligned}
(-1)^n n! R(\lambda, A)^{n+1}x &= R^{(n)}(\lambda, A)x \\
&= \sum_{k=0}^{n} C_n^k \frac{n!}{k!}\lambda^k \int_0^\infty (-t)^k e^{-\lambda t}S(t)x dt.
\end{aligned}$$

Since $\int_0^\infty \lambda^{n+1}\frac{t^n}{n!}e^{-\lambda t}dt = 1$, we have

$$\left\| n! R(\lambda, A)^{n+1}x \right\|$$

$$\leq M \|x\| n! \sum_{k=0}^{n} C_n^k \frac{1}{k!} \lambda^k \int_0^\infty t^k e^{-(\lambda-\omega)t} dt$$

$$\leq M \|x\| n! \sum_{k=0}^{n} C_n^k \frac{\lambda^k}{(\lambda-\omega)^{k+1}} \int_0^\infty (\lambda-\omega)^{k+1} \frac{t^k}{k!} e^{-(\lambda-\omega)t} dt$$

$$= M \|x\| n! \sum_{k=0}^{n} C_n^k \frac{\lambda^k}{(\lambda-\omega)^{k+1}}$$

$$= M \|x\| n! \frac{1}{\lambda-\omega} \sum_{k=0}^{n} C_n^k \frac{\lambda^k}{(\lambda-\omega)^k}$$

$$= M \|x\| n! \frac{1}{\lambda-\omega} (1 + \frac{\lambda}{\lambda-\omega})^n.$$

Hence, there is an $M_\omega > M$ such that $\left\| R(\lambda, A)^{n+1} \right\| \leq M_\omega \frac{1}{\lambda-\omega}$. Then for $\lambda > \omega + M_\omega \|B\|^n$,

$$\left\| B^n R(\lambda, A)^{n+1} \right\| \leq \|B\|^n M_\omega \frac{1}{\lambda-\omega} < 1.$$

Therefore, for $\lambda > \omega + M_\omega \|B\|^n, \lambda \in \rho(A + B)$. Furthermore, since for sufficiently large $\lambda, R(\lambda, A)$ and B commute, we have

$$R(\lambda, A + B)$$

$$= (I - BR(\lambda, A))^{-1} R(\lambda, A) = \sum_{k=0}^{\infty} B^k R(\lambda, A)^{k+1}$$

$$= \sum_{k=0}^{\infty} B^k (-1)^k \frac{1}{k!} R^{(k)}(\lambda, A)$$

$$= \sum_{k=0}^{\infty} B^k (-1)^k \frac{1}{k!} \frac{d^k}{d\lambda^k} \left(\lambda^n \int_0^\infty e^{-\lambda t} S(t) dt \right)$$

$$= \sum_{k=0}^{\infty} (-B)^k \frac{1}{k!} \sum_{i=0}^{\min\{k,n\}} C_k^i \frac{n!}{(n-i)!} \lambda^{n-i} \int_0^\infty (-t)^{k-i} e^{-\lambda t} S(t) dt$$

$$= \sum_{k=0}^{n} \frac{B^k}{k!} \sum_{i=0}^{k} C_k^i (-1)^i \frac{n!}{(n-i)!} \lambda^{n-i} \int_0^\infty t^{k-i} e^{-\lambda t} S(t) dt$$

$$+ \sum_{k=n+1}^{\infty} \frac{B^k}{k!} \sum_{i=0}^{n} C_k^i (-1)^i \frac{n!}{(n-i)!} \lambda^{n-i} \int_0^\infty t^{k-i} e^{-\lambda t} S(t) dt$$

$$= I_1 + I_2$$

where

$$
\begin{aligned}
I_1 &= \sum_{i=0}^{n}(-1)^i \frac{n!}{(n-i)!}\lambda^{n-i}\sum_{k=i}^{n}\frac{B^k}{k!}C_k^i\int_0^\infty t^{k-i}e^{-\lambda t}S(t)dt \\
&= \sum_{i=0}^{n}C_n^i(-B)^i\lambda^{n-i}\sum_{k=i}^{n}\frac{B^{k-i}}{(k-i)!}\int_0^\infty t^{k-i}e^{-\lambda t}S(t)dt \\
&= \sum_{i=0}^{n}C_n^i(-B)^i\lambda^{n-i}\sum_{k=0}^{n-i}\frac{B^k}{k!}\int_0^\infty t^k e^{-\lambda t}S(t)dt \\
I_2 &= \sum_{i=0}^{n}(-1)^i\frac{n!}{(n-i)!}\lambda^{n-i}\sum_{k=n+1}^{\infty}\frac{B^k}{k!}C_k^i\int_0^\infty t^{k-i}e^{-\lambda t}S(t)dt \\
&= \sum_{i=0}^{n}C_n^i(-B)^i\lambda^{n-i}\sum_{k=n+1}^{\infty}\frac{B^{k-i}}{(k-i)!}\int_0^\infty t^{k-i}e^{-\lambda t}S(t)dt \\
&= \sum_{i=0}^{n}C_n^i(-B)^i\lambda^{n-i}\sum_{k=n-i+1}^{\infty}\frac{B^k}{k!}\int_0^\infty t^k e^{-\lambda t}S(t)dt \\
&= \sum_{i=0}^{n}C_n^i(-B)^i\lambda^{n-i}\int_0^\infty [e^{Bt}-\sum_{k=0}^{n-i}\frac{B^k}{k!}t^k]e^{-\lambda t}S(t)dt \\
&= (\lambda-B)^n\int_0^\infty e^{Bt}e^{-\lambda t}S(t)dt - I_1.
\end{aligned}
$$

Therefore,

$$
\begin{aligned}
R(\lambda, A+B) &= (\lambda-B)^n\int_0^\infty e^{Bt}e^{-\lambda t}S(t)dt \\
&= \lambda^n\int_0^\infty e^{-\lambda t}\Big(\sum_{i=0}^{n}C_n^i(-B)^i\int_0^t\cdots\int_0^t e^{B\tau}S(\tau)(d\tau)^i\Big)dt.
\end{aligned}
\tag{2.150}
$$

By Definition 2.78, $A+B$ generates an n-times integrated semigroup $S_B(t)$ which is given by

$$
S_B(t) = \sum_{i=0}^{n}C_n^i(-B)^i\int_0^t\cdots\int_0^t e^{B\tau}S(\tau)(d\tau)^i.
\tag{2.151}
$$

Since $\|S(t)\| \le M_\omega e^{\omega t}$, we have

$$
\|S_B(t)\| \le \sum_{i=0}^{n}C_n^i\|B\|^i\int_0^t\cdots\int_0^t e^{\|B\|\tau}M_\omega e^{\omega\tau}(d\tau)^k
$$

$$= M_\omega \sum_{i=0}^{n} C_n^i \|B\|^i \sum_{k=0}^{\infty} \frac{(\|B\| + \omega)^k}{k!} \int_0^t \cdots \int_0^t \tau^k (d\tau)^i$$

$$= M_\omega \sum_{i=0}^{n} C_n^i \left(\frac{\|B\|}{\omega + \|B\|} \right) \sum_{k=i}^{\infty} \frac{[(\omega + \|B\|)t]^k}{k!}$$

$$\leq M_\omega \left(1 + \frac{\|B\|}{\omega + \|B\|} \right)^n e^{(\omega + \|B\|)t}$$

for all $t \geq 0$. This proves the theorem.   □

2.9   Nonlinear semigroups of contractions

Let A be a linear operator in a Banach space X. From the results in Section 2.3, we see that A generates a C_0-semigroup of contractions if and only if for every $x \in D(A)$ and for every $\lambda > 0$,

$$\|x\| \leq \|x - \lambda A x\|, \tag{2.152}$$
$$\mathcal{R}(I - \lambda A) = X, \tag{2.153}$$
$$\overline{D(A)} = X. \tag{2.154}$$

The goal of this section is to prove nonlinear versions of these theorems whenever they exist. We first give the definition of a nonlinear contraction semigroup.

Definition 2.97 *Let F be a nonempty closed subset of a Banach space X. A nonlinear contraction semigroup on F is a family of operators $T(t) : F \to F$ satisfying:*

(i) $T(t + s) = T(t)T(s)$ for every $s, t \geq 0, T(0) = I$ (identity on F).

(ii) $\|T(t)x - T(t)y\| \leq \|x - y\|$ for every $t \geq 0$ and every $x, y \in F$.

(iii) For every $x \in F, T(t)x \to x$ as $t \downarrow 0$.

Observe that if $T(t)$ satisfies (i)-(iii) then for every $x \in F, t \to T(t)x$ is continuous on $[0, \infty)$. As in semigroup theory for linear operators, we define the infinitesimal generator A of nonlinear semigroup $T(t)$ as

$$D(A) = \{x \in F | \lim_{h \downarrow 0} h^{-1}[T(h)x - x] \text{ exists}\},$$

$$Ax = \lim_{h \downarrow 0} h^{-1}[T(h)x - x], \text{ for every } x \in D(A). \tag{2.155}$$

Proposition 2.98 *Let A be the generator of a nonlinear contraction semigroup $T(t)$ defined on a closed subset F of a Banach space X. Then*

(i) $\langle Ax - Ay, f \rangle \leq 0$ for all $x, y \in D(A)$ and every $f \in F(x - y)$, i.e., A is dissipative.

(ii) For each $x \in D(A), T(t)x$ is Lipschitz continuous and

$$\|T(t + h)x - T(t)x\| \leq \|Ax\| h, \tag{2.156}$$

for all $t, h > 0$.

(iii) If X satisfies the Radon-Nikodym property (in particular if X is reflexive) then $t \rightarrow T(t)x$ is differentiable a.e. on $[0, \infty)$ for every $x \in D(A)$ and

$$\frac{d}{dt}T(t)x = AT(t)x \quad \text{for } t \in [0, \infty) \text{ a.e.}$$

Proof. (iii) follows from the well-known result that X satisfies the *Radon-Nikodym property* (in particular if X is reflexive) if and only if every Lipschitz continuous X-valued function is almost everywhere differentiable, and $\frac{d}{dt}T(t)x = AT(t)x$ whenever $\frac{d}{dt}T(t)x$ exists. To prove (i), we note that

$$\begin{aligned}
\langle Ax - Ay, f \rangle &= \lim_{h \downarrow 0} h^{-1}[\langle T(h)x - T(h)y, f \rangle - \|x - y\|^2] \\
&\leq \overline{\lim_{h \downarrow 0}} h^{-1}[\|T(h)x - T(h)y\| \|x - y\| - \|x - y\|^2] \leq 0,
\end{aligned}$$

for all $x, y \in D(A)$ and $f \in F(x - y)$. Now we consider (ii). Let $x \in D(A)$ and let $\{t_k\}$ be a monotonically decreasing sequence of positive numbers tending to zero as $k \rightarrow \infty$ and let $\epsilon > 0$ be such that

$$\|T(t_k)x - x\| \leq (\epsilon + \|Ax\|)t_k.$$

Let $h > 0$ and let n_k be a nonnegative integer such that $0 \leq h - n_k t_k < t_k$. Then

$$\begin{aligned}
\|T(t + h)x - T(t)x\| &\leq \|T(h)x - x\| \\
&= \|T(h - n_k t_k + n_k t_k)x - x\| \\
&\leq \|T(h - n_k t_k)x - x\| + (\epsilon + \|Ax\|)t_k n_k \\
&\leq \|T(h - n_k t_k)x - x\| + (\epsilon + \|Ax\|)h,
\end{aligned}$$

for every $t > 0$. Letting $k \rightarrow \infty$, we have

$$\|T(t + h)x - T(t)x\| \leq (\epsilon + \|Ax\|)h.$$

(ii) then follows from the arbitrariness of ϵ. $\qquad\square$

Remark 2.99 *Let*

$$D = \{x \in F, \lim_{t \downarrow 0} t^{-1} \|T(t)x - x\| < \infty\}. \tag{2.157}$$

Then it is obvious that $D(A) \subset D$ and the proof of (ii) of Proposition 2.98 shows that $T(t)x$ is also Lipschitz continuous for every $x \in D$.

Example 2.100 *Let $X = C[0,1]$. Define*

$$(T(t)x)(s) = f(t + f^{-}(x(s))), \; x \in X,$$

where

$$f(r) = \begin{cases} r, & \text{if } r \geq 0, \\ 2r, & \text{if } r \leq 0, \end{cases}$$

and f^{-} denotes the inverse function of f. It is easy to see that $T(t) : X \to X$ is a nonlinear contraction semigroup on X. Moreover,

$$h^{-1}[(T(h)x)(s) - x(s)] = \begin{cases} 1, & \text{if } x(s) \geq 0 \\ 1 - \dfrac{1}{2h}x(s), & \text{if } x(s) \leq 0. \end{cases}$$

This implies $h^{-1}[T(h)x - x]$ has no limit as $h \downarrow 0$ in X unless x is positive on [0,1]. That is, $D(A) = \{x(s) \in X | x(s) \geq 0 \text{ for all } s \in [0,1]\}$. A is not densely defined in X.

An interesting example was given in [40] which indicates that $D(A) = \emptyset$ may happen even if $F = X$.

Fortunately, the generator of a nonlinear contraction semigroup on Hilbert space is necessarily densely defined on the closed subset where the nonlinear semigroup is defined.

Theorem 2.101 *Let $T(t)$ be a nonlinear contraction semigroup defined on a closed convex subset F of a real Hilbert space H. Then the generator A of $T(t)$ is densely defined on F.*

We shall prove Theorem 2.101 by using the following lemmas.
Let $A_h : F \to H$ be the operator defined by

$$A_h x = h^{-1}[T(h)x - x], h > 0.$$

Since A_h is dissipative, we know that $(I - \lambda A_h)^{-1}$ exists and is non-expansive on $\mathcal{R}(I - \lambda A_h)$ for each λ and h positive. We can now verify

Lemma 2.102   *(i) $F \subset \cap_{\lambda, h > 0} \mathcal{R}(I - \lambda A_h)$.*

(ii) For every $x \in F$, the function $(\lambda, h) \to (I - \lambda A_h)^{-1}x$ is separately continuous.

Proof. Let $x \in F$. The equation $(I - \lambda A_h)y = x$ is equivalent to

$$y = \frac{h}{h + \lambda}x + \frac{\lambda}{h + \lambda}T(h)y. \tag{2.158}$$

It is easy to see that the mapping defined by the right-hand side of (2.158) maps F into itself since F is convex. Furthermore, the mapping is a contraction since

$$\left\| \frac{h}{h + \lambda}x + \frac{\lambda}{h + \lambda}T(h)y_1 - \frac{h}{h + \lambda}x - \frac{\lambda}{h + \lambda}T(h)y_2 \right\| \leq \frac{\lambda}{h + \lambda}\|y_1 - y_2\|.$$

By contraction mapping theorem, there is a unique solution $y_{\lambda,h} \in F$ to (2.158) given by

$$y_{\lambda,h} = (I - \lambda A_h)^{-1}x. \tag{2.159}$$

(ii) is an immediate consequence of the facts that $T(h)x$ is continuous in h and non-expansive in X. $\quad\square$

Lemma 2.103 *Let $\epsilon > 0$ and $0 < \delta < 1$ be such that $h \in (0, \delta)$ and $\|T(h)x - x\| \leq \epsilon$ for any $x \in F$. If $nh = t \in (0, \delta), n$ being a positive integer, then*

$$\|y_{\lambda,h} - y_{\lambda,t}\|^2 \leq 2\epsilon \|y_{\lambda,h} - x\|. \tag{2.160}$$

Proof. We may assume without loss of generality that $x = 0$. From (2.159), we have

$$(I - \lambda A_h)y_{\lambda,h} = 0$$

from which it follows that

$$T(h)y_{\lambda,h} = (1 + \frac{h}{\lambda})y_{\lambda,h}.$$

Consequently, for $i = 1, 2, \cdots, n$,

$$\|y_{\lambda,h} - T((i-1)h)y_{\lambda,t}\|^2$$
$$\geq \|T(h)y_{\lambda,h} - T(ih)y_{\lambda,t}\|^2$$
$$= \left\| (1 + \frac{h}{\lambda})y_{\lambda,h} - T(ih)y_{\lambda,t} \right\|^2$$
$$\geq \|y_{\lambda,h} - T(ih)y_{\lambda,t}\|^2 + \langle \frac{2h}{\lambda}y_{\lambda,h}, \ y_{\lambda,h} - T(ih)y_{\lambda,t}\rangle,$$

and also

$$\|y_{\lambda,h} - T(t)y_{\lambda,t}\|^2 = \left\| y_{\lambda,h} - (1 + \frac{t}{\lambda})y_{\lambda,t} \right\|^2$$
$$\geq \|y_{\lambda,h} - y_{\lambda,t}\|^2 + \langle \frac{2t}{\lambda}y_{\lambda,t}, \ y_{\lambda,t} - y_{\lambda,h}\rangle.$$

By summing these $n + 1$ inequalities, one finally obtains

$$\|y_{\lambda,h}\|^2 + \|y_{\lambda,t}\|^2 \leq \frac{1}{n} \sum_{i=1}^{n} \langle y_{\lambda,h}, T(ih)y_{\lambda,t} \rangle + \langle y_{\lambda,h}, y_{\lambda,t} \rangle.$$

Since $ih \in (0, \delta)$, we have

$$\|T(ih)y_{\lambda,t}\| \leq \|y_{\lambda,t}\| + \|T(ih)0\| \leq \|y_{\lambda,t}\| + \epsilon.$$

Thus,

$$\begin{aligned}
\|y_{\lambda,h}\|^2 + \|y_{\lambda,t}\|^2 &\leq \|y_{\lambda,h}\| [\|y_{\lambda,t}\| + \epsilon] + \langle y_{\lambda,h}, y_{\lambda,t} \rangle \\
&\leq \frac{1}{2}[\|y_{\lambda,h}\|^2 + \|y_{\lambda,t}\|^2] + \epsilon \|y_{\lambda,h}\| + \langle y_{\lambda,h}, y_{\lambda,t} \rangle.
\end{aligned}$$

Therefore,

$$\|y_{\lambda,h} - y_{\lambda,t}\|^2 \leq 2\epsilon \|y_{\lambda,h}\|$$

which is the desired result. $\square$

Lemma 2.104 *Let $x \in F$ and let ϵ and δ be as in Lemma 2.103. Then, for all $\lambda > 0$ and $h \in (0, \delta)$, the following inequality holds*

$$\|y_{\lambda,h} - x\| \leq 2\epsilon(1 + \frac{4\lambda}{\delta}). \tag{2.161}$$

In particular, for every $\lambda > 0$ the function $h \to y_{\lambda,h}$ is bounded on $(0, \delta)$.

Proof. We assume again, without loss of generality, that $x = 0$. If $h \in [\delta/2, \delta)$, then since $(I - \lambda A_h)y_{\lambda,h} = 0$, we have

$$\begin{aligned}
\|y_{\lambda,h}\| &= \lambda \|A_h y_{\lambda,h}\| = \frac{\lambda}{h} \|T(h)y_{\lambda,h} - y_{\lambda,h}\| \\
&= \frac{\lambda}{h} \|(I - \lambda A_h)^{-1}T(h)0 - (I - \lambda A_h)^{-1}0\| \\
&\leq \frac{\lambda}{h} \|T(h)0\| \leq \frac{\lambda}{h}\epsilon \leq \frac{2\lambda}{\delta}\epsilon. \tag{2.162}
\end{aligned}$$

If $h \in (0, \delta/2)$, there exists a positive integer n such that $nh = t \in [\delta/2, \delta)$. Thus, from Lemma 2.103 it follows that

$$\|y_{\lambda,h} - y_{\lambda,t}\|^2 \leq 2\epsilon \|y_{\lambda,h}\| \leq 2\epsilon \|y_{\lambda,h} - y_{\lambda,t}\| + 2\epsilon \|y_{\lambda,t}\|,$$

and by (2.162)

$$\|y_{\lambda,h} - y_{\lambda,t}\|^2 \leq 2\epsilon \|y_{\lambda,h} - y_{\lambda,t}\| + \frac{4\lambda}{\delta}\epsilon^2.$$

Hence

$$\|y_{\lambda,h} - y_{\lambda,t}\| \leq \epsilon[1 + (1 + 4\lambda/\delta)^{1/2}].$$

Together with (2.162), the above inequality implies (2.161). □

Lemma 2.105 *For all $\lambda > 0$,*

$$\lim_{h \downarrow 0} y_{\lambda,h} = y_\lambda$$

exists and belongs to $\overline{D(A)}$.

Proof. According to Lemma 2.104, there exists a positive constant M such that $\|y_{\lambda,h} - x\| \leq M$ for all $h, \lambda \in (0, \delta)$. Then inequality (2.160) implies that

$$\|y_{\lambda,h} - y_{\lambda,t}\| \leq (2M\epsilon)^{1/2}, \ t = nh \in (0, \delta).$$

Consequently,

$$\|y_{\lambda,h} - y_{\lambda,t}\| \leq 2(2M\epsilon)^{1/2} \tag{2.163}$$

for all $h, t \in (0, \delta)$ such that t/h is rational. Since for every $\lambda > 0$ the function $h \to y_{\lambda,h}$ is continuous on $(0, \infty)$, inequality (2.163) clearly extends to all $h, t > 0$. In particular, one concludes that $lim_{h \downarrow 0} y_{\lambda,h} = y_\lambda$ exists for all $\lambda > 0$.

In order to show that $y_\lambda \in \overline{D(A)}$, it suffices to show that

$$\|T(t)y_\lambda - y_\lambda\| \leq Mt \text{ for all } t \in [0, 1] \tag{2.164}$$

where M is independent of t. In fact, if (2.164) holds then $y_\lambda \in D$ by remark 2.99 and $T(t)y_\lambda$ is Lipschitz continuous. Since H is a Hilbert space, $T(t)y_\lambda$ is a.e. differentiable. $\frac{d}{dt}T(t)y_\lambda = AT(t)y_\lambda$ whenever $\frac{d}{dt}T(t)x$ exists. Hence, $T(t)y_\lambda \in D(A)$ for $t > 0$ a.e. and the strong continuity of $T(t)y_\lambda$ then implies that $y_\lambda \in \overline{D(A)}$.

Now, we show (2.164). Indeed, for $t = nh$, we have

$$\|T(t)y_{\lambda,h} - y_{\lambda,h}\| \leq \sum_{i=1}^{n} \|T((i-1)h)y_{\lambda,h} - T(ih)y_{\lambda,h}\|$$

$$\leq n\|y_{\lambda,h} - T(h)y_{\lambda,h}\| = \frac{t}{\lambda}\|x - y_{\lambda,h}\|.$$

Fix t and let $n \to \infty$. Then $y_{\lambda,h} \to y_\lambda$. The above inequality implies that

$$\|T(t)y_\lambda - y_\lambda\| \leq \frac{t}{\lambda}\|x - y_\lambda\| \text{ for all } \lambda > 0,$$

which completes the proof. □

Proof of Theorem 2.101. Given $x \in F$, we shall show that y_λ given by Lemma 2.105 converges strongly to x as $\lambda \to 0$. In fact, letting $h \to 0$ in estimation (2.161), one obtains

$$\|y_\lambda - x\| \leq 2\epsilon(1 + \frac{4\lambda}{\delta}) \text{ for all } \lambda > 0.$$

As ϵ is arbitrary, we deduce that $y_\lambda \to x$ as $\lambda \to 0$. This proves the result by Lemma 2.105.

$\square$

Although Theorem 2.101 shows that the generator of a nonlinear contraction semigroup defined on a closed convex subset of a Hilbert space is densely defined, the following example shows that the m-dissipativity does not hold in general.

Example 2.106 *Let $X = \mathbf{R}$ and let*

$$T(t)x = \begin{cases} (x - t)^+, & \text{if} \quad x \geq 0, \\ (x + t)^-, & \text{if} \quad x < 0, \end{cases} \tag{2.165}$$

where $a^+ = \max\{a, 0\}$ and $a^- = \min\{a, 0\}$. Then $T(t)$ is a contraction semigroup on $\mathbf{R}$ and

$$Ax = \begin{cases} 1, & \text{if} \quad x < 0, \\ 0, & \text{if} \quad x = 0, \\ -1, & \text{if} \quad x > 0. \end{cases} \tag{2.166}$$

A is dissipative in the sense that

$$\langle Ax - Ay, f \rangle \leq 0, \quad \text{for some } f \in F(x - y).$$

But $\mathcal{R}(I - hA) \neq \mathbf{R}$ for any $h > 0$, i.e. A is never m-dissipative.

The above example motivates us to consider semigroups generated by *multivalued differential operators*. For example, a multivalued operator $\mathcal{A} : \mathbf{R} \to 2^{\mathbf{R}}$ corresponding to the operator in Example 2.106 can be defined as

$$\mathcal{A}x = \begin{cases} 1, & \text{if} \quad x < 0, \\ [-1, 1], & \text{if} \quad x = 0, \\ -1, & \text{if} \quad x > 0. \end{cases} \tag{2.167}$$

Definition 2.107 *A multivalued operator $\mathcal{A} : D(\mathcal{A}) \subset X \to 2^X$ is called dissipative in a Banach space X if for every $x_i \in D(\mathcal{A})$ and every $y_i \in \mathcal{A}x_i, i = 1, 2$, there exists an $f \in F(x_1 - x_2)$ such that*

$$\langle y_1 - y_2, f \rangle \leq 0.$$

Or equivalently for every $x_i \in D(\mathcal{A})$ and every $y_i \in \mathcal{A}x_i$.

$$\|x_1 - x_2\| \leq \|x_1 - x_2 - \lambda(y_1 - y_2)\| \text{ for all } \lambda > 0.$$

A dissipative operator $\mathcal{A}$ is called m-dissipative if

$$\mathcal{R}(I - \mathcal{A}) = X.$$

It is now easy to verify that the multivalued operator $\mathcal{A}$ defined in (2.167) is dissipative, because for every $y_i \in \mathcal{A}x_i, i = 1, 2$, there exists $f \in F(x_1 - x_2)$ such that

$$\langle y_1 - y_2, f \rangle \leq 0.$$

Moreover, for every $\lambda > 0, \mathcal{R}(I - \lambda \mathcal{A}) = \mathbf{R}$. i.e., $\mathcal{A}$ is m-dissipative. Indeed,

$$(I - \lambda \mathcal{A})x = \begin{cases} x - \lambda, & \text{if} \quad x < 0, \\ [-\lambda, \lambda], & \text{if} \quad x = 0, \\ x + \lambda, & \text{if} \quad x > 0. \end{cases}$$

The relationship between operators defined by (2.166) and (2.167) can be expressed as

$$\mathcal{A}x = y, \quad \|y\| = |\mathcal{A}x| = \inf\{\|z\| \,\big|\, z \in \mathcal{A}x\}, \tag{2.168}$$

and the nonlinear semigroup defined by (2.165) satisfies

$$\frac{dT(t)x}{dt} = \mathcal{A}T(t)x \in \mathcal{A}T(t)x, \quad \forall x \in \mathbf{R}. \tag{2.169}$$

We shall see that there is a nice nonlinear analogue of nonlinear contraction semigroups for multivalued m-dissipative operators in Hilbert spaces.

Usually we identify a multivalued operator $\mathcal{A} : D(\mathcal{A}) \subset X \to 2^X$ with its graph in $X \times X$:

$$\mathcal{A} = \{(x, y) \mid y \in \mathcal{A}x\}. \tag{2.170}$$

Note that if $\mathcal{A}$ is dissipative, then for any $y \in \mathcal{R}(I - \lambda \mathcal{A})$ there is only one $x \in D(I - \lambda \mathcal{A})$ such that $y \in (I - \lambda \mathcal{A})x$. In fact, if $x_1 \neq x_2, x_i \in D(\mathcal{A})$ and $y \in (I - \lambda \mathcal{A})x_i, i = 1, 2$, then $y_i = \lambda^{-1}(x_i - y) \in \mathcal{A}x_i$ and

$$\|x_1 - x_2\| \leq \|x_1 - x_2 - \lambda(y_1 - y_2)\| = 0.$$

That is, $x_1 = x_2$. Thus, for any $y \in \mathcal{R}(I - \lambda \mathcal{A}), (I - \lambda \mathcal{A})^{-1}y$ is well defined and non-expansive on $\mathcal{R}(I - \lambda \mathcal{A})$, i.e.,

$$\left\|(I - \lambda \mathcal{A})^{-1}x - (I - \lambda \mathcal{A})^{-1}y\right\| \leq \|x - y\| \tag{2.171}$$

for any $x, y \in \mathcal{R}(I - \lambda \mathcal{A})$.

Proposition 2.108 *A multivalued m-dissipative operator $\mathcal{A} : D(\mathcal{A}) \subset X \to 2^X$ in a Banach space X is maximal dissipative, i.e., if there is another dissipative operator $\tilde{\mathcal{A}}$ such that $\mathcal{A} \subset \tilde{\mathcal{A}}$ then $\mathcal{A} = \tilde{\mathcal{A}}$.*

Proof. Suppose that $(x_0, y_0) \notin \mathcal{A}, (x_0, y_0) \in \tilde{\mathcal{A}}$. Then for any $(x, y) \in \mathcal{A}$, there exists $f \in F(x_0 - x)$ such that $\langle y_0 - y, f \rangle \leq 0$. Since $\mathcal{R}(I - \mathcal{A}) = X$, we may choose $(x_1, y_1) \in \mathcal{A}$ such that $x_1 - y_1 = x_0 - y_0$. Then

$$\langle y_0 - y_1, f \rangle \leq 0, f \in F(x_0 - x_1).$$

But, $\langle y_0 - y_1, f \rangle = \langle x_0 - x_1, f \rangle = \|x_0 - x_1\|^2$, so $x_0 = x_1$ and $y_0 = y_1$, which is a contradiction. Therefore, $\mathcal{A} = \tilde{\mathcal{A}}$. $\qquad\square$

Proposition 2.109 *Let $\mathcal{A}$ be a multivalued m-dissipative operator in a Banach space X. Then $\mathcal{A}$ is closed and for any $x \in D(\mathcal{A}), \mathcal{A}x$ is a closed convex subset of X.*

Proof. Let $y_n \in \mathcal{A}x_n$ be such that $x_n \to x_0, y_n \to y_0$ as $n \to \infty$. Since $\mathcal{A}$ is dissipative,

$$\|x_n - x\| \leq \|x_n - x - \lambda(y_n - y)\| \text{ for any } \lambda > 0 \text{ and } y \in \mathcal{A}x.$$

Letting $n \to \infty$ yields

$$\|x_0 - x\| \leq \|x_0 - x - \lambda(y_0 - y)\| \text{ for any } \lambda > 0 \text{ and } y \in \mathcal{A}x.$$

Thus, $\mathcal{A}_1 = (x_0, y_0) \bigcup \mathcal{A}$ is dissipative. Since $\mathcal{A}$ is maximal dissipative, we conclude that $(x_0, y_0) \in \mathcal{A}$. That is, $\mathcal{A}$ is closed.

Since $\mathcal{A}$ is maximal dissipative, for any $x_0 \in D(\mathcal{A}), \mathcal{A}x_0$ is given by

$$\mathcal{A}x_0 = \{y_0 \in X \big| \langle y - y_0, f \rangle \leq 0 \text{ for all } (x, y) \in \mathcal{A} \text{ and some } f \in F(x - x_0)\},$$

which implies that $\mathcal{A}x_0$ is closed and convex. $\qquad\square$

Recall that if C is a closed convex subset of a reflexive Banach space X, then for every $x \in X$ there exists at least one nearest point to x in C. Define the minimal section A for an m-dissipative multivalued operator $\mathcal{A}$ as in (2.168). If X is reflexive, then $D(A) = D(\mathcal{A})$. If X has more geometric properties (e.g., strictly convex) then A may be single-valued.

Now we state the nonlinear version of the Hille-Yosida theorem in Hilbert spaces for multivalued m-dissipative operators.

Theorem 2.110 *Let F be a nonempty closed convex subset of a Hilbert space H and let $T(t)$ be a nonlinear contraction semigroup defined on F. Then there exists a unique multivalued m-dissipative operator $\mathcal{A}$ such that the minimal section of A defined as in (2.168) is the generator of $T(t)$. Conversely, let $\mathcal{A}$ be an m-dissipative operator. Then there is a unique semigroup $T(t)$ defined on $\overline{D(\mathcal{A})}$ such that the minimal section A of $\mathcal{A}$ is the generator of $T(t)$.*

Proof. See [12](Chap.IV, Theorem 1.2). □

To associate nonlinear semigroups with nonlinear abstract Cauchy problems , we introduce the following definition

Definition 2.111 *A function $u(t)$ defined on $[0, \infty)$ with values in a Banach space X is said to be a solution of the following nonlinear abstract Cauchy problem (NACP):*

$$\begin{cases} \dfrac{du(t)}{dt} \in \mathcal{A}u(t), & t > 0, \\ u(0) = x, \end{cases} \qquad (2.172)$$

if $u(t)$ is continuous on $[0, \infty)$ and Lipschitz on every compact interval of $(0, \infty)$, $u(0) = x$, $u(t)$ is differentiable a.e. on $(0, \infty)$, $u(t) \in D(\mathcal{A})$ a.e. and $\frac{du(t)}{dt} \in \mathcal{A}u(t)$, a.e. for $t > 0$.

Theorem 2.112 *Let $\mathcal{A}$ be a multivalued dissipative operator in a Banach space X. Suppose that, for every $x \in D(\mathcal{A})$, (2.172) has at least one solution denoted by $u(t; x)$. Then*

(i) $\|u(t; x) - u(t; y)\| \leq \|x - y\|$, *for every $t \geq 0$ and $x, y \in D(\mathcal{A})$. Consequently, the solution is unique.*

(ii) *For every $t \geq 0$ define $T(t)x = u(t; x)$ and again denote by $T(t)$ the extension of $T(t)$ on $\overline{D(\mathcal{A})}$. Then $T(t)$ is a semigroup of nonlinear contractions on $\overline{D(\mathcal{A})}$.*

Proof. Let $x, y \in D(\mathcal{A})$. Then we have

$$\frac{d}{dt}[u(t; x) - u(t; y)] \in \mathcal{A}u(t; x) - \mathcal{A}u(t; y), \text{ for } t > 0 \text{ a.e.}$$

For any $h < 0$, and $f \in F(u(t; x) - u(t; y))$, since

$$\langle [u(t + h; x) - u(t + h; y)] - [u(t; x) - u(t; y)], f \rangle$$
$$\leq [\|u(t + h; x) - u(t + h; y)\| - \|u(t; x) - u(t; y)\|] \|f\|$$

and $\mathcal{A}$ is dissipative, we have

$$\frac{d}{dt} \|u(t; x) - u(t; y)\|^2 \leq 0, \text{ for } t > 0 \text{ a.e.}$$

which implies (i). It follows that $u(t; x)$ is unique. Define $T(t)x = u(t; x)$. Then the function $T(t)x$ is continuous in t and nonexpansive by (i). For every $t \geq 0$, $T(t)$ can be extended to a nonexpansive mapping on $\overline{D(\mathcal{A})}$. Obviously, $T(t)$ maps $\overline{D(\mathcal{A})}$ into itself and

$$\|T(t)x - T(t)y\| \leq \|x - y\|, \text{ for every } t \geq 0 \text{ and } x, y \in \overline{D(\mathcal{A})}.$$

The semigroup property $T(t+s) = T(t)T(s)$ is an immediate consequence of the fact that the solution $u(t; x)$ is unique for every $x \in D(\mathcal{A})$. $\quad\square$

Definition 2.113 *Let $T(t)$ be a nonlinear contraction semigroup defined on a closed subset F of a Banach space X. Let $\mathcal{A}$ be a multivalued dissipative operator. We say that $T(t)$ is generated by $\mathcal{A}$ if the following conditions hold:*

(i) $\overline{D(\mathcal{A})} = F$.

(ii) For every $x \in D(\mathcal{A}), u(t) = T(t)x$ is a solution (in the sense of definition 2.111) of (2.172).

We observe that if $T(t)$ is generated by some dissipative operators, then the generator of $T(t)$ is densely defined on the domain of $T(t)$. Example 2.100 shows that in a general Banach space, it may happen that there is a nonlinear semigroup of contractions, but no dissipative generator generates this semigroup.

By the above definitions and Theorem 2.112, we see that for a dissipative operator $\mathcal{A}$, if (NACP) has at least one (unique actually) solution for every $x \in D(\mathcal{A})$, then $\mathcal{A}$ generates a nonlinear contraction semigroup on $\overline{D(\mathcal{A})}$. Moreover, (ii) of Proposition 2.98, Theorems 2.101 and 2.110 show that the following result holds.

Proposition 2.114 *Let $\mathcal{A}$ be an m-dissipative multivalued operator in a Hilbert space H. Then $\mathcal{A}$ generates a unique nonlinear contraction semigroup $T(t)$ defined on $\overline{D(\mathcal{A})}$.*

The general result on Banach spaces can be stated as follows.

Theorem 2.115 (Crandall-Liggett) *Let X be a real Banach space and let $\mathcal{A}$ be a dissipative operator such that*

$$\overline{D(\mathcal{A})} \subset \mathcal{R}(I - \lambda\mathcal{A}), \tag{2.173}$$

for all sufficiently small λ. Then

$$T(t)x = \lim_{n\to\infty} (I - \frac{t}{n}\mathcal{A})^{-n}x \tag{2.174}$$

exists for all $x \in D(\mathcal{A})$, uniformly in t on every compact interval of $[0, \infty)$. Moreover, $T(t)$ defined by (2.174) is a nonlinear semigroup of contractions on $\overline{D(\mathcal{A})}$ and

$$\|T(t)x - T(s)x\| \le 2|t - s||\mathcal{A}x|, \tag{2.175}$$

where the norm $|\cdot|$ is defined as

$$|\mathcal{A}x| = \inf\{\|y\| : y \in \mathcal{A}x\},$$

so that $|\mathcal{A}x| = \|\mathcal{A}x\|$ when $\mathcal{A}$ is single-valued.

The proof of this theorem will be divided into several steps. We set

$$J_\lambda = (I - \lambda A)^{-1}, \lambda > 0.$$

The condition (2.173) implies $\overline{D(A)} \subset D(J_\lambda) = \mathcal{R}(I - \lambda A)$ for all sufficiently small λ and hence $x \in D(J_\lambda^n)$ for all $n \geq 1$ if $x \in D(J_\lambda)$.

Lemma 2.116 *Let* $\lambda > 0$. *Then the following statements hold*

(i) J_λ *is single-valued, and for every* $x, y \in D(J_\lambda)$,

$$\|J_\lambda x - J_\lambda y\| \leq \|x - y\|.$$

(ii) $\|J_\lambda x - x\| \leq \lambda |Ax|, \forall x \in D(A) \cap \mathcal{R}(I - \lambda A)$,

$$\|J_\lambda^n x - x\| \leq n \|J_\lambda x - x\|, \forall x \in D(J_\lambda), \ n = 1, 2, \cdots.$$

(iii) *If* $x \in \mathcal{R}(I - \lambda A)$ *and* $\mu > 0$, *then* $\frac{\mu}{\lambda} x + \frac{\lambda - \mu}{\lambda} J_\lambda x \in \mathcal{R}(I - \mu A)$ *and*

$$J_\lambda x = J_\mu \left(\frac{\mu}{\lambda} x + \frac{\lambda - \mu}{\lambda} J_\lambda x \right). \tag{2.176}$$

Proof. (i) has been explained in (2.171). Let

$$A_\lambda x = \lambda^{-1}(J_\lambda x - x), \ x \in \mathcal{R}(I - \lambda A).$$

Then for any $x \in \mathcal{R}(I - \lambda A)$,

$$A_\lambda x \in \lambda^{-1}(J_\lambda x - (I - \lambda A)J_\lambda x) = AJ_\lambda x.$$

Let $x \in D(A) \cap \mathcal{R}(I - \lambda A)$. Then $A_\lambda x = \lambda^{-1}(J_\lambda x - J_\lambda (I - \lambda A)x)$. Since J_λ is non-expansive on $\mathcal{R}(I - \lambda A)$, this implies that

$$\|A_\lambda x\| \leq \|y\|, \ \text{for any } y \in Ax,$$

i.e., $\|A_\lambda x\| \leq |Ax|$. Hence $\|J_\lambda x - x\| = \lambda \|A_\lambda x\| \leq \lambda |Ax|$. This is the first part of (ii). The second part follows from the iterative process:

$$\|J_\lambda^n x - x\| = \|J_\lambda^n x - J_\lambda^{n-1} x + J_\lambda^{n-1} x - x\| \leq \|J_\lambda x - x\| + \|J_\lambda^{n-1} x - x\|.$$

Now we shall prove (2.176) which is the nonlinear version of the resolvent formula. If $x \in \mathcal{R}(I - \lambda A)$, then there is $(x_0, y_0) \in A$ such that $x_0 - \lambda y_0 = x$. We can write

$$\frac{\mu}{\lambda} x + \frac{\lambda - \mu}{\lambda} J_\lambda x = \frac{\mu}{\lambda}(x_0 - \lambda y_0) + \frac{\lambda - \mu}{\lambda} x_0 = x_0 - \mu y_0.$$

Therefore,

$$\frac{\mu}{\lambda}x + \frac{\lambda - \mu}{\lambda}J_\lambda x \in (I - \mu\mathcal{A})x_0.$$

This means that

$$J_\mu(\frac{\mu}{\lambda}x + \frac{\lambda - \mu}{\lambda}J_\lambda x) = x_0 = J_\lambda x$$

as claimed.   □

Lemma 2.117 Let $\lambda \geq \mu > 0$ and $x \in D(J_\lambda^m) \cap D(J_\lambda^n)$ where m and n are positive integers satisfying $n \geq m$. Then

$$\|J_\mu^n x - J_\lambda^m x\|$$
$$\leq \sum_{i=0}^{m} C_n^i \alpha^i \beta^{n-i} \|J_\lambda^{m-i} x - x\| + \sum_{i=m}^{n} C_{m-1}^{i-1} \alpha^m \beta^{i-m} \|J_\mu^{n-i} x - x\|,$$
$$(2.177)$$

where $\alpha = \mu/\lambda$ and $\beta = (\lambda - \mu)/\lambda$.

Proof. For the positive integers i and k satisfying $0 \leq i \leq n$ and $0 \leq k \leq m$, set
$$a_{k,i} = \|J_\mu^i x - J_\lambda^k x\|.$$
Using the properties of J_λ listed in Lemma 2.116, we obtain

$$a_{k,i} = \left\|J_\mu^i x - J_\mu(\frac{\mu}{\lambda}J_\lambda^{k-1}x + \frac{\lambda - \mu}{\lambda}J_\lambda^k x)\right\|$$
$$\leq \frac{\mu}{\lambda}\|J_\mu^{i-1} x - J_\lambda^{k-1} x\| + \frac{\lambda - \mu}{\lambda}\|J_\mu^{i-1} x - J_\lambda^k x\|$$
$$= \alpha a_{k-1,i-1} + \beta a_{k,i-1}.$$

By solving the inequalities

$$a_{k,i} \leq \alpha a_{k-1,i-1} + \beta a_{k,i-1},$$

we obtain (2.177).   □

We also need the following combinatorial lemma.

Lemma 2.118 Let $n \geq m > 0$ be integers, and α, β be positive numbers satisfying $\alpha + \beta = 1$. Then

$$\sum_{i=0}^{m} C_n^i \alpha^i \beta^{n-i}(m - i) \leq [(n\alpha - m)^2 + n\alpha\beta]^{1/2}, \qquad (2.178)$$

$$\sum_{i=m}^{n} C_{m-1}^{i-1} \alpha^m \beta^{i-m}(n - i) \leq [m\beta/\alpha^2 + (m\beta/\alpha + m - n)^2]^{1/2}. \qquad (2.179)$$

Proof. In order to prove (2.178), note first that

$$\sum_{i=0}^{m} C_n^i \alpha^i \beta^{n-i} (m-i)$$

$$\leq \sum_{i=0}^{n} C_n^i \alpha^i \beta^{n-i} (m-i)$$

$$\leq [\sum_{i=0}^{n} C_n^i \alpha^i \beta^{n-i}]^{1/2} [\sum_{i=0}^{m} C_n^i \alpha^i \beta^{n-i} (m-i)^2]^{1/2}.$$

Combining this with the following relations

$$\sum_{i=0}^{n} C_n^i \alpha^i \beta^{n-i} = 2^n, \quad \sum_{i=0}^{n} i C_n^i \alpha^i \beta^{n-i} = 2^{n-1} n \alpha,$$

$$\sum_{i=0}^{n} i^2 C_n^i \alpha^i \beta^{n-i} = 2^{n-2} n \alpha [(n-1)\alpha + 2],$$

we obtain (2.178). (2.179) follows from

$$\sum_{i=m}^{n} C_{m-1}^{i-1} \alpha^m \beta^{i-m} (n-i)$$

$$\leq \sum_{i=m}^{\infty} C_{m-1}^{i-1} \alpha^m \beta^{i-m} |n-i|$$

$$\leq [\sum_{i=m}^{\infty} C_{m-1}^{i-1} \alpha^m \beta^{i-m}]^{1/2} [\sum_{i=m}^{\infty} C_{m-1}^{i-1} \alpha^m \beta^{i-m} (n-i)^2]^{1/2},$$

and the following identity

$$\sum_{i=m}^{\infty} C_{m-1}^{i-1} \beta^{i-m} = (1-\beta)^{-m}, \quad |\beta| < 1.$$

$\square$

Proof of Theorem 2.115. Let $x \in D(\mathcal{A})$ and let $\lambda \geq \mu > 0$ be sufficiently small. Let n and m be positive integers such that $n \geq m$. By assumption, $x \in D(J_\lambda^m) \cap D(J_\lambda^n)$. Then we combine (ii) of Lemma 2.116 and (2.177)-(2.179) to obtain

$$\|J_\mu^n x - J_\lambda^m x\| \leq \{(n\mu - m\lambda)^2 + n\mu(\lambda - \mu)^{1/2}$$
$$+ [m\lambda(\lambda - \mu) + (m\lambda - n\mu)^2]^{1/2}\}|\mathcal{A}x|. \quad (2.180)$$

Taking $\mu = \frac{t}{n}$ and $\lambda = \frac{t}{m}$ in (2.180), we obtain

$$\left\| J^n_{t/n}x - J^m_{t/m}x \right\| \leq 2t(\frac{1}{m} - \frac{1}{n})^{1/2}|\mathcal{A}x|. \qquad (2.181)$$

Therefore, $\lim_{n\to\infty} J^n_{t/n}x = T(t)x$ exists uniformly in t on every compact subset of $(0,\infty)$. Since $J^n_{t/n}$ is nonexpansive on $D(J^n_{t/n})$, we find that

$$\|T(t)x - T(t)y\| \leq \|x - y\| \text{ for every } t \geq 0 \text{ and } x, y \in D(\mathcal{A}).$$

This implies that $T(t)x = \lim_{n\to\infty} J^n_{t/n}x$ exists for $x \in \overline{D(\mathcal{A})}$ and $T(t)$ is nonexpansive on $\overline{D(\mathcal{A})}$. Moreover, for every $x \in D(\mathcal{A})$, $T(t)x$ is Lipschitz continuous in t on $[0,\infty)$. Indeed, taking limit in equality (2.180) with $n = m, \mu = t/n, \lambda = s/n$, where $0 \leq t \leq s$, we obtain (2.175),

$$\|T(t)x - T(s)x\| \leq 2|t - s||\mathcal{A}x|.$$

In particular, this shows that $T(t)x$ is continuous in t for every $x \in D(\mathcal{A})$. In order to complete the proof, we have to verify the semigroup property $T(t + s) = T(t)T(s)$. We note that the strong convergence, $T(t) = \lim_{n\to\infty} J^n_{t/n}$ together with (2.181), implies that

$$T(t)^m x = \lim_{n\to\infty} [J^n_{t/n}]^m x = \lim_{n\to\infty} [J^m_{t/n}]^n x.$$

Hence,

$$T(mt)x = \lim_{n\to\infty} J^n_{mt/n}x = \lim_{k\to\infty} J^{mk}_{mt/(mk)}x = \lim_{k\to\infty} [J^m_{t/k}]^k x = T(t)^m x.$$

Then by using a standard argument we can prove that $T(t + s) = T(t)T(s)$ for rational t and s. Since $T(t)x$ is continuous in t and Lipschitz continuous in x, this implies that $T(t + s) = T(t)T(s)$ for all t, s nonnegative real. This completes the proof.   □

In order to assure the existence of solution of the (NACP) with a dissipative operator $\mathcal{A}$, as given in (2.172), we have to impose further conditions on $\mathcal{A}$. More precisely, we shall need the following stronger version of condition (2.173), namely

$$\overline{conv D(\mathcal{A})} \subset \bigcap_{\lambda > 0} \mathcal{R}(I - \lambda\mathcal{A}) \qquad (2.182)$$

where $\overline{conv}$ denotes the closure of convex hull. It is obvious that condition (2.182) is trivially satisfied if $\mathcal{A}$ is m-dissipative.

Theorem 2.119 *Let X be a real Banach space and let $\mathcal{A}$ be a dissipative operator satisfying condition (2.182). Then for any $x \in D(\mathcal{A})$, if (NACP) given in (2.172) has a solution $u(t)$, then*

$$u(t) = T(t)x = \lim_{n\to\infty} (I - \frac{t}{n}\mathcal{A})^{-n}x, t \geq 0. \qquad (2.183)$$

Conversely, if $\mathcal{A}$ is closed and $T(t)x, x \in \overline{D(\mathcal{A})}$ defined by (2.183) is differentiable a.e. on $(0, \infty)$, then $u(t) = T(t)x$ is a solution of (NACP) (2.172).

Proof. See [12](Chap.III, Theorems 1.4 and 1.5). □

Since every Lipschitz continuous X-valued function is a.e. differentiable when X is reflexive, we have immediately the following result.

Corollary 2.120 *Let X be a reflexive Banach space and let $\mathcal{A}$ be a closed and dissipative operator satisfying (2.182) (in particular, $\mathcal{A}$ be m-dissipative operator). Then $\mathcal{A}$ generates a unique nonlinear semigroup of contractions on $\overline{D(\mathcal{A})}$.*

2.10 Notes and references

The classical theory of semigroups, in particular the theory of C_0−semigroups, is well developed and most of the results in Sections 2.1-2.7 can be found in standard textbooks such as Hille and Phillips [77] and Pazy [128]. The abstract theory of semigroups may be considered as a part of functional analysis and in that respect some textbooks on this subject such as Yosida [171] and Balakrishnan [6] are good sources to be consulted. For necessary background on operator theory, the reader is referred to textbooks such as Dunford and Schwartz [53] and Kato [84]. We also note that a different approach using sesquilinear form for semigroup generation, which is not included in this book, can be found in Lions [92] and Banks and Kunisch [9]. Except for Section 2.7, we do not consider evolution equations, for which the textbook by Tanabe [150] may be consulted. The theory of integrated semigroups, first established by Neubrander in his paper [123], is a relatively new topic in the research of semigroup theory. Since there are currently only a few references on this subject [1], we included a comprehensive and step-by-step treatment of the integrated semigroups in Section 2.8 by integrating the various theories in Arendt [4], Krein [89] and Zheng [175]. Most of the results on the theory of nonlinear semigroups are standard by now and the presentation in Section 2.9 is based on the textbook by Barbu [12].

Chapter 3

Stability of C_0-Semigroups

Stability analysis and feedback stabilization are issues of great importance in control system design. In this chapter, we study the stability of the following abstract Cauchy problem on a Banach space X:

$$\begin{cases} \dfrac{du(t)}{dt} = Au(t), \\ u(0) = x \in X, \end{cases} \qquad (3.1)$$

where $A : D(A)(\subset X) \to X$ is a linear operator which generates a C_0-semigroup $T(t)$ on X. The system stability is concerned with the behavior of the state $u(t)$ as time t evolves. As we saw in Chapter 2, the unique solution of (3.1) can be expressed as

$$u(t) = T(t)x,$$

which clearly shows that the time evolution of the C_0-semigroup $T(t)$ determines the stability of the solution $u(t)$. Therefore, this chapter is dedicated to the stability analysis of C_0-semigroups. We consider three types of stability as defined below:

Definition 3.1   (i) A C_0-semigroup, $T(t)$, is said to be weakly stable if for every $x \in X$ and $y \in X^*$, $\langle T(t)x, y \rangle \to 0$, as $t \to \infty$.

(ii) A C_0-semigroup, $T(t)$, is said to be asymptotically stable if for every $x \in X$, $\|T(t)x\| \to 0$, as $t \to \infty$.

(iii) A C_0-semigroup, $T(t)$, is said to be exponentially stable if there exist constants $M \geq 1$ and $\omega > 0$ such that $\|T(t)\| \leq M e^{-\omega t}$.

As seen in Chapter 1, the stability analysis for infinite dimensional systems is much more complicated than that for finite dimensional systems. To

demonstrate this, we provide two examples. The first shows that the semigroup is asymptotically stable, but not exponentially stable, while the second
shows that there is an operator whose spectrum is contained in the left-half
complex plane with $\mathrm{Re}\lambda \leq -1$, but the semigroup generated by the operator
is not exponentially stable.

Example 3.2 *Let $X = \ell^2$, the Hilbert space of all square-summable sequences, and define*

$$T(t)x = (e^{-t}x_1, e^{-t/2}x_2, \ldots, e^{-t/n}x_n, \ldots), \quad t \geq 0$$

*for all $x = (x_1, x_2, \ldots, x_n, \ldots) \in X$. Then $T(t)$ is a C_0-semigroup on X and
for every $x \in X$*

$$\|T(t)x\|^2 = \sum_{n=1}^{\infty} e^{-2t/n}x_n^2 \to 0 \ \ as \ t \to \infty.$$

Thus, $T(t)$ is asymptotically stable. However, for any $t \in [0, \infty)$,

$$\begin{aligned}\|T(t)\| &= \sup_{\|x\|=1} \|T(t)x\| \\ &= \sup_{\|x\|=1} \left(\sum_{n=1}^{\infty} e^{-2t/n}x_n^2\right)^{1/2} = \lim_{n\to\infty} e^{-t/n} = 1,\end{aligned}$$

*which indicates that $T(t)$ is not exponentially stable. The infinitesimal gen-
erator of $T(t)$ is found to be*

$$Ax = \frac{d^+}{dt}T(t)|_{t=0} = -(x_1, \frac{x_2}{2}, \ldots, \frac{x_n}{n}, \ldots),$$

and $\sigma(A) = \{-\frac{1}{n}|n \geq 1\}$.

Example 3.3 *[128, P.117] Let $X = \{f(x) \in L^p(0, \infty), 1 < p < \infty \mid |f|_1 = \int_0^{\infty} e^x|f(x)|dx < \infty\}$. X is a Banach space when it is equipped with the norm*

$$\|f\| = \|f\|_{L^p} + |f|_1.$$

Define a semigroup by

$$T(t)f(x) = f(x + t), \quad for \ all \ t \geq 0.$$

*It is easy to see that $T(t)$ is a C_0-semigroup of contractions on X. Consider
functions*

$$f_{t\epsilon}(x) = \begin{cases} 1, & x \in [t, t + \epsilon^p], \\ 0, & otherwise. \end{cases}$$

Then $\|f_{t\epsilon}\| = \epsilon + \int_t^{t+\epsilon^p} e^x dx = \epsilon + e^t[e^{\epsilon^p} - 1]$ *and*

$$T(t)f_{t\epsilon}(x) = \begin{cases} 1, & x \in [0, \epsilon^p], \\ 0, & otherwise. \end{cases}$$

Hence,

$$\|T(t)f_{t\epsilon}(x)\| = \epsilon + [e^{\epsilon^p} - 1] = \frac{\epsilon + [e^{\epsilon^p} - 1]}{\epsilon + e^t[e^{\epsilon^p} - 1]}\|f_{t\epsilon}\|.$$

Letting $\epsilon \to 0$, we see $\|T(t)\| \geq 1$, and so $\|T(t)\| = 1$, for all $t \geq 0$ because we already know that $T(t)$ is a contraction semigroup. The generator of $T(t)$ is given by

$$Af(x) = f'(x), \quad D(A) = \{f \text{ is absolutely continuous, } f' \in X\}.$$

We show that $\lambda \in \rho(A)$ when $Re\lambda > -1$ and

$$R(\lambda, A)f(x) = \int_x^\infty e^{-\lambda(s-x)}f(s)ds, \forall f \in X.$$

Indeed, let $u(x) = \int_x^\infty e^{-\lambda(s-x)}f(s)ds$. Then $\lambda u - Au = f$. It suffices to show that $u \in X$. If $Re\lambda \geq 0$ then

$$|u(x)| \leq \int_x^\infty e^{-Re\lambda(s-x)}e^{-s}e^s|f(s)|ds$$

$$\leq e^{-x}\int_x^\infty e^s|f(s)|ds \leq e^{-x}|f|_1.$$

If $-1 < Re\lambda < 0$ then

$$|u(x)| \leq e^{Re\lambda x}\int_x^\infty e^{-(Re\lambda+1)s}e^s|f(s)|ds$$

$$\leq e^{Re\lambda x}\int_x^\infty e^s|f(s)|ds \leq e^{Re\lambda x}|f|_1.$$

Therefore, $u(x) \in L^p(0, \infty)$ and

$$|u(x)|_1 \leq \int_0^\infty e^x \int_x^\infty e^{-Re\lambda(s-x)}|f(s)|dxds$$

$$= \int_0^\infty e^s|f(s)|ds \int_0^s e^{-(Re\lambda+1)(s-x)}dx.$$

$$= \int_0^\infty e^s|f(s)|ds(Re\lambda + 1)^{-1}(1 - e^{-(Re\lambda+1)s})$$

$$\leq (Re\lambda + 1)^{-1}|f|_1.$$

Hence, $\sigma(A) \subset \{\lambda| Re\lambda \leq -1\}$ while $\|T(t)\|$ does not decay exponentially.

In the rest of this chapter, we shall first prove spectral mapping theorems linking the spectrum of C_0-semigroups to that of their generators. These theorems will then be used to show that the spectrum-determined growth condition holds for C_0-semigroups which are continuous in the uniform operator topology, in particular, the compact semigroups, differentiable semigroups, and analytic semigroups. In Section 3.3, weak stability is addressed, and necessary and sufficient conditions, in terms of the spectrum of infinitesimal generators are derived for a C_0-semigroup to be asymptotically stable. Also, a relationship between the weak stability and asymptotic stability is established. In the three sections thereafter, we shall focus on the exponential stability. We derive some time domain criteria for the exponential stability of C_0-semigroups based on the information of the solution of (3.1). If we do not know about the solution, but instead we know the behavior of the resolvent of the system operator A, the frequency domain criteria can be invoked to test the exponential stability of C_0-semigroups. Finally, we shall discuss essential spectrum, compact perturbations to C_0-semigroups, and invariance principle for nonlinear semigroups.

The results of this chapter will be used frequently in subsequent chapters where we consider applications to specific problems in engineering.

3.1   Spectral mapping theorems

Let $T(t)$ be a C_0-semigroup on a Banach space with generator A. Let $\sigma(A)$ denote the spectrum set of A, and let $\sigma_p(A), \sigma_r(A)$, and $\sigma_c(A)$ denote, respectively, the point spectrum (eigenvalues), residual spectrum, and continuous spectrum of A. We have the following theorems.

Theorem 3.4 *Let $T(t)$ be a C_0-semigroup on a Banach space X with generator A. Then*

$$e^{t\sigma_p(A)} \subset \sigma_p(T(t)) \subset e^{t\sigma_p(A)} \cup \{0\}.$$

More precisely, if $\lambda \in \sigma_p(A)$, then $e^{\lambda t} \in \sigma_p(T(t))$, and if $e^{\lambda t} \in \sigma_p(T(t))$, then there exists an integer k such that $\lambda_k = \lambda + 2\pi ik/t \in \sigma_p(A)$.

Proof.   First, let $\lambda \in \sigma_p(A)$. Then there exists an $x_0 \neq 0$ such that

$$Ax_0 = \lambda x_0,$$

so the equation $\frac{du(t)}{dt} = Au(t)$ has a solution $u(t) = e^{\lambda t}x_0$ with $u(0) = x_0$. By uniqueness, we have $T(t)x_0 = e^{\lambda t}x_0$ and so $e^{\lambda t} \in \sigma_p(T(t))$. We show the second inclusion relation. Let $e^{\lambda t} \in \sigma_p(T(t))$ and let $x_0 \neq 0$ such that $T(t)x_0 = e^{\lambda t}x_0$, or equivalently $e^{-\lambda t}T(t)x_0 = x_0$. Since

$$e^{-\lambda(s+t)}T(s+t)x_0 = e^{-\lambda s}T(s)e^{-\lambda t}T(t)x_0 = e^{-\lambda s}T(s)x_0,$$

the continuous function $s \to e^{-\lambda s}T(s)x_0$ is periodic with period t. Since it does not vanish identically, one of its Fourier coefficients must be different from zero. Therefore, there is an integer k such that

$$x_k = \frac{1}{t}\int_0^t e^{-(2\pi ik/t)s}e^{-\lambda s}T(s)x_0 ds \neq 0.$$

We claim that $\lambda_k = \lambda + 2\pi ik/t \in \sigma_p(A)$ and $Ax_k = \lambda_k x_k$. In fact,

$$\frac{T(h) - I}{h}x_k = \frac{1}{ht}\int_0^t e^{-\lambda_k s}T(s+h)x_0 ds - \frac{1}{ht}\int_0^t e^{-\lambda_k s}T(s)x_0 ds$$

$$= \frac{1}{ht}\int_h^{t+h} e^{-\lambda_k(s-h)}T(s)x_0 ds - \frac{1}{ht}\int_0^t e^{-\lambda_k s}T(s)x_0 ds$$

$$= \frac{1}{ht}e^{\lambda_k h}\int_t^{t+h} e^{-\lambda_k s}T(s)x_0 ds - \frac{1}{ht}\int_0^t e^{-\lambda_k s}T(s)x_0 ds$$

$$+ \frac{1}{ht}e^{\lambda_k h}\int_h^t e^{-\lambda_k s}T(s)x_0 ds$$

$$= \frac{1}{ht}e^{\lambda_k h}\int_t^{t+h} e^{-\lambda_k s}T(s)x_0 ds - \frac{1}{ht}\int_0^h e^{-\lambda_k s}T(s)x_0 ds$$

$$+ \frac{1}{ht}(e^{\lambda_k h} - 1)\int_h^t e^{-\lambda_k s}T(s)x_0 ds$$

tends to

$$\frac{1}{t}e^{-\lambda_k t}T(t)x_0 - \frac{1}{t}x_0 + \frac{1}{t}\lambda_k \int_0^t e^{-\lambda_k s}T(s)x_0 ds$$

$$= \frac{1}{t}e^{-\lambda_k t}T(t)x_0 - \frac{1}{t}x_0 + \lambda_k x_k = \lambda_k x_k$$

as $h \to 0$. This means that $x_k \in D(A)$ and

$$Ax_k = \lambda_k x_k.$$

$\square$

Theorem 3.5 *Let $T(t)$ be a C_0-semigroup on a Banach space with generator A. Then*

$$e^{t\sigma(A)} \subset \sigma(T(t)).$$

Proof. Since $\rho(T(t)) \subset e^{t\rho(A)}$ implies $e^{t\sigma(A)} \subset \sigma(T(t))$, it is sufficient to show the former relation. Let $e^{\lambda t} \in \rho(T(t))$. Define

$$B_\lambda(t)x = \int_0^t e^{\lambda(t-s)} T(s)x\, ds.$$

Then a direct computation shows that

$$(\lambda - A)B_\lambda(t)x = e^{\lambda t}x - T(t)x, \text{ for all } x \in X.$$
$$B_\lambda(t)(\lambda - A)x = e^{\lambda t}x - T(t)x, \text{ for all } x \in D(A).$$

Hence, $\lambda \in \rho(A)$ and $R(\lambda, A) = R(e^{\lambda t}, T(t))B_\lambda(t)$. Unlike the finite dimensional case, $e^{t\sigma(A)} \subset \sigma(T(t))$ may hold strictly. Example 3.3 provides such an example. $\qquad\square$

3.2 Spectrum-determined growth condition

Let $T(t)$ be a C_0-semigroup on a Banach space with generator A. Let the growth rate ω_0 be as defined in Theorem 2.9. Denote the spectral bound of A by

$$S(A) = \sup\{\mathrm{Re}\lambda \mid \lambda \in \sigma(A)\}.$$

We say that the *spectrum-determined growth condition* holds if

$$\omega_0 = S(A). \tag{3.2}$$

From the Hille-Yosida theorem, we know that $S(A) \le \omega_0$ for any C_0-semigroup. However, in general, $\omega_0 \le S(A)$ is not true even for C_0-semigroups on Hilbert spaces. The following interesting example given originally by Zabczyk [173] (see also [78]) shows that, even for an operator with compact resolvent in a Hilbert space, the spectrum-determined growth condition may not hold.

Example 3.6 *Let $H_n = \mathrm{span}\{e_{ni}, i = 1, 2, \ldots, n\}$ be an n-dimensional Hilbert space spanned by standard basis $\{e_{ni}\}_1^n$ with all zero elements except the $i-th$ element which is 1. Let A_n denote a bounded linear operator on H_n specified by:*

$$A_n = \lambda_n + N_n = \begin{bmatrix} \lambda_n & 1 & 0 & \cdots & 0 \\ 0 & \lambda_n & 1 & \cdots & 0 \\ . & . & . & \cdots & . \\ 0 & 0 & 0 & \cdots & 1 \\ 0 & 0 & 0 & \cdots & \lambda_n \end{bmatrix},$$

where $\lambda_n = -\frac{1}{2} + i\omega_n, \omega_n \in \mathbf{R}$ and $|\omega_n| \to \infty$,

$$N_n e_{n1} = 0, \; N_n e_{ni} = e_{n(i-1)}, \; i = 2, 3, \ldots, n.$$

Let

$$H = \sum_{n=1}^{\infty} \oplus H_n, \ A = \sum_{n=1}^{\infty} \oplus A_n.$$

Then H is a Hilbert space with the norm

$$\|x\|^2 = \sum_{n=1}^{\infty} \sum_{m=1}^{n} |x_{nm}|^2$$

for any $x = \sum_{n=1}^{\infty} \sum_{m=1}^{n} x_{nm} e_{nm} \in H$. It can be verified that the resolvent of A is compact and $\sigma(A) = \{\lambda_n \mid n = 1, 2, \ldots\}$. Indeed, for any $\lambda \neq \lambda_n (n \geq 1)$ and any positive n, $\|N_n\| = 1, N_n^m = 0 (m \geq n)$ and hence

$$(\lambda - A_n)^{-1} = (\lambda - \lambda_n - N_n)^{-1} = \sum_{m=0}^{\infty} (\lambda - \lambda_n)^{-m-1} N_n^m = \sum_{m=0}^{n-1} (\lambda - \lambda_n)^{-m-1} N_n^m.$$

Therefore, if $\lambda \neq \lambda_n (n \geq 1)$, by $| \lambda - \lambda_n | \to \infty$ as $n \to \infty$, it can be easily shown that the series

$$\sum_{n=1}^{\infty} \oplus \sum_{m=0}^{n-1} (\lambda - \lambda_n)^{-m-1} N_n^m$$

converges strongly to a compact linear operator on H. The proof of convergence is simple; we show only the compactness. For any $y = \sum_{n=1}^{\infty} \oplus x_n \in H$ such that $\|y\| \leq C$, with C being a constant, we have

$$\left\| \sum_{n=N}^{\infty} \oplus \sum_{m=0}^{n-1} (\lambda - \lambda_n)^{-m-1} N_n^m y \right\|$$

$$\leq \sum_{n=N}^{\infty} \sum_{m=0}^{n-1} |\lambda - \lambda_n|^{-m-1} \|x_n\|$$

$$= \sum_{n=N}^{\infty} \frac{1 - |\lambda - \lambda_n|^{-n+1}}{|\lambda - \lambda_n| - 1} \|x_n\|$$

$$\leq C \sup_{n \geq N} \frac{1 - |\lambda - \lambda_n|^{-n+1}}{|\lambda - \lambda_n| - 1} \to 0 \quad as \ N \to \infty$$

which shows that $\sum_{n=1}^{\infty} \oplus \sum_{m=0}^{n-1} (\lambda - \lambda_n)^{-m-1} N_n^m$ is compact. Hence $\lambda \in \rho(A)$ and

$$(\lambda - A)^{-1} = \sum_{n=1}^{\infty} \oplus \sum_{m=0}^{n-1} (\lambda - \lambda_n)^{-m-1} N_n^m.$$

Consequently,

$$S(A) = \{Re\lambda \mid \lambda \in \sigma(A)\} = -\frac{1}{2}.$$

We claim that A is an infinitesimal generator of a C_0-semigroup on H. Indeed, denoting $A_0 = A - I$, we have $\mathcal{R}(A_0 - I) = \mathcal{R}(A - 2I) = H$. Moreover, for any $x \in H$,

$$x = \sum_{n=1}^{\infty} \sum_{m=1}^{n} C_{nm} e_{nm}$$

with

$$\|x\|^2 = \sum_{n=1}^{\infty} \sum_{m=1}^{n} |C_{nm}|^2$$

we have

$$
\begin{aligned}
&Re\langle A_0 x, x\rangle \\
&= Re\langle Ax, x\rangle - \|x\|^2 \\
&= Re \sum_{n=1}^{\infty} \langle (\lambda_n + N_n) \sum_{m=1}^{n} C_{nm} e_{nm}, \sum_{m=1}^{n} C_{nm} e_{nm}\rangle - \|x\|^2 \\
&\leq \sum_{n=1}^{\infty} \left[-\frac{1}{2} \sum_{m=1}^{n} |C_{nm}|^2 + \|N_n\| \left\| \sum_{m=1}^{n} C_{nm} e_{nm} \right\|^2 \right] - \|x\|^2 \\
&= -\frac{1}{2} \|x\|^2 + \|x\|^2 - \|x\|^2 \leq 0.
\end{aligned}
$$

It follows from the Lümer-Phillips theorem that A_0, and thus A, is an infinitesimal generator of a C_0-semigroup e^{tA}. We now show that the growth rate $\omega_0 = 1/2$. Note that

$$T(t) = \sum_{n=1}^{\infty} \oplus e^{tA_n}$$

is a C_0-semigroup on H since the right-hand side is strongly convergent. Clearly, $T(t)x = e^{tA}x$ for all $x \in \sum_{n=1}^{m} \oplus H_n$ which are dense in H. Hence, $T(t) = e^{tA}$. Let P_n be the orthogonal projection from H to H_n. Then for any $x \in H, x = \sum_{n=1}^{\infty} P_n x$. Since $\|e^{tN_n}\| \leq e^t$, we have

$$
\begin{aligned}
\|e^{tA}x\|^2 &= \sum_{n=1}^{\infty} \|e^{tA} P_n x\|^2 = \sum_{n=1}^{\infty} \|e^{\lambda_n t} e^{N_n t} P_n x\|^2 \\
&\leq e^{-t} \sum_{n=1}^{\infty} \|e^{tN_n}\|^2 \|P_n x\|^2 \leq e^{-t} \sum_{n=1}^{\infty} e^{2t} \|P_n x\|^2 = e^t \|x\|^2.
\end{aligned}
$$

Therefore, $\omega_0 \leq 1/2$. On the other hand, for

$$y_n = n^{-1/2} \sum_{i=1}^{n} e_{ni}, \quad n \geq 2,$$

we have $\|y_n\| = 1$ *and*

$$\left\| e^{tA} \right\|^2 \geq \left\| e^{tA} y_n \right\|^2 = \left\| e^{A_n t} y_n \right\|^2$$

$$= \left\| e^{\lambda_n t} e^{N_n t} (n^{-1/2} \sum_{i=1}^{n} e_{ni}) \right\|^2$$

$$= n^{-1} e^{-t} \left\| \sum_{m=0}^{n-1} (t^m N_n^m / m!) \sum_{i=1}^{n} e_{ni} \right\|^2$$

$$= n^{-1} e^{-t} \left\| \sum_{i=1}^{n} \sum_{m=0}^{i-1} (t^m / m!) e_{n(i-m)} \right\|^2$$

$$= n^{-1} e^{-t} \left\| \sum_{k=1}^{n} (\sum_{m=0}^{n-k} t^m / m!) e_{nk} \right\|^2$$

$$= n^{-1} e^{-t} \sum_{k=1}^{n} (\sum_{m=0}^{n-k} t^m / m!)^2 \geq n^{-1} e^{-t} \sum_{k=1}^{[n/2]} (\sum_{m=0}^{n-k} t^m / m!)^2$$

$$\geq n^{-1} e^{-t} \sum_{k=1}^{[n/2]} (\sum_{m=0}^{[n/2]} t^m / m!)^2 \geq n^{-1} e^{-t} [n/2] (\sum_{m=0}^{[n/2]} t^m / m!)^2$$

$$\geq \frac{1}{4} e^{-t} (\sum_{m=0}^{[n/2]} t^m / m!)^2,$$

where $[n/2]$ *denotes the largest integer not exceeding* $n/2$. *Letting* $n \to \infty$, *we obtain* $\left\| e^{tA} \right\|^2 \geq \frac{1}{4} e^t$ *and so* $\omega_0 \geq \frac{1}{2}$. *Therefore,* $\omega_0 = \frac{1}{2}$, *so the spectrum-determined growth condition does not hold.*

Remark 3.7 *If we consider* $A_n^\delta = \lambda_n + \delta N_n$, *for any* $\delta > 0$, *instead of* A_n *in Example 3.6, and if* A^δ *is similarly defined as* A, *then we can show that* $S(A^\delta) = -\frac{1}{2}$ *and* $\omega_0(A^\delta) = \delta - 1/2$. *This shows that even in Hilbert spaces, there exists an operator which has compact resolvent, but the gap between its growth rate and its spectral bound can be arbitrarily large.*

It is thus of interest to investigate what kind of C_0-semigroups satisfy the spectrum-determined growth condition. This is made clear in the following theorem.

Theorem 3.8 *Let* $T(t)$ *be a* C_0-*semigroup on a Banach space with generator* A. *Let the growth rate* ω_0, *the spectral radius* $r(T(t))$ *be defined as in Theorem 2.9, and the spectral bound* $S(A)$ *as defined above. If*

$$\sigma(T(t)) \backslash \{0\} \subset e^{t\sigma(A)}, \tag{3.3}$$

then $\omega_0 \leq S(A)$, *and hence the spectrum-determined growth condition holds.*

Proof. From (iii) of Theorem 2.9 we know that

$$r(T(t)) = e^{\omega_0 t}$$

for any $t > 0$. Therefore, $\sigma(T(t))\backslash\{0\} \subset e^{t\sigma(A)}$ implies $r(T(t)) \leq e^{tS(A)}$, or equivalently, $e^{t\omega_0} \leq e^{tS(A)}$. It follows that $\omega_0 \leq S(A)$. Hence, the spectrum-determined growth condition is satisfied because $S(A) \leq \omega_0$.   □

We have shown in Theorem 3.4 that (3.3) is satisfied for the point spectrum, and in Theorem 3.5 that the converse also holds. These results yield a direct consequence that for C_0-semigroups with pure point spectrum, condition (3.3) is always true; hence, the spectrum-determined growth condition holds. Compact C_0-semigroups belong to such a class of C_0-semigroups.

Theorem 3.9 *If $T(t)$ is compact for $t > t_0$ then the spectrum-determined growth condition (3.2) holds.*

Proof. It follows from Theorem 3.4 that if $T(t)$ is compact for $t > t_0$, then for every $t > t_0$

$$\sigma(T(t)) = \sigma_p(T(t)) \subset e^{t\sigma_p(A)} \cup \{0\} \subset e^{t\sigma(A)} \cup \{0\} \subset \sigma_p(T(t)) \cup \{0\}. \quad (3.4)$$

Therefore, (3.3) holds and so $\omega_0 = S(A)$.   □

We can generalize the results in Theorem 2.58 to semigroups which are compact for any $t > t_0$ with $t_0 > 0$.

Lemma 3.10 *Suppose that $T(t)$ is compact for $t > t_0$. Let A be its generator. Then $\sigma(A)$ consists of isolated eigenvalues with finite algebraic multiplicity, i.e., $\sigma(A) = \sigma_p(A)$.*

Proof. Since $T(t)$ is compact for every $t > t_0, \sigma(T(t))$ consists at most of countable eigenvalues. From (3.4), $\sigma(A)$ consists at most of countable points. Now we show that $\sigma(A) = \sigma_p(A)$. Indeed, let $\lambda \in \sigma(A)$. From (3.4), $e^{\lambda t} \in \sigma_p(T(t))$. By the spectral mapping theorem (Theorem 3.4), there is an integer $n(t)$ such that $\lambda + \frac{2\pi n(t)}{t} \in \sigma_p(A)$. If for some $t > t_0, n(t) = 0$ then $\lambda \in \sigma_p(A)$. Otherwise, if for all $t > t_0, n(t) \neq 0$ then

$$\{\lambda + \frac{2\pi n(t)}{t} | t > t_0\} \in \sigma_p(A).$$

The left-hand side above is an uncountable set, which is a contradiction. Therefore, $\lambda \in \sigma_p(A)$, i.e., $\sigma(A) = \sigma_p(A)$. Next, let $t > t_0$. According to the proof of Theorem 3.5, for all $x \in D(A), B_\lambda(t)x = \int_0^t e^{\lambda(t-s)} T(s)x\,ds$ satisfies

$$(\lambda - A)B_\lambda(t)x = B_\lambda(t)(\lambda - A)x = e^{\lambda t}x - T(t)x,$$

which implies that

$$B_\lambda{}^p(t)(\lambda - A)^p x = (e^{\lambda t} - T(t))^p x$$

holds for all $x \in D(A)$. Hence, the algebraic multiplicity of λ as the eigenvalues of A is less than that of $e^{\lambda t}$ as the eigenvalues of $T(t)$, which is finite by compactness of $T(t)$. □

Lemma 3.11 *Suppose $T(t)$ is compact for $t > t_0$. Let A be its generator. Then for any m and M, there are only a finite number of eigenvalues of A in the strip $\{\lambda | m \le Re\lambda \le M\}$.*

Proof. Note that for any m and M, the mapping

$$\lambda \to e^{\lambda t}$$

maps $\{\lambda | m \le Re\lambda \le M\}$ to a bounded set $\{\lambda | e^{mt} \le |\lambda| \le e^{Mt}\}$. If $\lambda \in \{\lambda | m \le Re\lambda \le M\} \cap \sigma(A)$, then $e^{\lambda t} \in \sigma_p(T(t))$ and $e^{\lambda t} \in \{\lambda | e^{mt} \le |\lambda| \le e^{Mt}\}$ for $t > t_0$, and the numbers of such $e^{\lambda t}$ are finite. Hence, there are only finite numbers of $\lambda \in \{\lambda | m \le Re\lambda \le M\} \cap \sigma(A)$ with different real parts. If there is a line paralleling the imaginary axis on which there are infinite numbers of λ such that $e^{\lambda t} = \mu$ which is a fixed complex number, then since $x \in \mathcal{N}(\lambda - A)$ implies $x \in \mathcal{N}(\mu - T(t))$, and since the eigenvectors corresponding to different eigenvalues are linearly independent, $\mathcal{N}(\mu - T(t))$ must be an infinite dimensional space, which contradicts the compactness of $T(t)(t > t_0)$. The proof is complete. □

Due to the above lemmas, for a C_0-semigroup $T(t)$ which is compact for $t > t_0$, we can arrange the eigenvalues $\{\lambda_i\}$ of its generator A as

$$\lambda_0 \to \lambda_1 \to \lambda_2 \to \cdots \to \lambda_n \to \cdots \tag{3.5}$$

with $Re\lambda_i \ge Re\lambda_{i+1}, i = 0, 1, 2, \ldots$. Let P_{λ_i} be the projection corresponding to λ_i

$$P_{\lambda_i} = \frac{1}{2\pi i} \int_\Gamma R(\lambda, A) d\lambda$$

where Γ is the circumference of a circle with center at λ_i inside which there are no other eigenvalues of A except λ_i.

Theorem 3.12 *Let $T(t)$ be a C_0-semigroup which is compact for $t > t_0$ in a Banach space with generator A. Let $\sigma(A) = \{\lambda_i\}$ be arranged as in (3.5). If there is an integer k such that $Re\lambda_k > Re\lambda_{k+1}$, then for any $\epsilon > 0$ such that $Re\lambda_k - \epsilon > Re\lambda_{k+1}$, there exists a constant $C(\epsilon, k)$, depending on ϵ and k, such that*

$$\|T(t) - T(t) \sum_{i=0}^{k} P_{\lambda_i}\| \le C(\epsilon, k) e^{(Re\lambda_k - \epsilon)t}.$$

Proof. It suffices to show the desired result for sufficiently large $t > t_0$. Since

$$T(t) - T(t) \sum_{i=0}^{k} P_{\lambda_i} = T(t)(I - \sum_{i=0}^{k} P_{\lambda_i}) = (I - \sum_{i=0}^{k} P_{\lambda_i})T(t),$$

it follows from (3.4) that

$$\sigma(T(t)(I - \sum_{i=0}^{k} P_{\lambda_i})) \subset \{e^{\lambda t}, \lambda \in \sigma(A) \backslash \{\lambda_i\}_0^k\} \cup \{0\}.$$

Hence,

$$\lim_{n \to \infty} \|T(nt_0)(I - \sum_{i=0}^{k} P_{\lambda_i})\|^{1/n} = \lim_{n \to \infty} \|[T(t_0)(I - \sum_{i=0}^{k} P_{\lambda_i})]^n\|^{1/n}$$

$$\leq e^{(\mathrm{Re}\lambda_k - \epsilon)t_0}.$$

Let $N > 0$ be such a number that for all $n \geq N$

$$\|T(nt_0)(I - \sum_{i=0}^{k} P_{\lambda_i})\| \leq e^{n(\mathrm{Re}\lambda_k - \epsilon)t_0}.$$

Let $t > Nt_0$ and select n such that $nt_0 \leq t < (n+1)t_0$. Then $n \geq N$ and so

$$\left\| T(t)(I - \sum_{i=0}^{k} P_{\lambda_i}) \right\| = \left\| T(t - nt_0)T(nt_0)(I - \sum_{i=0}^{k} P_{\lambda_i}) \right\|$$

$$\leq \sup_{0 \leq s \leq t_0} \|T(s)\| \, e^{n(\mathrm{Re}\lambda_k - \epsilon)t_0} e^{-(\mathrm{Re}\lambda_k - \epsilon)t} e^{(\mathrm{Re}\lambda_k - \epsilon)t}$$

$$\leq \sup_{0 \leq s \leq t_0} \|T(s)\| \sup_{0 \leq s \leq t_0} e^{-(\mathrm{Re}\lambda_k - \epsilon)s} e^{(\mathrm{Re}\lambda_k - \epsilon)t} = C(\epsilon, k)e^{(\mathrm{Re}\lambda_k - \epsilon)t}.$$

This is the desired result.   □

Actually, we can identify a larger class of C_0-semigroups which satisfy the spectrum-determined growth condition. This class is known as the semigroups which are continuous in the uniform operator topology.

It has been shown in Theorem 2.56 that a compact semigroup is continuous in the uniform operator topology. For a C_0-semigroup $T(t)$ which is continuous in the uniform operator topology, the following spectral mapping theorem is stated without proof (see, e.g., [46]).

Theorem 3.13 *Let $T(t)$ be a C_0-semigroup which is continuous for $t > t_0 \geq 0$ in the uniform operator topology on a Banach space with generator A. Then*

$$\sigma(T(t))\backslash\{0\} = e^{t\sigma(A)} \quad \text{for every } t \geq 0.$$

Consequently, the spectrum-determined growth condition holds.

If a C_0-semigroup $T(t)$ is differentiable for $t > t_0$, then in view of Theorem 2.45, $T'(t) = AT(t)$ is a linear bounded operator for every $t > t_0$. For any $t \geq s > t_0$, and every $x \in X$,

$$T(t)x - T(s)x = \int_s^t AT(\tau)x d\tau = \int_s^t T(\tau - s)AT(s)x d\tau.$$

Therefore,

$$\|T(t) - T(s)\| \leq \sup_{0 \leq \tau \leq t-s} \|T(\tau)\| \|AT(s)\| (t - s),$$

which implies that $T(t)$ is continuous in the uniform operator topology for $t > t_0$.

By Theorem 3.13, we have

Corollary 3.14 *For a C_0-semigroup $T(t)$ which is differentiable for $t > t_0 (\geq 0)$, the following holds.*

$$\sigma(T(t))\backslash\{0\} = e^{t\sigma(A)} \quad \text{for every } t \geq 0.$$

Therefore, the spectrum-determined growth condition is satisfied. In particular, analytic semigroups are differentiable, and hence the spectrum-determined growth condition also holds for analytic semigroups.

Remark 3.15 *It is also known that the spectrum-determined growth condition holds for C_0-semigroups generated by Riesz spectral operators. Details can be found in [41].*

Remark 3.16 *In [138], Renardy showed that the spectrum-determined growth condition is generally satisfied, or in other words, the spectrum-determined growth condition holds "most of the time". However, the counter-example of a lower order perturbation to the wave equation in [137] (see also Section 6.7) destroys all hope that practical application problems are well-behaved in such a way that the "pathological" cases where $\omega_0(A) > S(A)$ does not appear.*

3.3   Weak stability and asymptotic stability

This section is devoted to addressing the weak stability, establishing a relationship between the weak and asymptotic stabilities, and characterizing the necessary and sufficient conditions for a C_0−semigroup to be asymptotically stable in terms of the spectrum of its infinitesimal generator.

Definition 3.17 *Let H be a Hilbert space and $L \in \mathcal{L}(H)$. We say that a subspace Y of H reduces L if*

$$LY \subseteq Y \ and \ L^*Y \subseteq Y.$$

Recall that a linear operator L is unitary if and only if L is an isometric-onto-mapping on H.

Theorem 3.18 *[56] Let $T(t)$ be a C_0-semigroup of contractions on a Hilbert space H. Define the weakly stable subspace W of H with respect to $T(t)$ by*

$$W = \{x \in H \mid T(t)x \to 0 (weakly) \ as \ t \to \infty\}.$$

Then

(i) *W reduces $T(s)$ for each $s \geq 0$.*

(ii) *$T(t)$ is reduced to a unitary group on $W^{\perp}$, the orthogonal complement of W in H.*

Proof.   First we note that W is a closed subspace of H.
(i): Let $x \in W$. Then for any $s \geq 0, T(t)T(s)x = T(t+s)x \to 0 (weakly)$ as $t \to \infty$ and so $T(s)x \in W$. Now we prove $T^*(t)W \subseteq W$. Note that $T^*(t)$ is also a C_0-semigroup of contractions. Therefore, for any $x \in H, \|T^*(t)x\|^2$ is nonincreasing with respect to t; hence, converges as $t \to \infty$. Thus, for any fixed s,
$$f(t) = \|T^*(t)x\|^2 - \|T^*(t+s)x\|^2 \to 0 \text{ as } t \to \infty.$$
On the other hand,
$$\begin{aligned} f(t) &= \langle T^*(t)x, \ T^*(t)x\rangle - \langle T(s)T^*(s)T^*(t)x, \ T^*(t)x\rangle \\ &= \langle [I - T(s)T^*(s)]T^*(t)x, \ T^*(t)x\rangle \\ &= \left\| [I - T(s)T^*(s)]^{1/2}T^*(t)x \right\|^2 . \end{aligned}$$
Hence, $[I - T(s)T^*(s)]^{1/2}T^*(t)x \to 0$ as $t \to \infty$. Consequently,
$$[I - T(s)T^*(s)]T^*(t)x \to 0 \ \text{ as } t \to \infty \tag{3.6}$$

for any $x \in H$ and $s \geq 0$. Multiplying $T^*(s)$ on both sides above, we have

$$T^*(s)[I - T(s)T^*(s)]T^*(t)x = [I - T^*(s)T(s)]T^*(t+s)x \to 0, \text{ as } t \to \infty.$$

It follows that
$$[I - T^*(s)T(s)]T^*(t)x \to 0, \text{ as } t \to \infty. \qquad (3.7)$$

Next, we use an apparent fact that if $S(t)$ is a bounded operator such that $S(t)x \to 0$ for any $x \in H$, then $S^*(t)x \to 0(weakly)$ for any $x \in H$. Applying this fact to both (3.6) and (3.7), we obtain

$$T(t)[I - T(s)T^*(s)]x \to 0(weakly), \text{ as } t \to \infty, \qquad (3.8)$$

$$T(t)[I - T^*(s)T(s)]x \to 0(weakly), \text{ as } t \to \infty. \qquad (3.9)$$

Now we are in a position to finish the proof of (i). For any $x \in W$, since $T(t)x \to 0(weakly)$, we have, from (3.8), that

$$-T(t)T(s)T^*(s)x = -T(t+s)T^*(s)x \to 0(weakly), \text{ as } t \to \infty.$$

It follows that

$$T(t)T^*(s)x \to 0(weakly), \text{ as } t \to \infty$$

which shows that $T^*(s)x \in W$ and (i) is proved. Consequently, $W^\perp$ also reduces $T(s)$ for any s.

(ii): Interpreting (3.8) and (3.9) from a different point of view, we have

$$\mathcal{R}(I - T(s)T^*(s)) \subseteq W, \quad \mathcal{R}(I - T^*(s)T(s)) \subseteq W \qquad (3.10)$$

for any $s \geq 0$. Since both $I - T(s)T^*(s)$ and $I - T^*(s)T(s)$ are self-adjoint, we have

$$\mathcal{R}(I - T(s)T^*(s))^\perp = \mathcal{N}(I - T(s)T^*(s)),$$
$$\mathcal{R}(I - T^*(s)T(s))^\perp = \mathcal{N}(I - T^*(s)T(s)).$$

Therefore, (3.10) reads

$$W^\perp \subseteq \mathcal{N}(I - T(s)T^*(s)) \cap \mathcal{N}(I - T^*(s)T(s)),$$

and hence

$$W^\perp \subseteq \bigcap_{s \geq 0} \mathcal{N}(I - T(s)T^*(s)) \cap \mathcal{N}(I - T^*(s)T(s))$$

which implies that for all $x \in W^\perp$ and all $s \geq 0$,

$$T(s)T^*(s)x = T^*(s)T(s)x = x.$$

Therefore, $T(t)$ is reduced to a unitary C_0-group on $W^\perp$. □

For any self-adjoint operator Q in a Hilbert space H, denote by

$$H_{ac}(Q) = \{x \in H \mid x \text{ is absolutely continuous with respect to } Q\}$$

the spectrally absolutely continuous part [84, p.516]. Using Theorem 3.18, we can derive a condition for the weak stability of C_0-semigroups of contractions in Hilbert spaces in terms of the absolutely continuous spectrum of the imaginary component of their infinitesimal generators.

Corollary 3.19 *Let $T(t)$ be a C_0-semigroup of contractions on a Hilbert space H with generator A. Then $T(t)$ is weakly stable if $H_{ac}(A_I) = H$, where $A_I = (A - A^*)/(2i)$ is the imaginary component of A.*

Proof. By Theorem 3.18, $T_\perp(t) = T(t)|_{W^\perp}$ is a unitary C_0-semigroup on $W^\perp$. It follows from Stone's theorem that there is a self-adjoint operator Q in $W^\perp$ such that $T_\perp = e^{iQt}$ and $A|_{W^\perp} = iQ$. Hence $(A|_{W^\perp})^* = A^*|_{W^\perp} = -iQ$, which gives $Q = A_I|_{W^\perp}$. By assumption, we have $H_{ac}(Q) = W^\perp$. Furthermore, using the spectral family E_λ of Q, we can express the self-adjoint operator Q as

$$Qx = \int_{-\infty}^{\infty} \lambda dE_\lambda x, \quad \forall x \in W^\perp.$$

Since the condition $H_{ac}(Q) = W^\perp$ implies that $\langle E_\lambda x, y \rangle$ is absolutely continuous for any $x, y \in W^\perp$ [84, p.519], we have

$$\langle T_\perp(t)x, y \rangle = \int_{-\infty}^{\infty} e^{i\lambda t} d\langle E_\lambda x, y \rangle = \int_{-\infty}^{\infty} e^{i\lambda t} \psi(\lambda) d\lambda$$

for some $\psi \in L^1(-\infty, \infty)$. By the Riemann-Lebesgue lemma, we obtain $\langle T_\perp(t)x, y \rangle \to 0$ as $t \to \infty$ for all $x, y \in W^\perp$, which shows $W^\perp = \{0\}$. Therefore, $T(t)$ is weakly stable. □

Remark 3.20 *The condition $H_{ac}(Q) = H$ for a self-adjoint operator Q in H can be characterized by the resolvent $R(\lambda, Q)$ of Q; see, e.g., [135, pp.137-138]. In Corollary 3.19, since $A|_{W^\perp} = iQ$, and only $H_{ac}(Q) = W^\perp$ is needed in our proof, the condition on $R(\lambda, Q)$ to guarantee $H_{ac}(Q) = W^\perp$ can also be given in terms of $R(\lambda, A)$.*

We now establish a relationship between weak stability and asymptotic stability.

Proposition 3.21 *Let $X = H$ be a Hilbert space. Suppose that $T(t)$ is a weakly stable C_0-semigroup on H, i.e., $\langle T(t)x, y \rangle \to 0$ as $t \to \infty$ for all $x, y \in H$. If its infinitesimal generator A has compact resolvent, then $T(t)$ is asymptotically stable, i.e., $\|T(t)z\| \to 0$ as $t \to \infty$ for all $z \in H$.*

Proof. Since a weakly convergent sequence in a Hilbert space is bounded, for any $y \in H$, there exists a constant $M_0(y) > 0$ such that $\|T(n)y\| \leq M_0(y)$, uniformly for $n = 1, 2, \cdots$. Applying the uniform boundedness theorem, we obtain $\|T(n)\| \leq M_1$ for $n = 1, 2, \cdots$. Since any $t > 0$ can be written as $t = n + r(t)$ for some $0 \leq r(t) < 1$ and $\|T(t)\| \leq M_2 e^{\omega t}$ for some $M_2 > 0, \omega \in \mathbf{R}$, it follows that

$$\|T(t)\| = \|T(n + r(t))\| \leq \|T(n)\| \, \|T(r(t))\| \leq M_1 M_2 \max(1, e^{\omega r(t)}) < \infty,$$

which proves that $T(t)$ is uniformly bounded for any $t \geq 0$.

Since we have assumed that there exists a λ such that $(\lambda - A)^{-1}$ is compact, and since $T(n)y$ is weakly convergent, we have a subsequence n_i of n such that $(\lambda - A)^{-1}T(n_i)y$ strongly converges to zero as $i \to \infty$. The uniform boundedness of $T(t)$ thus shows that $(\lambda - A)^{-1}T(t)y \to 0$ as $t \to \infty$. Now, for any $y \in H$, let $x = (\lambda - A)^{-1}y \in D(A)$. Then

$$T(t)x = T(t)(\lambda - A)^{-1}y = (\lambda - A)^{-1}T(t)y \to 0, \quad \text{as } t \to \infty,$$

which holds for any $x \in D(A)$. Since $D(A)$ is dense in H and $T(t)$ is uniformly bounded, it is easily shown that $T(t)$ is asymptotically stable for any $z \in H$, i.e., $\|T(t)z\| \to 0$ as $t \to \infty$. $\qquad\square$

Since in most practical problems the generators of C_0-semigroups do have compact resolvents, it turns out that the notion of the weak stability is not so attractive due to the above lemma. For this reason, we shall concentrate on asymptotic stability or exponential stability in the sequel. However, it should be recognized that the relationship established in the above lemma is very useful in stability analysis.

We now turn to the study of the asymptotic stability of C_0-semigroups. Let $T(t)$ be a C_0-semigroup generated by operator A in a Banach space X. If $T(t)$ is asymptotically stable, then by the uniform boundedness theorem there is a constant $M \geq 1$ such that $\|T(t)\| \leq M$ for all $t \geq 0$, which in turn implies that $\text{Re}\lambda \leq 0$ for all $\lambda \in \sigma(A)$. To prove the main results of this section, we need the following basic properties of C_0-semigroup of isometrics on Banach spaces.

A C_0-semigroup on a Banach space X is called a C_0-*semigroup of isometrics* if

$$\|T(t)x\| = \|x\|, \quad \text{for all } x \in X \text{ and } t \geq 0.$$

Lemma 3.22 *Let $T(t)$ be a C_0-semigroup of isometrics on a Banach space X with generator A. Then*

(i) If $\text{Re}\lambda < 0$ then

$$\|(\lambda - A)x\| \geq |\text{Re}\lambda| \|x\| \qquad (3.11)$$

for all $x \in D(A)$.

*(ii) If $T(t)$ does not extend to a C_0-group of isometrics on X, then $\lambda \in \sigma(A)$
for all λ with $\mathrm{Re}\lambda \leq 0$ and $\lambda \in \sigma_r(A)$ if $\mathrm{Re}\lambda < 0$.*

*(iii) If $X \neq \{0\}$, and $T(t)$ is a C_0-group of isometrics on X, then $\sigma(A) \cap
i\mathbf{R} \neq \emptyset$.*

Proof. (i): Let $\lambda = -\sigma + i\tau, \sigma > 0$ and let $u(t) = e^{-\lambda t}T(t)x$, $x \in D(A)$.
Obviously,

$$\|u(t)\| = e^{\sigma t}\|x\|. \tag{3.12}$$

On the other hand, since

$$u(t) = x + \int_0^t \frac{du(\tau)}{d\tau}d\tau = x + \int_0^t e^{-\lambda\tau}T(\tau)(Ax - \lambda x)d\tau$$

we have

$$\|u(t)\| \leq \|x\| + \frac{e^{\sigma t} - 1}{\sigma}\|Ax - \lambda x\|. \tag{3.13}$$

Comparing (3.12) and (3.13) gives (3.11).
(ii): By (3.11) and the closedness of operator A, for any λ with $\mathrm{Re}\lambda <
0, \mathcal{R}(\lambda - A)$ is a closed subspace of X. If there is a λ_0 with $\mathrm{Re}\lambda_0 < 0$ such
that $\mathcal{R}(\lambda_0 - A) = X$, then mimicking the proof of Theorem 2.27 we can
show that $\mathcal{R}(\lambda - A) = X$ for all λ with $\mathrm{Re}\lambda < 0$. It follows from (3.11) that
$\lambda \in \rho(A)$ for all λ with $\mathrm{Re}\lambda < 0$ and $\|(\lambda - A)^{-1}\| \leq -1/\mathrm{Re}\lambda$. This shows
that $-A$ is the generator of a C_0-semigroup by the Lümer-Phillips theorem.
Therefore, $T(t)$ extends to a C_0-group of isometrics on X, contradicting the
assumption. Therefore, $\lambda \in \sigma(A)$ for all λ with $\mathrm{Re}\lambda < 0$. Since $\sigma(A)$ is closed,
we have $\{\lambda| \mathrm{Re}\lambda \leq 0\} \subset \sigma(A)$. In particular, the above discussion shows that
$\lambda \in \sigma_r(A)$ if $\mathrm{Re}\lambda < 0$.
(iii): Suppose $\sigma(A) \cap i\mathbf{R} = \emptyset$. Let $f(t) \in L^1(\mathbf{R})$ and let $\hat{f}(\tau)$ be its Fourier
transform given by

$$\hat{f}(\tau) = \frac{1}{\sqrt{2\pi}} \int_{-\infty}^{\infty} f(t)e^{-it\tau}dt.$$

Then by the classical inverse Fourier transform,

$$f(t) = \frac{1}{\sqrt{2\pi}} \int_{-\infty}^{\infty} \hat{f}(\tau)e^{it\tau}d\tau$$

provided that $\hat{f}(\tau) \in L^1(\mathbf{R})$. Let $\Omega = \{f(t) \in L^1(\mathbf{R}) \mid \mathrm{supp}\hat{f}$ is compact in
$\mathbf{R}\}$. Let S be the Schwartz space which consists of all functions $g(t) \in C^\infty(\mathbf{R})$
with

$$\lim_{|t|\to\infty} |t^m g^{(n)}(t)| = 0$$

for all integers $m, n \geq 0$. It is known that Fourier transform is an isomorphism from S to S [122]. Therefore, Ω is dense in S and hence in $L^1(\mathbf{R})$. For any $f \in L^1(\mathbf{R})$, define $\Pi(f) : X \to X$ by

$$\Pi(f)x = \int_{-\infty}^{\infty} f(t)T(t)x\, dt.$$

Then $\Pi(f)$ is a linear bounded operator on X. The proof is finished if we can prove that for any given $x \in X$ and all $f \in L^1(\mathbf{R})$, $\Pi(f)x = 0$. Indeed, take $f_\epsilon(t) = 1/\epsilon$ in $[0, \epsilon]$ and $f_\epsilon(t) = 0$ outside $[0, \epsilon]$. Then

$$0 = \Pi(f_\epsilon)x = \frac{1}{\epsilon}\int_0^{\epsilon} T(t)x\, dt \to T(0)x = x, \quad \text{as } \epsilon \to 0.$$

It follows that $X = \{0\}$, a contradiction.

Now, we show that $\Pi(f)x = 0$ for any given $x \in H$ and all $f \in L^1(\mathbf{R})$. By density argument, we need only to show this for $f \in \Omega$. For $\delta > 0$, define

$$f_\delta^+(t) = \frac{1}{\sqrt{2\pi}}\int_{-\infty}^{\infty} \hat{f}(\tau)e^{it\tau - \delta t}\, d\tau, \quad t > 0,$$

$$f_\delta^-(t) = \frac{1}{\sqrt{2\pi}}\int_{-\infty}^{\infty} \hat{f}(\tau)e^{it\tau + \delta t}\, d\tau, \quad t < 0.$$

Then

$$f_\delta^+(t) - f(t) = f(t)[e^{-\delta t} - 1], \quad t > 0,$$

$$f_\delta^-(t) - f(t) = f(t)[e^{\delta t} - 1], \quad t < 0.$$

By the Lebesgue's dominated convergence theorem,

$$\int_0^{\infty} \mid f_\delta^+(t) - f(t) \mid dt \to 0, \quad \int_{-\infty}^0 \mid f_\delta^-(t) - f(t) \mid dt \to 0,$$

as $\delta \to 0$. Therefore, for any given $\epsilon > 0$, there is a $\delta_0 > 0$ such that

$$\int_0^{\infty} |f_\delta^+(t) - f(t)|dt < \epsilon/2, \quad \int_{-\infty}^0 |f_\delta^-(t) - f(t)|dt < \epsilon/2$$

for all $0 < \delta < \delta_0$. Then

$$\|\Pi(f)x\| = \left\|\int_{-\infty}^{\infty} f(t)T(t)x\, dt\right\|$$

$$\leq \epsilon\|x\| + \left\|\int_0^{\infty} f_\delta^+(t)T(t)x\, dt + \int_{-\infty}^0 f_\delta^-(t)T(t)x\, dt\right\|$$

$$= \epsilon\|x\| + \frac{1}{\sqrt{2\pi}}\left\|\int_{-\infty}^{\infty} \hat{f}(\tau)d\tau\left[\int_0^{\infty} e^{it\tau - \delta t}T(t)x\, dt\right.\right.$$

$$+ \int_{-\infty}^{0} e^{it\tau + \delta t} T(t)x \, dt \Big] \Big\|$$

$$= \epsilon \|x\| + \frac{1}{\sqrt{2\pi}} \left\| \int_{-\infty}^{\infty} \hat{f}(\tau)[R(\delta - i\tau, A) - R(-\delta + i\tau, A)] d\tau x \right\|.$$

Since $\operatorname{supp} \hat{f}$ is compact in $\mathbf{R}$, $R(\delta - i\tau, A)$ and $R(-\delta - i\tau, A)$ are uniformly bounded for all $0 < \delta < \delta_0$ and $\tau \in \operatorname{supp} \hat{f}$. By the Lebesgue's dominated convergence theorem,

$$\left\| \int_{-\infty}^{\infty} \hat{f}(\tau)[R(\delta - i\tau, A) - R(-\delta - i\tau, A)] d\tau \right\| \to 0$$

as $\delta \to 0$. Therefore, $\Pi(f)x = 0$. The proof is complete. $\square$

Remark 3.23 *The conclusion (iii) of Lemma 3.22 holds for any C_0-group $T(t)$ satisfying*

$$\int_{-\infty}^{\infty} \frac{\log \|T(t)\|}{1 + t^2} dt < \infty.$$

The general discussion can be found in [129]

An operator $Q \in \mathcal{L}(X)$ is said to be a *Hermitian operator* if

$$\|e^{i\alpha Q}\| = 1, \text{ for all } \alpha \in \mathbf{R}.$$

Note that this definition is slightly different from those given in many textbooks.

Lemma 3.24 *For any Hermitian operator $Q, \sigma(Q)$ is real and $\|Q\| = r(Q)$.*

Proof.   We follow the proof in [52]. Let $U = e^{iQ}$. Then for any integer n,

$$\|U^n\| = 1,$$

and so $r(U) = 1$. If $|\lambda| < 1$, then a direct computation shows that $\lambda \in \rho(U)$ and

$$(\lambda - U)^{-1} = -\sum_{n=0}^{\infty} \lambda^n U^{-n-1}.$$

Thus, $\sigma(U) \subset \{z | |z| = 1\}$. By the spectral mapping theorem [Theorem 3.5], $\sigma(Q)$ is real. To prove $r(Q) = \|Q\|$, we may assume $r(Q) < \pi/2$, by the positive homogeneity of the spectral radius and the norm. Then $\sigma(Q) \subset (-\pi/2, \pi/2)$. It suffices to show that $\|Q\| \leq \pi/2$. Indeed, if $r(Q) < \|Q\|$, then taking $0 < \delta < \pi/2(\|Q\|/r(Q) - 1), \alpha = (\pi/2 + \delta)/\|Q\|$, we have that $\|\alpha Q\| = \pi/2 + \delta > \pi/2$ but $r(\alpha Q) < \pi/2$. This is a contradiction since αQ

is also a Hermitian operator. Now, for $|t| \le 1$, it is well known that $\arcsin t$ can be expanded as

$$\arcsin t = t + \frac{1}{2}\frac{t^3}{3} + \frac{1 \cdot 3}{2 \cdot 4}\frac{t^5}{5} + \frac{1 \cdot 3 \cdot 5}{2 \cdot 4 \cdot 6}\frac{t^7}{7} + \cdots,$$

which is rewritten as

$$\arcsin t = \sum_{i=1}^{\infty} a_i t^i, \quad |t| \le 1$$

with $a_i \ge 0$ denoting the coefficients of t^i. Using this, we can express t as

$$t = \sum_{i=1}^{\infty} a_i (\sin t)^i.$$

Define

$$F_n(z) = \sum_{i=1}^{n} a_i (\sin z)^i, \; z \in \mathbf{C}.$$

Since $\sigma(Q) \subset (-\pi/2, \pi/2), |\sin \lambda| < 1$ for $\lambda \in \sigma(Q)$. For each $\lambda \in \sigma(Q)$, there is a bounded open ball O_λ of $\mathbf{C}$ centered at λ such that $|\sin z| \le 1$ for all $z \in O_\lambda$. By compactness of $\sigma(Q)$ in $\mathbf{C}$, we can find finite numbers of $O_{\lambda_i}, i = 1, 2, ..., N$ such that $\sigma(Q) \subset O = \cup_{i=1}^{N} O_{\lambda_i}$. Obviously, O is an open bounded subset of $\mathbf{C}$ with $|\sin z| \le 1 (z \in O)$. Furthermore, since each a_i is nonnegative, $a_i |\sin z|^i \le a_i$ on O, and $\sum_{i=1}^{\infty} a_i = \pi/2 < \infty$, the Weierstrass M-test is applicable to show that $\sum_{i=1}^{\infty} a_i (\sin z)^i$ converges uniformly on O, so that the limit function is analytic on O. From functional calculus, we know that

$$Q = \lim_{n \to \infty} F_n(Q) = \sum_{i=1}^{\infty} a_i (\sin Q)^i$$

with the convergence being in the norm of $\mathcal{L}(X)$. By noting that

$$\| \sin Q \| \le \frac{1}{2}\|e^{iQ}\| + \frac{1}{2}\|e^{-iQ}\| = 1,$$

we have

$$\|Q\| \le \sum_{i=1}^{\infty} a_i = \frac{\pi}{2}.$$

The desired result follows. $\qquad\qquad\qquad\qquad\qquad\qquad\qquad\qquad\qquad\qquad$ □

Lemma 3.25 *Let $Q \in \mathcal{L}(X)$ be a Hermitian operator. If $\sigma(Q) = \{\lambda_0\}$ is a single point set, then $Q = \lambda_0 I$.*

Proof. It follows from Lemma 3.24 that λ_0 is real and so $\lambda_0 - Q$ is Hermitian. Since $\sigma(\lambda_0 - Q) = \{0\}$, it follows from Lemma 3.24 that $\|\lambda_0 - Q\| = 0$ and so $Q = \lambda_0 I$. $\square$

Now we prove the main result of this section.

Theorem 3.26 *Let $T(t)$ be a uniformly bounded C_0-semigroup on a Banach space X and let A be its generator. Then*

(i) *If $T(t)$ is asymptotically stable then $\sigma(A) \cap i\mathbf{R} \subset \sigma_c(A)$, the continuous spectrum of A.*

(ii) *If $\sigma(A) \cap i\mathbf{R} \subset \sigma_c(A)$, and $\sigma_c(A)$ is countable, then $T(t)$ is asymptotically stable.*

(iii) *If $R(\lambda, A)$ is compact, then $T(t)$ is asymptotically stable if and only if $\mathrm{Re}\lambda < 0$ for all $\lambda \in \sigma(A)$.*

Proof. We only need to prove (i) and (ii) since (iii) is a direct consequence of (i) and (ii).

(i): Since $T(t)$ is a bounded C_0-semigroup, $\mathrm{Re}\lambda \leq 0$ for all $\lambda \in \sigma(A)$. If $T(t)$ is asymptotically stable, we claim that A has no eigenvalue and residual spectrum on the imaginary axis. Indeed, if $x \neq 0$ such that $Ax = i\beta x, \beta \in \mathbf{R}$ then $T(t)x = e^{i\beta t}x$, which contradicts the assumption that $T(t)$ is asymptotically stable. Suppose $(i\beta - A)^{-1}$ exists for some $\beta \in \mathbf{R}$ and let $S(t) = T(t)e^{-i\beta t}$ be the C_0-semigroup generated by $A - i\beta$. Then for any $x \in X$, it follows from the semigroup property that $\int_0^t S(\tau)x d\tau \in D(A - i\beta) = D(A)$, and

$$(A - i\beta) \int_0^t S(\tau)x d\tau = S(t)x - x,$$

which means $S(t)x - x \in \mathcal{R}(A - i\beta)$. Since $T(t)$ is asymptotically stable, $T(t)x \to 0$ as $t \to \infty$, so does $S(t)x \to 0$ as $t \to \infty$. Thus $S(t)x - x \to -x$ as $t \to \infty$, which shows that $\mathcal{R}(A - i\beta)$ is dense in X. That is, $i\beta$ is not in the residual spectrum of A.

(ii): We may assume, without loss of generality, that $T(t)$ is a semigroup of contractions, i.e., $\|T(t)\| \leq 1$ (Since otherwise we can introduce an equivalent new norm $|x| = \sup_{t \geq 0} \|T(t)x\|$, as we did in the proof of Theorem 2.37, such that $|T(t)| \leq 1$). Thus, for any $x \in X$, $\|T(t)x\|$ is nonincreasing with respect to t. Define a seminorm ℓ on X by

$$\ell(x) = \lim_{t \to \infty} \|T(t)x\|.$$

Let $X_0 = \{x \in X | \ell(x) = 0\}$. Then X_0 is a closed subspace of X by the uniform boundedness of $T(t)$. We show that $X_0 = X$. Let

$$\tilde{X} = X/X_0 = \{[x] \big| [x] = x + X_0\}$$

be the quotient space. A norm $\tilde{\ell}$ in $\hat{X}$ induced by ℓ is given by

$$\tilde{\ell}[x] = \ell(x).$$

And a semigroup $\tilde{T}(t)$ is induced by $T(t)$ in the following way

$$\tilde{T}(t)[x] = [T(t)x].$$

Since $\ell(T(t)x) = \ell(x)$ for every $x \in X$ and all $t \geq 0$, $\tilde{T}(t)$ is an isometric semigroup in $\hat{X}$ with respect to norm $\tilde{\ell}$, namely

$$\tilde{\ell}(\tilde{T}(t)[x]) = \tilde{\ell}([x]) \text{ for all } t \geq 0 \text{ and } [x] \in \hat{X}.$$

Furthermore, by

$$
\begin{aligned}
\tilde{\ell}(\tilde{T}(t)[x] - [x]) &= \tilde{\ell}([T(t)x - x]) = \ell(T(t)x - x) \\
&= \lim_{s \to \infty} \|T(s)[T(t)x - x]\| \leq \|T(t)x - x\| \to 0 \text{ as } t \to 0.
\end{aligned}
$$

we see that $\tilde{T}(t)$ is strongly continuous.

Let X_1 be the completion of $\hat{X}$ with respect to the norm $\tilde{\ell}$. Then the extension of $\tilde{T}(t)$ by continuity produces a C_0-semigroup of isometrics $T_1(t)$. Let A_1 be the generator of $T_1(t)$. We show that

$$\sigma(A_1) \subset \sigma(A).$$

Let $\lambda \in \rho(A)$. Then for every $[x] \in \hat{X}$, since

$$
\begin{aligned}
\tilde{\ell}(R(\lambda, A)[x]) &= \tilde{\ell}([R(\lambda, A)x]) = \ell(R(\lambda, A)x) \\
&= \lim_{t \to \infty} \|R(\lambda, A)T(t)x\| \leq \|R(\lambda, A)\|\ell(x) = \|R(\lambda, A)\|\tilde{\ell}[x],
\end{aligned}
$$

$R(\lambda, A)$ has a natural bounded extension R_λ in X_1. If $\operatorname{Re}\lambda > 0$, since

$$R(\lambda, A)x = \int_0^\infty e^{-\lambda t} T(t)x\, dt$$

for every $x \in X$, we have

$$R_\lambda x_1 = \int_0^\infty e^{-\lambda t} T_1(t)x_1\, dt$$

for every $x \in X_1$. Therefore, $R_\lambda = R(\lambda, A_1)$ for all λ with $\operatorname{Re}\lambda > 0$. By identity

$$R_\mu - R_\lambda = (\lambda - \mu)R_\lambda R_\mu, \quad \lambda, \mu \in \rho(A),$$

we have

$$R_\mu - R(\lambda, A_1) = (\lambda - \mu)R(\lambda, A_1)R_\mu, \text{ for } \operatorname{Re}\lambda > 0, \mu \in \rho(A).$$

Therefore, $\mathcal{R}(R_\mu) \subset D(A_1)$ and

$$(\lambda - A_1)R_\mu = I + (\lambda - \mu)R_\mu,$$

which implies that $(\mu - A_1)R_\mu = I$. Similarly, $R_\mu(\mu - A_1) = I|_{D(A_1)}$. Therefore $\mu \in \rho(A_1)$ or $\rho(A) \subset \rho(A_1)$ and so $\sigma(A_1) \subset \sigma(A)$.

Since $\sigma(A) \cap i\mathbf{R}$ is countable, it follows that $\sigma(A_1) \cap i\mathbf{R}$ is at most countable. Hence, by virtue of (ii) of Lemma 3.22, $T_1(t)$ extends to a C_0-group of isometrics on X_1 and so $\sigma(A_1) \subset i\mathbf{R}$ and $\sigma(A_1)$ is at most a countable closed subset of $i\mathbf{R}$. If $X_1 \neq \{0\}$, then by (iii) of Lemma 3.22 it follows that $\sigma(A_1) \neq \emptyset$. Therefore, $\sigma(A_1)$ contains an isolated point $\lambda_0 = i\omega, \omega \in \mathbf{R}$. Let $\rho > 0$ be sufficiently small such that $C(\rho) = \{\lambda \in \mathbf{C} \mid |\lambda - \lambda_0| \leq \rho\} \cap \sigma(A_1) = \{\lambda_0\}$. Let

$$P_{\lambda_0} = \frac{1}{2\pi i} \int_{|\lambda - \lambda_0| = \rho} R(\lambda, A_1) d\lambda$$

be the eigen-projection. Then $X_1 = Y_1 \oplus Y_2$, where $Y_1 = P_{\lambda_0} X_1, Y_2 = (I - P_{\lambda_0})X_1$, and

$$A_1 P_{\lambda_0} = \frac{1}{2\pi i} \int_{|\lambda - \lambda_0| = \rho} A_1 R(\lambda, A_1) d\lambda = \frac{1}{2\pi i} \int_{|\lambda - \lambda_0| = \rho} \lambda R(\lambda, A_1) d\lambda$$

belongs to $\mathcal{L}(X)$. And also for $\rho_1 < \rho$,

$$
\begin{aligned}
&P_{\lambda_0} A_1 P_{\lambda_0} \\
&= \frac{1}{2\pi i} \int_{|\lambda - \lambda_0| = \rho} R(\lambda, A_1) d\lambda \Big[\frac{1}{2\pi i} \int_{|\mu - \lambda_0| = \rho_1} \mu R(\mu, A_1) d\mu \Big] \\
&= \frac{1}{2\pi i} \int_{|\lambda - \lambda_0| = \rho} d\lambda \Big[\frac{1}{2\pi i} \int_{|\mu - \lambda_0| = \rho_1} \frac{\mu}{\lambda - \mu} [R(\mu, A_1) - R(\lambda, A_1)] d\mu \Big] \\
&= \frac{1}{2\pi i} \int_{|\lambda - \lambda_0| = \rho} d\lambda \Big[\frac{1}{2\pi i} \int_{|\mu - \lambda_0| = \rho_1} \frac{\mu}{\lambda - \mu} R(\mu, A_1) d\mu \Big] \\
&= \frac{1}{2\pi i} \int_{|\mu - \lambda_0| = \rho_1} \mu R(\mu, A_1) d\mu \Big[\frac{1}{2\pi i} \int_{|\lambda - \lambda_0| = \rho} \frac{1}{\lambda - \mu} d\lambda \Big] \\
&= \frac{1}{2\pi i} \int_{|\mu - \lambda_0| = \rho_1} \mu R(\mu, A_1) d\mu = A_1 P_{\lambda_0}.
\end{aligned}
$$

Thus Y_1 is invariant for A_1 and $A_1|_{Y_1} \in \mathcal{L}(Y_1)$. Since $P_{\lambda_0} = P_{\lambda_0}^2$ is commutative with A_1 and with all $T_1(t)$, we see that $T_{\lambda_0}(t) = T_1(t)|_{Y_1}$ is a C_0-group of isometrics on Y_1 with generator $A_1|_{Y_1}$. It follows that $iA_1|_{Y_1}$ is Hermitian. Since $\sigma(iA_1|_{Y_1}) = i\lambda_0$ is a single point, we know by Lemma 3.25 that $A_1|_{Y_1} = \lambda_0 I$ or $A_1 P_{\lambda_0} = \lambda_0 P_{\lambda_0}$. It follows that $A_1^* f_1^* = \overline{\lambda_0} f_1^*$ for all $f_1^* \in Y_1^*$. Since for every $f_1^* \in Y_1^*, f^* = f_1^* \cdot P_{\lambda_0} \in X_1^*$, we have $A_1^* f^* = \overline{\lambda_0} f^*$. So there exists a nonzero $f^* \in X_1^*$ such that $T_1^*(t)f^* = e^{\overline{\lambda_0}t}f^*$. Let $Q : X \to X_1$ be the natural map and $g^* = Q^*f^*$. Then $T^*(t)g^* = e^{\overline{\lambda_0}t}g^*$. Since $\overline{\lambda_0} \in i\mathbf{R}$, by referring to the process of proof of Theorem 3.4 we see that there is a $\hat{\lambda}_0 \in \sigma_p(A^*) \cap i\mathbf{R}$. Therefore, $\hat{\lambda}_0 \in \{\sigma_p(A) \cup \sigma_r(A)\} \cap i\mathbf{R} = \emptyset$, which is a contradiction; hence, $X_1 = \{0\}$. Consequently, $\|T(t)x\| \to 0$ for each x in X. $\square$

Corollary 3.27 *Let $T(t)$ be a C_0-semigroup on a Banach space X and A be its generator. Suppose that $\sigma(A) \cap i\mathbf{R} \subset \sigma_c(A)$ and $\sigma_c(A)$ is countable, then $T(t)$ is weakly stable if and only if $T(t)$ is asymptotically stable.*

Proof. Suppose that $T(t)$ is weakly stable. From the proof of Proposition 3.21 it follows that $T(t)$ is bounded. Theorem 3.26 then implies that $T(t)$ is asymptotically stable. $\square$

3.4 Exponential stability — time domain criteria

In control system analysis, it would be much more convenient if we know that the semigroup generated by the underlying system operator was exponentially stable. It is also highly desirable to design a feedback control such that the closed-loop system generates an exponentially stable semigroup which guarantees the convergence rate of the system solutions.

An important criterion for the exponential stability is the following.

Theorem 3.28 *Let A be the infinitesimal generator of a C_0-semigroup $T(t)$ on a Banach space X. If for some $p \geq 1$,*

$$\int_0^\infty \|T(t)x\|^p dt < \infty \text{ for every } x \in X, \tag{3.14}$$

then $T(t)$ is exponentially stable.

Proof. Define a mapping $S : X \to L^p(0, \infty; X)$ by

$$Sx = T(t)x, \quad \forall t \geq 0.$$

Then $D(S) = X$. It is not difficult to see that S is closed. Indeed, let $x_n \to x$ in X and $\int_0^\infty \|T(t)x_n - y(t)\|^p dt \to 0$. Then for any $a, b < \infty$,

$$\int_a^b \|T(t)x_n - T(t)x\|^p dt \to 0 \text{ and } \int_a^b \|T(t)x_n - y(t)\|^p dt \to 0.$$

Hence, $y(t) = T(t)x$. By the closed graph theorem, S is bounded, i.e., there exists an $M > 0$ such that

$$\int_0^\infty \|T(t)x\|^p dt \le M\|x\|^p.$$

Let $\omega > 0$ be a constant such that $\|T(t)\| \le M_1 e^{\omega t}$ for some $M_1 > 0$. Then it follows from

$$\frac{1 - e^{-p\omega t}}{p\omega}\|T(t)x\|^p = \int_0^t e^{-p\omega(t-s)}\|T(t)x\|^p ds$$

$$\le \int_0^t e^{-p\omega(t-s)}\|T(t-s)\|^p\|T(s)x\|^p ds$$

$$\le M_1^p \int_0^t \|T(s)x\|^p ds \le M_1^p M\|x\|^p$$

that $T(t)$ is uniformly bounded, i.e., $\|T(t)\| \le C$ for some $C > 0$. Therefore,

$$t\|T(t)x\|^p = \int_0^t \|T(t)x\|^p ds \le \int_0^t \|T(t-s)\|^p \|T(s)x\|^p ds$$

$$\le C^p \int_0^t \|T(s)x\|^p ds \le C^p M \|x\|^p.$$

Consequently, $\|T(t)\| \le (C^p M/t)^{1/p}$, and hence there exists a t_0 such that $\|T(t_0)\| < 1$. It follows from Corollary 2.11 that $T(t)$ is exponentially stable. $\square$

Remark 3.29 *We say that $T(t)$ is exponentially asymptotically stable if for every $x \in X$, there exist $M_x, \omega_x > 0$ depending on x such that*

$$\|T(t)x\| \le M_x e^{-\omega_x t}.$$

Theorem 3.28 shows that a linear C_0-semigroup is exponentially asymptotically stable if and only if it is exponentially stable.

There are some alternative criteria for the exponential stability of C_0-semigroups which can be very useful.

Theorem 3.30 *Let $T(t)$ be a C_0-semigroup with infinitesimal generator A. The following statements are equivalent.*

(i) $T(t)$ is exponentially stable, i.e., $\|T(t)\| \le M e^{-\omega t}$, for $M \ge 1$, $\omega > 0$;

(ii) $\lim_{t\to\infty} \|T(t)\| = 0$;

(iii) there exists a $t_0 > 0$ such that

$$\|T(t_0)\| < 1.$$

Proof. It is obvious that (i) implies (ii), and (ii) implies (iii). Moreover, (iii) implies (i) by Corollary 2.11. □

3.5 Exponential stability — frequency domain criteria

The time domain criteria allow us to conclude the exponential stability when either the system operator generates some specific semigroups or the explicit expressions of the semigroups are available. In some cases, however, we may not have any information about the semigroups. Instead, we always know the system operators from which we can estimate the bound of their resolvents. The question is now whether we can derive conditions on resolvents which guarantee the exponential stability of the corresponding semigroups; the answer is yes. We will show, following the ideas in [168], how to do this as follows.

We assume, in this section, that $X = H$ is a Hilbert space with the inner product $\langle \cdot, \cdot \rangle$ and the induced norm $\|\cdot\|$. Recall that if A generates a C_0-semigroup $T(t)$ on H with $\|T(t)\| \leq Me^{\omega t}$, then for all λ with $\mathrm{Re}\lambda > \omega$,

$$R(\lambda, A)x = \int_0^\infty e^{-\lambda t} T(t)x dt \tag{3.15}$$

for every $x \in H$ (Theorem 2.39). Let $\lambda = \sigma + i\tau$. Then (3.15) becomes

$$R(\sigma + i\tau, A)x = \int_0^\infty e^{-i\tau t} e^{-\sigma t} T(t)x dt = \frac{1}{\sqrt{2\pi}} \int_{-\infty}^\infty e^{-i\tau t} f_\sigma(t)x dt \tag{3.16}$$

where

$$f_\sigma(t)x = \begin{cases} \sqrt{2\pi} e^{-\sigma t} T(t)x, & t \geq 0, \\ 0, & t < 0. \end{cases} \tag{3.17}$$

Formally, $R(\sigma+i\tau, A)x$, as an H-valued function of τ, is the Fourier transform of the H-valued function $f_\sigma(t)x$. So that by studying the behavior of the resolvent, we are able to know the stability of the semigroup by the inverse Fourier transform.

Before deriving conditions on $R(\lambda, A)$ to guarantee the exponential stability of $T(t)$, we need some basic results about the Fourier transform of H−valued functions.

For any $p \geq 1$, define

$$L^p(\mathbf{R}; H) = \{f(t) : \mathbf{R} \to H | \int_{-\infty}^\infty \|f(t)\|^p dt < \infty\}, \text{ if } p < \infty;$$

$$L^\infty(\mathbf{R}; H) = \{f(t) : \mathbf{R} \to H | \text{ ess sup} \|f(t)\| < \infty\}.$$

$L^p(\mathbf{R}; H)$ is a Banach space for every $1 \le p \le \infty$. When $p = 2, L^2(\mathbf{R}; H)$ is a Hilbert space with the inner product

$$\langle f, g \rangle = \int_{-\infty}^{\infty} \langle f(t), g(t) \rangle dt.$$

We use $\|f\|_{L^p(\mathbf{R};H)}$ or $\|f\|_{L^p}$ to denote the norm of f in $L^p(\mathbf{R}; H)$ whenever there is no confusion.

For $f \in L^2(\mathbf{R}; H) \cap L^1(\mathbf{R}; H)$, denote the Fourier transform of f by

$$\mathcal{F}f(\tau) = \hat{f}(\tau) = \frac{1}{\sqrt{2\pi}} \int_{-\infty}^{\infty} f(t)e^{-it\tau} dt. \tag{3.18}$$

It is easily seen that for $f \in L^2(\mathbf{R}; H) \cap L^1(\mathbf{R}; H)$,

$$\|\hat{f}(\tau)\| \le \frac{1}{\sqrt{2\pi}} \int_{-\infty}^{\infty} \|f(t)\| dt = \frac{1}{\sqrt{2\pi}} \|f\|_{L^1},$$

hence $\hat{f} \in L^{\infty}(-\infty, \infty)$. Furthermore, since

$$\|\hat{f}(\tau + h) - \hat{f}(\tau)\| \le \frac{1}{\sqrt{2\pi}} \int_{-\infty}^{\infty} |e^{iht} - 1| \|f(t)\| dt$$

is independent of τ, it follows from the Lebesgue's dominated convergence theorem that $\hat{f}(\tau)$ is uniformly continuous with respect to τ.

Now we state a result in classical Fourier analysis.

Theorem 3.31 *Let $H = \mathbf{R}$. Then the Fourier transform operator $\mathcal{F}$ defined by (3.18) can be extended continuously to a unitary operator on the entire space $L^2(-\infty, \infty)$. More precisely, for every $f(t) \in L^2(-\infty, \infty)$, there exists an $\hat{f}(\tau) \in L^2(-\infty, \infty)$ such that*

$$\left\| \frac{1}{\sqrt{2\pi}} \int_a^b f(t)e^{-it\tau} dt - \hat{f}(\tau) \right\|_{L^2} \to 0,$$

$$\left\| \frac{1}{\sqrt{2\pi}} \int_a^b \hat{f}(\tau)e^{it\tau} d\tau - f(t) \right\|_{L^2} \to 0,$$

as $a \to -\infty, b \to \infty$, and

$$\|f\|_{L^2} = \|\hat{f}\|_{L^2}.$$

Furthermore, if $f, \hat{f} \in L^1(-\infty, \infty)$ then

$$\frac{1}{\sqrt{2\pi}} \int_{-\infty}^{\infty} \hat{f}(\tau)e^{it\tau} d\tau = f(t) \quad \text{almost everywhere.}$$

Proof.   See [54].   □

Generalizing the above result to H-valued functions, we have

Theorem 3.32 *The Fourier transform $\mathcal{F}$ defined by (3.18) can be extended continuously to a unitary operator on $L^2(\mathbf{R}; H)$.*

To prove this result, we need the following classical lemma which can be found in [54].

Lemma 3.33 *Let $f \in L^1(-\infty, \infty) \cap L^\infty(-\infty, \infty)$ and $\hat{f}(\tau) \geq 0$ for all $\tau \geq 0$. Then $\hat{f} \in L^1(-\infty, \infty)$.*

Lemma 3.34 *For any $f \in L^1(\mathbf{R}; H) \cap L^\infty(\mathbf{R}; H), \|f\|_{L^2(\mathbf{R};H)} = \|\hat{f}\|_{L^2(\mathbf{R};H)}$.*

Proof.   Let $f \in L^1(\mathbf{R}; H) \cap L^\infty(\mathbf{R}; H)$ and define

$$h_f(t) = \int_{-\infty}^{\infty} \langle f(t-s), \; f(-s) \rangle ds, \forall t \in \mathbf{R}.$$

We claim that $h_f(t) \in L^1(-\infty, \infty) \cap L^\infty(-\infty, \infty)$. Indeed, $|h_f(t)| \leq \|f\|_\infty \|f\|_{L^1}$ and so $h_f(t) \in L^\infty(-\infty, \infty)$. Meanwhile,

$$\int_{-\infty}^{\infty} |h_f(t)| dt \leq \int_{-\infty}^{\infty} \int_{-\infty}^{\infty} \|f(t-s)\| \|f(-s)\| ds dt = \|f\|_{L^2(\mathbf{R};H)}^2$$

implies that $h_f \in L^1(-\infty, \infty)$. In addition, by Theorem 3.31,

$$
\begin{aligned}
\hat{h}_f(\tau) &= \frac{1}{\sqrt{2\pi}} \int_{-\infty}^{\infty} h_f(t) e^{-it\tau} dt \\
&= \frac{1}{\sqrt{2\pi}} \int_{-\infty}^{\infty} e^{-it\tau} \int_{-\infty}^{\infty} \langle f(t-s), \; f(-s) \rangle ds dt. \\
&= \int_{-\infty}^{\infty} \langle \hat{f}(\tau), \; e^{is\tau} f(-s) \rangle ds = \sqrt{2\pi} \|\hat{f}(\tau)\|^2 \geq 0, \forall \tau \geq 0,
\end{aligned}
$$

it follows from Lemma 3.33 that $\hat{h}_f \in L^1(-\infty, \infty)$ and so $\hat{f}(\tau) \in L^2(\mathbf{R}; H)$. Since $\hat{h}_f(\tau)$ is continuous in τ, by the inversion of Fourier transform, we have

$$\frac{1}{\sqrt{2\pi}} \int_{-\infty}^{\infty} \hat{h}_f(\tau) e^{it\tau} d\tau = h_f(t).$$

Letting in particular $t = 0$ we have

$$\frac{1}{\sqrt{2\pi}} \int_{-\infty}^{\infty} \hat{h}_f(\tau) d\tau = h_f(0).$$

Therefore, $\|f\|_{L^2(\mathbf{R};H)} = \|\hat{f}\|_{L^2(\mathbf{R};H)}$. □

Proof of Theorem 3.32. Let $S = \{f(t) \in C^\infty(\mathbf{R};H) \mid \|t^m f^{(n)}(t)\| \to 0$ as $|t| \to \infty\}$ be the H-valued Schwartz space, where $f^{(n)}(t)$ denotes the n-th Frechét derivative of f with respect to t. Clearly, $S \subset L^1(\mathbf{R};H) \cap L^\infty(\mathbf{R};H)$, and hence by Lemma 3.34,

$$\|f\|_{L^2(\mathbf{R};H)} = \|\hat{f}\|_{L^2(\mathbf{R};H)} \qquad \text{for all } f \in S.$$

Let $f \in S$. We show that $\hat{f} \in S$. Indeed, by

$$\hat{f}(\tau) = \frac{1}{\sqrt{2\pi}} \int_{-\infty}^\infty e^{-it\tau} f(t) dt,$$

we have

$$\begin{aligned}
\hat{f}^{(n)}(\tau) &= \frac{1}{\sqrt{2\pi}} \int_{-\infty}^\infty e^{-it\tau}(-it)^n f(t) dt \\
&= \frac{1}{\sqrt{2\pi}} \int_{-\infty}^\infty e^{-it\tau} g(t) dt,
\end{aligned}$$

where $g(t) = (-it)^n f(t) \in S$ by assumption. By induction, we can easily show that

$$\begin{aligned}
(-i\tau)^m \hat{f}^{(n)}(\tau) &= \frac{1}{\sqrt{2\pi}} \int_{-\infty}^\infty e^{-it\tau} g^{(m)}(t) dt \\
&= -\frac{1}{i\tau} \frac{1}{\sqrt{2\pi}} \int_{-\infty}^\infty e^{-it\tau} g^{(m+1)}(t) dt.
\end{aligned}$$

Therefore,

$$\left\| \tau^m \hat{f}^{(n)}(\tau) \right\| \le \frac{1}{|\tau|} \frac{1}{\sqrt{2\pi}} \left\| g^{(m+1)} \right\|_{L^1} \to 0 \quad \text{as } |\tau| \to \infty$$

for any integers n and m. Thus $\hat{f} \in S$.

Conversely, for any $\hat{f} \in S$, define

$$f(t) = \frac{1}{\sqrt{2\pi}} \int_{-\infty}^\infty \hat{f}(\tau) e^{it\tau} d\tau.$$

Similarly, we can show that $f \in S$. Repeating the proof procedure in Lemma 3.34, one has

$$\|f\|_{L^2(\mathbf{R};H)} = \|\hat{f}\|_{L^2(\mathbf{R};H)},$$

and

$$\hat{f}(\tau) = \frac{1}{\sqrt{2\pi}} \int_{-\infty}^\infty f(t) e^{-it\tau} dt.$$

Hence, $\mathcal{F}$ is an isomorphism from S to S. Since S is dense in $L^2(\mathbf{R}; H)$, we first extend $\mathcal{F}$ to an isometric operator on $L^2(\mathbf{R}; H)$ and then by

$$L^2(\mathbf{R}; H) = \overline{S} \subset \overline{\mathcal{R}(\mathcal{F})} = \mathcal{R}(\mathcal{F}) \subset L^2(\mathbf{R}; H),$$

we obtain $\mathcal{R}(\mathcal{F}) = L^2(\mathbf{R}; H)$. Consequently, $\mathcal{F}$ is unitary on $L^2(\mathbf{R}; H)$. □

With these preparations, we can prove the following theorem, which allows us to conclude the exponential stability of C_0-semigroups in terms of the behavior of the resolvents of the generators.

Theorem 3.35 *Let $T(t)$ be a C_0-semigroup on a Hilbert space H with generator A. Then $T(t)$ is exponentially stable if and only if $\{\lambda \mid Re\lambda \geq 0\} \subset \rho(A)$ and*

$$\|R(\lambda, A)\| \leq M$$

for all λ with $Re\lambda \geq 0$ and some constant $M > 0$.

Proof. The necessity is trivial by the Hille-Yosida theorem. We prove the sufficiency. Let ω_0 be the growth rate of $T(t)$. Then (3.16) and (3.17) hold for all $\sigma > \omega_0$ and any $x \in H$. Since for every $\sigma_0 > \omega_0$, $f_{\sigma_0}x$, defined by (3.17), belongs to $L^2(\mathbf{R}; H)$, it follows from Theorem 3.32 that $R(\sigma_0 + i\cdot, A)x$, as the Laplace transform of $f_{\sigma_0}x$, belongs to $\in L^2(\mathbf{R}; H)$ and

$$\|f_{\sigma_0}x\|_{L^2} = \|R(\sigma_0 + i\cdot, A)x\|_{L^2}. \tag{3.19}$$

By the resolvent identity,

$$R(\sigma + i\tau, A) = R(\sigma_0 + i\tau, A) + (\sigma_0 - \sigma)R(\sigma + i\tau, A)R(\sigma_0 + i\tau, A)$$

and the assumption, we know that $R(\sigma + i\cdot, A)x \in L^2(\mathbf{R}; H)$ for every $\sigma \geq 0$. It follows from Theorem 3.32 that there exists a $g_\sigma(t)x \in L^2(\mathbf{R}; H)$ such that

$$R(\sigma + i\tau, A)x = \frac{1}{\sqrt{2\pi}} \int_{-\infty}^{\infty} e^{-i\tau t} g_\sigma(t)x dt = \hat{g}_\sigma(\tau)x.$$

Suppose that $\omega_0 \geq 0$. Then we have in particular

$$R(\omega_0 + i\tau, A)x = \frac{1}{\sqrt{2\pi}} \int_{-\infty}^{\infty} e^{-i\tau t} g_{\omega_0}(t)x dt.$$

But $R(\sigma + i\tau, A)x = \frac{1}{\sqrt{2\pi}} \int_{-\infty}^{\infty} e^{-i\tau t} f_\sigma(t)x dt$ for every $\sigma > \omega_0$. It follows from the Lebesgue's dominated convergence theorem that

$$\|R(\sigma + i\tau, A)x - R(\omega_0 + i\tau, A)x\|_{L^2} \to 0,$$

as $\sigma \to \omega_0$; hence, by (3.19)

$$\|f_\sigma(t)x - g_{\omega_0}(t)x\|_{L^2} \to 0,$$

as $\sigma \to \omega_0$, which implies that

$$f_\sigma(t)x - g_{\omega_0}(t)x \to 0,$$

as $\sigma \to \omega_0$ for almost all t. On the other hand, as $\sigma \to \omega_0$,

$$f_\sigma(t)x - f_{\omega_0}(t)x \to 0$$

for all t, we have

$$g_{\omega_0}(t) = f_{\omega_0}(t).$$

Now $f_{\omega_0}(t)x \in L^2(\mathbf{R}; H)$ means that

$$\int_0^\infty \|e^{-\omega_0 t}T(t)x\|^2 dt < \infty$$

for all $x \in H$. It follows from Theorem 3.28 that there exist $M > 0, \mu > 0$ such that

$$e^{-\omega_0 t}\|T(t)\| \le Me^{-\mu t},$$

that is, $\|T(t)\| \le Me^{(\omega_0 - \mu)t}$, contradicting the definition of ω_0. Therefore, $\omega_0 < 0$ which shows that $T(t)$ is exponentially stable. $\qquad\square$

Corollary 3.36 *Let $T(t)$ be a uniformly bounded C_0-semigroup on a Hilbert space H with generator A. Then $T(t)$ is exponentially stable if and only if $i\mathbf{R} \subset \rho(A)$ and*

$$M_0 := \sup_{\tau \in \mathbf{R}} \|R(i\tau, A)\| < \infty.$$

Proof. The necessity is a direct consequence of Theorem 3.35. We show the sufficiency. By assumption, ω_0 is nonpositive. According to the Hille-Yosida theorem, there exists a constant $M > 0$ such that for all $\sigma > 0$ and $\tau \in \mathbf{R}$,

$$\|R(\sigma + i\tau, A)\| \le \frac{M}{\sigma}.$$

This together with the resolvent identity

$$R(\sigma + i\tau, A) = R(i\tau, A) - \sigma R(\sigma + i\tau, A)R(i\tau, A)$$

gives

$$
\begin{aligned}
\|R(\sigma + i\tau, A)\| &\le \|R(i\tau, A)\| + \|\sigma R(\sigma + i\tau, A)\|\|R(i\tau, A)\| \\
&\le (1 + M)\|R(i\tau, A)\| \\
&\le (1 + M)\sup_{\tau \in \mathbf{R}} \|R(i\tau, A)\| \\
&= (1 + M)M_0.
\end{aligned}
$$

By virtue of Theorem 3.35, $T(t)$ is exponentially stable.   □

This is the so-called frequency domain test method for the exponential stability of C_0-semigroups on Hilbert spaces. It has proved to be a very effective method even for problems where other test methods, such as the energy multiplier method, do not work.

In what follows, we use a similar idea as above to prove the well-known Paley-Wiener theorem in Fourier analysis for H-valued functions in $L^2(\mathbf{R}; H)$, and to characterize the growth rate ω_0 of C_0-semigroups on Hilbert spaces.

For any $\beta \in \mathbf{R}$, define

$$\Im^2(\beta) = \Big\{ g(\lambda) : \mathbf{R} \to H \mid g(\lambda) \text{ is analytic in } \mathrm{Re}\lambda > \beta \text{ and}$$

$$\sup_{\sigma > \beta} \int_{-\infty}^{\infty} \|g(\sigma + i\tau)\|^2 d\tau < \infty \Big\}. \tag{3.20}$$

Obviously, $f(\lambda) = g(\lambda + \beta)$ is analytic in $\mathrm{Re}\lambda > 0$ for every $g(\lambda) \in \Im^2(\beta)$.

Theorem 3.37 (Paley-Wiener theorem in $L^2(\mathbf{R}; H)$) *[162] Let $\Im^2(\beta)$ be defined as in (3.20). Then $g(\lambda) \in \Im^2(\beta)$ if and only if there exists a $G(t) \in L^2(\mathbf{R}; H)$ such that*

$$g(\lambda) = \int_0^{\infty} e^{-(\lambda-\beta)t} G(t) dt$$

for all λ with $\mathrm{Re}\lambda > \beta$ where the integral is taken in the sense of Bochner integration.

Proof. The range of $g(\lambda)$ is contained in a subspace H_0 of H with a countable orthonormal basis $\{x_k\}_1^{\infty}$ because all λ with $\mathrm{Re}\lambda > \beta$ can be approximated by rational numbers and hence $g(\lambda)$ is separable. Let $g(\lambda) = \sum_{k=1}^{\infty} g_k(\lambda) x_k$ with $g_k(\lambda) = \langle g(\lambda), x_k \rangle$ analytic in $\mathrm{Re}\lambda > \beta$. Then

$$\|g(\lambda)\|^2 = \sum_{k=1}^{\infty} |g_k(\lambda)|^2.$$

By the classical Paley-Wiener theorem in Fourier analysis (see e.g. [54]), each $g_k(\lambda)$ is the Laplace transform of a function $G_k(t) \in L^2(-\infty, \infty)$, i.e.,

$$g_k(\lambda) = \int_0^{\infty} e^{-(\lambda-\beta)t} G_k(t) dt = \int_0^{\infty} e^{-i\tau t} e^{-(\sigma-\beta)t} G_k(t) dt.$$

Hence,

$$g(\lambda) = \int_0^{\infty} e^{-(\lambda-\beta)t} \sum_{k=1}^{\infty} G_k(t) x_k dt = \int_0^{\infty} e^{-(\lambda-\beta)t} G(t) dt$$

where $G(t) = \sum_{k=1}^{\infty} G_k(t)x_k$. The series is convergent in $L^2(\mathbf{R}; H_0)$ by the orthogonality of $\{x_k\}$ and the isometric property of Fourier transform in $L^2(-\infty, \infty)$,

$$\int_0^{\infty} e^{-2(\sigma-\beta)t} \sum_{k=1}^{\infty} |G_k(t)|^2 dt = \int_{-\infty}^{\infty} \sum_{k=1}^{\infty} |g_k(\sigma+i\tau)|^2 d\tau = \int_{-\infty}^{\infty} \|g(\sigma+i\tau)\|^2 d\tau.$$

Applying Fatou's lemma, we have

$$\int_0^{\infty} \sum_{k=1}^{\infty} |G_k(t)|^2 dt \leq \inf_{\sigma \to \beta} \lim \int_0^{\infty} e^{-2(\sigma-\beta)t} \sum_{k=1}^{\infty} |G_k(t)|^2 dt$$

$$\leq \inf_{\sigma \to \beta} \lim \int_{-\infty}^{\infty} \|g(\sigma+i\tau)\|^2 d\tau$$

$$\leq \sup_{\sigma > \beta} \int_{-\infty}^{\infty} \|g(\sigma+i\tau)\|^2 d\tau < \infty.$$

Thus, $G(t) \in L^2(\mathbf{R}; H)$. The proof is complete.   □

As an application of Theorem 3.37, we discuss characterizations of the growth rate of C_0-semigroups on Hilbert spaces.

Theorem 3.38 *Let ω_0 be the growth rate of a C_0-semigroup $T(t)$ on a Hilbert space H. Let A be its generator. Then*

$$\omega_0 = \inf\{\omega > S(A) | \ R(\lambda, A)x \in \mathfrak{I}^2(\omega) \ \text{for any } x \in H\}.$$

Proof. For any $\omega > \omega_0$, take σ_0 such that $\omega > \sigma_0 > \omega_0$. Then there is an $M_0 > 0$ such that

$$\|T(t)\| \leq M_0 e^{\sigma_0 t}.$$

By (3.16) and (3.17) for all $\lambda = \sigma + i\tau$ with $\mathrm{Re}\lambda = \sigma > \omega$ and all $x \in H$,

$$R(\lambda, A)x = R(\sigma + i\tau, A)x = \frac{1}{\sqrt{2\pi}} \int_{-\infty}^{\infty} e^{-i\tau t} f_\sigma(t) x dt$$

where $f_\sigma(t)x$ is as defined in (3.17). We have seen before that $f_\sigma(t)x \in L^2(\mathbf{R}; H)$ for all $x \in H$ and $\sigma > \omega$. Since

$$\int_{-\infty}^{\infty} \|R(\sigma + i\tau, A)x\|^2 d\tau = 2\pi \int_0^{\infty} e^{-2\sigma t} \|T(t)x\|^2 dt$$

$$\leq 2\pi \int_0^{\infty} e^{-2\omega t} M_0{}^2 e^{2\sigma_0 t} \|x\|^2 dt < \infty,$$

it follows that $R(\lambda, A)x \in \Im^2(\omega)$ which means that

$$\inf\{\omega > S(A)|\ R(\lambda, A)x \in \Im^2(\omega)\ \text{for any}\ x \in H\} \leq \omega_0. \qquad (3.21)$$

Conversely, for any $\omega > S(A)$ such that $R(\lambda, A)x \in \Im^2(\omega)$ for any $x \in H$, it follows from Theorem 3.37 that there is a $G_x(t) \in L^2(0, \infty; H)$ such that

$$R(\lambda, A)x = \int_0^\infty e^{-(\lambda-\omega)t} G_x(t) dt$$

for all λ with $\text{Re}\lambda > \omega$ and $x \in H$. Since for $\text{Re}\lambda > \omega_0$,

$$R(\lambda, A)x = \int_0^\infty e^{-\lambda t} T(t)x dt,$$

we have $e^{\omega t} G_x(t) = T(t)x$ for all $t \geq 0$ from the uniqueness of Laplace transforms. Therefore,

$$\int_0^\infty e^{-2\omega t} \|T(t)x\|^2 dt = \int_0^\infty \|G_x(t)\|^2 dt < \infty.$$

By Theorem 3.28, there exist $M, \mu > 0$ such that

$$e^{-\omega t} \|T(t)\| \leq M e^{-\mu t}.$$

Thus, $\omega > \omega_0$ which implies that

$$\inf\{\omega > S(A)|\ R(\lambda, A) \in \Im^2(\omega)\ \text{for any}\ x \in H\} \geq \omega_0. \qquad (3.22)$$

The desired result then follows from (3.21) and (3.22).   □

Theorem 3.39 *Let ω_0 be the growth rate of a C_0-semigroup $T(t)$ on a Hilbert space H. Let A be its generator. Then*

$$\omega_0 = \inf\{\omega > S(A) \mid \exists M_\omega\ \text{s.t.} \sup_{\tau \in \mathbf{R}, \sigma \geq \omega} \|R(\sigma + i\tau, A)\| < M_\omega\}.$$

Proof. Let

$$r_0 = \inf\{\omega > S(A) \mid \exists M_\omega\ \text{s.t.} \sup_{\tau \in \mathbf{R}, \sigma \geq \omega} \|R(\sigma + i\tau, A)\| < M_\omega\}.$$

For any $\omega > \omega_0$, take σ_0 so that $\omega > \sigma_0 > \omega_0$. Then by the Hille-Yosida theorem, $\sigma_0 > S(A)$, and there exists an $M > 0$ such that

$$\|R(\sigma + i\tau, A)\| \leq \frac{M}{\sigma - \sigma_0}$$

for all $\sigma > \sigma_0$ and $\tau \in \mathbf{R}$. Hence,

$$\sup_{\tau \in \mathbf{R}, \sigma \geq \omega} \|R(\sigma + i\tau, A)\| \leq \frac{M}{\omega - \sigma_0}.$$

Consequently, $\omega \geq r_0$; hence, $\omega_0 \geq r_0$. Conversely, for any $\omega > r_0$, by definition

$$\sup_{\tau \in \mathbf{R}, \sigma \geq \omega} \|R(\sigma + i\tau, A)\| < M_\omega.$$

Choose $\omega_1 > \max\{\omega, \omega_0\}$. Then $R(\omega_1 + i\tau, A)x \in L^2(\mathbf{R}; H)$ for any $x \in H$. For any $\omega < \sigma \leq \omega_1$, from the resolvent identity,

$$R(\sigma + i\tau, A)x = R(\omega_1 + i\tau, A)x + (\omega_1 - \sigma)R(\omega_1 + i\tau, A)R(\sigma + i\tau, A)x$$

for all $x \in H$, we have

$$\|R(\sigma + i\tau, A)x\| \leq [1 + |\omega_1 - \sigma|M_\omega]\|R(\omega_1 + i\tau, A)x\|, \ \forall \omega_1 \geq \sigma > \omega.$$

Hence, $R(\sigma + i\tau, A)x \in \Im^2(\omega)$ for any $\sigma > \omega$ and all $x \in H$. It follows from Theorem 3.38 that $\omega \geq \omega_0$. Therefore, $r_0 \geq \omega_0$ and so $r_0 = \omega_0$.   □

Corollary 3.40 Let ω_0 be the growth rate of a C_0-semigroup $T(t)$ on a Hilbert space H. Let A be its generator. Then

$$\omega_0 = \inf\{\omega > S(A)| \sup_{\tau \in \mathbf{R}} \|R(\sigma + i\tau, A)\| < M_\sigma < \infty, \forall \sigma \geq \omega\}.$$

Proof. It is sufficient to show, under the assumption, that M_σ can be chosen to be independent of σ. Let $\omega > S(A)$ and $\sigma \geq \omega$ such that $\sup_{\tau \in \mathbf{R}} \|R(\sigma + i\tau, A)\| < M_\sigma < \infty$. For $\sigma \geq \omega$, define

$$f(\sigma) = \sup_{\tau \in \mathbf{R}} \|R(\sigma + i\tau, A)\|.$$

Choose $\omega_1 > \max\{\omega, \omega_0 + \epsilon\}$ where $\epsilon > 0$ is a small number. Then by the Hille-Yosida theorem, there exists an $M > 0$ such that

$$f(\sigma) \leq \frac{M}{\sigma - \omega_0 - \epsilon} \leq \frac{M}{\omega_1 - \omega_0 - \epsilon}, \text{ for all } \sigma \geq \omega_1.$$

We need only to prove

$$f(\sigma) \leq M \text{ for all } \sigma \in [\omega, \omega_1],$$

or sufficiently, $f(\sigma)$ is continuous in $[\omega, \omega_1]$. For any $\omega_2 \in [\omega, \omega_1]$, by the resolvent identity

$$R(\sigma + i\tau, A) = R(\omega_2 + i\tau, A) + (\omega_2 - \sigma)R(\omega_2 + i\tau, A)R(\sigma + i\tau, A),$$

we have

$$\|R(\sigma + i\tau, A)\| \le f(\omega_2) + |\omega_2 - \sigma| f(\omega_2) f(\sigma),$$

which in turn implies that

$$f(\sigma) \le f(\omega_2) + |\omega_2 - \sigma| f(\omega_2) f(\sigma).$$

Similarly, we can obtain

$$f(\omega_2) \le f(\sigma) + |\omega_2 - \sigma| f(\omega_2) f(\sigma).$$

Therefore,

$$|f(\omega_2) - f(\sigma)| \le |\omega_2 - \sigma| f(\omega_2) f(\sigma).$$

If in a neighborhood of ω_2, $f(\sigma)$ is not bounded, then there exists a sequence $\sigma_n \to \omega_2$ such that $f(\sigma_n) \to \infty$, and hence

$$|f(\omega_2)/f(\sigma_n) - 1| \le |\omega_2 - \sigma_n| f(\omega_2) \to 0$$

which is a contradiction. Hence $f(\sigma)$ is bounded in a neighborhood of ω_2, and is therefore continuous. Thus we have proved that

$$\sup_{\sigma \ge \omega} f(\sigma) = \sup_{\tau \in \mathbf{R}, \sigma \ge \omega} \|R(\sigma + i\tau, A)\| < \infty.$$

Applying Theorem 3.39 completes the proof. □

Remark 3.41 *It should be understood that the conclusions drawn in this section are not generally true on Banach spaces. Example 3.3 is one of such examples.*

3.6 Essential spectrum and compact perturbations

From the previous sections, we have seen that there is a gap between the growth rate ω_0 of a C_0-semigroup $T(t)$ and its spectral bound $S(A)$. The trouble arises when the spectral mapping relation $e^{\sigma(A)t} = \sigma(T(t))\backslash\{0\}$ does not hold. In this section, we shall show that it is those non-isolated spectra and the isolated eigenvalues with infinite algebraic multiplicity of $T(t)$, in addition to $S(A)$, that determine the growth rate ω_0. This part of the spectrum is called the *essential spectrum*, which usually arises from the spectral mapping $\lambda \in \sigma(A) \to e^{\lambda t} \in \sigma(T(t))$, even if $\sigma(A)$ consists only of isolated eigenvalues with finite algebraic multiplicity of A.

Recall that, for a linear closed operator L in a Banach space X and for each isolated point λ_0 in the spectrum $\sigma(L)$, the resolvent $R(\lambda, L)$ can be expanded in a Laurent series about λ_0 as follows:

$$R(\lambda, L) = \sum_{n=0}^{\infty}(\lambda - \lambda_0)^n L_n + \sum_{n=1}^{\infty}(\lambda - \lambda_0)^{-n} B_n, \qquad (3.23)$$

where for each n,

$$L_n = \frac{1}{2\pi i}\int_{\Gamma}(\lambda - \lambda_0)^{-n-1} R(\lambda, L)d\lambda,$$

$$B_n = \frac{1}{2\pi i}\int_{\Gamma}(\lambda - \lambda_0)^{n-1} R(\lambda, L)d\lambda \qquad (3.24)$$

are bounded linear operators on X, and Γ is a positively oriented circle of sufficiently small radius such that no other points of $\sigma(A)$ except λ_0 lie on or inside Γ. From (3.24), we have

$$(L - \lambda_0)L_0 = B_1 - I, \ (L - \lambda_0)^n L_n = L_0, \ n \geq 1,$$
$$B_{n+1} = (L - \lambda_0)^n B_1, \ B_{n+1}B_{m+1} = B_{n+m+1}, \ n, m \geq 0. \quad (3.25)$$

If there is a $p > 0$ such that $B_p \neq 0$ (consequently, $B_n \neq 0$ for all $n \leq p$), while $B_n = 0$ for all $n > p$, then the point λ_0 is called a pole of $R(\lambda, L)$ of order p. In view of (3.25), this is true if $B_p \neq 0$ and $B_{p+1} = 0$. In this case,

$$B_p = \lim_{\lambda \to \lambda_0}(\lambda - \lambda_0)^p R(\lambda, L).$$

In particular, B_1 is called the residue of $R(\lambda, L)$ at λ_0. Our next result is stated without proof (see [35] and Theorem 9.2 in [151]).

Theorem 3.42 *Let λ_0 be an isolated spectral point of the closed operator L in a Banach space X. Then*

(i). B_1 is a projection on X (i.e. $B_1^2 = B_1$), $\mathcal{R}(B_1)$ and $\mathcal{R}(I - B_1)$ are closed, and the restriction of L to $\mathcal{R}(B_1)$ is bounded and has spectrum $\{\lambda_0\}$.

(ii). If $\mathcal{R}(B_1)$ has finite dimension, then λ_0 is a pole of $R(\lambda, L)$.

(iii). If $\mathcal{R}(B_1)$ has finite dimension, then $\mathcal{R}(\lambda_0 - L)$ is closed.

(iv). If λ_0 is a pole of $R(\lambda, L)$ of order $p < \infty$, then λ_0 is an eigenvalue of L,

$$\mathcal{R}(B_1) = \mathcal{N}((\lambda_0 - L)^p) = \mathcal{N}((\lambda_0 - L)^{p+1}) = \ldots,$$
$$\mathcal{R}(I - B_1) = \mathcal{R}((\lambda_0 - L)^p) = \mathcal{R}((\lambda_0 - L)^{p+1}) = \ldots,$$

and

$$X = \mathcal{N}((\lambda_0 - L)^p) \oplus \mathcal{R}((\lambda_0 - L)^p).$$

We call $m_a = \dim \mathcal{R}(B_1)$ the algebraic multiplicity of λ_0, and $m_g = \dim \mathcal{N}(\lambda_0 - L)$ the geometric multiplicity of λ_0.

Theorem 3.43 *[35] Let λ_0 be an isolated spectral point of the closed operator L in a Banach space X. If $\mathcal{R}(\lambda_0 - L)$ is closed in X, then*

$$m_a = \dim M_{\lambda_0}, \tag{3.26}$$

where

$$M_{\lambda_0} = \bigcup_{n=1}^{\infty} \mathcal{N}((\lambda_0 - L)^n) \tag{3.27}$$

is called the generalized eigenspace of L associated with λ_0.

Proof. If $\dim \mathcal{R}(B_1) < \infty$, then from Theorem 3.42, λ_0 is a pole of $R(\lambda, L)$ of order p and so $\mathcal{R}(B_1) = \mathcal{N}((\lambda_0 - L)^p) = \mathcal{N}((\lambda_0 - L)^{p+1})$. Hence, $\dim M_{\lambda_0} = \dim \bigcup_{n=1}^{p} \mathcal{N}((\lambda_0 - L)^n) = \dim \mathcal{N}((\lambda_0 - L)^p) = \dim \mathcal{R}(B_1) < \infty$. That is, $\dim M_{\lambda_0} = \infty$ implies that $\dim \mathcal{R}(B_1) = \infty$. Now suppose $\dim M_{\lambda_0} < \infty$. Then it follows from (3.27) that $M_{\lambda_0} = \mathcal{N}((\lambda_0 - L)^p)$ for some finite integer $p \geq 0$. We show that $M_{\lambda_0} \subset \mathcal{R}(B_1)$. Obviously, $\mathcal{N}((\lambda_0 - L)^0) = \{0\} \subset \mathcal{R}(B_1)$. Let $m \geq 0$ and suppose that $\mathcal{N}((\lambda_0 - L)^m) \subset \mathcal{R}(B_1)$. Let $x \in \mathcal{N}((\lambda_0 - L)^{m+1})$ and $y = (\lambda_0 - L)x$, then $y \in \mathcal{N}((\lambda_0 - L)^m) \subset \mathcal{R}(B_1)$. Let Γ be the contour in (3.24). If $\lambda \in \Gamma$, then

$$y = (\lambda_0 - L)x = (\lambda_0 - \lambda)x + (\lambda - L)x$$

and so

$$R(\lambda, L)x = (\lambda_0 - \lambda)^{-1} R(\lambda, L)y - (\lambda_0 - \lambda)^{-1} x.$$

Since $y \in \mathcal{R}(B_1)$ we have $y = B_1 z$ for some $z \in X$; hence,

$$R(\lambda, L)y = R(\lambda, L)B_1 z = B_1 R(\lambda, L)z.$$

Integrating along the contour yields

$$\begin{aligned}
B_1 x &= \frac{1}{2\pi i} \int_\Gamma R(\lambda, L)x d\lambda \\
&= \frac{1}{2\pi i} \int_\Gamma (\lambda_0 - \lambda)^{-1} R(\lambda, L)y d\lambda - \frac{1}{2\pi i} \int_\Gamma (\lambda_0 - \lambda)^{-1} x d\lambda \\
&= \frac{1}{2\pi i} \int_\Gamma (\lambda_0 - \lambda)^{-1} R(\lambda, L)y d\lambda - x.
\end{aligned}$$

Hence,

$$\begin{aligned}
x &= \frac{1}{2\pi i} \int_\Gamma (\lambda_0 - \lambda)^{-1} R(\lambda, L)y d\lambda - B_1 x \\
&= B_1 \left(\frac{1}{2\pi i} \int_\Gamma (\lambda_0 - \lambda)^{-1} R(\lambda, L)z d\lambda - x \right),
\end{aligned}$$

which says that $x \in \mathcal{R}(B_1)$, and by mathematical induction we have proved $M_{\lambda_0} \subset \mathcal{R}(B_1)$. Next, we show $M_{\lambda_0} = \mathcal{R}(B_1)$. Define $\tilde{L} = (L - \lambda_0)|_{\mathcal{R}(B_1)}$. It follows from Theorem 3.42 that $\tilde{L}$ is bounded and $\sigma(\tilde{L}) = \{0\}$. Let $X_0 = \mathcal{R}(B_1)/M_{\lambda_0}$, with the usual Banach space structure, and let $\hat{L}$ be the operator on X_0 induced by $\tilde{L}$, i.e., $\hat{L}(x + M_{\lambda_0}) = \tilde{L}x + M_{\lambda_0}$. For every $k \geq 1, \hat{L}^k$ is the operator on X_0 induced by $\tilde{L}^k$, and $\left\|\hat{L}^k\right\| \leq \left\|\tilde{L}^k\right\|$. Therefore, $r(\hat{L}) \leq r(\tilde{L}) = 0$. We want to show that $\mathcal{R}(\hat{L})$ is closed in X_0. But this follows from the following three points.

a) $\mathcal{R}(\tilde{L})$ is closed in $\mathcal{R}(B_1)$. This follows from the assumption that $\mathcal{R}(\lambda_0 - L)$ is closed in X and Problem 2 in Taylor [152, P.335].

b) $\mathcal{R}(\tilde{L}) + M_{\lambda_0}$ is closed. This is because $\dim M_{\lambda_0} < \infty$ and $\mathcal{R}(\tilde{L})$ is closed in X (Theorem 5.3, P.73 in [152]).

c) Note that $\mathcal{R}(\hat{L}) = \mathcal{R}(\tilde{L}) + M_{\lambda_0}$. It follows from a), b) and Theorem 5.2 (P.72 in [152]) that $\mathcal{R}(\hat{L})$ is closed in $\mathcal{R}(B_1)/M_{\lambda_0} = X_0$.

Finally, $\hat{L}(x + M_{\lambda_0}) = M_{\lambda_0}$ implies that $\tilde{L}x \in M_{\lambda_0}$, or $(\lambda_0 - L)^{p+1}x = 0$ for some $p \geq 0$, and so $x \in M_{\lambda_0}$. This shows that $\hat{L}$ is a one-to-one mapping of X_0 onto the closed subspace $\mathcal{R}(\hat{L})$ of X_0. If $X_0 \neq \{0\}$, then by the open mapping theorem there is a constant $c > 0$ such that

$$\left\|\hat{L}x\right\| \geq c\left\|x\right\|, \quad x \in X_0.$$

This implies, however, that $\left\|\hat{L}^k x\right\| \geq c^k\left\|x\right\|$ and hence $r(\hat{L}) \geq c$, contradicting $r(\hat{L}) = 0$. Therefore, $X_0 = \{0\}$ or $\mathcal{R}(B_1) = M_{\lambda_0}$.   □

If λ_0 is a pole of $R(\lambda, L)$ of order p, then $m_a = \dim \mathcal{N}(\lambda_0 - L)^p$, and it can be verified easily that [119]

$$\max\{m_g, p\} \leq m_a \leq p \cdot m_g.$$

In particular, $p = 1$ implies that $m_a = m_g$. If λ_0 is not an eigenvalue, then $m_g = 0$ and $p = \infty$ by Theorem 3.42. In this case, the above inequality still holds by regulating $0 \cdot \infty = \infty$ and so $m_a = \infty$. Therefore, the necessary and sufficient condition for $m_a < \infty$ is that λ_0 is a pole and $m_g < \infty$. It is possible that λ_0 is a pole but $m_g = \infty$ (for instance, if $L = I$ and $\dim X = \infty$). In the case $m_a = 1$, we say that λ_0 is algebraically simple.

Now, we introduce the concept of (Brower) essential spectrum $\sigma_{ess}(L)$ of operator L, which is quite useful in understanding the relationship between $S(A)$ and ω_0 in semigroup theory.

Definition 3.44 *Let L be a closed operator in a Banach space X. The complex number λ belongs to $\sigma_{ess}(L)$, the essential spectrum of L, if at least one of the following conditions is satisfied:*

(i) λ is a limit of $\sigma(L)$.

(ii) $\mathcal{R}(\lambda - L)$ is not closed.

(iii) $dim M_\lambda = \infty$.

Theorem 3.45 *Let L be a closed operator in a Banach space X. Then $\sigma(L)\backslash$ $\sigma_{ess}(L)$ consists of all isolated eigenvalues of L with finite algebraic multi-plicities, or equivalently, all poles of $R(\lambda, L)$ with finite rank of the residue B_1.*

Proof. By Theorem 3.43 and Definition 3.44, $\sigma(L)\backslash\sigma_{ess}(L)$ consists of all isolated eigenvalues λ of L with finite algebraic multiplicity and $\mathcal{R}(\lambda - L)$ is closed. The equivalent statement is a consequence of the finiteness of algebraic multiplicity by (iii) of Theorem 3.42. This proves the result. □

For $L \in \mathcal{L}(X)$, we define the *essential spectral radius* of L by

$$r_{ess}(L) = \sup\{|\lambda| \,\big|\, \lambda \in \sigma_{ess}(L)\}. \qquad (3.28)$$

In order to characterize $r_{ess}(L)$ in a similar way to the spectral radius $r(L)$, we need the concept of the measure of *noncompactness* .

For a bounded subset Ω of X we define the (Kuratowski-) measure of noncompactness $\alpha(\Omega)$ by

$$\begin{aligned} \alpha(\Omega) \;=\; &\inf\{d > 0 \mid \text{there exist finite subsets } \Omega_1, \ldots, \Omega_n \\ &\text{of } X \text{ with diameters of } \Omega_i \text{ less than } d \text{ such that} \\ &\Omega \subset \bigcup_{i=1}^{n} \Omega_i\}. \end{aligned}$$

$$(3.29)$$

The following elementary properties of α can be easily verified [35]:

Lemma 3.46 *Let $\Omega, \Omega_1, \Omega_2$ be bounded subsets of X. Then*

(i) $\alpha(\Omega) = \alpha(\overline{\Omega})$,

(ii) $\alpha(\Omega) = 0$ if and only if $\overline{\Omega}$ is compact,

(iii) $\alpha(\lambda\Omega) = |\lambda|\alpha(\Omega), \forall \lambda \in \mathbf{C}$,

(iv) $\alpha(\Omega_1 \cup \Omega_2) = \max\{\alpha(\Omega_1), \alpha(\Omega_2)\}$.

(v) $\alpha(\Omega_1 \cap \Omega_2) \leq \min\{\alpha(\Omega_1), \alpha(\Omega_2)\}$.

(vi) $\alpha(\Omega_1 + \Omega_2) \leq \alpha(\Omega_1) + \alpha(\Omega_2)$.

For $L \in \mathcal{L}(X)$, its measure of noncompactness $|L|_\alpha$ is defined by

$$|L|_\alpha = \inf\{k > 0 \mid \alpha(L\Omega) \leq k\alpha(\Omega) \text{ for every bounded subset } \Omega \text{ of } X\}. \tag{3.30}$$

The following lemma follows immediately from Lemma 3.46.

Lemma 3.47 *Let* $L, K, L_1, L_2 \in \mathcal{L}(X)$. *Then*

(i) $|L|_\alpha \leq \|L\|$,

(ii) $|\lambda L|_\alpha = |\lambda||L|_\alpha$, $\forall \lambda \in \mathbf{C}$,

(iii) $|L|_\alpha = 0$ *if and only if* L *is compact,*

(iv) $|L + K|_\alpha = |L|_\alpha$ *for any compact* K,

(v) $|L_1 + L_2|_\alpha \leq |L_1|_\alpha + |L_2|_\alpha$,

(vi) $|L_1 L_2|_\alpha \leq |L_1|_\alpha |L_2|_\alpha$.

The following result is proved by Nussbaum [126].

Theorem 3.48 *Let* $L \in \mathcal{L}(X)$. *Then*

$$r_{ess}(L) = \lim_{n\to\infty} |L^n|_\alpha^{1/n}. \tag{3.31}$$

As a consequence of Theorem 3.48, if L is a compact operator, we have that $r_{ess}(L) = 0$, which means $\sigma_{ess}(L) = \{0\}$. This also follows from the well-known Riesz-Schauder theory for compact operators.

In the sequel, we associate the concepts introduced above with the C_0-semigroup $T(t)$ and its infinitesimal generator A.

Proposition 3.49 *Let* $T(t)$ *be a* C_0-*semigroup on a Banach Space* X *with generator* A. *Then for every* $t \geq 0$,

$$e^{t\sigma_{ess}(A)} \subset \sigma_{ess}(T(t)).$$

Proof. Fix $t > 0$ and suppose that $e^{\lambda t} \in \sigma(T(t)) \backslash \sigma_{ess}(T(t))$ for some $\lambda \in$ **C**. We show that $\lambda \in \sigma(A) \backslash \sigma_{ess}(A)$. By Theorem 3.45, $e^{\lambda t}$ is an isolated eigenvalue of $T(t)$. We claim that λ is not a limit of $\sigma(A)$. Indeed, suppose $\lambda_k \in \sigma(A), \lambda_k \neq \lambda, k = 1, 2, \ldots$, and, $\lambda_k \to \lambda$ as $k \to \infty$. Then $e^{\lambda_k t} \to e^{\lambda t}$ as $k \to \infty$ and $e^{\lambda_k t} \neq e^{\lambda t}$ whenever k is large enough, for $e^{\lambda_k t} = e^{\lambda t}$ implies that $Re\lambda_k = Re\lambda$ and $Im\lambda_k = Im\lambda + 2m\pi/t$ for some integer m. This is impossible for $\lambda_k \to \lambda$ as $k \to \infty$. Next, we show that $\mathcal{R}(\lambda - A)$ is closed. Let Y be a subspace of X such that

$$X = \mathcal{N}(e^{\lambda t} - T(t)) \oplus Y.$$

Since $\dim \mathcal{N}(e^{\lambda t} - T(t)) < \infty, Y$ is closed. Because $(\lambda - A)(\mathcal{N}(e^{\lambda t} - T(t)) \cap D(A))$ is finite dimensional and thus closed, it suffices to show that $(\lambda - A)(Y \cap D(A))$ is closed. Note that the restriction of $e^{\lambda t} - T(t)$ to Y is an isomorphism of Y onto $\mathcal{R}(e^{\lambda t} - T(t))$ which is closed. From the open mapping theorem, it follows that there is a constant $c > 0$ such that

$$\left\| (e^{\lambda t} - T(t))x \right\| \geq c \left\| x \right\| \tag{3.32}$$

for every $x \in Y$. From the proof of Theorem 3.5, we know that

$$B_\lambda(t)(\lambda - A)x = e^{\lambda t}x - T(t)x \tag{3.33}$$

for all $x \in D(A)$ where

$$B_\lambda(t)x = \int_0^t e^{\lambda(t-s)} T(s)x \, ds.$$

Hence, there is a constant $M > 0$ which may depend on λ such that

$$\left\| e^{\lambda t}x - T(t)x \right\| \leq M \left\| \lambda x - Ax \right\|$$

for all $x \in D(A)$. Combining this inequalities with (3.32) gives

$$\left\| \lambda x - Ax \right\| \geq c/M \left\| x \right\|, \quad \text{for all } x \in Y \cap D(A),$$

and from the fact that $\lambda - A$ is closed, we conclude that $\mathcal{R}((\lambda - A)|_Y)$ is closed. Thus, $\mathcal{R}(\lambda - A)$ also is closed.

Finally, it follows from (3.33) that for all integers k and m,

$$\mathcal{N}((\lambda_m - A)^k) \subset \mathcal{N}((e^{\lambda t} - T(t))^k) \tag{3.34}$$

where $\lambda_m = \lambda + \frac{2m\pi}{t} i$. Therefore, $M_\lambda^A \subset M_\lambda^T$, where

$$M_\lambda^A = \bigcup_{k=1}^{\infty} \mathcal{N}((\lambda - A)^k),$$

$$M_\lambda^T = \bigcup_{k=1}^{\infty} \mathcal{N}((e^{\lambda t} - T(t))^k).$$

It follows that $\dim M_\lambda^A \leq \dim M_\lambda^T < \infty$. Summarizing, we have proved that $\lambda \in \sigma(A) \backslash \sigma_{ess}(A)$. $\qquad\qquad\qquad\qquad\qquad\qquad\qquad\qquad\qquad\qquad\qquad\qquad\qquad\square$

Remark 3.50 *It follows from (3.34) that if there are an infinite number of integers m such that $\lambda_m = \lambda + i2m\pi/t \in \sigma_p(A)$, then $e^{\lambda t}$ must belong to the essential spectrum of $T(t)$. Actually, Nussbaum proved in [127] that*

$$\mathcal{N}((e^{\lambda t} - T(t))^k) = \bigvee_{m=-\infty}^{\infty} \mathcal{N}((\lambda_m - A)^k) \qquad (3.35)$$

where $\bigvee_{m=-\infty}^{\infty} \mathcal{N}((\lambda_m - A)^k)$ denotes the smallest closed linear subspace of X such that $\mathcal{N}((\lambda_m - A)^k) \subset \bigvee_{m=-\infty}^{\infty} \mathcal{N}((\lambda_m - A)^k)$.

Like (iii) of Theorem 2.9, we have

$$r_{ess}(T(t)) = e^{\omega_{ess} t}, \quad t \geq 0 \qquad (3.36)$$

where $\omega_{ess} = \omega_{ess}(A)$ is called the essential type or *essential growth rate* of $T(t)$, which is defined by

$$\omega_{ess} = \inf_{t>0} \frac{1}{t} \log |T(t)|_\alpha = \lim_{t\to\infty} \frac{1}{t} \log |T(t)|_\alpha. \qquad (3.37)$$

Since $|T(t)|_\alpha \leq \|T(t)\|$, it follows immediately that

$$\omega_{ess} \leq \omega_0. \qquad (3.38)$$

If $T(t)$ is compact for $t > t_0 \geq 0$, then $r_{ess}(T(t)) = 0$ and hence $\omega_{ess} = -\infty$. Define

$$S_1(A) = \sup\{Re\lambda \mid \lambda \in \sigma(A) \backslash \sigma_{ess}(A)\}. \qquad (3.39)$$

We have the following inequality

$$S_1(A) \leq S(A) \leq \omega_0 \qquad (3.40)$$

where $S(A)$ is the spectral bound of A.

Theorem 3.51 *Let $T(t)$ be a C_0-semigroup on a Banach Space X with generator A. Then*

(i) $\sup\{Re\lambda \mid \lambda \in \sigma_{ess}(A)\} \leq \omega_{ess}.$

(ii) $\omega_0 = \max\{S(A), \omega_{ess}\} = \max\{S_1(A), \omega_{ess}\}.$

Proof.   (i): Assume $\lambda \in \sigma_{ess}(A)$, then by Proposition 3.49, $e^{\lambda t} \in \sigma_{ess}(T(t))$; hence $e^{\omega_{ess}t} = r_{ess}(T(t)) \geq e^{Re\lambda t}$ which proves (i).

(ii): Set $p = \max\{S_1(A), \omega_{ess}\}$. Then $p \leq \omega_0$. To prove $p \geq \omega_0$, it suffices to show that $r(T(t)) \leq e^{pt}$. Let $\mu \in \sigma(T(t))$ with $\mu \neq 0$. If $\mu \in \sigma_{ess}(T(t))$, then $|\mu| \leq r_{ess}(T(t)) = e^{\omega_{ess}t} \leq e^{pt}$. If $\mu \notin \sigma_{ess}(T(t))$, then μ is an isolated eigenvalue of $T(t)$ with finite algebraic multiplicity by virtue of Theorem 3.45. In particular, Theorem 3.4 now gives $\mu = e^{\lambda t}$ for some $\lambda \in \sigma_p(A)$. If $\lambda \in \sigma_{ess}(A)$, then $e^{\lambda t} \in \sigma_{ess}(T(t))$ by Proposition 3.49, which contradicts the assumption. Therefore, $\lambda \in \sigma(A)\backslash\sigma_{ess}(A)$ and so $|\mu| = e^{Re\lambda t} \leq e^{S_1(A)t} \leq e^{pt}$. This proves $p = \omega_0$. Since $S_1(A) \leq S(A) \leq \omega_0$, the first equality of (ii) follows immediately.   □

From (ii) of Theorem 3.51 and (3.36), we see that ω_0 is determined by the spectral bound $S(A)$ and essential spectrum $\sigma_{ess}(T(t))$. Remark 3.50 tells us that even all eigenvalues of the operator A are isolated with finite algebraic multiplicities, the mapping: $\lambda \in \sigma_p(A) \rightarrow e^{\lambda t} \in \sigma_p(T(t))$ may produce essential spectrum of $T(t)$. For instance, in Example 3.6, we have seen that $\sigma(A) = \{\lambda_n \mid n = 1, 2, \cdots\}$ and $S(A) = \{Re\lambda \mid \lambda \in \sigma(A)\} = -1/2$, but $\omega_0 = \omega_{ess} = 1/2$ by Theorem 3.51, or $r_{ess}(e^{At}) = e^{t/2}$. This clearly shows the gap between $e^{t\sigma(A)}$ and $\sigma(e^{At})$.

Using the essential spectrum results developed above, we now turn to consider the properties of solutions of the following control problems on a Banach space X with compact feedback:

$$\frac{dx(t)}{dt} = Ax(t) + u(t).$$

Suppose the operator A generates a C_0-semigroup $T(t)$. Let B be a linear operator on X with $D(B) \supset D(A)$. The closed-loop system by state feedback $u(t) = Bx(t)$ will take the form:

$$\frac{dx(t)}{dt} = (A+B)x(t).$$

In the context of control study for dynamic systems, one wants to select feedback operator B to force the closed-loop system to possess stability properties that is not enjoyed by the original system. One important class in physical applications is that of compact operators $K \in \mathcal{L}(X)$. We will give several results in this respect which are frequently used in applications. The characterization (3.37) of the essential growth rate, in terms of the measure of noncompactness, gives the first result on compact perturbations of C_0-semigroups.

Theorem 3.52 *Let $T_i(t)$ be C_0-semigroups on a Banach space X with generators $A_i, i = 1, 2$. If $T_1(t_0) - T_2(t_0)$ is compact for some $t_0 > 0$, then*

$$\omega_{ess}(A_1) = \omega_{ess}(A_2).$$

Proof. From (3.31), (3.36), and (iv) of Lemma 3.47, we have

$$
\begin{aligned}
\omega_{ess}(A_1) &= \frac{1}{t_0} \log r_{ess}(T_1(t_0)) \\
&= \frac{1}{t_0} \log r_{ess}(T_1(t_0) - T_2(t_0) + T_2(t_0)) \\
&= \frac{1}{t_0} \log \lim_{n \to \infty} |(T_2(t_0) + T_1(t_0) - T_2(t_0))^n|_\alpha^{1/n} \\
&= \frac{1}{t_0} \log \lim_{n \to \infty} |T_2(t_0)^n|_\alpha^{1/n} \\
&= \omega_{ess}(A_2).
\end{aligned}
$$

$\square$

Theorem 3.52 is very useful in studying the exponential stability of linear dynamical systems with compact perturbations in Banach spaces. The next result is one of its consequences.

Theorem 3.53 *Let $T(t)$ be a C_0-semigroup on a Banach space X with generator A. If $K \in \mathcal{L}(X)$ is compact, then*

$$
\omega_{ess}(A + K) = \omega_{ess}(A).
$$

Proof. Let $x(t)$ be the solution of the abstract Cauchy problem associated with $A + K$:

$$
\frac{dx(t)}{dt} = Ax(t) + Kx(t).
$$

Then the variation of constant gives

$$
\begin{aligned}
x(t) &= T(t)x(0) + \int_0^t T(t-s)Kx(s)ds \\
&= T(t)x(0) + \int_0^t T(s)Kx(t-s)ds
\end{aligned}
$$

so that for all $x \in X$,

$$
e^{(A+K)t}x - T(t)x = \int_0^t T(s)Ke^{(A+K)(t-s)}xds.
$$

By Theorem 3.52, it suffices to show that $S(t)$ defined by

$$
S(t)x = \int_0^t T(s)Ke^{(A+K)(t-s)}xds
$$

is compact for each fixed $t > 0$.

Let $\Omega_1 = \{e^{(A+K)(t-s)}x \mid x \in X, \|x\| \leq 1, 0 \leq s \leq t\}$, which is a bounded subset of X. Then $\Omega_2 = K(\Omega_1)$ is precompact in X. We show that $\Omega_3 = \{T(s)x \mid x \in \Omega_2, 0 \leq s \leq t\}$ is precompact. Let C be a constant such that $\|T(s)\| \leq C$ for all $0 \leq s \leq t$. Let $\epsilon > 0$. There exist $x_1, x_2, \ldots, x_n \in \Omega_2$ such that if $x \in \Omega_2$ then $\|x - x_i\| \leq \epsilon/2C$ for some i, $1 \leq i \leq n$. Since $T(s)$ is strongly continuous, there exists $s_1^i, \ldots, s_{k_i}^i \in [0, t]$ such that if $s \in [0, t]$, then $\left\| T(s_{k_j}^i)x_i - T(s)x_i \right\| < \epsilon/2$ for some j. Then for any $x \in \Omega_2$, there exist i and j such that

$$\left\| T(s)x - T(s_{k_j}^i)x_i \right\| \leq \|T(s)x - T(s)x_i\| + \left\| T(s)x_i - T(s_{k_j}^i)x_i \right\|$$
$$\leq \epsilon/2 + \epsilon/2 = \epsilon.$$

Therefore, Ω_3 is precompact. By Mazur's theorem, the convex hull $\overline{conv}(\Omega_3)$ also is precompact. Note that

$$S(t)x = \int_0^t T(s)Ke^{(A+K)(t-s)}x\,ds \in \overline{tconv}(\Omega_3)$$

for all $x \in X$ with $\|x\| \leq 1$, so $S(t)$ is compact. $\qquad\square$

Corollary 3.54 *Let $T(t)$ be an exponentially stable C_0-semigroup on a Banach space X with generator A. Let $K \in \mathcal{L}(X)$ be compact. Then $e^{(A+K)t}$ is exponentially stable if and only if $S(A+K) < 0$ or $S_1(A+K) < 0$.*

Proof. By Theorem 3.51 and Theorem 3.53,

$$\omega_0(A+K) = \max\{S(A+K), \omega_{ess}(A)\} = \max\{S_1(A+K), \omega_{ess}(A)\}.$$

Since $\omega_{ess}(A) \leq \omega_0(A) < 0$ by assumption, we see that $\omega_0(A+K) < 0$ if and only if $S(A+K) < 0$ or $S_1(A+K) < 0$. $\qquad\square$

Theorem 3.55 *Let $T_i(t)$ be C_0-semigroups on a Banach space X with generators A_i, $i = 1, 2$. Assume the following:*

(i) $T_1(t)$ is asymptotically stable,

(ii) $T_1(t_0) - T_2(t_0)$ is compact for some $t_0 > 0$,

(iii) $T_2(t)$ is exponentially stable.

Then $T_1(t)$ is exponentially stable.

Proof. By assumption (ii), Theorem 3.51, and Theorem 3.52, we know that $\omega_{ess}(A_1) = \omega_{ess}(A_2) < 0$. Therefore, it is sufficient to show that $S_1(A_1) < 0$. Since T_1 is asymptotically stable, $S_1(A_1) \leq 0$. Suppose $S_1(A_1) = 0$. Then there are $\{\lambda_n\}_1^\infty \subset \sigma(A_1)\backslash\sigma_{ess}(A_1)$ such that $\operatorname{Re}\lambda_{n+1} > \operatorname{Re}\lambda_n$ for all $n \geq 1$ and $\operatorname{Re}\lambda_n \to 0$ as $n \to \infty$. Because λ_n are eigenvalues of A_1, it follows from (i) and Theorem 3.26 that $\operatorname{Re}\lambda_n < 0$. Clearly, $e^{\lambda_n t_0}$ are eigenvalues of $T_1(t_0)$ satisfying $|e^{\lambda_n t_0}| < 1$ for $n \geq 1$ and $|e^{\lambda_n t_0}| \to 1$ as $n \to \infty$. Let z be an accumulation point of $\{e^{\lambda_n t_0}\}_1^\infty$ in the complex plane $\mathbf{C}$, then $z \in \sigma_{ess}(T_1(t_0))$ and $|z| = 1$. Consequently, $r_{ess}(T_1(t_0)) = 1$. On the other hand,

$$r_{ess}(T_1(t_0)) = e^{\omega_{ess}(A_1)t_0} = e^{\omega_{ess}(A_2)t_0} \leq e^{\omega_0(A_2)t_0} < 1.$$

This contradiction leads to $S_1(A_1) < 0$. $\square$

Corollary 3.56 *Let $T(t)$ be a C_0-semigroup on a Banach space X with generator A. Let $K \in \mathcal{L}(X)$ be compact. Assume the following:*

(i) $T(t)$ is asymptotically stable.

(ii) The semigroup $e^{(A+K)t}$ is exponentially stable.

Then $T(t)$ is exponentially stable.

Proof. From the proof of Theorem 3.53, $e^{(A+K)t} - T(t)$ is compact for any $t > 0$. The conclusion then follows from Theorem 3.55. $\square$

Theorem 3.57 *[140] Let X be an infinite dimensional Banach space and let A be the generator of a C_0-group $T(t)$. Then there cannot exist compact operators K_1 and K_2, positive numbers $\tau > 0$ and $0 \leq \gamma < 1$ such that the groups $T_1(t), T_2(t)$ generated by $A + K_1, A + K_2$, respectively, satisfy*

$$\|T_1(\tau)\| \leq 1, \ \|T_2(-\tau)\| \leq \gamma, \tag{3.41}$$

or

$$\|T_1(\tau)\| \leq \gamma, \ \|T_2(-\tau)\| \leq 1. \tag{3.42}$$

Proof. Consider the Cauchy problems associated with $A + K_1$ and $A + K_2$. The variation of constant gives

$$T_1(t) = T(t) + \int_0^t T(t-s)K_1 T_1(s)ds$$

and

$$T_2(t) = T(t) + \int_0^t T(t-s)K_2 T_2(s)ds.$$

Therefore,

$$
T_1(\tau)T_2(-\tau) = \left[T(\tau) + \int_0^\tau T(\tau - s)K_1 T_1(s)ds \right]
$$

$$
\times \left[T(-\tau) + \int_0^{-\tau} T(-\tau - s)K_2 T_2(s)ds \right]
$$

$$
= I + \int_0^{-\tau} T(-s)K_2 T_2(s)ds + \int_0^\tau T(\tau - s)K_1 T_1(s)T(-\tau)ds
$$

$$
+ \int_0^\tau T(\tau - s)K_1 T_1(s)ds \int_0^{-\tau} T(-\tau - s)K_2 T_2(s)ds
$$

$$
= I + K,
$$

with K representing the apparent integration terms. Since K_1 and K_2 are compact, we know from the proof of Theorem 3.53, that $K \in \mathcal{L}(X)$ is compact. Recall the properties of the measure of noncompactness, we have

$$
\|I + K\| \geq |I + K|_\alpha = |I|_\alpha = 1.
$$

The last equality $|I|_\alpha = 1$ holds because the unit ball of X is not compact for its infinite dimensionality. Therefore,

$$
\|T_1(\tau)T_2(-\tau)\| \geq 1. \tag{3.43}
$$

This shows that neither (3.41) nor (3.42) can be true. $\qquad\square$

Corollary 3.58 *Let X be an infinite dimensional Banach space and let A be the generator of a C_0-group $T(t)$. If $\|T(-t)\| \leq 1$ for all $t \geq 0$, then $e^{(A+K)t}$ cannot be exponentially stable for any compact $K \in \mathcal{L}(X)$.*

Proof. Take $K_1 = K, K_2 = 0$ in Theorem 3.57. If $e^{(A+K)t}$ is exponentially stable, then there exists a γ with $0 \leq \gamma < 1$ such that $\left\|e^{(A+K)t}\right\| < \gamma$, for all t sufficiently large, which contradicts (3.42) by assumption $\|T(-t)\| \leq 1$. This proves the result. $\qquad\square$

3.7 Invariance principle for nonlinear semigroups

In this section, we consider the stability of a continuous nonlinear semigroup of contractions $T(t)$ on a closed subset F of a real Banach space X. For every $x \in F$, denote by

$$
\gamma(x) = \bigcup_{t \geq 0} T(t)x \tag{3.44}
$$

the orbit through x and by

$$\omega(x) = \{y \in F | y = \lim_{n \to \infty} T(t_n)x \text{ with } t_n \to \infty \text{ as } n \to \infty\} \qquad (3.45)$$

the (possibly empty) ω-limit set of x. In case $T(t)$ is generated by a multival-ued operator $\mathcal{A} : X \to 2^X$, the structure of ω-limit set is certainly responsible for the asymptotic behavior of solution of the nonlinear evolution equation:

$$\frac{du(t)}{dt} \in \mathcal{A}u(t). \qquad (3.46)$$

$\omega(x)$ is always *positively invariant*, i.e., $T(t)\omega(x) \subset \omega(x)$.

Proposition 3.59 $\omega(x)$ *is closed.*

Proof. Consider a sequence $\{y_m\} \subset \omega(x)$ such that $\|y_m - y_0\| < \frac{1}{m}$ for $m = 1, 2, \ldots$ and some $y_0 \in F$. To each y_m we can find a sequence $\{t_{nm}\}$ such that $t_{nm} \to \infty$ as $n \to \infty$ and $\|T(t_{nm})x - y_m\| < \frac{1}{n}$. Hence,

$$\|T(t_{nn})x - y_0\| \leq \|T(t_{nn})x - y_n\| + \|y_n - y_0\| \leq \frac{2}{n} \to 0 \text{ as } n \to \infty.$$

This implies that $y_0 \in \omega(x)$. $\qquad\qquad\square$

It is easily seen that

$$\omega(x) = \bigcap_{\tau \geq 0} \overline{\gamma(T(\tau)x)}. \qquad (3.47)$$

Proposition 3.60 *If* $\omega(x)$ *is nonempty, then*

(i) $\omega(x) = \omega(y) = \overline{\gamma(y)}$ *for every* $y \in \omega(x)$.

(ii) *The mapping* $T(t) : \omega(x) \to \omega(x)$ *is onto and isometric.*

(iii) *If* w *is a fixed point of* $T(t) : T(t)w = w$ *for all* $t \geq 0$, *then*

$$\omega(x) \subset \{z \mid \|z - w\| = r\},$$

with $r \leq \|x - w\|$.

Proof. Fix $y \in \omega(x)$, say $y = \lim_n T(t_n)x$ with $t_n \to \infty$ as $n \to \infty$. For any $t \geq 0, T(t)y = \lim_n T(t + t_n)x \in \omega(x)$; i.e. $\overline{\gamma(y)} \subset \omega(x) = \omega(x)$. Suppose now $z \in \omega(x)$, say $z = \lim_n T(\tau_n)x, \tau_n \to \infty$ as $n \to \infty$. We may assume without loss of generality that $s_n = \tau_n - t_n \geq n, n = 1, 2, \ldots$. Since

$$\begin{aligned} \|T(s_n)y - z\| &\leq \|T(s_n)y - T(s_n + t_n)x\| + \|T(\tau_n)x - z\| \\ &\leq \|y - T(t_n)x\| + \|T(\tau_n)x - z\| \to 0 \end{aligned}$$

as $n \to \infty$, we have $z \in \omega(y)$ so that $\omega(x) \subset \omega(y) \subset \overline{\gamma(y)}$. Therefore,

$$\omega(x) = \omega(y) = \overline{\gamma(y)}. \tag{3.48}$$

This is (i). Next we show (ii). For any $y \in \omega(x)$, since $\omega(x) = \omega(y)$, there is $\sigma_n \to \infty$ as $n \to \infty$, such that $y = \lim_n T(\sigma_n)y$. For any $z \in \omega(x)$, if $z = \lim_n T(s_n)y$, then

$$
\begin{aligned}
\|T(\sigma_n)z - z\| &\leq \|T(\sigma_n)z - T(s_n + \sigma_n)y\| \\
&\quad + \|T(s_n + \sigma_n)y - T(s_n)y\| + \|T(s_n)y - z\| \\
&\leq 2\|z - T(s_n)y\| + \|T(\sigma_n)y - y\| \to 0 \text{ as } n \to \infty.
\end{aligned}
$$

Therefore,

$$\|y - z\| \geq \|T(t)y - T(t)z\| \geq \lim_n \|T(\sigma_n)y - T(\sigma_n)z\| = \|y - z\|.$$

Thus,

$$\|T(t)y - T(t)z\| = \|y - z\|, \tag{3.49}$$

i.e., $T(t)$ is an isometry on $\omega(x)$. Now we show that $T(t)$ is onto $\omega(x)$. Fix $z \in \omega(x)$ and suppose $z = \lim_n T(\sigma_n)z$. For large k, m, n,

$$
\begin{aligned}
\|T(\sigma_n - t)z - T(\sigma_m - t)z\| &\leq \|T(\sigma_n - t)z - T(\sigma_n + \sigma_k - t)z\| \\
&\quad + \|T(\sigma_n + \sigma_k - t)z - T(\sigma_m + \sigma_k - t)z\| \\
&\quad + \|T(\sigma_m + \sigma_k - t)z - T(\sigma_m - t)z\| \\
&\leq 2\|T(\sigma_k)z - z\| + \|T(\sigma_m)z - T(\sigma_n)z\|.
\end{aligned}
$$

Now, $T(\sigma_k)z - z \to 0$ and $\{T(\sigma_n)z\}$ is a Cauchy sequence. It follows that $\{T(\sigma_n - t)z\}$ is a Cauchy sequence and let y be its limit. Then $T(t)y = \lim_n T(\sigma_n)z = z$. Obviously, $y \in \omega(z) = \omega(x)$ and so $T(t)$ is onto on $\omega(x)$. Finally, assume w is a fixed point of $T(t)$. Then $\|T(t)x - w\|$ is nonincreasing and tends, as $t \to \infty$, to a limit $r \leq \|x - w\|$. If $y \in \omega(x)$, say $y = \lim_n T(t_n)x, t_n \to \infty$ as $n \to \infty$, then $\|y - w\| = \lim_n \|T(t_n)x - w\| = r$. Thus

$$\omega(x) \subset \{z \mid \|z - w\| = r\}.$$

The proof is complete. $\qquad\qquad\square$

To guarantee $\omega(x)$ to be nonempty, a sufficient condition is that the orbit $\gamma(x)$ is precompact, a well-known result of classical topological dynamics.

Theorem 3.61 *If $x \in F$, and $\gamma(x)$ is precompact, then $\omega(x)$ is nonempty, compact, connected, and moreover*

$$\lim_{t \to \infty} d(T(t)x, \omega(x)) = 0. \tag{3.50}$$

Here, for $y \in X$, and $\Omega \subset X, d(y, \Omega)$ denotes the distance from y to Ω. i.e., $d(y, \Omega) = \inf_{w \in \Omega} \|y - w\|$. In fact, $\omega(x)$ is the smallest closed set that $T(t)x$ approaches: if $T(t)x \to \mathcal{E} \subset F$ as $t \to \infty$, then $\omega(x) \subset \overline{\mathcal{E}}$.

Proof. As $\gamma(x)$ is precompact, $\omega(x) = \overline{\gamma(x)}$ is compact and it follows that there is a Cauchy sequence $\{T(t_n)x\}$ with $t_n \to \infty$ as $n \to \infty$. The limit of this sequence belongs to $\omega(x)$, i.e. $\omega(x)$ is nonempty.

Suppose (3.50) is not true, i.e., there exists a $\epsilon > 0$ and a sequence $\{T(t_n)x\}, t_n \to \infty$ as $n \to \infty$ such that

$$d(T(t_n)x, \omega(x)) > \epsilon, n = 1, 2, \dots . \tag{3.51}$$

Then, by precompactness of $\{T(t_n)x\}$, there is a convergent subsequence of $\{T(t_n)x\}$, its limit belongs to $\omega(x)$. As this contradicts (3.51), we see that (3.50) holds.

Suppose that $\omega(x)$ is not connected, that is, $\omega(x) = \Omega_1 \cup \Omega_2$ with $\Omega_1 \cap \Omega_2 = \emptyset$, where Ω_1 and Ω_2 are closed subset of $\omega(x)$. As $\omega(x)$ is compact, so are Ω_1 and Ω_2, hence

$$\inf_{y \in \Omega_1, z \in \Omega_2} \|y - z\| = \delta > 0.$$

Given $y \in \Omega_1, z \in \Omega_2$, there exist sequences $\{T(t_n)x\}$ and $\{T(\tau_n)x\}, \tau_n > t_n \to \infty$ as $n \to \infty$ such that $\lim_n T(t_n)x = y, \lim_n T(\tau_n)x = z$. Therefore, there exists N such that $d(T(t_n)x, \Omega_1) < \delta/2$ and $d(T(\tau_n)x, \Omega_2) < \delta/2$ for all $n > N$. However, by the continuity of $T(\cdot)x : \mathbf{R}^+ \to X$, there exists a sequence $\{s_n\}, \tau_n > s_n > t_n$, such that $d(T(s_n)x, \Omega_1) \geq \delta/2$ and $d(T(s_n)x, \Omega_2) \geq \delta/2$ for all $n > N$. This means that $d(T(s_n)x, \Omega) \geq \delta/2$ for all $n > N$, contradicting (3.50). Hence, $\omega(x)$ is connected. $\square$

Theorem 3.61 can be used to characterize the asymptotic behavior of $u(t, x) = T(t)x$ provided that one is able to prove precompactness of the orbit $\gamma(x)$ and to determine the ω-limit set of x. The first respect motivates the search for compactness criteria of orbits of contraction semigroups and as to the second aspect, *Lyapunov functions* turn out to be extremely useful.

Definition 3.62 *Let* $\mathcal{V} : F \to \mathbf{R}$ *be a continuous function. For* $x \in F$ *we define*

$$\dot{\mathcal{V}}(x) = \lim_{t \downarrow 0} \frac{1}{t}\{\mathcal{V}(T(t)x) - \mathcal{V}(x)\} \tag{3.52}$$

where it is allowable that $\dot{\mathcal{V}}(x) = -\infty$. *The function* $\mathcal{V}$ *is called a (continuous) Lyapunov function for* $T(t)$ *on* F *if*

$$\dot{\mathcal{V}}(x) \leq 0, \text{ for all } x \in F. \tag{3.53}$$

Lemma 3.63 *Let* $\mathcal{V}$ *be a Lyapunov function. Then for every* $x \in F, f(t) = \mathcal{V}(T(t)x) : \mathbf{R}^+ \to \mathbf{R}$ *is continuously nonincreasing, and hence differentiable almost everywhere.*

Proof. It is seen that $f(t) : \mathbf{R}^+ \to R$ is a continuous function. Choosing some $\epsilon > 0$, define $g(t) = f(t) - \epsilon t$. Then $g(t)$ is continuous and

$$\lim_{h \downarrow 0} \frac{1}{h}\{g(t+h) - g(t)\} \leq -\epsilon t \text{ for every } t \geq 0. \tag{3.54}$$

Let $t_1 \geq 0$ be fixed. If there exists $t_2 > t_1 \geq 0$ such that $g(t_1) < g(t_2)$, then the continuity of $g(t)$ implies that there exists $t_3 \in [t_1, t_2)$ such that $g(t_3) \leq g(t_1)$ and $g(t) > g(t_1)$ for all $t \in (t_3, t_2]$. This leads to the contradiction

$$\lim_{h \downarrow 0} \frac{1}{h}\{g(t_3 + h) - g(t_3)\} \geq 0.$$

We conclude that for any $t > t_1, g(t) \leq g(t_1)$, and, as $\epsilon > 0$ is arbitrary, the same is true for f, i.e., $f(t) \leq f(t_1)$ for all $t \geq t_1$. Hence, $f(t)$ is nonincreasing by arbitrariness of t_1. □

The following result is called the *invariance principle*.

Theorem 3.64 (LaSalle's Invariance Principle) *Let $\mathcal{V}$ be a continuous Lyapunov function for $T(t)$ on F and let $\mathcal{E}$ be the largest invariant subset of*

$$\{x| \quad \dot{\mathcal{V}}(x) = 0\}. \tag{3.55}$$

If $x \in F$ and $\gamma(x)$ is precompact, then

$$\lim_{t \to \infty} d(T(t)x, \mathcal{E}) = 0. \tag{3.56}$$

Here, by invariance of $\mathcal{E}$ under $T(t)$, we mean $T(t)\mathcal{E} = \mathcal{E}$ for all $t \geq 0$.

Proof. For every x, it follows from Lemma 3.63 that $\mathcal{V}(T(t)x)$ is continuously nonincreasing and so $\mathcal{V}(T(t)x) \to \beta < \infty$ as $t \to \infty$, where $\beta = \inf_{t \geq 0} \mathcal{V}(T(t)x)$. It follows from the definition of $\omega(x)$ that $\mathcal{V}(y) = \beta$ for every $y \in \omega(x)$. As $\omega(x)$ is positively invariant, we have $\mathcal{V}(T(t)y) = \beta$ for every $y \in \omega(x)$. Hence, $\dot{\mathcal{V}}(y) = 0$ for every $y \in \omega(x)$. Thus, $\omega(x) \subset \{z| \quad \dot{\mathcal{V}}(z) = 0\}$. Other conclusions follow from Theorem 3.61 and (ii) of Proposition 3.60. □

Theorem 3.64 is a special case of the general result proved in [46, Theorem 4.4.2].

Finally, we give a condition which characterizes the compactness of the orbit and is easy to use in applications. Recall Crandall-Liggett theorem (Theorem 2.115) which states that if a dissipative operator $\mathcal{A}$ in a real Banach space X satisfies

$$\overline{D(\mathcal{A})} \subset \mathcal{R}(I - \lambda\mathcal{A}), \tag{3.57}$$

for all sufficiently small λ, then there is a nonlinear semigroup of contractions on $\overline{D(\mathcal{A})}$ defined by

$$T(t)x = \lim_{n \to \infty} \left(I - \frac{t}{n}\mathcal{A}\right)^{-n} x \tag{3.58}$$

and

$$\|T(t)x - T(s)x\| \leq 2|t - s||\mathcal{A}x|, \tag{3.59}$$

where $|\mathcal{A}x| = \min\{y|\ y \in \mathcal{A}x\}$ is the minimal section of $\mathcal{A}x$.

Theorem 3.65 *Let $\mathcal{A}$ be a (multivalued) dissipative operator $\mathcal{A}$ in a real Banach space X such that (3.57) holds for all sufficiently small λ, and let $T(t)$ be the contraction semigroup defined by (3.58). Assume that $0 \in \mathcal{R}(\mathcal{A})$ and $(I - \lambda\mathcal{A})^{-1}$ is compact for some $\lambda > 0$. Then $\gamma(x)$ is precompact for any $x \in \overline{D(\mathcal{A})}$.*

Proof. For $\lambda > 0$ let $J_\lambda = (I - \lambda\mathcal{A})^{-1}$. It was proved in Lemma 2.116 that J_λ is single-valued and $J_\mu x = J_\lambda(\frac{\lambda}{\mu}x + \frac{\mu-\lambda}{\mu}J_\mu x)$ for all $x \in \mathcal{R}(I - \mu\mathcal{A})$ for any $\mu > 0$. It follows that the compactness of J_λ implies that J_μ is compact for every $\mu > 0$. Moreover, for all $\lambda > \mu > 0$,

$$\|x - J_\lambda x\| \leq \frac{\lambda}{\mu}\|x - J_\mu x\| \tag{3.60}$$

for all $x \in D(J_\lambda) \cap D(J_\mu)$. Indeed, let $\mathcal{A}_\lambda = \lambda^{-1}(J_\lambda x - x)$ for $x \in D(J_\lambda)$. From the proof of Lemma 2.116 (i) it follows that $\mathcal{A}_\lambda x \in \mathcal{A}J_\lambda x$. Since $\mathcal{A}$ is dissipative, we have

$$\langle \mathcal{A}_\lambda x - \mathcal{A}_\mu x,\ f \rangle \leq 0$$

for any $f \in F(J_\lambda x - J_\mu x)$. It follows that

$$\begin{aligned}
\|J_\lambda x - J_\mu x\|^2 &= \langle J_\lambda x - J_\mu x,\ f \rangle = \langle \lambda\mathcal{A}_\lambda x - \mu\mathcal{A}_\mu x,\ f \rangle \\
&\leq \frac{\lambda - \mu}{\mu}\langle J_\mu x - x,\ f \rangle \\
&\leq \frac{\lambda - \mu}{\mu}\|J_\mu x - x\|\,\|J_\lambda x - J_\mu x\|.
\end{aligned}$$

Therefore,

$$\|J_\lambda x - J_\mu x\| \leq \frac{\lambda - \mu}{\mu}\|J_\mu x - x\|.$$

On the other hand, since

$$\|J_\lambda x - J_\mu x\| \geq \|J_\lambda x - x\| - \|J_\mu x - x\|,$$

we finally get $\|J_\lambda x - x\| \leq \frac{\lambda}{\mu}\|J_\mu x - x\|$ as desired.

It follows from (3.60) that

$$
\begin{aligned}
\|(I - J_\lambda)J_{t/n}^n x\| &\leq \frac{n\lambda}{t}\|J_{t/n}^n x - J_{t/n}^{n+1}x\| \\
&= \frac{n\lambda}{t}\|J_{t/n}^{n+1}(I - \frac{t}{n}\mathcal{A})x - J_{t/n}^{n+1}x\| \leq \lambda|\mathcal{A}x|
\end{aligned}
$$

for $n = 1, 2, \ldots$, $0 \leq t < n\lambda$, and $x \in D(\mathcal{A})$ (and hence $x \in D(J_\lambda)$ for λ sufficiently small by (3.57)). Thus, we see that $\|T(t)x - J_\lambda T(t)x\| \leq \lambda|\mathcal{A}x|$ for all $x \in D(\mathcal{A}), t \geq 0$. In particular, letting $\lambda = 1/n$, we have

$$
\|T(t)x - J_{1/n}T(t)x\| \leq 1/n \mid \mathcal{A}x \mid . \tag{3.61}
$$

Noting that $J_{1/n}$ is compact for all sufficiently large n, we claim that $\gamma(x)$ is precompact provided that $\gamma(x)$ is bounded. Indeed, for any given sequence $\{\tau_m\} \subset R^+$, Tychonov's theorem [171] implies the existence of a subsequence $\{t_m\}$ such that for each n, $\|J_{1/n}T(t_{m+1})x - J_{1/n}T(t_m)x\| < 1/m$ for all $m \geq 1$. Noting that

$$
\begin{aligned}
\|T(t_{m+1})x - T(t_m)x\| &\leq \|T(t_{m+1})x - J_{1/n}T(t_{m+1})x\| \\
&\quad + \|J_{1/n}T(t_{m+1})x - J_{1/n}T(t_m)x\| \\
&\quad + \|J_{1/n}T(t_m)x - T(t_m)x\| \\
&\leq 2/n|\mathcal{A}x| + 1/m
\end{aligned}
$$

and letting $n \to \infty$, we see that $\{T(t_m)x\}$ is a Cauchy sequence. Therefore, $\gamma(x)$ is precompact.

Next, since $0 \in \mathcal{R}(\mathcal{A})$, there exists $x^* \in D(\mathcal{A})$ such that $0 \in \mathcal{A}x^*$, and $u(t) = x^*$ is a strong solution of (3.46). By Theorem 2.119, $T(t)x^* = x^*$ for all $t \geq 0$. As $\|T(t)x - T(t)x^*\| \leq \|x - x^*\|$, we see that $\|T(t)x\| \leq \|x^*\| + \|x - x^*\|$ and so $\gamma(x)$ is bounded. Therefore, $\gamma(x)$ is precompact for every $x \in D(\mathcal{A})$. Finally, let $y \in \overline{D(\mathcal{A})}$, there exists a sequence $\{x_n\} \subset D(\mathcal{A})$ such that $\|T(t)y - T(t)x_n\| \leq \|y - x_n\| < \frac{1}{n}$ for all $t \geq 0, n \geq 1$. As $\gamma(x_n)$ is precompact, Tychonov's theorem again implies that any given sequence $\{\tau_m\} \subset R^+$ admits a subsequence $\{t_m\}$ such that $\{T(t_m)x_n\}_{m=1,2,\ldots}$ is Cauchy, uniformly in $n = 1, 2, \ldots$. Noting that

$$
\begin{aligned}
\|T(t_k)y - T(t_m)y\| &\leq \|T(t_k)y - T(t_k)x_n\| + \|T(t_k)x_n - T(t_m)x_n\| \\
&\quad + \|T(t_m)x_n - T(t_m)y\| \\
&\leq \frac{2}{n} + \|T(t_k)x_n - T(t_m)x_n\|
\end{aligned}
$$

and letting $n \to \infty$, it follows that $\{T(t_m)y\}$ is a Cauchy sequence. Hence, $\gamma(y)$ is precompact and the proof is complete. $\qquad\square$

3.8   Notes and references

The spectral mapping theorem in Section 3.1 and the spectrum-determined growth condition in Section 3.2 are based on the books by Pazy [128] and by Nagel [119]. Example 3.6 can be found in Zabczyk [173] and Huang [78]. The asymptotic expansion Theorem 3.12 for compact semigroups are from [172]. The weak stability results in Section 3.3 are from Foguel [56]. The characterizations of the asymptotic stability of C_0−semigroups by the spectrum of their generators (Theorem 3.26) are mainly from Lyubich and Phong [105] and Batty [14]. The time domain criteria for the exponential stability of C_0−semigroups can be found in many reference books (see, for instance, [32], [41], and [18]). The frequency domain criteria in Section 3.5 was first worked out by Prüss [133], Huang [78], and Weiss [162], and can also be found in [32]. Our proof, however, is much simpler. The characterizations of the growth rate of C_0−semigroups in Hilbert spaces, described in a number of theorems in the latter part of Section 3.5, are contributed by Yao [168] and Guo [63]. References for Section 3.6 are Clément et al. [35], Webb [159], Russell [140], and Gibson [60] and Guo [64]. Section 3.7 is based on Davies [46] and Dafermos and Slemrod [42].

Chapter 4

Static Sensor Feedback Stabilization of Euler-Bernoulli Beam Equations

In this chapter, we consider stabilization problems of Euler-Bernoulli beam equations arising in the area of space and industrial robots with lightweight and flexible arms, as well as in the area of flexible space structures. We shall first derive a general model for a Euler-Bernoulli beam with a rigid tip body. The model is more general than those models in the existing literature in the sense that both bending and torsional vibrations of the beam will be considered, and the tip body is allowed to be a rigid body. Although the derived model looks complicated, it can be reformulated into a simple abstract equation in some appropriately defined Hilbert spaces. For this reason, in the sections which follow, we shall consider a simplified version of this model by assuming a free beam (with no tip bodies) without loss of generality.

For the derived Euler-Bernoulli beam model, we first show that it is impossible to exponentially stabilize the equation using direct velocity feedback which results in a compact feedback operator, as shown at the end of Section 4.1. This motivates us to consider higher order spatial derivative feedback such as strain feedback and shear force feedback.

In Section 4.2, we introduce the concept of A-dependent operators which enables us to exploit, in a unified manner, the semigroup generation property and asymptotic stability of the strain and shear force feedback controlled closed-loop equations for the Euler-Bernoulli model of rotating beams without damping. In practice, however, all physical systems do possess damping, no matter how small, so a good model that reflects the physical evidence should include damping.

In Section 4.3, we analyze the semigroup generation property and exponential stability of the strain and shear force feedback controlled Euler-Bernoulli beam equations with viscous, Kelvin-Voigt and structural damping.

In Section 4.4, we turn back to consider the undamped strain and shear force feedback controlled closed-loop equations with the purpose of investigating their exponential stabilities. By transforming these equations into boundary control problems, and by invoking the time and frequency domain criteria for the exponential stability of C_0-semigroups developed in Chapter 3, we are able to show that these equations are exponentially stable. Furthermore, it is shown that the eigenvalues with large moduli of the closed-loop equation approach a vertical line paralleling the imaginary axis in the left-half complex plane. This result demonstrates that the strain feedback, which is unbounded, is more powerful than bounded feedback in the sense that they can uniformly shift all the closed-loop eigenvalues in the left-half complex plane.

Section 4.5 is devoted to the stability analysis for the shear force feedback controlled Euler-Bernoulli equation of rotating beams. The exponential stability of this equation is not easily proved by using the A-dependent operator concept or the energy multiplier method. However, by estimating the resolvent bound of the operator associated with the closed-loop equation, and by applying the Paley-Wiener theorem explained in Chapter 3, we are able to prove the exponential stability of this equation. The spectral analysis result, when compared with those in the last section, reveals that the higher the order of derivative of the feedback operators, the more powerful they are, in the sense that the closed-loop eigenvalues, especially those with large moduli, can be bent further to the left-half complex plane.

Section 4.6 is concerned with the stability analysis of a hybrid system consisting of a coupled partial differential equation and an ordinary differential equation. Using the essential spectrum theory developed in Section 3.6, Chapter 3, we derive conditions on the feedback gains which guarantee the closed-loop stability of the hybrid system.

The final section of this chapter presents stability results for a nonlinear Euler-Bernoulli beam equation which arises from gain adaptive strain feedback control. The nonlinear semigroup theory stated in Section 2.9 is incorporated to show the existence, uniqueness, and the exponential stability of the solutions.

4.1   Modeling of a rotating beam with a rigid tip body

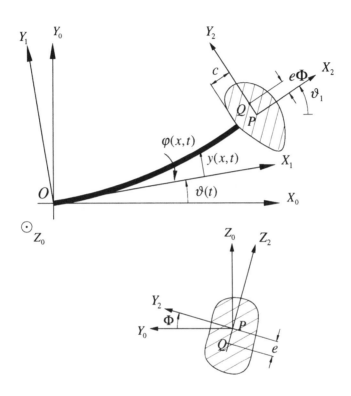

Figure 4.1: Vibrations of a Flexible Arm with a Rigid Tip Body

We consider a flexible beam, one end of which is clamped to a control motor shaft and rotated by the motor at an angular velocity $\dot{\theta}(t)$ in the horizontal plane. We will assume that the beam satisfies the Euler-Bernoulli hypothesis, i.e., the rotary inertia and shear deformation are negligible, and that the beam is of constant cross section, having length ℓ, uniform linear mass density ρ, uniform flexural rigidity EI, and uniform torsional rigidity GJ. An arbitrarily shaped rigid tip body of mass m is securely attached to the free end. Due to the existence of this rigid body, two kinds of vibrations — bending and torsion — can be observed in general, and the two vibrations are coupled through motion of the rigid tip body. It is the purpose of this section to derive a dynamic model which governs the coupled vibrations. We do this by applying the Hamilton's principle.

Let X_0, Y_0, Z_0 designate the inertial Cartesian coordinate axes, where X_0 and Y_0 axes span a horizontal plane, and Z_0 axis is taken so that it coincides with the vertical rotation shaft of the motor. Let $X_1, Y_1, Z_1 (Z_1 = Z_0)$ denote coordinate axes rotating with the motor, and let $\theta(t)$ be the angle of rotation of the motor. Let Q denote the mass center of the rigid tip body, and let P denote the intersection of the beam's tip tangent with a perpendicular plane passing through the mass center Q. Let c be the distance between the beam's tip point to point P. It is further assumed that points P and Q lie on the same vertical line in the equilibrium state. The distance between P and Q are denoted by e.

Let X_2, Y_2, Z_2 be another Cartesian coordinate fixed in the tip body with X_2 being the beam's tip tangent. The axis X_2 still lies in the horizontal plane and is obtained by rotating X_1 axis by an angle of θ_1 due to bending of the beam. We will work under the physically plausible assumption that the tip body oscillates like a pendulum about a shear center axis PX_2, and the change of potential due to the up and down motion of the mass center can be neglected.

Let Φ be the angle of rotation of the tip body about PX_2. The axes Y_2 and Z_2 also oscillate together with the tip body. Since the tip body is a rigid body, it is characterized by two moments of inertia, J_o and J_e, where J_o is with respect to the line passing through the mass center Q and parallel to the axis PZ_2, and J_e is with respect to the line passing through the mass center Q and parallel to the axis PX_2.

Now let $y(x,t)$ and $\phi(x,t)$ denote the transverse displacement of the beam in the rotating frame $X_1 Y_1$ and the angle of twist of the beam, respectively, at location $x(0 < x < \ell)$ and time t. Note that both y and ϕ are assumed to be small. Under this assumption, it is well known, from the elastic mechanics, that the kinetic energy and potential energy associated with the motion of the rotating beam is given, respectively, by

$$T_b(t) = \frac{\rho}{2} \int_0^\ell \left[x\dot{\theta}(t) + \frac{\partial y(x,t)}{\partial t} \right]^2 dx + \frac{\rho\kappa^2}{2} \int_0^\ell \left[\frac{\partial \phi(x,t)}{\partial t} \right]^2 dx \qquad (4.1)$$

and

$$V_b(t) = \frac{EI}{2} \int_0^\ell \left[\frac{\partial^2 y(x,t)}{\partial x^2} \right]^2 dx + \frac{GJ}{2} \int_0^\ell \left[\frac{\partial \phi(x,t)}{\partial x} \right]^2 dx, \qquad (4.2)$$

where $\rho\kappa^2$ is the mass polar moment of inertia per unit length of the beam.

We now need to derive the expression of the kinetic energy of the tip body. In Figure 4.1, let $\vec{R}(t)$ denote the three-dimensional position vector of the mass center Q of the tip body with respect to the inertial coordinate system $X_0 Y_0 Z_0$. Let $\vec{\xi}_1, \vec{\eta}_1,$ and $\vec{\zeta}_1$ be the unit vectors along the axes $OX_1, OY_1,$ and OZ_1. Let $\vec{\xi}_2, \vec{\eta}_2,$ and $\vec{\zeta}_2$ be the unit vectors along the axes $PX_2, PY_2,$ and PZ_2, respectively. Then it can be easily seen from Figure 4.1 that

$$\vec{R}(t) = \vec{\xi}_1(t)\ell + \vec{\eta}_1(t)y(\ell,t) + \vec{\xi}_2(t)c - \vec{\zeta}_2(t)e.$$

Since $y(x,t)$, $\phi(x,t)$ and their derivatives $y_x(x,t)$ and $\phi_x(x,t)$ are assumed to be small, the following relation holds

$$
\begin{bmatrix} \vec{\xi}_2 \\ \vec{\eta}_2 \\ \vec{\zeta}_2 \end{bmatrix} = \begin{bmatrix} 1 & y_x(\ell,t) & 0 \\ -y_x(\ell,t) & 1 & \phi(\ell,t) \\ 0 & -\phi(\ell,t) & 1 \end{bmatrix} \begin{bmatrix} \vec{\xi}_1 \\ \vec{\eta}_1 \\ \vec{\zeta}_1 \end{bmatrix}.
$$

Substituting this equation into the expression of $\vec{R}(t)$ yields

$$
\vec{R}(t) = \vec{\xi}_1(t)(\ell + c) + \vec{\eta}_1(t)[y(\ell,t) + cy_x(\ell,t) + e\phi(\ell,t)] - \vec{\zeta}_1(t)e.
$$

Since the angular velocity $\vec{\omega}(t)$ of the coordinate system $X_1 Y_1 Z_1$ with respect to the inertial coordinate system $X_0 Y_0 Z_0$ can be written as $\vec{\omega}(t) = \dot{\theta}(t)\vec{\zeta}_1$, we obtain

$$
\frac{d}{dt}\vec{\xi}_1 = \vec{\omega} \times \vec{\xi}_1 = \dot{\theta}(t)\vec{\eta}_1(t).
$$

Similarly,

$$
\frac{d}{dt}\vec{\eta}_1 = -\dot{\theta}(t)\vec{\xi}_1(t), \quad \frac{d}{dt}\vec{\zeta}_1 = 0.
$$

Using these relations, we see that

$$
\begin{aligned}
\frac{d}{dt}\vec{R}(t) = {} & \vec{\eta}_1(t)[\dot{y}(\ell,t) + c\dot{y}_x(\ell,t) + e\dot{\phi}(\ell,t) + (\ell + c)\dot{\theta}(t)] \\
& -\vec{\xi}_1(t)\dot{\theta}(t)[y(\ell,t) + cy_x(\ell,t) + e\phi(\ell,t)].
\end{aligned}
$$

Ignoring the higher order terms $\dot{\theta}y$, $\dot{\theta}y_x$, and $\dot{\theta}\phi$, we obtain

$$
\left\| \frac{d}{dt}\vec{R} \right\|^2 = [(\ell + c)\dot{\theta}(t) + \dot{y}(\ell,t) + c\dot{y}_x(\ell,t) + e\dot{\phi}(\ell,t)]^2.
$$

Thus, the total kinetic energy of the tip body is given by

$$
\begin{aligned}
T_t = {} & \frac{1}{2}J_e[\dot{\phi}(\ell,t)]^2 + \frac{1}{2}J_0[\dot{\theta}(t) + \dot{y}_x(\ell,t)]^2 \\
& + \frac{1}{2}m[(\ell + c)\dot{\theta}(t) + \dot{y}(\ell,t) + c\dot{y}_x(\ell,t) + e\dot{\phi}(\ell,t)]^2. \tag{4.3}
\end{aligned}
$$

The first term in this equation represents the energy of rotation about the axis passing through the mass center Q and parallel to PX_2, and the second term represents the energy of rotation about the axis PZ_2. The third term represents the kinetic energy of the translational motion of the mass center.

Hamilton's principle states that the variation of the total kinetic energy and potential energy of the system considered during any time interval t_0 to t_1 must equal zero. Hence, in the usual notation,

$$
\delta \int_{t_0}^{t_1} (T_b + T_t - V_b)dt = 0.
$$

Substituting the expressions in (4.1), (4.2), and (4.3) into this equation and using standard variational arguments, we obtain the resulting Euler-Lagrange equations governing the motion of the rotating beam with a rigid tip body:

$$\frac{\partial^2 y(x,t)}{\partial t^2} + \frac{EI}{\rho}\frac{\partial^4 y(x,t)}{\partial x^4} = -x\ddot{\theta}(t), \qquad (4.4)$$

which describes the bending vibration of the beam,

$$\frac{\partial^2 \phi(x,t)}{\partial t^2} - \frac{GJ}{\rho\kappa^2}\frac{\partial^2 \phi(x,t)}{\partial x^2} = 0, \qquad (4.5)$$

which describes the torsional vibration of the beam, and

$$
\begin{aligned}
EIy_{xxx}(\ell,t) &= m[(\ell+c)\ddot{\theta}(t) + \ddot{y}(\ell,t) + c\ddot{y}_x(\ell,t) + e\ddot{\phi}(\ell,t)], & (4.6)\\
-EIy_{xx}(\ell,t) &= mc[(\ell+c)\ddot{\theta}(t) + \ddot{y}(\ell,t) + c\ddot{y}_x(\ell,t) + e\ddot{\phi}(\ell,t)]\\
&\quad + J_o[\ddot{\theta}(t) + \ddot{y}_x(\ell,t)], & (4.7)\\
-GJ\phi_x(\ell,t) &= me[(\ell+c)\ddot{\theta}(t) + \ddot{y}(\ell,t) + c\ddot{y}_x(\ell,t) + e\ddot{\phi}(\ell,t)]\\
&\quad + J_e\ddot{\phi}(\ell,t), & (4.8)
\end{aligned}
$$

which describes motion of the tip body. In addition to these equations, the beam is subject to the clamped boundary conditions at $x = 0$ which are given by

$$y(0,t) = y_x(0,t) = 0, \ \ \phi(0,t) = 0. \qquad (4.9)$$

Finally, the equation of motion of the control motor can be written as

$$J_m\ddot{\theta}(t) = \tau(t) + EIy_{xx}(0,t), \qquad (4.10)$$

where J_m is the moment of inertia of the motor and τ is the torque developed by the motor; $EIy_{xx}(0,t)$ represents the reaction torque exerted on the motor shaft by the beam.

The complete dynamic model which governs the coupled bending and torsional vibrations of the rotating flexible beam is thus given by (4.4) and (4.5) subject to the essential boundary conditions (4.9) and the natural boundary conditions (4.6)-(4.8) with the initial conditions

$$y(x,0) = y_0(x), \ \ \dot{y}(x,0) = y_1(x), \qquad (4.11)$$
$$\phi(x,0) = \phi_0(x), \ \ \dot{\phi}(x,0) = \phi_1(x). \qquad (4.12)$$

In order to investigate the properties that this model possesses, it is desirable to write the boundary-initial value problem described by the above mentioned equations as an abstract equation.

Let us first introduce a Hilbert space $H = L^2(0,\ell) \times L^2(0,\ell) \times \mathbf{C}^3$ as the state space and

$$u(t) = (y(\cdot,t), \phi(\cdot,t), y(\ell,t), y_x(\ell,t), \phi(\ell,t))$$

as the state variable. The inner product in H is defined by

$$\langle (u_1, u_2, u_3, u_4, u_5),\ (v_1, v_2, v_3, v_4, v_5) \rangle$$

$$= \rho \int_0^\ell [u_1(x)\overline{v_1(x)} + \kappa^2 u_2(x)\overline{v_2(x)}]dx$$

$$+ \left\langle M \begin{bmatrix} u_3 \\ u_4 \\ u_5 \end{bmatrix},\ \begin{bmatrix} v_3 \\ v_4 \\ v_5 \end{bmatrix} \right\rangle_{\mathbf{C}^3}, \qquad (4.13)$$

where

$$M = \begin{bmatrix} m & mc & me \\ mc & J_o + mc^2 & mce \\ me & mce & J_e + me^2 \end{bmatrix}. \qquad (4.14)$$

Define the operator $\Sigma : D(\Sigma) \subset H \to H$ by

$$\Sigma u = \left(\frac{EI}{\rho} u_1''''(\cdot),\ -\frac{GJ}{\rho \kappa^2} u_2''(\cdot),\ -EI u_1'''(\ell),\ EI u_1''(\ell),\ GJ u_2'(\ell) \right)^T, \qquad (4.15)$$

$$D(\Sigma) = \{u = (u_1, u_2, u_3, u_4, u_5) | u_1 \in H^4(0, \ell),\ u_2 \in H^2(0, \ell)$$
$$u_3 = u_1(\ell),\ u_4 = u_1'(\ell),\ u_5 = u_2(\ell)$$
$$u_1(0) = 0,\ u_1'(0) = 0,\ u_2(0) = 0\}, \qquad (4.16)$$

where $D(\Sigma)$ denotes domain of the operator Σ. Let Λ denote

$$\Lambda = \left[\begin{array}{cc|c} 1 & 0 & 0 \\ 0 & 1 & \\ \hline & 0 & M \end{array} \right]$$

and

$$\Omega = -(x,\ 0,\ m(\ell + c),\ J_o + mc(\ell + c),\ me(\ell + c))^T \in H. \qquad (4.17)$$

With these preparations, we can rewrite (4.4), (4.5), (4.6)-(4.8), (4.9), (4.11), and (4.12) as the following abstract equation on H:

$$\begin{cases} \Lambda \ddot{u}(t) + \Sigma u(t) = \Omega \ddot{\theta}(t), \\ u(0) = u_0,\ \dot{u}(0) = u_1, \end{cases} \qquad (4.18)$$

where

$$u_0 = (y_0,\ \phi_0,\ y_0(\ell),\ y_0'(\ell),\ \phi_0(\ell))^T,$$
$$u_1 = (y_1,\ \phi_1,\ y_1(\ell),\ y_1'(\ell),\ \phi_1(\ell))^T.$$

Since the inertia matrix M is always positive, Λ is invertible. If we define an operator B in H by

$$Bu = \Lambda^{-1}\Sigma u, \quad D(B) = D(\Sigma), \qquad (4.19)$$

then (4.18) is equivalently written as

$$\ddot{u}(t) + Bu(t) = \Lambda^{-1}\Omega f(t) \qquad (4.20)$$

where $f(t) = \ddot{\theta}(t)$ is considered as the control input. Equation (4.20) is a second order evolution equation without damping in the Hilbert space H. The vibration suppression problem is to appropriately choose the control input as a function of $u(t)$ and $\dot{u}(t)$ such that the closed-loop solution is stable either in the exponential convergence sense or in the asymptotic convergence sense. Before we discuss control problems, we wish to explore what properties the operator B possesses.

Proposition 4.1 *Let A be a closed symmetric operator in a Hilbert space H. Then the following holds.*

(i) A is densely defined and self-adjoint if and only if $\sigma(A)$ is confined to the real axis.

(ii) If A is densely defined and $0 \in \rho(A)$, then A is self-adjoint.

Proof. For (i), we refer the reader to Taylor [151, p.385]. For (ii), since A is symmetric, $D(A) \subset D(A^*)$. If we can show $D(A^*) \subset D(A)$, then A is self-adjoint. To this end, let $y \in D(A^*)$ be arbitrary. Since $0 \in \rho(A)$, there exists an $x_0 \in D(A)$ such that $Ax_0 = A^*y$. From the identities

$$\langle Ax, \, y \rangle = \langle x, \, A^*y \rangle = \langle x, \, Ax_0 \rangle = \langle Ax, \, x_0 \rangle$$

for all $x \in D(A)$, we see that

$$\langle Ax, \, y - x_0 \rangle = 0$$

for all $x \in D(A)$. Hence, $y = x_0, i.e., y \in D(A)$, which shows that $D(A^*) \subset D(A)$. $\qquad \square$

Theorem 4.2 *The operator B given in (4.19) is densely defined and symmetric in H. The inverse B^{-1} of B exists and is compact. Hence, B is self-adjoint and positive definite.*

Proof. The denseness is obvious. We first show that B is symmetric. For any $u, v \in D(B)$, we see that

$$
\begin{aligned}
\langle Bu, v \rangle &= \langle \Lambda^{-1} \Sigma u, v \rangle \\
&= \int_0^\ell EI u_1''''(x) \overline{v_1(x)} dx - \int_0^\ell GJ u_2''(x) \overline{v_2(x)} dx \\
&\quad - EI u_1'''(\ell) \overline{v_1(\ell)} + EI u_1''(\ell) \overline{v_1'(\ell)} + GJ u_2'(\ell) \overline{v_2(\ell)} \\
&= EI \int_0^\ell u_1''(x) \overline{v_1''(x)} dx + GJ \int_0^\ell u_2'(x) \overline{v_2'(x)} dx, \qquad (4.21)
\end{aligned}
$$

where the last equality is derived by integration by parts and by employing the boundary conditions

$$
u_1(0) = u_1'(0) = u_2(0) = 0. \qquad (4.22)
$$

Equation (4.21) implies that B is symmetric.

Next, we show that $0 \in \rho(B)$. It is sufficient to show that $0 \in \rho(\Sigma)$. Given $v = (f, g, a, b, c) \in H$, solving $\Sigma u = v$, we find that this equation has a unique solution $u(x) = (u_1(x), u_2(x), u_1(\ell), u_1'(\ell), u_2(\ell))$ where

$$
\begin{aligned}
u_1(x) &= -\frac{1}{6EI}[a\ell^3 + \rho \int_0^\ell \tau^3 f(\tau) d\tau] + \frac{1}{2EI}[a\ell^2 + \rho \int_0^\ell \tau^2 f(\tau) d\tau]x \\
&\quad + \frac{b}{2EI}x^2 - \frac{1}{6EI}(x-\ell)^3 + \frac{\rho}{EI}\int_x^\ell \frac{(\tau - x)^3}{6} f(\tau) d\tau, \\
u_2(x) &= cx + \frac{\rho \kappa^2}{GJ} \int_0^\ell \tau g(\tau) d\tau - \frac{\rho \kappa^2}{GJ} \int_x^\ell (\tau - x) g(\tau) d\tau.
\end{aligned}
$$

It is obvious from these expressions that $0 \in \rho(B)$ and B^{-1} is compact. By Proposition 4.1, we see that B is self-adjoint. Finally, from (4.21), it is clear that $\langle Bu, u \rangle \geq 0$ for all $u \in D(B)$, which together with $0 \in \rho(B)$ shows that B is positive definite. $\qquad \square$

Before we proceed, we wish to introduce the concept of A-dependent operators and to study their properties.

Definition 4.3 (A-dependent operators) *Let A be an unbounded, self-adjoint, and positive definite operator on the complex Hilbert space H. Let Π be an operator with domain $D(\Pi) \supset D(A)$ and satisfies*

i) Π is A-bounded, i.e., there exist positive constants a and b such that

$$
\|\Pi u\| \leq a \|u\| + b \|Au\|, \quad \forall u \in D(A);
$$

ii) Π *is A-symmetric, i.e., for any* $u, v \in D(A)$,

$$\langle \Pi u, \ Av \rangle = \langle Au, \ \Pi v \rangle;$$

iii) Π *is A-positive semidefinite, i.e., for any* $u \in D(A)$

$$\langle \Pi u, \ Au \rangle \geq 0.$$

Then Π *is said to be A-dependent.*

We state, in the following, a theorem which holds for A-dependent operators.

Theorem 4.4 *An A-dependent operator* Π *can be decomposed as* $\Pi = QA$ *with* Q *being a bounded, self-adjoint, and positive semidefinite operator on* H.

Proof. Let $Q = \Pi A^{-1}$. Then $\Pi = QA$. The proof is complete if we can show that Q satisfies all the properties required in this theorem. For any $u \in H$, since Π is A-bounded, i.e., there exist constants a, b, and c such that

$$
\begin{aligned}
\|Qu\| &= \ \left\|\Pi A^{-1} u\right\| \leq a \left\|A^{-1} u\right\| + b \left\|AA^{-1} u\right\| \\
&= \ b \|u\| + a \left\|A^{-1} u\right\| \\
&\leq \ b \|u\| + ac \|u\| = (b + ac) \|u\|,
\end{aligned}
$$

where the boundedness of A^{-1} has been used, which shows that Q is bounded. For any $u, v \in H$, set $y_1 = A^{-1} u \in D(A)$ and $y_2 = A^{-1} v \in D(A)$. Since Π is A-symmetric, we have

$$
\begin{aligned}
\langle Qu, v \rangle &= \ \langle \Pi A^{-1} u, \ v \rangle = \langle \Pi y_1, \ Ay_2 \rangle \\
&= \ \langle Ay_1, \ \Pi y_2 \rangle = \langle u, \ \Pi A^{-1} v \rangle \\
&= \ \langle u, \ Qv \rangle,
\end{aligned}
$$

which shows Q is self-adjoint. Finally, for any $u \in H$, set $y = A^{-1} u$. Then

$$\langle Qu, u \rangle = \langle \Pi A^{-1} u, u \rangle = \langle \Pi y, \ Ay \rangle \geq 0,$$

which shows that Q is positive semidefinite. $\square$

The next theorem compares the eigenvalues of the operator $A + \Pi$ with those of A.

Theorem 4.5 *Let* $\tilde{\lambda}_i$ *and* λ_i *be the i-th eigenvalues of the operator* $A + \Pi$ *and* A, *respectively. Let* Π *be an A-dependent operator. Then* $\tilde{\lambda}_i$ *are real and positive. Moreover* $\tilde{\lambda}_i \geq \lambda_i$ *for* $i = 1, 2, \cdots$.

Proof. Decompose Π as $\Pi = Q_0 A$, where Q_0 is bounded, self-adjoint and positive semidefinite by Theorem 4.4. It is clear that $A + \Pi = (I + Q_0)A$, where I denotes the identity operator on H. Putting $Q = I + Q_0$, we know that Q is bounded, self-adjoint, and positive definite. The following facts can easily be proved: 1) the spectral set of $A + \Pi$ consists only of its eigenvalues, since $(A + \Pi)^{-1} = A^{-1}Q^{-1}$ is compact; 2) QA and $Q^{1/2}AQ^{1/2}$ have the same set of eigenvalues; 3) $Q^{-1/2}A^{-1}Q^{-1/2}$ is compact. Let $\hat{\lambda}_i$ denote the ith eigenvalue of $Q^{-1/2}A^{-1}Q^{-1/2}$. Using the Courant-Fischer's min-max theorem for self-adjoint and compact operators [32], we have

$$\hat{\lambda}_i = \min_{\dim \mathcal{M}=i-1} \max_{\substack{x \perp \mathcal{M} \\ \|x\| \le 1}} \langle Q^{-1/2}A^{-1}Q^{-1/2}x,\, x \rangle$$

$$= \min_{\dim \mathcal{M}=i-1} \max_{\substack{y \perp Q^{1/2}\mathcal{M} \\ \|Q^{1/2}y\| \le 1}} \langle A^{-1}y,\, y \rangle$$

where $\mathcal{M}$ denotes any $(i-1)$-dimensional Euclidean space. It is obvious that $\dim \mathcal{M} = i - 1$ is equivalent to $\dim Q^{1/2}\mathcal{M} = i - 1$. Let m denote the minimal eigenvalue of Q. Then $m \ge 1$ from the definition of Q. From this, we see that $\sqrt{m}\,\|y\| \le \|Q^{1/2}y\|$ for any $y \in H$. If $\|Q^{1/2}y\| \le 1$, then $\|y\| \le \frac{1}{\sqrt{m}} \le 1$. Thus $\hat{\lambda}_i$ can be bounded by

$$\hat{\lambda}_i \le \min_{\dim \mathcal{M}=i-1} \max_{\substack{y \perp \mathcal{M} \\ \|y\| \le 1}} \langle A^{-1}y,\, y \rangle = \frac{1}{\lambda_i},$$

which implies that

$$\tilde{\lambda}_i = \frac{1}{\hat{\lambda}_i} \ge \lambda_i.$$

That is, the i-th eigenvalue of the perturbed operator $A + \Pi$ is greater than or equal to the i-th eigenvalue of the operator A. □

Equation (4.20) represents a general vibration model for a wide class of systems. For some particular cases, simple models can be directly derived from this model. For example, when we consider the tip body of the flexible beam as a concentrated mass, instead of a rigid body, only bending vibration has to be considered, and the dynamic equation reduces to

$$
\begin{cases}
\dfrac{\partial^2 y(x,t)}{\partial t^2} + \dfrac{EI}{\rho}\dfrac{\partial^4 y(x,t)}{\partial x^4} = -x\ddot{\theta}(t), \quad 0 < x < \ell,\ t > 0, \\[2mm]
y(0,t) = y_x(0,t) = 0, \\[2mm]
EIy_{xx}(\ell,t) = 0, \\[2mm]
m\ddot{y}(\ell,t) - EIy_{xxx}(\ell,t) = -\ell\ddot{\theta}(t), \\[2mm]
y(x,0) = y_0(x),\ \dot{y}(x,0) = y_1(x)
\end{cases}
\qquad (4.23)
$$

which can be formulated as the abstract equation (4.20) if we properly define the operator A, and choose the Hilbert space H as $H = L^2(0, \ell) \times C$ and $u(t) = (y(\cdot, t), y(\ell, t))$ as the state variable.

Based on this observation, we will consider the most simple case where the tip end of the flexible beam is free, i.e., no tip body. Also, for notational simplicity, we assume from now on that all the physical variables in the beam dynamic model are nondimensionalized such that ρ, EI and ℓ can be taken to be 1, without loss of generality. In this case, the dynamic model of a rotating beam is given by

$$
\begin{cases}
\dfrac{\partial^2 y(x, t)}{\partial t^2} + \dfrac{\partial^4 y(x, t)}{\partial x^4} = -x\ddot{\theta}(t), \quad 0 < x < 1, \ t > 0, \\[2mm]
y(0, t) = y_x(0, t) = 0, \\[2mm]
y_{xx}(1, t) = 0, \\[2mm]
-y_{xxx}(1, t) = 0, \\[2mm]
y(x, 0) = y_0(x), \ \dot{y}(x, 0) = y_1(x).
\end{cases}
\tag{4.24}
$$

Here again we assume that there is no need to consider torsional vibrations. The operator A corresponding to (4.24) is reduced to

$$
\begin{cases}
D(A) = \{y(x) \in H^4(0, 1) | y(0) = y'(0) = y''(1) = y'''(1) = 0\}, \\[2mm]
Ay(x) = y''''(x), \quad \forall y \in D(A)
\end{cases}
\tag{4.25}
$$

and A shares all the properties which the operator B possesses in Theorem 4.2. We use this A throughout this chapter unless otherwise specified. The results in the following lemma will be used frequently in the sequel.

Lemma 4.6 *Again, let $H = L^2(0, 1)$. Let A be the operator defined in (4.25). Let $\{(\lambda_n, \phi_n)\}_{n=1}^{\infty}$ be the eigenvalues and eigenvectors of A. Then*

(i) $\lambda_n = \beta_n^4$ with $\beta_n = O(n)$, where β_n are the solutions of

$$
1 + \cos(\beta_n) \cosh(\beta_n) = 0;
$$

(ii) $\{\phi_n\}$ forms an orthogonal basis of H and can be expressed as

$$
\phi_n(x) = -\frac{1 + \gamma_n}{2} \exp(\beta_n x) - \frac{1 - \gamma_n}{2} \exp(-\beta_n x)
$$
$$
+ \gamma_n \sin(\beta_n x) + \cos(\beta_n x),
$$
$$
\gamma_n = -\frac{\exp(\beta_n) - \sin(\beta_n) + \cos(\beta_n)}{\exp(\beta_n) + \sin(\beta_n) + \cos(\beta_n)} \to -1, \ as \ n \to \infty;
$$

(iii) $\gamma_n < 0$ for each n;

(iv) Let $x = \sum_{n=1}^{\infty} b_n \phi_n(x) \in L^2(0,1)$. Then

$$\phi_n''(0) = -2\beta_n^2, \ b_n = -2\beta_n^{-2} \|\phi_n\|^{-2}, \ \|\phi_n\| = O(1)$$

and

$$\phi_n'''(0) = -2\gamma_n \beta_n^3;$$

(v) Let $1 = \sum_{n=1}^{\infty} c_n \phi_n(x) \in H$. Then

$$c_n = \frac{\langle 1, \ \phi_n \rangle}{\|\phi_n\|^2} = O(\beta_n^{-1}).$$

Proof. The formulations of λ_n and $\phi_n(x)$ can be verified by directly solving $A\phi_n(x) = \lambda_n \phi_n(x)$, $\phi_n(x) \in D(A)$. Since it can be shown that A^{-1} is compact, the spectrum of A contains only eigenvalues. Again since A^{-1} is compact and self-adjoint, $\{\phi_n(x)\}$ is complete. Thus, (i) and (ii) are proved. To show (iii), notice that

$$
\begin{aligned}
\gamma_n &= -\frac{\exp(\beta_n) - \sin(\beta_n) + \cos(\beta_n)}{\exp(\beta_n) + \sin(\beta_n) + \cos(\beta_n)} \\
&= -\frac{[\exp(\beta_n) - \sin(\beta_n) + \cos(\beta_n)][\exp(\beta_n) + \sin(\beta_n) + \cos(\beta_n)]}{[\exp(\beta_n) + \sin(\beta_n) + \cos(\beta_n)]^2}.
\end{aligned}
$$

Thus, it is only needed to show that

$$
\begin{aligned}
g(x) &= (\exp(x) - \sin(x) + \cos(x))(\exp(x) + \sin(x) + \cos(x)) \\
&= \exp(2x) + 2\exp(x)\cos(x) + \cos^2(x) - \sin^2(x) > 0
\end{aligned}
$$

for all x satisfying $1 + \cos(x)\cosh(x) = 0$, which is left as an exercise. Note that the numerator of γ_n is exactly $g(\beta_n)$. (iv) follows from the fact that

$$
\begin{aligned}
b_n \|\phi_n\|^2 &= \langle x, \ \phi_n \rangle = \frac{1}{\lambda_n} \int_0^1 x\phi_n''''(x)dx \\
&= -\frac{1}{\lambda_n} \int_0^1 \phi_n'''(x)dx = \frac{1}{\lambda_n}\phi_n''(0) = -2\beta_n^{-2}.
\end{aligned}
$$

$\phi_n''(0)$ and $\phi_n'''(0)$ can be directly calculated from the expression of $\phi_n(x)$. Similarly, since

$$
\begin{aligned}
c_n \|\phi_n\|^2 &= \langle 1, \ \phi_n \rangle = \frac{1}{\lambda_n} \int_0^1 \phi_n''''(x)dx \\
&= -\frac{1}{\lambda_n}\phi_n'''(0),
\end{aligned}
$$

it is seen that $c_n = O(\beta_n^{-1})$ from the results in (iv). $\qquad\square$

Using A and letting $y(t) = y(\cdot, t)$, we can now write (4.24)as the following abstract equation

$$
\begin{cases}
\ddot{y}(t) + Ay(t) = -\Omega\ddot{\theta}(t), \\
y(0) = y_0, \ \dot{y}(0) = y_1
\end{cases}
\tag{4.26}
$$

on the Hilbert space $L^2(0,1)$, where $\Omega = x \in L^2(0,1)$.

So far, we have derived dynamic models for the rotating beam with or without tip bodies. We next consider a flexible beam driven by a moving XY table. Suppose the flexible beam is aligned along the Z-axis and one end is rigidly attached to the moving table of the robot while the other end is free. We shall call this a *translating beam* because the beam and the robot are configured in a Cartesian coordinate. For such beams, vibrations can occur in both the X-axis direction and the Y-axis direction. Since vibrations in these two directions are decoupled, we only consider the vibration in the X-axis direction. Denote the moving distance of the moving table in the X-axis direction by $s(t)$. Then $\ddot{s}(t)$ is the linear acceleration of the moving table. Let all the other variables and constants be the same as in the rotating beam, and again the Euler-Bernoulli beam is assumed and the effect of gravity is to be ignored. Then, by a similar argument as before, it is easy to see that the dynamic model for the beam vibration is governed by

$$
\begin{cases}
\dfrac{\partial^2 y(x,t)}{\partial t^2} + \dfrac{\partial^4 y(x,t)}{\partial x^4} = -\ddot{s}(t), \quad 0 < x < 1, \ t > 0, \\[2mm]
y(0,t) = y_x(0,t) = 0, \\[2mm]
y_{xx}(1,t) = 0, \\[2mm]
-y_{xxx}(1,t) = 0, \\[2mm]
y(x,0) = y_0(x), \ \dot{y}(x,0) = y_1(x).
\end{cases}
\tag{4.27}
$$

Equation (4.27) should be compared to (4.24). The only difference between these two models is that it is the angular acceleration $\ddot{\theta}(t)$ that causes vibration in the rotating beam, while the vibration in the translating beam is caused by the linear acceleration $\ddot{s}(t)$ of the moving table to which the beam is fixed. Again, by using the operator A defined in (4.25), we can write (4.27) as an abstract equation similar to (4.26) on the Hilbert space $L^2(0,1)$.

A common technique to stabilize (4.26) is to use velocity feedback to introduce damping into the system. We show that bounded velocity feedback cannot exponentially stabilize (4.26) by using the compact perturbation results stated in Chapter 3.

Let F be a bounded linear functional on H. Choose the control input $f(t) = \ddot{\theta}(t)$ as

$$
f(t) = -F\dot{y}(t).
\tag{4.28}
$$

Substituting this into (4.26) yields

$$\begin{cases} \ddot{y}(t) + B\dot{y}(t) + Ay(t) = 0, \\[2mm] y(0) = y_0, \ \dot{y}(0) = y_1, \end{cases} \tag{4.29}$$

where $B = \Omega F$ is a linear compact operator on H because its range is of one dimension. Let

$$z_1 = A^{1/2}y, \quad z_2 = \dot{y},$$

$$C = \begin{bmatrix} 0 & A^{1/2} \\ -A^{1/2} & 0 \end{bmatrix}, \ D = \begin{bmatrix} 0 & 0 \\ 0 & B \end{bmatrix}.$$

Then, (4.29) can be written as

$$\begin{bmatrix} \dot{z}_1 \\ \dot{z}_2 \end{bmatrix} = \begin{bmatrix} 0 & A^{1/2} \\ -A^{1/2} & B \end{bmatrix} \begin{bmatrix} z_1 \\ z_2 \end{bmatrix} = (C+D) \begin{bmatrix} z_1 \\ z_2 \end{bmatrix} \tag{4.30}$$

on the Hilbert space $Y = H \times H$. Clearly, $C^* = -C$, and hence by Stone's theorem C generates a unitary C_0−group $T(t)$ and $C+D$ generates C_0−group $T_D(t)$, respectively. The solution of (4.30) can be expressed as

$$\begin{bmatrix} z_1(t) \\ z_2(t) \end{bmatrix} = T_D(t) \begin{bmatrix} z_1(0) \\ z_2(0) \end{bmatrix}. \tag{4.31}$$

Since $\|T(t)\| = 1$ for all $t \in (-\infty, \infty)$, by virtue of Corollary 3.58 in Chapter 3, we know that $T_D(t)$ is not exponentially stable.

Therefore, the solution of (4.29) is not exponentially stable. That is, system (4.26) is not exponentially stabilizable using bounded velocity feedback (4.28).

This negative result motivates us to consider unbounded feedback to achieve exponential stabilizability of Euler-Bernoulli beam equations, which is discussed below.

4.2 Stabilization using strain or shear force feedback

Since the velocity feedback with bounded feedback operator cannot exponentially stabilize (4.26), we try to use *strain feedback* to stabilize the rotating beam equation and *shear force feedback* to stabilize the translating beam equation. From the elastic mechanics, we know that $y_{xx}(0, t)$ corresponds to the strain of the beam at the point $x = 0$, which can be measured by cementing strain gauge foils at the clamped end of the beam, and that $y_{xxx}(0, t)$ corresponds to the shear force of the beam, which can be measured by either properly arranged strain gauges or load cells. Suppose now that we can control the driving motor such that its angular velocity is proportional to the strain $y_{xx}(0, t)$, i.e.,

$$\dot{\theta}(t) = ky_{xx}(0, t), \tag{4.32}$$

where $k > 0$ is a feedback gain. Then, the time derivative of both sides of the above equation leads to

$$\ddot{\theta}(t) = ky_{xxt}(0, t). \tag{4.33}$$

Substituting this into (4.24) yields the strain feedback controlled closed-loop equation for the rotating beam:

$$\begin{cases} \dfrac{\partial^2 y(x, t)}{\partial t^2} + kx\dfrac{\partial^3 y(0, t)}{\partial t \partial x^2} + \dfrac{\partial^4 y(x, t)}{\partial x^4} = 0, \quad 0 < x < 1, \ t > 0, \\[2mm] y(0, t) = y_x(0, t) = 0, \\[1mm] y_{xx}(1, t) = y_{xxx}(1, t) = 0, \\[1mm] y(x, 0) = y_0(x), \ \dot{y}(x, 0) = y_1(x). \end{cases} \tag{4.34}$$

Similarly, for the translating beam, suppose we can control the motion of the driving motor such that the linear velocity of the moving table is proportional to the shear force $y_{xxx}(0, t)$, i.e.,

$$\dot{s}(t) = -ky_{xxx}(0, t). \tag{4.35}$$

Substituting this into (4.27) gives the shear force feedback controlled closed-loop equation for the translating beam:

$$\begin{cases} \dfrac{\partial^2 y(x, t)}{\partial t^2} - k\dfrac{\partial^4 y(0, t)}{\partial t \partial x^3} + \dfrac{\partial^4 y(x, t)}{\partial x^4} = 0, \quad 0 < x < 1, \ t > 0, \\[2mm] y(0, t) = y_x(0, t) = 0, \\[1mm] y_{xx}(1, t) = y_{xxx}(1, t) = 0, \\[1mm] y(x, 0) = y_0(x), \ \dot{y}(x, 0) = y_1(x). \end{cases} \tag{4.36}$$

The question is now whether equations (4.34) and (4.36) are exponentially stable, or in other words, whether the terms $kxy_{xxt}(0, t)$ and $-ky_{xxxt}(0, t)$ introduce damping. We use the A-dependent operator method to answer this question.

For this purpose, define the operators Π_0 and Π_1 as follows:

$$\Pi_0 u(x) = xu''(0), \quad \Pi_1 u(x) = -u'''(0), \ \forall u \in D(A). \tag{4.37}$$

It is seen that both Π_0 and Π_1 are unbounded operators, but neither Π_0 nor Π_1 is self-adjoint. However, we have the following.

Lemma 4.7 *Both Π_0 and Π_1 are A-dependent for the operator A specified by (4.25).*

Proof. We only show that Π_0 satisfies all three conditions in Definition 4.3. The proof for Π_1 is similar. i) For any $u \in D(A)$, we have

$$
\begin{aligned}
\|\Pi_0 u\|^2 &= \int_0^1 x^2 |u''(0)|^2 dx \\
&= \frac{1}{3} \left| \int_0^1 u'''(x) dx \right|^2 \\
&= \frac{1}{3} \left| \int_0^1 \int_x^1 u''''(\tau) d\tau dx \right|^2 \\
&\leq \frac{1}{3} \int_0^1 \int_0^1 |u''''(\tau)|^2 d\tau dx \\
&\leq C^2 \|Au\|^2
\end{aligned}
$$

where $C = \sqrt{1/3}$. ii) For any $u, v \in D(A)$, it is seen that

$$
\begin{aligned}
\langle \Pi_0 u, \ Av \rangle &= \int_0^1 x u''(0) \overline{v}''''(x) dx \\
&= -\int_0^1 u''(0) \overline{v}'''(x) dx \\
&= u''(0) \overline{v}''(0) = \langle Au, \ \Pi_0 v \rangle.
\end{aligned}
$$

iii) Setting $v = u$ in above equation leads to

$$
\langle \Pi_0 u, \ Au \rangle = |u''(0)|^2 \geq 0,
$$

which completes the proof. □

With the aid of this lemma, we are now able to show the existence, uniqueness, semigroup generation property, and stability of the solutions of the strain feedback controlled closed-loop equation (4.34) and the shear force feedback controlled closed-loop equation (4.36). From the arguments before, we see that both (4.34) and (4.36) can be written as

$$
\begin{cases}
\ddot{y}(t) + QA\dot{y}(t) + Ay(t) = 0, \quad t > 0, \\
y(0) = y_0, \ \dot{y}(0) = y_1
\end{cases}
\tag{4.38}
$$

with some appropriate Q which is bounded, self-adjoint, and positive semidefinite. Consider a product Hilbert space $\mathcal{H} = H \times H$. Let $z_1 = Ay, z_2 = A^{1/2}\dot{y}$ and

$$
\mathcal{A} = \begin{bmatrix} A^{1/2} & 0 \\ 0 & -A^{1/2} \end{bmatrix} \begin{bmatrix} 0 & I \\ I & QA^{1/2} \end{bmatrix},
\tag{4.39}
$$

$$
D(\mathcal{A}) = \{(z_1, z_2)^T | z_2 \in D(A^{1/2}), z_1 + QA^{1/2}z_2 \in D(A^{1/2})\}.
\tag{4.40}
$$

Then (4.38) can be written as

$$\frac{d}{dt}\left[\begin{array}{c} z_1(t) \\ z_2(t) \end{array}\right] = \mathcal{A}\left[\begin{array}{c} z_1(t) \\ z_2(t) \end{array}\right]. \tag{4.41}$$

We have the following theorem.

Theorem 4.8 *The operator $\mathcal{A}$ defined above is the infinitesimal generator of a C_0-semigroup of contractions $T(t)$ on $\mathcal{H}$, and hence, there exists a unique mild solution to (4.38) which is expressed as*

$$\left[\begin{array}{c} y(t) \\ \dot{y}(t) \end{array}\right] = \left[\begin{array}{cc} A^{-1} & 0 \\ 0 & A^{-1/2} \end{array}\right] T(t) \left[\begin{array}{cc} A & 0 \\ 0 & A^{1/2} \end{array}\right] \left[\begin{array}{c} y_0 \\ y_1 \end{array}\right]$$

$$= \bar{T}(t) \left[\begin{array}{c} y_0 \\ y_1 \end{array}\right]$$

for any initial conditions satisfying $y_0 \in D(A), y_1 \in D(A^{1/2})$ and $(y, \dot{y})^T \in C(0, \infty; D(\mathcal{A}))$ when $y_1 \in D(A), A y_0 + Q A y_1 \in D(A^{1/2})$, where $\bar{T}(t)$ is a C_0-semigroup of contractions on $D(A) \times D(A^{1/2})$. Thus, the solution is stable in the Lyapunov sense.

Proof. For any $\Phi = (\phi_1, \phi_2)^T \in D(\mathcal{A})$,

$$\begin{aligned} \mathrm{Re}\langle \mathcal{A}\Phi, \, \Phi \rangle_{\mathcal{H}} &= \mathrm{Re}\langle A^{1/2}\phi_2, \, \phi_1 \rangle - \mathrm{Re}\langle A^{1/2}(\phi_1 + Q A^{1/2}\phi_2), \, \phi_2 \rangle \\ &= \mathrm{Re}\langle A^{1/2}\phi_2, \, \phi_1 \rangle - \mathrm{Re}\langle \phi_1, \, A^{1/2}\phi_2 \rangle \\ &\quad - \mathrm{Re}\langle Q A^{1/2}\phi_2, \, A^{1/2}\phi_2 \rangle \\ &= -\mathrm{Re}\langle Q A^{1/2}\phi_2, \, A^{1/2}\phi_2 \rangle \le 0, \end{aligned}$$

which shows that $\mathcal{A}$ is dissipative. Since $P = I + A^{-1} + Q$ is self-adjoint and positive definite on H, it has bounded inverse P^{-1}. A direct calculation shows that $I - \mathcal{A}$ also has bounded inverse given by

$$(I - \mathcal{A})^{-1} = \left[\begin{array}{cc} I - P^{-1} & P^{-1}A^{-1/2} \\ -A^{-1/2}P^{-1} & A^{-1/2}P^{-1}A^{-1/2} \end{array}\right],$$

which is bounded on $H \times H$. This shows that $\mathcal{A}$ is m-dissipative. Thus, by the Lümer-Phillips theorem, we know that $\mathcal{A}$ is the infinitesimal generator of a C_0-semigroup of contractions $T(t)$ on $\mathcal{H} = H \times H$. It remains to be shown that $\bar{T}(t)$ is a C_0-semigroup on $D(A) \times D(A^{1/2})$, which can be shown easily. $\square$

Note that we were unable to show that the solution of (4.38) is asymptotically stable because Q is positive semidefinite. A sharp result, which shows that the unique solutions of (4.34) and (4.36) are indeed exponentially stable, can be obtained by using the energy multiplier method. This will be explained in Section 4.4.

A-dependent operators have many applications when studying the existence and uniqueness of some nonstandard second order equations. Consider, for example, the following partial differential equation:

$$\begin{cases} \dfrac{\partial^2 y(x,t)}{\partial t^2} + \dfrac{\partial^4 y(x,t)}{\partial x^4} - \dfrac{\partial^3 y(0,t)}{\partial x^3} = 0, \\[2mm] y(0,t) = y_x(0,t) = y_{xx}(1,t) = y_{xxx}(1,t) = 0, \\[2mm] y(x,0) = y_0(x), \quad \dot{y}(x,0) = y_1(x), \end{cases} \tag{4.42}$$

which is the dynamic model of a flexible beam attached to a free moving base in space where there is no gravity [103]. Physically, the perturbation term $-\dfrac{\partial^3 y(0,t)}{\partial x^3}$ represents the reaction force acting on the flexible beam due to the base movement. Our concern here is to show the existence and uniqueness of the solutions of (4.42).

Let A and $\Pi_1 y = -y'''(0)$ be the operators defined before. Then (4.42) can be written as the following abstract equation on the Hilbert space H.

$$\begin{cases} \ddot{y}(t) + (A + \Pi_1)y(t) = 0, \\[2mm] y(0) = y_0, \quad \dot{y}(0) = y_1, \end{cases} \tag{4.43}$$

where $\Pi_1 = Q_1 A$ is an A-dependent operator as shown before. If Π_1 were only A-bounded, but not A-symmetric (thus not A-dependent), then in general only a sufficient condition for the existence and uniqueness of solutions of (4.43) could be obtained, for small bounding constants a and b in Definition 4.3. Here we stress that for *any bounded* a and b there exists a unique solution of (4.43), since Π_1 is an A-dependent operator which has special structures.

Let $Q = I + Q_1$, $A_1 = Q^{1/2} A Q^{1/2}$, and $w(t) = Q^{-1/2} y(t)$. Then $w(t)$ satisfies the following equation:

$$\begin{cases} \ddot{w}(t) + A_1 w(t) = 0, \\[2mm] w(0) = Q^{-1/2} y_0, \quad \dot{w}(0) = Q^{-1/2} y_1. \end{cases} \tag{4.44}$$

It is obvious that A_1 is symmetric and $A_1^{-1} = Q^{-1/2} A^{-1} Q^{-1/2}$. By Proposition 4.1, A_1 is self-adjoint and positive definite. Hence, we can define the square root $A_1^{1/2}$ which is also self-adjoint and positive definite. Introduce a Hilbert space $\bar{H} = D(A_1^{1/2}) \times H$ with the inner product

$$\langle h, \bar{h} \rangle_{\bar{H}} = \langle A_1^{1/2} h_0, A_1^{1/2} \bar{h}_0 \rangle + \langle h_1, \bar{h}_1 \rangle$$
$$\forall h = \begin{bmatrix} h_0 \\ h_1 \end{bmatrix}, \quad \bar{h} = \begin{bmatrix} \bar{h}_0 \\ \bar{h}_1 \end{bmatrix} \in \bar{H}.$$

Putting $z(t) = \begin{bmatrix} w(t) \\ \dot{w}(t) \end{bmatrix}$, we can rewrite (4.44) as a first order abstract equation on $\bar{H}$

$$\begin{cases} \dot{z}(t) = \bar{A}z(t), \\[2mm] z(0) = \begin{bmatrix} Q^{-1/2}y_0 \\ Q^{-1/2}y_1 \end{bmatrix}, \end{cases} \qquad (4.45)$$

where $\bar{A}$ is defined by

$$D(\bar{A}) = D(A_1) \times D(A_1^{1/2}), \quad \bar{A} = \begin{bmatrix} 0 & I \\ -A_1 & 0 \end{bmatrix}.$$

Thus defined operator $\bar{A}$ is obviously a closed operator in $\bar{H}$ and its adjoint operator $\bar{A}^*$ is uniquely determined by $\bar{A}^* = -\bar{A}$. It follows from Stone's theorem that $\bar{A}$ generates a unitary C_0-group $\bar{T}(t)$ on $\bar{H}$. Define

$$T(t) = \begin{bmatrix} Q^{1/2} & 0 \\ 0 & Q^{1/2} \end{bmatrix} \bar{T}(t) \begin{bmatrix} Q^{-1/2} & 0 \\ 0 & Q^{-1/2} \end{bmatrix}. \qquad (4.46)$$

We wish to show that $T(t)$ is a uniformly bounded C_0-semigroup on $\mathcal{H} = D(A^{1/2}) \times H$ with the infinitesimal generator

$$\mathcal{A} = \begin{bmatrix} 0 & I \\ -(A+\Pi_1) & 0 \end{bmatrix} = \begin{bmatrix} 0 & I \\ -QA & 0 \end{bmatrix}, \quad D(\mathcal{A}) = D(A) \times D(A^{1/2}).$$

Let the Hilbert space $\mathcal{H} = D(A^{1/2}) \times H$ be endowed with the analogous inner product as in $\bar{H}$. It is important to note that

$$\begin{bmatrix} Q^{-1/2} & 0 \\ 0 & Q^{-1/2} \end{bmatrix} : D(\mathcal{A}) = D(A) \times D(A^{1/2}) \to D(\bar{A}) = D(A_1) \times D(A_1^{1/2})$$

is an isomorphism. To see this, one only needs to show that $D(A_1^{1/2}) = Q^{-1/2}D(A^{1/2})$ since $D(A_1) = Q^{-1/2}D(A)$ is already known. Writing

$$A_1 = Q^{1/2}AQ^{1/2} = Q^{1/2}A^{1/2}A^{1/2}Q^{1/2} = (A^{1/2}Q^{1/2})^*(A^{1/2}Q^{1/2}),$$

and using the polar decomposition of the operator $A^{1/2}Q^{1/2}$ [84, p.334], we have $UA_1^{1/2} = A^{1/2}Q^{1/2}$, where U is a partially isometric operator. Hence, $D(A_1^{1/2}) = Q^{-1/2}D(A^{1/2})$ and $\left\| A_1^{1/2}Q^{-1/2}x \right\| = \left\| A^{1/2}x \right\|$ for any $x \in D(A^{1/2})$. We can thus consider $T(t)$ as a uniformly bounded semigroup on $\mathcal{H}$, instead of on $\bar{H}$, since it can be checked straightforwardly that

- $\|T(t)\|_{\mathcal{H}} \leq M$ for some constant M;

- $T(t+s) = T(t)T(s), \forall t, s \geq 0$;

- $T(0) = I$;

- $\lim_{t \to 0} \|T(t)z - z\| = 0, \forall z \in \mathcal{H}$.

We now show that $\mathcal{A}$ is the infinitesimal generator of $T(t)$. Let $\tilde{\mathcal{A}}$ denote the infinitesimal generator of $T(t)$ on $\mathcal{H}$. Note that

$$
\int_0^t T(\tau)\mathcal{A}z d\tau
$$
$$
= \int_0^t \begin{bmatrix} Q^{1/2} & 0 \\ 0 & Q^{1/2} \end{bmatrix} \bar{T}(\tau) \begin{bmatrix} Q^{-1/2} & 0 \\ 0 & Q^{-1/2} \end{bmatrix} \begin{bmatrix} 0 & I \\ -QA & 0 \end{bmatrix} z d\tau
$$
$$
= \begin{bmatrix} Q^{1/2} & 0 \\ 0 & Q^{1/2} \end{bmatrix} \int_0^t \bar{T}(\tau) \begin{bmatrix} 0 & I \\ -A_1 & 0 \end{bmatrix} \begin{bmatrix} Q^{-1/2} & 0 \\ 0 & Q^{-1/2} \end{bmatrix} z d\tau
$$
$$
= T(t)z - z, \quad \forall z \in D(\mathcal{A}), \tag{4.47}
$$

where the relation $\int_0^t \bar{T}(\tau)\bar{A}\bar{z}d\tau = \bar{T}(t)\bar{z} - \bar{z}, \forall \bar{z} \in D(\bar{A})$ has been used. This relation implies that $\mathcal{A} \subset \tilde{\mathcal{A}}$. Since it is easy to show

1) $D(\mathcal{A})$ is $T(t)$-invariant;

2) $D(\mathcal{A})$ is dense in $\mathcal{H}$;

3) $\mathcal{A}$ is a closed operator,

we see, by (4.47), that for any $z \in D(\mathcal{A})$

$$
T(t)z - z = \int_0^t T(\tau)\mathcal{A}z d\tau
$$
$$
= \int_0^t T(\tau)\tilde{\mathcal{A}}z d\tau
$$
$$
= \int_0^t \tilde{\mathcal{A}}T(\tau)z d\tau
$$
$$
= \int_0^t \mathcal{A}T(\tau)z d\tau
$$
$$
= \mathcal{A}\int_0^t T(\tau)z d\tau.
$$

Now, for any $y \in D(\tilde{\mathcal{A}})$, since $D(\mathcal{A})$ is dense in $\mathcal{H}$ by 1), there exist $y_n \in D(\mathcal{A})$ such that $y_n \to y$. It follows that

$$
\int_0^t T(\tau)y_n d\tau \to \int_0^t T(\tau)y d\tau,
$$
$$
T(t)y_n - y_n \to T(t)y - y.
$$

By the closedness of $\mathcal{A}$, we have

$$\int_0^t T(\tau)y d\tau \in D(\mathcal{A}), \quad \text{and } \mathcal{A}\int_0^t T(\tau)y d\tau = T(t)y - y.$$

Thus,

$$\mathcal{A}\left(\frac{1}{t}\int_0^t T(\tau)y d\tau\right) = \frac{T(t)y - y}{t} \to \tilde{\mathcal{A}}y.$$

Since $\frac{1}{t}\int_0^t T(\tau)y d\tau \to y$, again by the closedness of $\mathcal{A}$, we have $y \in D(\mathcal{A})$ and $\mathcal{A}y = \tilde{\mathcal{A}}y$, which shows that $\mathcal{A}$ is the infinitesimal generator of $T(t)$ on $\mathcal{H}$.

We have actually proved the following theorem:

Theorem 4.9 *The nonstandard second order abstract equation (4.43) is well-posed in the sense that the operator*

$$\mathcal{A} = \begin{bmatrix} 0 & I \\ -(A + \Pi_1) & 0 \end{bmatrix}$$

generates a C_0-semigroup $T(t)$ on $\mathcal{H} = D(A^{1/2}) \times H$. The solution of (4.43) can be expressed as

$$\begin{bmatrix} y(t) \\ \dot{y}(t) \end{bmatrix} = T(t)\begin{bmatrix} y_0 \\ y_1 \end{bmatrix},$$

for any initial conditions $(y_0, y_1) \in \mathcal{H}$ and $(y(t), \dot{y}(t)) \in C^1([0, \infty); \mathcal{H})$ when $(y_0, y_1) \in D(\mathcal{A}) = D(A) \times D(A^{1/2})$.

4.3 Damped second order systems

In the dynamic models we have discussed so far, the damping effect has been ignored on purpose. The reason for doing this is that a good controller must introduce damping even without considering natural damping in the models. In other words, if we can design a good controller without considering natural damping, then the performance should be better when natural damping are considered. The concept of undamped systems also helps us to better understand the intrinsic properties of infinite dimensional systems. For example, it was shown in Section 4.1 that undamped second order systems cannot be exponentially stabilized by compact velocity feedbacks, which is a unique property for undamped systems. In practice, however, if a system is set in motion and allowed to vibrate freely, the vibration will eventually die out; the rate of decay depends on the amount of damping. Therefore, a good model that reflects the physical evidence should include damping. There are several mathematical models to represent damping. Among these are:

- **Viscous damping.** In this case, air or fluid damping is usually assumed to be proportional to the velocity of displacement. This is also called "external damping" because it models external friction forces.

- **Kelvin-Voigt damping.** This damping originates from the internal friction of the material of the vibrating structures and is thus one type of "internal dampings". In this damping model, the damping moment is postulated as being proportional to the the strain rate, and mathematically the damping operator is a differential operator with the same order as the system stiffness operator. This is the strongest damping model among widely used damping models in literature.

- **Structural damping.** In this damping model, the damping operator is assumed to be proportional to the square root of the system stiffness operator.

There also are other damping mechanisms, such as spatial and time hysteresis dampings which can be found in [8][9][10][11]. It should be noted that damping mechanism is very complicated and is far from being fully understood. We shall be interested in the three damping models mentioned above. We show semigroup generation properties for abstract differential equations on Hilbert spaces. In particular, we compare results for the cases with and without strain or shear force feedback for damped second order systems.

Before studying systems equations with specific damping models, we consider the following general equation:

$$\begin{cases} \ddot{y}(t) + QA\dot{y}(t) + Ay(t) = 0, & t > 0, \\ y(0) = y_0, \ \dot{y}(0) = y_1, \end{cases} \tag{4.48}$$

where A is the same operator as defined in (4.25) and thus has the same properties as before; Q is assumed to be bounded, self-adjoint and positive definite. Note that the operator Q was positive semidefinite in previous sections.

Again, let the underlying Hilbert space be $\mathcal{H} = H \times H$ and let the operator $\mathcal{A} : D(\mathcal{A}) \to \mathcal{H}$ be defined as in (4.39). Define $z_1 = Ay, z_2 = A^{1/2}\dot{y}$. Then (4.48) can be written as an abstract equation on $\mathcal{H}$ as (4.41). We have shown in Theorem 4.8 that $\mathcal{A}$ generates a C_0-semigroup $T(t)$ on $\mathcal{H}$.

Theorem 4.10 *Let $\sigma_p(\mathcal{A}), \sigma_c(\mathcal{A})$, and $\sigma_r(\mathcal{A})$ denote, respectively, the point spectrum, continuous spectrum, and residual spectrum of the operator $\mathcal{A}$. Then,*

(i) $\lambda \in \sigma_p(\mathcal{A})$ if and only if there exists a $\phi \in D(A), \phi \neq 0$ satisfying

$$\lambda^2\phi + \lambda QA\phi + A\phi = 0.$$

In this case,

$$Re\lambda \le -\frac{\delta}{2}\lambda_1,$$

where $\lambda_1 > 0$ is the smallest eigenvalue of A and

$$\delta = \min_{\psi \in \mathcal{H}, \psi \neq 0} \frac{\langle Q\psi, \psi \rangle}{\|\psi\|^2}.$$

Moreover,

$$|\lambda + \frac{1}{\delta}| \le \frac{1}{\delta}, \quad \text{for any } \lambda \in \sigma_p(\mathcal{A}), Im\lambda \neq 0;$$

(ii) If $\lambda \in \sigma_r(\mathcal{A})$, then $\bar{\lambda} \in \sigma_p(\mathcal{A})$;

(iii) $\sigma_c(\mathcal{A}) \subset \mathbf{R}$ and $\omega \le -1/\delta$ for any $\omega \in \sigma_c(\mathcal{A})$.

Proof. If $\lambda \in \sigma_p(\mathcal{A})$, let $\Phi = (\phi_1, \phi_2)^T \in D(\mathcal{A})$ be the corresponding eigenvector. Then, $\phi = A^{-1/2}\phi_2$ satisfies $\lambda^2\phi + \lambda QA\phi + A\phi = 0$. Conversely, if $\lambda^2\phi + \lambda QA\phi + A\phi = 0$, then $\lambda \neq 0$, and $(\phi_1, \phi_2) = (A\phi/\lambda, A^{1/2}\phi)$ is the corresponding eigenvector of $\mathcal{A}$. Taking inner product with $A\phi$ on both sides of $\lambda^2\phi + \lambda QA\phi + A\phi = 0$, we obtain

$$\lambda^2 \langle \phi, A\phi \rangle + \lambda\langle QA\phi, A\phi \rangle + \|A\phi\|^2 = 0.$$

Let $\lambda = a + bi$. Then the above equation means that

$$(a^2 - b^2)\left\|A^{1/2}\phi\right\|^2 + a\langle QA\phi, A\phi \rangle + \|A\phi\|^2 = 0 \tag{4.49}$$

and

$$2ab\left\|A^{1/2}\phi\right\|^2 + b\langle QA\phi, A\phi \rangle = 0. \tag{4.50}$$

If $b = 0$, then

$$a^2\left\|A^{1/2}\phi\right\|^2 + a\langle QA\phi, A\phi \rangle + \|A\phi\|^2 = 0,$$

so

$$a \le -\frac{\langle QA\phi, A\phi \rangle}{2\left\|A^{1/2}\phi\right\|^2} \le -\frac{\delta}{2}\lambda_1.$$

If $b \neq 0$, then in view of (4.50) we have

$$a = -\frac{\langle QA\phi, A\phi \rangle}{2\left\|A^{1/2}\phi\right\|^2} \le -\frac{\delta}{2}\lambda_1$$

and

$$-b^2 \left\| A^{1/2} \phi \right\|^2 - \frac{\langle QA\phi, \, A\phi \rangle^2}{4 \left\| A^{1/2} \phi \right\|^2} + \left\| A\phi \right\|^2 = 0,$$

from which we obtain

$$b^2 = \frac{\left\| A\phi \right\|^2}{\left\| A^{1/2} \phi \right\|^2} - \frac{\langle QA\phi, \, A\phi \rangle^2}{4 \left\| A^{1/2} \phi \right\|^4}.$$

A simple calculation shows that

$$
\begin{aligned}
\left| \lambda + \frac{1}{\delta} \right|^2 &= \left(a + \frac{1}{\delta} \right)^2 + b^2 \\
&= a^2 + b^2 + \frac{2}{\delta} a + \frac{1}{\delta^2} \\
&= \frac{\left\| A\phi \right\|^2}{\left\| A^{1/2} \phi \right\|^2} - \frac{\langle QA\phi, \, A\phi \rangle}{\delta \left\| A^{1/2} \phi \right\|^2} + \frac{1}{\delta^2} \leq \frac{1}{\delta^2}, \quad (4.51)
\end{aligned}
$$

which proves (i).

Now note that the adjoint operator $\mathcal{A}^*$ of $\mathcal{A}$ is given by

$$\mathcal{A}^* = \begin{bmatrix} -A^{1/2} & 0 \\ 0 & A^{1/2} \end{bmatrix} \begin{bmatrix} 0 & I \\ I & -QA^{1/2} \end{bmatrix}.$$

It is easy to see that $\sigma_p(\mathcal{A}) = \sigma_p(\mathcal{A}^*)$ since if $\Phi = (\phi_1, \phi_2)^T$ is an eigenvector of $\mathcal{A}$ corresponding to $\lambda \in \sigma_p(\mathcal{A})$, then $\hat{\Phi} = (-\phi_1, \phi_2)^T$ is an eigenvector of $\mathcal{A}^*$ corresponding to the same λ, and vice versa. Now, let $\lambda \in \sigma_r(\mathcal{A})$ which means that $\mathcal{R}(\lambda - \mathcal{A})$ is not dense in $\mathcal{H}$. This in turn implies that there exists a $y \in \mathcal{H}$ with $y \neq 0$ such that

$$\langle (\lambda - \mathcal{A})x, \, y \rangle = 0$$

holds for all $x \in D(\mathcal{A})$. Rearranging this equation, we see that $y \in D(\mathcal{A}^*)$ and $\mathcal{A}^* y = \bar{\lambda} y$, which clearly shows that $\bar{\lambda} \in \sigma_p(\mathcal{A}^*) = \sigma_p(\mathcal{A})$. Thus, (ii) is proved.

Finally, let $\lambda \in \sigma_c(\mathcal{A})$. Then there exists a sequence of eigenvectors $\Phi_n = (\phi_{1n}, \phi_{2n})^T, \|\Phi_n\| = 1$ such that

$$(\lambda - \mathcal{A})\Phi_n \to 0 \text{ as } n \to \infty,$$

which is equivalently written as

$$\lambda \phi_{1n} - A^{1/2} \phi_{2n} \to 0, \quad (4.52)$$

$$\lambda \phi_{2n} + A^{1/2} (\phi_{1n} + QA^{1/2} \phi_{2n}) \to 0. \quad (4.53)$$

Since $\{\lambda \phi_{1n}\}$ is a bounded set and $A^{-1/2}$ is compact, one can find a subsequence of $\{\lambda A^{-1/2} \phi_{1n}\}$, still denoted by $\{\lambda A^{-1/2} \phi_{1n}\}$, which converges to ϕ_2. Hence, $\phi_{2n} \to \phi_2$ by (4.52). If $\phi_2 = 0$, then $\|\phi_{1n}\| = 1$ and $A^{1/2}(\phi_{1n} + QA^{1/2}\phi_{2n}) \to 0$ by (4.53). Thus,

$$\begin{aligned}
&\langle A^{1/2}(\phi_{1n} + QA^{1/2}\phi_{2n}), \phi_{2n} \rangle \\
&= \langle \phi_{1n} + QA^{1/2}\phi_{2n}, A^{1/2}\phi_{2n} \rangle \\
&= \langle \phi_{1n}, A^{1/2}\phi_{2n} - \lambda\phi_{1n} \rangle + \langle QA^{1/2}\phi_{2n}, A^{1/2}\phi_{2n} \rangle + \bar{\lambda}\|\phi_{1n}\|^2 \to 0,
\end{aligned}$$

which implies that

$$\langle QA^{1/2}\phi_{2n}, A^{1/2}\phi_{2n} \rangle + \bar{\lambda} \to 0.$$

This shows that λ must be a negative real number which is denoted by ω. Since

$$\delta \left\| A^{1/2}\phi_{2n} \right\|^2 + \omega \le \langle QA^{1/2}\phi_{2n}, A^{1/2}\phi_{2n} \rangle + \bar{\lambda} \to 0,$$

it follows from (4.52) that

$$|\omega|^2 \delta + \omega = \delta \left\| A^{1/2}\phi_{2n} \right\|^2 + \omega \le 0,$$

from which we obtain $\omega \le -1/\delta$ since $\omega = \lambda$ is nonzero.

On the other hand, if $\phi_2 \ne 0$, then by (4.53),

$$\lambda A^{-1/2}\phi_{2n} + \phi_{1n} + QA^{1/2}\phi_{2n} \to 0$$

or equivalently,

$$\lambda^2 A^{-1/2}\phi_{2n} + A^{1/2}\phi_{2n} + \lambda QA^{1/2}\phi_{2n} \to 0,$$

which is

$$(I + \lambda Q)A^{1/2}\phi_{2n} \to -\lambda^2 A^{-1/2}\phi_2.$$

Since Q is self-adjoint, $\sigma(Q)$ is real. For both the case $\text{Im}\lambda \ne 0$ and the case $\lambda = \omega > -1/\delta$, the operator $(I + \lambda Q)^{-1}$ exists and hence $(I + \lambda Q)A^{1/2}$ is closed, which in turn implies that $\phi_2 \in D(A^{1/2})$. Setting $\phi = A^{-1/2}\phi_2$, we see that ϕ satisfies

$$\lambda^2 \phi + \lambda QA\phi + A\phi = 0,$$

which implies that $\lambda \in \sigma_p(\mathcal{A})$, contradicting the assumption that $\lambda \in \sigma_c(\mathcal{A})$. Therefore, a point in $\sigma_c(\mathcal{A})$ must be a real number smaller than $-1/\delta$. The proof is complete. $\square$

Theorem 4.11 *The semigroup $T(t)$ generated by $\mathcal{A}$ is analytic with exponential decay. That is, for any $\varepsilon > 0$, $\omega_0 + \varepsilon < 0$, there exists a constant $M \ge 1$ such that*

$$\|T(t)\| \le Me^{(\omega_0 + \varepsilon)t}$$

where $\omega_0 = S(\mathcal{A})$ is the growth rate of $T(t)$.

Proof. Define $P_\lambda = I + \lambda^2 A^{-1} + \lambda Q$. P_λ is a bounded linear operator for any complex λ. We first show that $0 \in \rho(P_\lambda)$ and

$$\left\| P_\lambda^{-1} \right\| \le \frac{1}{\delta |\lambda|}, \ \forall \, \mathrm{Re}\lambda \ge 0, \ \lambda \neq 0, \tag{4.54}$$

where δ is the same as defined in Theorem 4.10. In fact, for any $\phi \in H$, it is easy to verify that

$$\left\| \frac{P_\lambda}{\lambda} \phi \right\| \|\phi\| \ge \mathrm{Re}\langle \frac{P_\lambda}{\lambda}\phi, \ \phi \rangle \ge \mathrm{Re}\langle Q\phi, \ \phi \rangle \ge \delta \|\phi\|^2 ,$$

which implies $\mathcal{R}(P_\lambda/\lambda)$, the range of P_λ/λ, is a closed subset of H. If there is a $\phi_0 \in H$ such that

$$\langle \frac{P_\lambda}{\lambda}\phi, \ \phi_0 \rangle = 0 \text{ for all } \phi \in H,$$

then, in particular, $\langle (P_\lambda/\lambda)\phi_0, \ \phi_0 \rangle = 0$. But we have shown that $\langle (P_\lambda/\lambda)\phi_0, \ \phi_0 \rangle \ge \delta \|\phi_0\|^2$. Thus, $\phi_0 = 0$. Consequently, $\mathcal{R}(P_\lambda/\lambda) = H$. That is, $0 \in \rho(P_\lambda)$ and $\left\| P_\lambda^{-1} \right\| \le 1/(\delta|\lambda|)$. It can be verified that

$$(\lambda - \mathcal{A})^{-1} = \begin{bmatrix} \frac{1}{\lambda} - \frac{1}{\lambda}P_\lambda^{-1} & P_\lambda^{-1}A^{-1/2} \\ -A^{-1/2}P_\lambda^{-1} & \lambda A^{-1/2}P_\lambda^{-1}A^{-1/2} \end{bmatrix}. \tag{4.55}$$

For any $\mathrm{Re}\lambda \ge 0$, the set

$$\{\phi \,|\, A^{-1/2}P_\lambda^{-1}\phi, \ \phi \in D(A^{1/2})\}$$

is dense in $\mathcal{H}$, and hence $A^{1/2}P_\lambda A^{1/2}$ is a densely defined closed operator with bounded inverse $A^{-1/2}P_\lambda^{-1}A^{-1/2}$. Furthermore, it can be shown that

$$\mathrm{Re}\langle \lambda^{-2}A^{1/2}P_\lambda A^{1/2}\phi, \ \phi \rangle$$
$$= \|\phi\|^2 + \mathrm{Re}\frac{1}{\lambda}\langle QA^{1/2}\phi, \ A^{1/2}\phi \rangle + \mathrm{Re}\frac{1}{\lambda^2}\left\| A^{1/2}\phi \right\|^2$$
$$\ge \|\phi\|^2 + \mathrm{Re}\left(\frac{\delta}{\lambda} + \frac{1}{\lambda^2} \right)\left\| A^{1/2}\phi \right\|^2$$
$$= \|\phi\|^2 + \mathrm{Re}\left(\frac{\bar{\lambda}|\lambda|^2\delta + \bar{\lambda}^2}{|\lambda|^4} \right)\left\| A^{1/2}\phi \right\|^2$$
$$\ge \|\phi\|^2, \text{ for } \mathrm{Re}\lambda \ge \frac{1}{\delta}.$$

Hence,

$$\left\| \lambda^2 A^{-1/2}P_\lambda^{-1}A^{-1/2} \right\| \le 1 \text{ for } \mathrm{Re}\lambda \ge 1/\delta. \tag{4.56}$$

Combining (4.54)-(4.56), we have

$$\|R(\lambda, \mathcal{A})\| \leq \max\left\{\frac{1}{|\lambda|} + \frac{1}{\delta|\lambda|^2}, \frac{\|A^{-1/2}\|}{\delta|\lambda|}, \frac{1}{|\lambda|}\right\}$$

$$\leq \frac{1}{|\lambda|}\max\left\{2, \frac{\|A^{-1/2}\|}{\delta}\right\} \text{ for } \operatorname{Re}\lambda \geq \frac{1}{\delta}. \qquad (4.57)$$

By Theorem 2.48, $T(t)$ is an analytic semigroup on $\mathcal{H}$. Furthermore, because the spectrum-determined growth condition holds for analytic semigroups and because $\sigma(\mathcal{A})$ is contained in the open left-half complex plane, $\omega_0 = \sup\{\operatorname{Re}\lambda \mid \lambda \in \sigma(\mathcal{A})\} \leq \max\{-\frac{1}{\delta}, -\frac{1}{\delta}\lambda_1\} < 0$, that is, the semigroup $T(t)$ decays exponentially with decay rate ω_0. The proof is complete. $\qquad \square$

Now we consider some special cases.

• **Beam equation with Kelvin-Voigt damping:** We consider the following initial-boundary value problem:

$$\begin{cases} \dfrac{\partial^2 y(x,t)}{\partial t^2} + \delta\dfrac{\partial^5 y(x,t)}{\partial t \partial x^4} + \dfrac{\partial^4 y(x,t)}{\partial x^4} = 0, \quad 0 < x < 1, \ t > 0, \\ y(0,t) = y_x(0,t) = y_{xx}(1,t) = y_{xxx}(1,t) = 0, \\ y(x,0) = y_0(x), \ \dot{y}(x,0) = y_1(x), \end{cases} \qquad (4.58)$$

where $\delta > 0$ is the damping constant. If we define the operator A as in (4.25), then (4.58) can be rewritten as the abstract equation on the Hilbert space H as follows:

$$\begin{cases} \ddot{y}(t) + \delta A\dot{y}(t) + Ay(t) = 0, \\ y(0) = y_0, \ \dot{y}(0) = y_1, \end{cases} \qquad (4.59)$$

which is a special case of (4.48) with $Q = \delta I$.

The solution properties of (4.59) is extensively studied by Sakawa [142]. It is shown that the system operator $\mathcal{A}$ associated with (4.59), if defined as in (4.39), generates an analytic semigroup, and all the complex eigenvalues of $\mathcal{A}$ lie on the circle with center at point $-1/\delta$ and radius $1/\delta$, as shown in Figure 4.2. This result also can be directly derived from (4.51) in the proof of (i) of Theorem 4.10. Because in this case $Q = \delta I$, so (4.51) reduces to

$$\left(a + \frac{1}{\delta}\right)^2 + b^2 = \frac{\|A\phi\|^2}{\|A^{1/2}\phi\|^2} - \frac{\langle QA\phi, \phi\rangle}{\|A^{1/2}\phi\|^2} + \frac{1}{\delta^2} = \frac{1}{\delta^2}.$$

• **Strain feedback controlled beam equation with Kelvin-Voigt damping:** The closed-loop equation for an Euler-Bernoulli beam with strain feed-

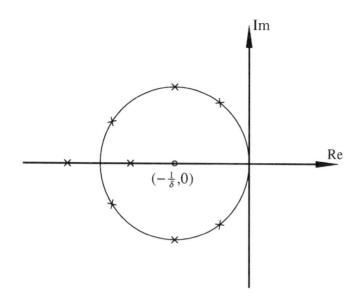

Figure 4.2: Spectrum of the Second Order Equation with Kelvin-Voigt Damping

back and Kelvin-Voigt damping is given by

$$
\begin{cases}
\dfrac{\partial^2 y(x,t)}{\partial t^2} + \delta \dfrac{\partial^5 y(x,t)}{\partial t \partial x^4} + kx \dfrac{\partial^3 y(0,t)}{\partial t \partial x^2} + \dfrac{\partial^4 y(x,t)}{\partial x^4} = 0, \\
\qquad 0 < x < 1,\ t > 0, \\
y(0,t) = y_x(0,t) = y_{xx}(1,t) = y_{xxx}(1,t) = 0, \\
y(x,0) = y_0(x),\ \dot{y}(x,0) = y_1(x).
\end{cases}
\tag{4.60}
$$

Let the operator A be the same as defined before. Put

$$
Q = \delta + k\Pi A^{-1}, \quad \Pi\phi(x) = x\phi''(0), \forall \phi \in D(\Pi).
$$

Clearly, Q is positive definite and (4.60) can be written as (4.48). The corresponding system operator $\mathcal{A}$ has no residual spectrum as is implied by (ii) of Theorem 4.10, since in this case $\lambda \in \sigma_p(\mathcal{A})$ if and only if $\bar{\lambda} \in \sigma_p(\mathcal{A})$. Since it is easy to verify that the point $-1/\delta$ is not in the point spectrum, we show that it is in the continuous spectrum of $\mathcal{A}$ for any $k \geq 0$. In fact, letting $Y = (y_1, 0)^T$ and solving for $\Phi = (\phi_1, \phi_2)^T$ from $(-1/\delta - \mathcal{A})\Phi = Y$, i.e.,

$$
-\frac{1}{\delta}\phi_1 - A^{1/2}\phi_2 = y_1,
$$

$$
-\frac{1}{\delta}\phi_2 + A^{1/2}(\phi_1 + QA^{1/2}\phi_2) = 0,
$$

we have

$$A^{-1/2}\phi_2 = \delta k x \phi''(0) - \delta^2 y_1 \in D(A^{1/2}) \subset H^2(0,1)$$

which implies that at least $y_1 \in H^2(0,1)$. Thus, $-1/\delta$ does not belong to the resolvent set of $\mathcal{A}$ because y_1 cannot take values in $H = L^2(0,1)$. It follows that $-1/\delta \in \sigma_c(\mathcal{A})$.

Now we wish to study the locations of the complex point spectrum. Let $\lambda \in \sigma_p(\mathcal{A})$. Then there should exist a nonzero $\phi(x)$ such that

$$\begin{cases} \lambda^2\phi(x) + (1+\delta\lambda)\phi''''(x) + kx\lambda\phi''(0) = 0, \\ \phi(0) = \phi'(0) = \phi''(1) = \phi'''(1) = 0. \end{cases} \tag{4.61}$$

Let $\psi(x) = \lambda\phi(x) + kx\phi''(0)$, then $\psi(x) \neq 0$ and the following equation is satisfied

$$\begin{cases} \lambda^2\psi(x) + (1+\delta\lambda)\psi''''(x) = 0, \\ \psi(0) = \psi''(1) = \psi'''(1) = 0, \ \psi'(0) = \frac{k}{\lambda}\psi''(0). \end{cases}$$

Taking the inner product with $\psi(x)$ on both sides of the first equation, we obtain

$$\frac{\lambda}{k}|\psi'(0)|^2 + \|\psi''\| + \frac{\lambda^2}{1+\delta\lambda}\|\psi\|^2 = 0. \tag{4.62}$$

Rearranging terms leads to

$$p_1\lambda^2 + p_2\lambda + p_3 = 0, \tag{4.63}$$

where

$$p_1 = \delta|\psi'(0)|^2 + \|\psi\|^2,$$
$$p_2 = |\psi'(0)|^2 + k\delta\|\psi''\|^2,$$
$$p_3 = k\|\psi''\|^2.$$

Let a and b denote, respectively, the real and imaginary parts of λ, that is, $\lambda = a + bi$. Substituting this into (4.63) and letting the real and imaginary parts be equal to zero yields

$$\begin{cases} (a^2 - b^2)p_1 + ap_2 + p_3 = 0, \\ 2abp_1 + bp_2 = 0. \end{cases}$$

Therefore, if $b \neq 0$, then

$$a = -\frac{1}{2}\frac{p_2}{p_1}, \quad b^2 = \frac{p_3}{p_1} - \frac{1}{4}\frac{p_2^2}{p_1^2},$$

from which we see that

$$
\begin{aligned}
\left| \lambda + \frac{1}{\delta} \right|^2 &= a^2 + b^2 + \frac{2a}{\delta} + \frac{1}{\delta^2} \\
&= -\frac{|\psi'(0)|^2}{\delta^2 |\psi'(0)|^2 + \delta \, \|\psi\|^2} + \frac{1}{\delta^2} \\
&< \frac{1}{\delta^2}.
\end{aligned}
$$

The last inequality comes from the fact that $\psi'(0) \neq 0$ because otherwise it can be inferred that $\psi''(0) = 0$, which in turn implies that $\psi(x) = 0$.

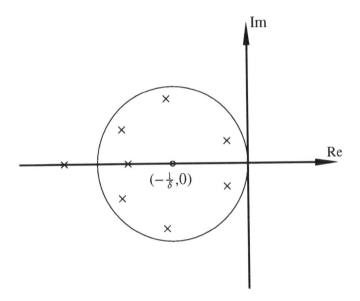

Figure 4.3: Spectrum of the Strain Feedback Controlled Beam Equation with Kelvin-Voigt Damping

The above analysis implies that all the complex eigenvalues of $\mathcal{A}$ lie *inside* the circle with center at $-1/\delta$ and radius equal to $1/\delta$, as depicted in Figure 4.3. This result should be compared with that in the case $k = 0$, that is, no strain feedback, as discussed before (see Figure 4.2). It is seen that strain feedback can shift the eigenvalues to the left-half complex plane, thus can enhance damping.

Finally, we study the analytic semigroup generation property of a beam equation with structural natural damping and strain feedback. Namely, we consider the following equation on H:

$$\begin{cases} \ddot{y}(t) + \delta A^{1/2}\dot{y}(t) + k\Pi\dot{y}(t) + Ay(t) = 0, \quad t > 0, \\ y(0) = y_0, \ \dot{y}(0) = y_1, \end{cases} \quad (4.64)$$

where A is the operator defined in (4.25) and Π is defined by

$$\Pi\phi(x) = x \cdot B\phi(x) = x\phi''(0), \quad \forall \phi \in D(\Pi).$$

Define $Q = \delta A^{-1/2} + k\Pi A^{-1}$. We see that (4.64) can be written as (4.48), but Q is only positive semidefinite. So whether there exists an analytic semigroup for system (4.64) is not clear from the previous results. It is already known, however, that system (4.64) generates an analytic semigroup when $k = 0$. We show here, by using compact perturbation theory, that system (4.64) still generates an analytic semigroup for $k > 0$.

Lemma 4.12   (i) $x \in D(A^{1/4})$;

(ii) $BA^{-3/4}$ is a linear bounded functional on H.

Proof.   Let $x = \sum_{n=1}^{\infty} b_n\phi_n(x)$ as in Lemma 4.6. Since it has been shown that $b_n = O(\beta_n^{-2}) = O(n^{-2})$ and $\lambda_n = O(n^4)$ in Lemma 4.6, there is a constant C such that

$$\sum_{n=1}^{\infty} b_n^2\lambda_n^{1/2} \leq C\sum_{n=1}^{\infty} n^{-2} < \infty,$$

which implies $x \in D(A^{1/4})$ and $A^{1/4}x = \sum_{n=1}^{\infty} b_n\lambda_n^{1/4}\phi_n(x)$.

To show the boundedness of $BA^{-3/4}$ on H, let $\phi \in H$ be expressed as $\phi = \sum_{n=1}^{\infty} a_n\phi_n(x)$. From the definition of B, we have

$$\begin{aligned} |BA^{-3/4}\phi| &= |\sum_{n=1}^{\infty} a_n\lambda_n^{-3/4}\phi_n''(0)| \\ &\leq \sum_{n=1}^{\infty} |a_n\lambda_n^{-3/4}\phi_n''(0)| \\ &\leq \left(\sum_{n=1}^{\infty} |a_n|^2\right)^{1/2}\left(\sum_{n=1}^{\infty} |\lambda_n^{-3/2}||\phi_n''(0)|^2\right)^{1/2} \\ &\leq 2\left(\sum_{n=1}^{\infty} |a_n|^2\right)^{1/2}\left(\sum_{n=1}^{\infty} |\lambda_n^{-1/2}|\right)^{1/2} \\ &\leq C\left(\sum_{n=1}^{\infty} n^{-2}\right)^{1/2}\|\phi\|, \end{aligned}$$

from which we see that $BA^{-3/4}$ is a bounded functional. $\qquad\qquad$ □

Now, we introduce a new variable, $w(t) = A^{1/4}y(t)$, with which we can rewrite (4.64) as

$$\begin{cases} \ddot{w}(t) + \delta A^{1/2}\dot{w}(t) + kA^{1/4}\Pi A^{-1/4}\dot{w}(t) + Aw(t) = 0, & t > 0, \\ w(0) = A^{1/4}y_0, \quad \dot{w}(0) = A^{1/4}y_1. \end{cases} \tag{4.65}$$

Define an operator Q_k by

$$Q_k = kA^{1/4}\Pi A^{-1}A^{1/4} = kA^{1/4}x \cdot BA^{-3/4}.$$

We see that Q_k is a self-adjoint operator because ΠA^{-1} is, as shown in Theorem 4.4. Moreover, Q_k is a compact operator on H because Q_k is bounded (since $BA^{-3/4}$ is bounded) and is of rank one from lemma 4.12.

Let $z_1(t) = A^{1/2}w(t), z_2(t) = \dot{w}(t)$. Then (4.65) can be written as an abstract equation on the Hilbert space $H \times H$:

$$\frac{d}{dt}\begin{bmatrix} z_1 \\ z_2 \end{bmatrix} = \begin{bmatrix} 0 & A^{1/2} \\ -A^{1/2} & -(\delta + Q_k)A^{1/2} \end{bmatrix}\begin{bmatrix} z_1 \\ z_2 \end{bmatrix} = \mathcal{A}_k\begin{bmatrix} z_1 \\ z_2 \end{bmatrix} \tag{4.66}$$

with initial conditions $z_1(0) = z_{10}$, $z_2(0) = z_{20}$, and the domain of $\mathcal{A}_k$ being defined as $D(\mathcal{A}_k) = D(A^{1/2}) \times D(A^{1/2})$.

Theorem 4.13 *The operator $\mathcal{A}_k$ defined in (4.66) generates an analytic semigroup $T_k(t)$ on $H \times H$. Thus system (4.65) admits a unique classical solution which is given by*

$$\begin{bmatrix} z_1 \\ z_2 \end{bmatrix} = T_k(t)\begin{bmatrix} z_{10} \\ z_{20} \end{bmatrix}$$

for any initial conditions $(z_{10}, z_{20})^T \in D(\mathcal{A}_k)$.

Remark 4.14 *Once z_1 and z_2 are obtained, the solution of (4.64) can be expressed as $y = A^{-3/4}z_1, \dot{y} = A^{-1/4}z_2$.*

Proof. First, notice that $\mathcal{A}_k$ can be decomposed as $\mathcal{A}_k = \mathcal{A}_0 + P\mathcal{A}_0$ where

$$\mathcal{A}_0 = \begin{bmatrix} 0 & A^{1/2} \\ -A^{1/2} & -\delta A^{1/2} \end{bmatrix}, \quad P = \begin{bmatrix} 0 & 0 \\ -Q_k & 0 \end{bmatrix}.$$

It is well known that the operator $\mathcal{A}_0$ is the infinitesimal generator of an analytic semigroup of contractions on $H \times H$ [31]. Since P is a bounded operator with finite rank, by Theorem 2.54 we conclude that $\mathcal{A}_k$ generates an analytic semigroup. $\qquad\qquad$ □

4.4   Exponential stability and spectral analysis

The A-dependent operators have been successfully applied to the study of
well-posedness of the abstract equations corresponding to (4.34) and (4.36).
When there existed Kelvin-Voigt damping or structural damping in the Euler-
Bernoulli equations, we were also able to show that the strain or shear
force feedback could enhance damping. However, for the undamped Euler-
Bernoulli equation, the closed-loop exponential stability of strain or shear
force feedback was not shown because the operator Q in (4.38) was only
positive semidefinite.

In this section, we shall study the exponential stability of the strain feedback controlled closed-loop equation

$$
\begin{cases}
\dfrac{\partial^2 y(x,t)}{\partial t^2} + kx\dfrac{\partial^3 y(0,t)}{\partial t \partial x^2} + \dfrac{\partial^4 y(x,t)}{\partial x^4} = 0, & 0 < x < 1,\ t > 0, \\[2mm]
y(0,t) = y_x(0,t) = 0, \\[2mm]
y_{xx}(1,t) = y_{xxx}(1,t) = 0, \\[2mm]
y(x,0) = y_0(x),\ \dot{y}(x,0) = y_1(x)
\end{cases}
\tag{4.67}
$$

for the rotating Euler-Bernoulli beam, and the shear force feedback controlled
closed-loop equation

$$
\begin{cases}
\dfrac{\partial^2 y(x,t)}{\partial t^2} - k\dfrac{\partial^4 y(0,t)}{\partial t \partial x^3} + \dfrac{\partial^4 y(x,t)}{\partial x^4} = 0, & 0 < x < 1,\ t > 0, \\[2mm]
y(0,t) = y_x(0,t) = 0, \\[2mm]
y_{xx}(1,t) = y_{xxx}(1,t) = 0, \\[2mm]
y(x,0) = y_0(x),\ \dot{y}(x,0) = y_1(x)
\end{cases}
\tag{4.68}
$$

for the translating beam. The main idea is to introduce the variable transformations such that (4.67) and (4.68) can be transformed into initial-boundary
value problems with boundary dampings for which we can use energy multiplier method or frequency domain test method to derive their exponential
stabilities. We shall also analyze the asymptotic behavior of eigenvalues of
the strain feedback controlled closed-loop equation (4.67) (the analysis can
be done similarly for shear force feedback controlled closed-loop equation
(4.68)). One interesting result is that the eigenvalues of the system (4.67),
with large moduli, approaches the vertical line $\mathrm{Re}\lambda = -2k$ in the complex
plane. This implies that the system operator associated with (4.67) cannot
generate an analytic semigroup.

4.4.1   Exponential stability

Let us introduce a new variable

$$w(x,t) = y_{xx}(1-x,t). \tag{4.69}$$

Suppose the initial conditions associated with (4.67) are sufficiently smooth such that the solution admits continuous spatial derivatives up to the sixth order. Then taking the spatial derivative of both sides of (4.67) twice yields

$$\frac{\partial^2 w(x,t)}{\partial t^2} + \frac{\partial^4 w(x,t)}{\partial x^4} = 0, \quad 0 < x < 1, \ t > 0. \tag{4.70}$$

Also, from the boundary conditions of (4.67), we see that $\ddot{y}(0,t) = \ddot{y}'(0,t) = 0$. Therefore, we obtain the following boundary conditions on $w(x,t)$:

$$\begin{cases} w(0,t) = y_{xx}(1,t) = 0, \\ w_x(0,t) = -y_{xxx}(1,t) = 0, \\ w_{xx}(1,t) = y_{xxxx}(0,t) = 0, \\ w_{xxx}(1,t) = -y_{xxxxx}(0,t) = k\ddot{y}_{xx}(0,t) = k\dot{w}(1,t), \end{cases} \tag{4.71}$$

where we have used the first equation of (4.67) and its spatial derivative in obtaining the last two equalities. Summarizing, we get an initial-boundary value problem for $w(x,t)$ which is characterized by

$$\begin{cases} \dfrac{\partial^2 w(x,t)}{\partial t^2} + \dfrac{\partial^4 w(x,t)}{\partial x^4} = 0, \quad 0 < x < 1, \ t > 0, \\ w(0,t) = w_x(0,t) = 0, \\ w_{xx}(1,t) = 0, \\ w_{xxx}(1,t) = k\dot{w}(1,t), \\ w(x,0) = y_0''(1-x), \ \dot{w}(x,0) = y_1''(1-x). \end{cases} \tag{4.72}$$

The transformation (4.69) establishes a connection between the strain feedback controlled closed-loop equation (4.67) and the direct boundary velocity controlled closed-loop equation (4.72). For the latter, we can apply the energy multiplier method to derive its exponential stability [28]. We have the following theorem.

Theorem 4.15 *There exists a unique solution to (4.72). Moreover, the energy of vibration*

$$E(t) = \frac{1}{2} \int_0^1 \dot{w}^2(x,t)dx + \frac{1}{2} \int_0^1 w_{xx}^2(x,t)dx \tag{4.73}$$

decays exponentially, that is, there exist constants $M \geq 1$ and $\beta > 0$ such that

$$E(t) \leq M e^{-\beta t} E(0), \tag{4.74}$$

where $E(0)$ stands for the initial energy.

Proof. We first define an appropriate product Hilbert space and rewrite (4.72) as the first order evolution equation on this product space. Then we show that the system operator is m-dissipative, thus generating a C_0-semigroup of contractions on the underlying space according to the Lümer-Phillips Theorem.

Let us consider a function space

$$\mathcal{H} = \left\{ (h_1, h_2)^T \,|\, h_1 \in H^2(0,1), h_2 \in L^2(0,1), h_1(0) = h_1'(0) = 0 \right\}.$$

It can be easily verified that this space, when equipped with the following inner product

$$\left\langle \begin{bmatrix} h_1 \\ h_2 \end{bmatrix}, \begin{bmatrix} \tilde{h}_1 \\ \tilde{h}_2 \end{bmatrix} \right\rangle_{\mathcal{H}} = \int_0^1 [h_1''(x)\overline{\tilde{h}_1''(x)} + h_2(x)\overline{\tilde{h}_2(x)}]dx$$

becomes a Hilbert space. Define

$$z_1(\cdot, t) = w(\cdot, t), \quad z_2(\cdot, t) = \dot{w}(\cdot, t)$$

and

$$\mathcal{A} \begin{bmatrix} h_1 \\ h_2 \end{bmatrix} = \begin{bmatrix} h_2 \\ -h_1'''' \end{bmatrix} \tag{4.75}$$

with domain

$$D(\mathcal{A}) = \left\{ \begin{bmatrix} h_1 \\ h_2 \end{bmatrix} \middle| h_1 \in H^4(0,1), h_2 \in H^2(0,1), h_1(0) = h_1'(0) = 0, \right.$$
$$\left. h_2(0) = h_2'(0) = 0, h''(1) = 0, h_1'''(1) = kh_2(1) \right\}. \tag{4.76}$$

Then (4.72) can be written as

$$\frac{d}{dt} \begin{bmatrix} z_1(\cdot, t) \\ z_2(\cdot, t) \end{bmatrix} = \mathcal{A} \begin{bmatrix} z_1(\cdot, t) \\ z_2(\cdot, t) \end{bmatrix}. \tag{4.77}$$

It is routine to verify that $\mathcal{A}$ is densely defined and closed. Also, for any $[h_1, h_2]^T \in \mathcal{H}$, because

$$\left\langle \mathcal{A} \begin{bmatrix} h_1 \\ h_2 \end{bmatrix}, \begin{bmatrix} h_1 \\ h_2 \end{bmatrix} \right\rangle_{\mathcal{H}} = -k|h_2(1)|^2 \leq 0,$$

we see that $\mathcal{A}$ is dissipative. To show that $\mathcal{A}$ generates a C_0-semigroup on $\mathcal{H}$, we have to show that there exists a $\lambda_0 \geq 0$ such that $\mathcal{R}(\lambda_0 - \mathcal{A})$, the range of $(\lambda_0 - \mathcal{A})$, is the whole space $\mathcal{H}$, or sufficiently, $\mathcal{A}^{-1}$ exists and is bounded, which is shown below. For any $[f, g]^T \in \mathcal{H}$, the solution of

$$\begin{cases} [h_1, h_2] \in D(\mathcal{A}), \\ \mathcal{A}\begin{bmatrix} h_1 \\ h_2 \end{bmatrix} = -\begin{bmatrix} f \\ g \end{bmatrix} \in \mathcal{H}, \end{cases}$$

or equivalently,

$$\begin{cases} h_1''''(x) = g(x), \\ h_1(0) = 0, \\ h_1'(0) = 0, \\ h_1''(1) = 0, \\ h_1'''(1) = kh_2(1), \\ h_2(x) = -f(x), \\ h_2(0) = h_2'(0) = 0 \end{cases}$$

is given by

$$h_1(x) = k\left(\frac{x^2}{2} - \frac{x^3}{6}\right)f(1) + \int_0^x \int_0^{x_4} \int_{x_3}^1 \int_{x_2}^1 g(x_1)dx_1 dx_2 dx_3 dx_4 \in H^4(0, 1),$$
$$h_2(x) = -f(x) \in H^2(0, 1),$$

which shows that $\mathcal{A}^{-1}$ exists and is bounded on $\mathcal{H}$. Therefore, by the Lümer-Phillips theorem, $\mathcal{A}$ generates a C_0-semigroup of contractions on $\mathcal{H}$. Moreover, $\mathcal{A}^{-1}$ is compact by Sobolev imbedding theorem since $\mathcal{A}^{-1}\mathcal{H} \subset H^4(0, 1) \times H^2(0, 1)$. System (4.72) is therefore well posed.

It remains to show the stability of (4.72). We notice that the system energy $E(t)$ defined in (4.73) can be written as

$$E(t) = \frac{1}{2}\int_0^1 \dot{w}^2(x, t)dx + \frac{1}{2}\int_0^1 w_{xx}^2(x, t)dx = \frac{1}{2}\left\|\begin{bmatrix} w(x, t) \\ \dot{w}(x, t) \end{bmatrix}\right\|_{\mathcal{H}}^2, \quad (4.78)$$

in terms of the solution of (4.72). This explains why we take $\mathcal{H}$ as the state space. Taking the time derivative of $E(t)$ along the solution of (4.77) yields

$$\dot{E}(t) = -w_{xxx}(1, t)\dot{w}(1, t) = -k\dot{w}^2(1, t) \leq 0. \quad (4.79)$$

Thus, the energy of system (4.72) is dissipative. In this sense, we say that $(w_{xxx}(1, t), w(1, t))$ is an adjoint pair. Using the energy multiplier method, we can further show that $E(t)$ decays exponentially. For the concept of the energy multiplier and a systematic way to find multipliers for a given specific system, the reader is referred to the book by Komornik [86].

Introduce a function

$$V(t) = E(t) + \varepsilon\rho(t), \qquad (4.80)$$

where the constant ε is to be determined, and

$$\rho(t) = \int_0^1 x\dot{w}(x,t)w_x(x,t)dx.$$

Calculating the time derivative of $\rho(t)$ along the solutions of (4.72) and making full use of the boundary conditions in (4.72) yields

$$
\begin{aligned}
\dot{\rho}(t) &= \int_0^1 x\ddot{w}(x,t)w_x(x,t)dx + \int_0^1 x\dot{w}(x,t)\dot{w}_x(x,t)dx \\
&= -\int_0^1 xw_{xxxx}(x,t)w_x(x,t)dx + \int_0^1 x\dot{w}(x,t)\dot{w}_x(x,t)dx \\
&= -k\dot{w}(1,t)w_x(1,t) + \int_0^1 xw_{xx}(x,t)w_{xxx}(x,t)dx \\
&\quad + \int_0^1 w_{xxx}(x,t)w_x(x,t)dx + \int_0^1 x\dot{w}(x,t)\dot{w}_x(x,t)dx, \\
&= -k\dot{w}(1,t)w_x(1,t) - \frac{3}{2}\int_0^1 w_{xx}^2(x,t)dx + \int_0^1 x\dot{w}(x,t)\dot{w}_x(x,t)dx,
\end{aligned}
$$
$$(4.81)$$

where we have used the relation

$$\int_0^1 xw_{xx}(x,t)w_{xxx}(x,t)dx = -\frac{1}{2}\int_0^1 w_{xx}^2(x,t)dx, \qquad (4.82)$$

which is easily verified by integration by parts. Now, again by integration by parts, and by using the Cauchy-Schwartz inequality and the inequality

$$2ab \le \varepsilon a^2 + \frac{1}{\varepsilon}b^2, \quad \text{for any real numbers } a \text{ and } b, \text{ and } \varepsilon > 0,$$

we can bound the terms on the right-hand side of (4.81) as follows:

$$
\begin{aligned}
-k\dot{w}(1,t)w_x(1,t) &= -k\dot{w}(1,t)\int_0^1 w_{xx}(x,t)dx \\
&\le \frac{k}{2}\left[\frac{1}{\varepsilon}\dot{w}^2(1,t) + \varepsilon\left(\int_0^1 w_{xx}(x,t)dx\right)^2\right] \\
&\le \frac{k}{2}\left[\frac{1}{\varepsilon}\dot{w}^2(1,t) + \varepsilon\int_0^1 w_{xx}^2(x,t)dx\right], \qquad (4.83)
\end{aligned}
$$

$$\int_0^1 x\dot{w}(x,t)\dot{w}_x(x,t)dx = \dot{w}^2(1,t) - \int_0^1 (x\dot{w}_x(x,t) + \dot{w}(x,t))\dot{w}(x,t)dx$$

$$= \frac{1}{2}\dot{w}^2(1,t) - \frac{1}{2}\int_0^1 \dot{w}^2(x,t)dx. \qquad (4.84)$$

Substituting the expressions in (4.83) and (4.84) into (4.81), we have

$$2\dot{\rho}(t) \leq -(3-k\varepsilon)\int_0^1 w_{xx}^2(x,t)dx - \int_0^1 \dot{w}^2(x,t)dx.$$
$$+(\frac{k}{\varepsilon}+1)\dot{w}^2(1,t),$$

that is,

$$\dot{\rho}(t) \leq -C_1 E(t) + C_2 \dot{w}^2(1,t), \qquad (4.85)$$

for some positive constants C_1 and C_2. Note that, for C_1 to be positive, ε must be chosen such that $\varepsilon < 3/k$.

On the other hand, using integration by parts, it is easy to show that there exists a positive constant C_3 such that

$$|\rho(t)| \leq C_3 E(t), \qquad (4.86)$$

from which we have

$$(1-C_3\varepsilon)E(t) \leq V(t) \leq (1+C_3\varepsilon)E(t).$$

Consequently, if we choose ε such that $\varepsilon \leq \min\{3/k, k/C_2, 1/C_3\}$, then

$$\dot{V}(t) = \dot{E}(t) + \varepsilon\dot{\rho}(t)$$
$$\leq -\varepsilon C_1 E(t) - (k - \varepsilon C_2)\dot{w}^2(1,t)$$
$$\leq -\frac{\varepsilon C_1}{1+\varepsilon C_3}V(t).$$

Therefore,

$$V(t) \leq \exp(-\frac{\varepsilon C_1}{1+\varepsilon C_3}t)V(0) \leq (1+\varepsilon C_3)\exp(-\frac{\varepsilon C_1}{1+\varepsilon C_3}t)E(0),$$

which means

$$E(t) \leq \frac{1+\varepsilon C_3}{1-\varepsilon C_3}\exp(-\frac{\varepsilon C_1}{1+\varepsilon C_3}t)E(0).$$

That is, the energy decays exponentially. In view of (4.78), we also see that the solution of (4.72) is exponentially stable because

$$\left\|\begin{bmatrix} w(x,t) \\ \dot{w}(x,t) \end{bmatrix}\right\|_{\mathcal{H}} \leq \sqrt{\frac{2(1+\varepsilon C_3)}{1-\varepsilon C_3}}\exp\left(-\frac{\varepsilon C_1}{2(1+\varepsilon C_3)}t\right)\left\|\begin{bmatrix} w(x,0) \\ \dot{w}(x,0) \end{bmatrix}\right\|_{\mathcal{H}}.$$

We have thus completed the proof of the exponential stability of the strain feedback controlled closed-loop system (4.67). □

As we have seen, the success of the energy multiplier method relies heavily on the selection of the function $\rho(t)$. We show how to choose $\rho(t)$ by considering $xw_x(x,t)$ as a multiplier. Multiplying $xw_x(x,t)$ on both sides of $\ddot{w}(x,t) + w_{xxxx}(x,t) = 0$, and integrating from 0 to 1 with respect to x, we obtain

$$
\begin{aligned}
0 &= \int_0^1 (xw_x(x,t)\ddot{w}(x,t) + xw_x(x,t)w_{xxxx}(x,t))dx \\
&= \frac{d}{dt}\int_0^1 xw_x(x,t)\dot{w}(x,t)dx - \int_0^1 x\dot{w}(x,t)\dot{w}_x(x,t)dx \\
&\quad - \int_0^1 [w_x(x,t) + xw_{xx}(x,t)]w_{xxx}(x,t)dx + kw_x(1,t)\dot{w}(1,t) \\
&= \frac{d}{dt}\int_0^1 xw_x(x,t)\dot{w}(x,t)dx - \frac{1}{2}\dot{w}^2(1,t) + \frac{1}{2}\int_0^1 \dot{w}^2(x,t)dx \\
&\quad + \frac{3}{2}\int_0^1 w_{xx}^2(x,t)dx + kw_x(1,t)\dot{w}(1,t),
\end{aligned}
$$

where we have used (4.84) and (4.82). Hence,

$$
\begin{aligned}
\frac{d}{dt}\int_0^1 x\dot{w}(x,t)w_x(x,t)dx &= -\frac{1}{2}\int_0^1 \dot{w}^2(x,t)dx - \frac{3}{2}\int_0^1 w_{xx}^2(x,t)dx \\
&\quad + \frac{1}{2}\dot{w}^2(1,t) - kw_x(1,t)\dot{w}(1,t).
\end{aligned}
$$

Since $w_x(1,t) = \int_0^1 w_{xx}(s,t)ds$, we have

$$
\frac{d}{dt}\int_0^1 x\dot{w}(x,t)w_x(x,t)dx \le -C_1 E(t) + C_2 \dot{w}^2(1,t).
$$

This is why we chose $\rho(t) = \int_0^1 xw_x(x,t)\dot{w}(x,t)dx$.

Now that we have shown the exponential stability of (4.72), we are able to show the exponential stability of (4.67), which is our ultimate purpose.

Theorem 4.16 *The solution $y(x,t)$ of (4.67) is exponentially stable in the sense that there exist positive constants M and ω such that*

$$
\int_0^1 [y_{xxxx}^2(x,t) + y_{txx}^2(x,t)]dx \le Me^{-\omega t}\int_0^1 \{[y_0''''(x)]^2 + [y_1''(x)]^2\}dx,
$$

or

$$
\left\| \begin{bmatrix} y(x,t) \\ \dot{y}(x,t) \end{bmatrix} \right\|_{D(A)\times D(A^{1/2})}^2 \le Me^{-\omega t} \left\| \begin{bmatrix} y_0 \\ y_1 \end{bmatrix} \right\|_{D(A)\times D(A^{1/2})}^2,
$$

where $D(A)$ is defined in (4.25).

Proof. In Theorem 4.15, we have shown that

$$\int_0^1 [w_{xx}^2(x,t) + w_t^2(x,t)]dx \le Me^{-\omega t} \int_0^1 \{w_{xx}^2(x,0) + \dot{w}^2(x,0)\}dx \quad (4.87)$$

for some $M > 0$ and $\omega > 0$. Since $w(x,t) = y''(1-x,t)$, (4.87) implies that

$$\int_0^1 [y_{xxxx}^2(x,t) + y_{txx}^2(x,t)]dx \le Me^{-\omega t} \int_0^1 \{[y_0''''(x)]^2 + [y_1''(x)]^2\}dx. \quad (4.88)$$

On the other hand, we know from Theorem 4.8 that the unique solution of (4.67) can be expressed as $(y, y_t)^T = T(t)(y_0, y_1)^T$, where $T(t)$ is a C_0-semigroup on $D(A) \times D(A^{1/2})$. From the arguments in Example 2.36, Chapter 2, we see that $D(A) \times D(A^{1/2})$ can be characterized as

$$D(A) \times D(A^{1/2}) = \{(f,g) \in H^2 \times L^2 \mid f(0) = f'(0) = f''(1) = f'''(1) = 0,$$
$$g(0) = g'(0) = 0\}.$$

The norm in $D(A) \times D(A^{1/2})$ is thus equivalent to

$$\left(\int_0^1 [f_{xxxx}^2(x) + g_{xx}^2(x)]dx \right)^{1/2}.$$

This fact, together with (4.88), implies that

$$\left\| \begin{bmatrix} y(x,t) \\ \dot{y}(x,t) \end{bmatrix} \right\|_{D(A) \times D(A^{1/2})}^2 \le Me^{-\omega t} \left\| \begin{bmatrix} y_0 \\ y_1 \end{bmatrix} \right\|_{D(A) \times D(A^{1/2})}^2.$$

This is the desired result. □

For the shear force feedback controlled closed-loop equation (4.68), we also can use the same variable transformation as before to transfer (4.68) into a boundary controlled Euler-Bernoulli beam equation like (4.72). Again, let $w(x,t) = y_{xx}(1-x,t)$. It is easy to see that (4.68) becomes

$$\begin{cases} \dfrac{\partial^2 w(x,t)}{\partial t^2} + \dfrac{\partial^4 w(x,t)}{\partial x^4} = 0, \quad 0 < x < 1,\ t > 0, \\ w(0,t) = w_x(0,t) = 0, \\ w_{xx}(1,t) = -k\dot{w}_x(1,t), \\ w_{xxx}(1,t) = 0, \\ w(x,0) = y_0''(1-x),\ \dot{w}(x,0) = y_1''(1-x). \end{cases} \quad (4.89)$$

This equation differs from (4.72) in the boundary conditions. Using the same argument above, it can be shown that the system operator associated with

(4.89) generates a C_0-semigroup which guarantees the well-posedness of the equation. Also, the energy function associated with (4.89) is given by

$$E(t) = \frac{1}{2} \int_0^1 \dot{w}^2(x,t)dx + \frac{1}{2} \int_0^1 w_{xx}^2(x,t)dx,$$

which is the same as above. Since the time derivative of $E(t)$ along the solution of (4.89) is given by

$$\dot{E}(t) = w_{xx}(1,t)\dot{w}_x(1,t) = -k\dot{w}_x^2(1,t) \leq 0,$$

energy is also dissipative. Again, $(w_{xx}(1,t), w_x(1,t))$ forms an adjoint pair. It might be conceived that the exponential stability of (4.89) also can be proved in parallel using the energy multiplier methods. Unfortunately, it is reported that the energy multiplier method did not work in this case, no matter how many different multipliers were tried [30]. The exponential stability was finally proved by the frequency test method, as discussed in Chapter 3. Namely, if we can show that the resolvent operator $(\lambda - \mathcal{A})^{-1}$ is uniformly bounded for all $\lambda = i\omega$ on the imaginary axis, then the semigroup generated by $\mathcal{A}$ is exponentially stable, where $\mathcal{A}$ is the system operator when (4.89) is written as a first order equation on an appropriate Hilbert space. Since the proof of this is rather lengthy, it will not be included here. The interested reader will find a similar proof in the next section. There, we use the frequency test method to prove the exponential stability of a different Euler-Bernoulli beam equation with boundary control.

4.4.2   Spectral analysis

In what follows, we turn to analyze the asymptotic behavior of the eigenvalues of the operator $\mathcal{A}$ defined in (4.75) and (4.76). Recall that all these eigenvalues lie on the left-half complex plane because (4.72) is already shown to be exponentially stable. We further show below that the eigenvalues with large moduli approach the vertical line specified by $\operatorname{Re}\lambda = -2k$ in the complex plane for finite feedback gain k. Changing the feedback gain can change the location of all closed-loop poles in strain feedback control, which is impossible with compact or bounded feedback control. This result also shows that the system operator $\mathcal{A}$ does not generate an analytic semigroup because its eigenvalues do not lie in a cone sector on the complex plane, as required by the generation of analytic semigroups.

Obviously, a complex number λ is an eigenvalue of $\mathcal{A}$ if and only if there exists a $\phi(x) \neq 0$ satisfying

$$\begin{cases} \lambda^2\phi(x) + kx\lambda\phi''(0) + \phi''''(x) = 0, & 0 < x < 1, \\ \phi(0) = \phi'(0) = 0, \\ \phi''(1) = \phi'''(1) = 0, \end{cases} \qquad (4.90)$$

which is obtained by assuming that $y(x,t) = e^{\lambda t}\phi(x)$ is a solution of (4.67). The boundary value problem (4.90) can be written as

$$\lambda^2 \phi(x) + k\lambda\Pi\phi(x) + A\phi(x) = 0, \tag{4.91}$$

where A is defined in (4.25) and $\Pi\phi(x) = x\phi''(0)$.

Lemma 4.17 λ *is an eigenvalue of* A *if and only if it satisfies*

$$F(\lambda) = 1 + k\lambda \int_0^1 x[A(\lambda^2 + A)^{-1}x]dx$$

$$= 1 + k\lambda \sum_{n=1}^{\infty} \frac{\lambda_n}{\lambda^2 + \lambda_n} b_n^2 \|\phi_n\|^2 = 0, \tag{4.92}$$

where λ_n, b_n, *and* $\phi_n(x)$ *are as defined in Lemma 4.6.*

Proof. Since all the eigenvalues of A have negative real parts, it is seen that $\lambda^2 \in \rho(-A)$, the resolvent set of $-A$. Solving for $\phi(x)$ from (4.91), we get

$$\phi(x) = -k\lambda\phi''(0)(\lambda^2 + A)^{-1}x. \tag{4.93}$$

On the other hand, taking L^2 inner product with x on both sides of (4.91), we obtain

$$\phi''(0)(1 + \frac{k\lambda}{3}) + \lambda^2 \int_0^1 x\phi(x)dx = 0. \tag{4.94}$$

Substituting (4.93) into (4.94) yields

$$\phi''(0)\left[1 + \frac{k\lambda}{3} - k\lambda^3 \int_0^1 x[(\lambda^2 + A)^{-1}x]dx\right]$$

$$= \phi''(0)\left[1 + k\lambda \int_0^1 x[A(\lambda^2 + A)^{-1}x]dx\right]$$

$$= 0.$$

If $\phi''(0) = 0$, then $\phi(x) = 0, \forall x \in (0,1)$ by (4.93). Thus, the eigenvalues of A should satisfy

$$1 + k\lambda \int_0^1 x[A(\lambda^2 + A)^{-1}x]dx = 0,$$

which is the first equality in (4.92). The second equality is a direct application of the eigenvector expansion. $\square$

Lemma 4.18 *The meromorphic function $F(\lambda)$ in Lemma 4.17 can be alternatively expressed as*

$$F(\lambda) = 1 + 2k\frac{f'(\sqrt{2\lambda})}{f(\sqrt{2\lambda})}, \qquad (4.95)$$

where

$$f(\lambda) = 4 + \exp(\lambda i) + \exp(-\lambda i) + \exp(\lambda) + \exp(-\lambda). \qquad (4.96)$$

Moreover, the zeros of $F(\lambda)$ consist of all those λ which satisfy

$$f(\sqrt{2\lambda}) + 2kf'(\sqrt{2\lambda}) = 0,$$

and the poles of $F(\lambda)$ consist of all the points $\pm i\beta_n^2 = \pm i\sqrt{\lambda_n}$, $n = 1, \cdots$.

Proof. Let $y(x) = (\lambda^2 + A)^{-1}x$, then

$$\begin{cases} \lambda^2 y(x) + y''''(x) = x, \\ y(0) = y'(0) = y''(1) = y'''(1) = 0. \end{cases}$$

Taking inner product with x on both sides of the first equation, we obtain

$$\lambda^2 \int_0^1 xy(x)dx + y''(0) = \frac{1}{3}.$$

Therefore,

$$\begin{aligned} F(\lambda) &= 1 + \frac{k\lambda}{3} - k\lambda^3 \int_0^1 xy(x)dx \\ &= 1 + k\lambda y''(0) = 1 + k\frac{g''(0)}{\lambda}, \qquad (4.97) \end{aligned}$$

where $g(x) = \lambda^2 y(x) - x$ and $g(x)$ satisfies

$$\begin{cases} \lambda^2 g(x) + g''''(x) = 0, \\ g(0) = g''(1) = g'''(1) = 0, \ g'(0) = -1. \end{cases} \qquad (4.98)$$

For any complex number $\lambda = \rho e^{i\theta}$, $-\pi < \theta \le \pi$, let $\sqrt{\lambda}$ denote the positive branch of λ. Then

$$\sqrt{\lambda} = \sqrt{\rho}e^{i\theta/2}, \quad \sqrt{\lambda i} = \sqrt{\lambda}e^{i\pi/4}, \quad \sqrt{2\lambda} = \sqrt{2}\sqrt{\lambda}.$$

The solution of (4.98) can be expressed as

$$g(x) = a_1 \exp(\sqrt{\lambda i}x) + a_2 \exp(-\sqrt{\lambda i}x) + a_3 \sin(\sqrt{\lambda i}x) + a_4 \cos(\sqrt{\lambda i}x),$$

where a_i, $i = 1, \cdots, 4$ are determined by the boundary conditions in (4.98). That is, a_i are the solutions to the following linear algebraic equation:

$$
\begin{bmatrix}
1 & 1 & 0 & 1 \\
1 & -1 & 1 & 0 \\
\exp(\sqrt{\lambda}i) & \exp(-\sqrt{\lambda}i) & -\sin(\sqrt{\lambda}i) & -\cos(\sqrt{\lambda}i) \\
\exp(\sqrt{\lambda}i) & -\exp(-\sqrt{\lambda}i) & -\cos(\sqrt{\lambda}i) & \sin(\sqrt{\lambda}i)
\end{bmatrix}
\begin{bmatrix}
a_1 \\ a_2 \\ a_3 \\ a_4
\end{bmatrix}
=
\begin{bmatrix}
0 \\ -\frac{1}{\sqrt{\lambda}i} \\ 0 \\ 0
\end{bmatrix},
$$

from which we get

$$
a_4 =
$$

$$
\frac{1}{2\sqrt{\lambda}i} \frac{\sin(\sqrt{\lambda}i)\left[\exp(\sqrt{\lambda}i) + \exp(-\sqrt{\lambda}i)\right] - \cos(\sqrt{\lambda}i)\left[\exp(\sqrt{\lambda}i) - \exp(-\sqrt{\lambda}i)\right]}{2 + \cos(\sqrt{\lambda}i)\left[\exp(\sqrt{\lambda}i) + \exp(-\sqrt{\lambda}i)\right]}.
$$

Using the identities

$$
\sin(\sqrt{\lambda}i) = [\exp(i\sqrt{\lambda}i) - \exp(-i\sqrt{\lambda}i)]/(2i),
$$
$$
\cos(\sqrt{\lambda}i) = [\exp(i\sqrt{\lambda}i) + \exp(-i\sqrt{\lambda}i)]/2,
$$
$$
\frac{i+1}{\sqrt{i}} = \sqrt{2}, \quad \frac{1-i}{\sqrt{i}} = -\sqrt{2}i, \quad (1+i)\sqrt{\lambda}i = i\sqrt{2\lambda}, \quad (1-i)\sqrt{\lambda}i = \sqrt{2\lambda},
$$

we are able to show that

$$
a_4 = \frac{1}{\sqrt{2\lambda}} \frac{-\left[\exp(i\sqrt{2\lambda}) - \exp(-i\sqrt{2\lambda})\right] + i\left[\exp(\sqrt{2\lambda}) - \exp(-\sqrt{2\lambda})\right]}{4 + \exp(i\sqrt{2\lambda}) + \exp(-i\sqrt{2\lambda}) + \exp(\sqrt{2\lambda}) + \exp(-\sqrt{2\lambda})}.
$$

Consequently,

$$
\begin{aligned}
F(\lambda) &= 1 + k\frac{g''(0)}{\lambda} = 1 + \frac{k}{\lambda}(-2a_4\lambda i) = 1 - 2ka_4 i \\
&= 1 + \frac{2k}{\sqrt{2\lambda}} \frac{i\left[\exp(i\sqrt{2\lambda}) - \exp(-i\sqrt{2\lambda})\right] + \left[\exp(\sqrt{2\lambda}) - \exp(-\sqrt{2\lambda})\right]}{4 + \exp(i\sqrt{2\lambda}) + \exp(-i\sqrt{2\lambda}) + \exp(\sqrt{2\lambda}) + \exp(-\sqrt{2\lambda})} \\
&= 1 + 2k\frac{f'(\sqrt{2\lambda})}{f(\sqrt{2\lambda})},
\end{aligned}
$$

which is exactly (4.95). Note that $f'(\sqrt{2\lambda})$ denotes the derivative with respect to λ. Let $\sqrt{2\lambda} = a + bi$. It follows that

$$
\begin{aligned}
f(\sqrt{2\lambda}) = \ & 4 + (e^b + e^{-b})\cos(a) + (e^a + e^{-a})\cos(b) \\
& +i[(e^{-b} - e^b)\sin(a) + (e^a - e^{-a})\sin(b)], \quad (4.99) \\
\sqrt{2\lambda}f'(\sqrt{2\lambda}) = \ & -(e^b + e^{-b})\sin(a) + (e^a - e^{-a})\cos(b) \\
& +i[(e^{-b} - e^b)\cos(a) + (e^a + e^{-a})\sin(b)]. \quad (4.100)
\end{aligned}
$$

For $\lambda = i\beta_n^2$, $\sqrt{2\lambda} = \beta_n + \beta_n i$, i.e., $a = b = \beta_n$. Thus, in view of (4.99),

$$
\begin{aligned}
f(\sqrt{2\beta_n^2 i}) &= 4 + 2[\exp(\beta_n) + \exp(-\beta_n)]\cos(\beta_n) \\
&= 1 + \cos(\beta_n)\cosh(\beta_n) = 0.
\end{aligned}
$$

The last equality follows from (i) of Lemma 4.6. This shows that $+i\beta_n^2$, $n = 1, \cdots, \infty$ are the zeros of $f(\sqrt{2\lambda})$. If we can show that $f'(\sqrt{2\beta_n^2 i}) \neq 0$, that is, there is no zero point cancellation between $f(\sqrt{2\lambda})$ and $f'(\sqrt{2\lambda})$, then clearly $+i\beta_n^2$, $n = 1, \cdots, \infty$, are the poles of $F(\lambda)$. Suppose $f'(\sqrt{2\beta_n^2 i}) = 0$, then from (4.100),

$$
\begin{aligned}
&-[\exp(\beta_n) + \exp(-\beta_n)]\sin(\beta_n) + [\exp(\beta_n) - \exp(-\beta_n)]\cos(\beta_n) \\
&+i\{[\exp(\beta_n) + \exp(-\beta_n)]\sin(\beta_n) - [\exp(\beta_n) - \exp(-\beta_n)]\cos(\beta_n)\} = 0,
\end{aligned}
$$

and hence, the real part must be zero, i.e.,

$$
[\exp(\beta_n) + \exp(-\beta_n)]\sin(\beta_n) = [\exp(\beta_n) - \exp(-\beta_n)]\cos(\beta_n). \tag{4.101}
$$

Again, from (i) of Lemma 4.6, we have

$$
\cos(\beta_n) = -\frac{2}{\exp(\beta_n) + \exp(-\beta_n)}.
$$

Substituting this into (4.101) gives

$$
\sin(\beta_n) = -\frac{2[\exp(\beta_n) - \exp(-\beta_n)]}{[\exp(\beta_n) + \exp(-\beta_n)]^2}.
$$

These two expressions imply that

$$
1 = \sin^2(\beta_n) + \cos^2(\beta_n) = \left(\frac{2}{\exp(\beta_n) + \exp(-\beta_n)}\right)^2 < 1,
$$

which is a contradiction. Hence, we have shown $f'(\sqrt{2\beta_n^2 i}) \neq 0$ and $i\beta_n^2$, $n = 1, \cdots, \infty$ are the poles of $F(\lambda)$. Since $F(\lambda)$ is a meromorphic function with real coefficients, $-i\beta_n^2$, $n = 1, \cdots, \infty$ also must be the poles of $F(\lambda)$. $\quad\square$

We shall use the expressions of $f(\sqrt{2\lambda})$ and $f'(\sqrt{2\lambda})$ in (4.95) to analyze the asymptotic behavior of the zeros of $F(\lambda)$ which are the eigenvalues of $\mathcal{A}$. The method is based on [30].

Theorem 4.19 *Let λ_n and their conjugates $\bar{\lambda}_n$, $n = 1, 2, \cdots$, be the eigenvalues of the operator $\mathcal{A}$, i.e., the zeros of $F(\lambda) = 0$. Then the following asymptotic expression holds*

$$
\lambda_n = [-2k + O(\frac{1}{n^2})] + i\{[(n - \frac{1}{2})\pi]^2 + O(\frac{1}{n})\}.
$$

as $n \to \infty$.

Remark 4.20 *The result in this theorem only applies to the case where the feedback gain k is finite. When $k \to \infty$, the eigenvalues of $\mathcal{A}$ approach the zeros of $f'(\sqrt{2\lambda})$, as can be seen from the expression of $F(\lambda) = 0$.*

Remark 4.21 *It is interesting to observe that the strain feedback can shift all the eigenvalues of $\mathcal{A}$ by changing the feedback gain k, which demonstrates the power of unbounded feedback. Note that bounded feedback cannot shift all the eigenvalues of the beam equation.*

Remark 4.22 *Theorem 4.19 shows that $\mathcal{A}$ does not generate an analytic semigroup.*

Proof. Let λ be an eigenvalue of $\mathcal{A}$, i.e., a zero of $F(\lambda)$. Denote

$$\lambda = |\lambda|e^{i\theta}. \tag{4.102}$$

Since all the eigenvalues locate on the open left-half complex plane, and since λ is symmetric with respect to the real axis, we need only to consider the case where $\pi/2 < \theta \leq \pi$ in (4.102). We divide the discussion into two different cases. First, suppose that there exists a $\delta > 0$ such that

$$\frac{\pi}{2} < \theta < \pi - \delta, \text{ for } |\lambda| \text{ sufficiently large.}$$

Since $\sqrt{\lambda} = \sqrt{|\lambda|}(\cos\frac{\theta}{2} + i\sin\frac{\theta}{2})$, we see that

$$e^{-\sqrt{2\lambda}} = O(e^{-\gamma\sqrt{|\lambda|}}), \quad e^{i\sqrt{2\lambda}} = O(e^{-\gamma\sqrt{|\lambda|}})$$

for some $\gamma > 0$. Thus, for large $|\lambda|$, $F(\lambda) = 0$, given in (4.95) and (4.96), can be interpreted as

$$4\sqrt{\lambda} + e^{-i\sqrt{2\lambda}}[\sqrt{\lambda} - i\sqrt{2}k] + e^{\sqrt{2\lambda}}[\sqrt{\lambda} + \sqrt{2}k] = O(|\lambda|e^{-\gamma\sqrt{|\lambda|}}).$$

Dividing both sides of this equation by $e^{-i\sqrt{2\lambda}}(\sqrt{\lambda} + \sqrt{2}k)$ and rearranging terms yields

$$
\begin{aligned}
e^{(1+i)\sqrt{2\lambda}} &= -\frac{\sqrt{\lambda} - i\sqrt{2}k}{\sqrt{\lambda} + \sqrt{2}k} + O(|\lambda|e^{-\gamma\sqrt{|\lambda|}}) \\
&= -1 + (1+i)\frac{\sqrt{2}k}{\sqrt{\lambda}} - \frac{2(1+i)k^2}{\lambda} + O(|\lambda|^{-3/2}) \quad (4.103)
\end{aligned}
$$

where the second equality is obtained from

$$-\frac{\sqrt{\lambda} - i\sqrt{2}k}{\sqrt{\lambda} + \sqrt{2}k} = -1 + \frac{(1+i)\sqrt{2}k}{\sqrt{\lambda} + \sqrt{2}k} = -1 + \frac{(1+i)\sqrt{2}k/\sqrt{\lambda}}{1 + \sqrt{2}k/\sqrt{\lambda}}$$

and the Taylor series expansion of

$$\frac{1}{1 + \sqrt{2}k/\sqrt{\lambda}} = 1 - \frac{\sqrt{2}k}{\sqrt{\lambda}} + O(|\lambda|^{-1}).$$

Consequently,

$$e^{\sqrt{2|\lambda|}(\cos\frac{\theta}{2} - \sin\frac{\theta}{2})} = -[1 - (1+i)\frac{\sqrt{2}k}{\sqrt{|\lambda|}}e^{-i\frac{\theta}{2}}]e^{-i\sqrt{2|\lambda|}(\cos\frac{\theta}{2} + \sin\frac{\theta}{2})} + O(|\lambda|^{-1}).$$

$$(4.104)$$

Since the left-hand side of (4.104) is a positive real number, we have

$$e^{\sqrt{2|\lambda|}(\cos\frac{\theta}{2} - \sin\frac{\theta}{2})} = |-1 + (1+i)\frac{\sqrt{2}k}{\sqrt{|\lambda|}}e^{-i\frac{\theta}{2}}| + O(|\lambda|^{-1})$$

$$= 1 - \frac{\sqrt{2}k}{\sqrt{|\lambda|}}(\cos\frac{\theta}{2} + \sin\frac{\theta}{2}) + O(|\lambda|^{-1}). \quad (4.105)$$

The last equality is obtained from the relation

$$\left| -1 + (a + bi)|\lambda|^{-1/2} \right| = \sqrt{(-1 + a|\lambda|^{-1/2})^2 + b^2|\lambda|^{-1}}$$

$$= \sqrt{1 - 2a|\lambda|^{-1/2} + O(|\lambda|^{-1})}$$

$$= 1 - a|\lambda|^{-1/2} + O(|\lambda|^{-1}),$$

with $a = \sqrt{2}k(\cos\frac{\theta}{2} + \sin\frac{\theta}{2})$, $b = \sqrt{2}k(\sin\frac{\theta}{2} - \sin\frac{\theta}{2})$. Notice that we have used the Taylor series expansion of $\sqrt{1+x}$ at $x = 0$. Performing logarithm operation on both sides of (4.105) and expanding the right-hand side of the resultant equation in the Taylor series leads to

$$\sqrt{2|\lambda|}(\cos\frac{\theta}{2} - \sin\frac{\theta}{2}) = -\frac{\sqrt{2}k}{\sqrt{|\lambda|}}(\cos\frac{\theta}{2} + \sin\frac{\theta}{2}) + O(|\lambda|^{-1}).$$

Taking the square of both sides of this equation yields

$$|\lambda|(1 - \sin\theta) = \frac{k^2}{|\lambda|}(1 + \sin\theta) + O(|\lambda|^{-3/2}),$$

from which we have

$$\sin\theta = 1 - \frac{2k^2}{|\lambda|^2} + O(|\lambda|^{-5/2}). \quad (4.106)$$

The real part of λ is thus given by

$$\mathrm{Re}\lambda = |\lambda|\cos\theta = -|\lambda|\sqrt{1 - \sin^2\theta} = -2k + O(|\lambda|^{-1/2}). \quad (4.107)$$

We now calculate $|\lambda|$ and $\mathrm{Im}\lambda$. From (4.104) and (4.105), it follows that

$$e^{-i\sqrt{2|\lambda|}(\cos\frac{\theta}{2}+\sin\frac{\theta}{2})} = -1 + O(|\lambda|^{-1/2}), \tag{4.108}$$

which implies that

$$\sqrt{2|\lambda|}(\cos\frac{\theta}{2} + \sin\frac{\theta}{2}) = (2n-1)\pi + O(|\lambda|^{-1/2})$$

where n is a large positive integer. This shows that $|\lambda| = O(n^2)$ and $2|\lambda|(1+\sin\theta) = [(2n-1)\pi]^2 + O(1)$. In view of (4.106), we have

$$|\lambda| = [(n-\frac{1}{2})\pi]^2 + O(1) \tag{4.109}$$

and hence,

$$\mathrm{Im}\lambda = |\lambda|\sin\theta = |\lambda| + O(|\lambda|^{-1}) = [(n-\frac{1}{2})\pi]^2 + O(1). \tag{4.110}$$

Let

$$\lambda_n = -2k + i[(n-\frac{1}{2})\pi]^2 + O(1). \tag{4.111}$$

We claim that such λ_n exists. To show this, we first observe that

$$e^{(1+i)\mu} = -1$$

has the following solutions

$$\mu_n = (n-\frac{1}{2})\pi(1+i)$$

for every positive integer n. Let $\mathcal{O}_n$ be the circle with center at μ_n and radius $\alpha|\mu_n|^{-1}$ where $\alpha > 2k$ is a constant, that is,

$$\mathcal{O}_n = \{\lambda \mid |\lambda - \mu_n| \le \alpha|\mu_n|^{-1}\}.$$

Then for all λ located on the circumference of $\mathcal{O}_n$, we see that

$$
\begin{aligned}
|e^{(1+i)\lambda} + 1| &= |e^{(1+i)\lambda} - e^{(1+i)\mu_n}| \\
&= |e^{(1+i)(\lambda-\mu_n)} - 1| \\
&= |e^{(1+i)\alpha|\mu_n|^{-1}e^{i\theta}} - 1| \\
&= |\alpha(1+i)|\mu_n|^{-1}e^{i\theta} + O(|\mu_n|^{-2})| \\
&= \alpha|1+i||\mu_n|^{-1} + O(|\mu_n|^{-2}) \\
&= \alpha|1+i||\lambda|^{-1} + O(|\lambda|^{-2}) \\
&> 2k|1+i||\lambda|^{-1} + O(|\lambda|^{-2})
\end{aligned}
$$

holds for all sufficiently large n. By Rouché's theorem, there exists a unique solution σ_n to the equation

$$e^{(1+i)\lambda} = -1 + 2k(1+i)\lambda^{-1} + O(|\lambda|^{-2})$$

inside $\mathcal{O}_n$, such that

$$|\sigma_n - \mu_n| \le \alpha |\mu_n|^{-1}.$$

Hence,

$$
\begin{aligned}
|\sigma_n^2 - \mu_n^2| &= |\sigma_n - \mu_n||\sigma_n + \mu_n| \\
&\le \alpha|\mu_n|^{-1}\left(2|\mu_n| + \alpha|\mu_n|^{-1}\right) \\
&= 2\alpha + \alpha^2|\mu_n|^{-2}.
\end{aligned}
$$

Let $\hat{\lambda}_n = \frac{1}{2}\sigma_n^2$. Then

$$\left|\hat{\lambda}_n - \frac{1}{2}\mu_n^2\right| \le \alpha + \alpha^2/2|\mu_n|^{-2}.$$

Since $\frac{1}{2}\mu_n^2 = [(n - \frac{1}{2})\pi]^2 i$, and $\alpha > 2k$, we see that λ_n defined by (4.111) is a solution of (4.103).

It remains to consider the case that $\theta \to \pi$. Suppose that $|\theta - \pi| < \delta$ for $\delta > 0$ sufficiently small. In this case, it is easily checked that

$$|e^{-\sqrt{2\lambda}}| \le C, \quad |e^{i\sqrt{2\lambda}}| \le C$$

for some $C > 0$ so that $F(\lambda) = 0$ can be written as

$$\sqrt{2}(1 - ik)e^{\sqrt{2|\lambda|}(\sin\frac{\theta}{2} - i\cos\frac{\theta}{2})} = \sqrt{\lambda}(1 + \sqrt{2})ke^{\sqrt{2|\lambda|}(\cos\frac{\theta}{2} + i\sin\frac{\theta}{2})} + O(|\lambda|). \tag{4.112}$$

It is seen that for θ in the range of interest, $\sin\frac{\theta}{2} > \cos\frac{\theta}{2}$ and in particular $\sin\frac{\theta}{2} > 1/2$. Thus, the modulus of the left-hand side of (4.112) is much larger than that of the right-hand side for $|\lambda|$ large enough. Hence it has no solution for large $|\lambda|$.

Finally, we derive the asymptotic estimation of λ_n given in the Theorem which is more accurate than the estimation shown in (4.111).

From (4.111), it is easy to see, by the Taylor series expansion, that

$$
\begin{aligned}
\sqrt{2\lambda_n} &= \sqrt{2i[(n - 1/2)\pi]^2 + O(1)} \\
&= \sqrt{2i[(n - 1/2)\pi]^2[1 + O(\frac{1}{n^2})]} \\
&= \pm(1 + i)(n - 1/2)\pi + O(n^{-1}). \tag{4.113}
\end{aligned}
$$

Since the sign on the right-hand side does not affect the value of λ_n, we consider only the positive branch of $\sqrt{2\lambda_n}$. Substituting (4.113) into (4.103), we see that the left-hand side of (4.103) is equal to

$$-e^{(1+i)O(n^{-1})} = -1 - (1+i)O(n^{-1}) + O(n^{-2})$$

and the right-hand side of (4.103) is equal to

$$-1 + (1+i)\frac{\sqrt{2}k}{\sqrt{\lambda_n}} + O(n^{-2})$$

$$= -1 + (1+i)\frac{2k}{(1+i)(n-1/2)\pi}[1 + O(n^{-2})] + O(n^{-2})$$

$$= -1 + \frac{2k}{(n-1/2)\pi} + O(n^{-2}).$$

By equating the two sides, we arrive at

$$O(|n|^{-1}) = -\frac{k(1-i)}{(n-1/2)\pi} + O(n^{-2}).$$

Thus (4.113) can be reformulated as

$$\sqrt{2\lambda_n} = (1+i)(n-1/2)\pi - \frac{k(1-i)}{(n-1/2)\pi} + O(n^{-2}). \qquad (4.114)$$

Substituting (4.114) into (4.103) again and using the Taylor series expansion, we see that the left-hand side of (4.103) is equal to

$$-\exp\left(-\frac{2k}{(n-1/2)\pi}\right)e^{(1+i)O(n^{-2})}$$

$$= \left(-1 + \frac{2k}{(n-1/2)\pi} - \frac{2k^2}{[(n-1/2)\pi]^2} + O(n^{-3})\right)$$

$$\left(1 + (1+i)O(n^{-2}) + O(n^{-3})\right)$$

$$= -1 + \frac{2k}{(n-1/2)\pi} - \frac{2k^2}{[(n-1/2)\pi]^2} - (1+i)O(n^{-2}) + O(n^{-3})$$

and the right-hand side of (4.103) is equal to

$$-1 + (1+i)\frac{\sqrt{2}k}{\sqrt{\lambda_n}} - \frac{2(1+i)k^2}{\lambda_n} + O(|\lambda_n|^{-3/2})$$

$$= -1 + (1+i)\frac{2k}{(1+i)(n-1/2)\pi}[1 + O(n^{-2})] - \frac{4(1+i)k^2}{[(1+i)(n-1/2)\pi]^2}$$

$$+ O(n^{-3})$$

$$= -1 + \frac{2k}{(n-1/2)\pi} - \frac{4k^2}{(1+i)[(n-1/2)\pi]^2} + O(n^{-3}).$$

By equating the two expressions above, we have

$$O(n^{-2}) = -(1+i)\frac{k^2}{[(n-1/2)\pi]^2} + O(n^{-3}).$$

Putting this into (4.114) yields

$$\sqrt{2\lambda_n} = (1+i)(n-1/2)\pi - \frac{k(1-i)}{(n-1/2)\pi} - (1+i)\frac{k^2}{[(n-1/2)\pi]^2} + O(n^{-3}),$$

or

$$2\lambda_n = [-4k + O(n^{-2})] + i\{2[(n-1/2)\pi]^2 + O(n^{-1})\}$$

which is the desired result.   □

4.5   Shear force feedback control of a rotating beam

In the previous section, we discussed strain feedback control for a rotating beam and shear force feedback control for a translating beam. We have seen that both of them introduce damping into beam vibrations, resulting in the closed-loop equations which are exponentially stable. A natural question then, is whether the shear force feedback works for rotating beam. The motivation for us to consider shear force feedback for rotating beam is threefold.

1) Practically, there are cases in which it is easier to measure shear force than bending strain. For example, load cells are easily attached to the flexible arm to measure the shear force needed for feedback, especially for robots with a flexible arm whose length can vary (imagining a polar robot) where it is impossible to cement strain gauge foils to measure the bending strain.

2) Hopefully, we can get better control performance by using higher order derivative feedback. This is because shear force information contains third order spatial derivative while strain feedback contains only second order spatial derivative. As we saw in the last section, the closed-loop eigenvalues of the strain feedback control of the rotating beam cannot be shifted into a cone sector, which might be improved by higher order derivative feedback.

3) Theoretically, the shear force feedback control for rotating beam produces a closed-loop equation which is similar to, but different from the equations we encountered in the previous section. Issues such as the well-posedness and stability of this new equation have not been solved to date. Thus, it is of great interest to clarify these points.

In the sequel, we first formulate the closed-loop equation into an abstract equation on an appropriate Hilbert space. After deriving some preliminary results, we shall estimate the norm bound of the resolvent of the system operator. This estimation makes it possible to show the existence and uniqueness of the solutions of the closed-loop equation, and to show the exponential stability of this solution using the results developed in Chapter 3. These results imply that the system operator generates a 1-time integrated semigroup. The asymptotic behavior of the spectrum of the system operator is then analyzed, which clearly demonstrates that the closed-loop eigenvalues with large moduli are shifted to the left-half complex plane.

4.5.1   Well-posedness and exponential stability

Consider the dynamic model of the rotating Euler-Bernoulli beam

$$
\begin{cases}
\dfrac{\partial^2 y(x,t)}{\partial t^2} + \dfrac{\partial^4 y(x,t)}{\partial x^4} = -x\ddot{\theta}(t), & 0 < x < 1,\ t > 0, \\[2mm]
y(0,t) = y_x(0,t) = 0, \\[2mm]
y_{xx}(1,t) = y_{xxx}(1,t) = 0, \\[2mm]
y(x,0) = y_0(x),\ \dot{y}(x,0) = y_1(x).
\end{cases}
\tag{4.115}
$$

Suppose the shear force at the root end of the beam is available from the measurement, and suppose we can control the rotating motor such that the angular velocity is proportional to the shear force, i.e.,

$$
\dot{\theta}(t) = -k y_{xxx}(0,t), \quad k > 0.
$$

Taking the time derivative on both sides of the above equality yields

$$
\ddot{\theta}(t) = -k y_{xxxt}(0,t).
$$

Substituting this into (4.115) leads to the shear force feedback controlled closed-loop system equation:

$$
\begin{cases}
\dfrac{\partial^2 y(x,t)}{\partial t^2} - kx\dfrac{\partial^4 y(0,t)}{\partial t \partial x^3} + \dfrac{\partial^4 y(x,t)}{\partial x^4} = 0, & 0 < x < 1,\ t > 0, \\[2mm]
y(0,t) = y_x(0,t) = 0, \\[2mm]
y_{xx}(1,t) = y_{xxx}(1,t) = 0, \\[2mm]
y(x,0) = y_0(x),\ \dot{y}(x,0) = y_1(x).
\end{cases}
\tag{4.116}
$$

Our primary concern is whether there exists a unique solution and whether the solution is exponentially stable. It should be noted that (4.116) is different from (4.67) and (4.68). And we cannot use the A-dependent concept to discuss the issues of our concern.

As before, let us introduce a new variable $w(x,t) = y_{xx}(1-x,t)$. It can be seen that, in terms of the new variable $w(x,t)$, (4.116) becomes

$$
\begin{cases}
\dfrac{\partial^2 w(x,t)}{\partial t^2} + \dfrac{\partial^4 w(x,t)}{\partial x^4} = 0, & 0 < x < 1,\ t > 0, \\[2mm]
w(0,t) = w_x(0,t) = 0, \\[2mm]
w_{xx}(1,t) = 0, \\[2mm]
w_{xxx}(1,t) = kw_{xt}(1,t), \\[2mm]
w(x,0) = y_0''(1-x),\ \dot{w}(x,0) = y_1''(1-x).
\end{cases}
\tag{4.117}
$$

This equation is different from (4.72) and (4.89). Since the pair (w_{xxx}, w_x) is not adjoint, it seems difficult, if not impossible, to find an energy function for (4.117). Therefore, we have to seek other ways to investigate the existence, uniqueness and stability of the solutions to (4.117).

For this purpose, let us first rewrite (4.117) as an abstract first order equation on Hilbert space

$$
\mathcal{H} = \{(f(x), g(x)) \in H^2(0,1) \times L^2(0,1) \mid f(0) = f'(0) = 0\}.
$$

The inner product in $\mathcal{H}$ is defined as

$$
\left\langle \begin{bmatrix} f \\ g \end{bmatrix}, \begin{bmatrix} \tilde{f} \\ \tilde{g} \end{bmatrix} \right\rangle_{\mathcal{H}} = \int_0^1 (f''(x)\overline{\tilde{f}''(x)} + g(x)\overline{\tilde{g}(x)})dx
$$

and the induced norm is given by

$$
\|(f,g)\|_{\mathcal{H}}^2 = \int_0^1 (|f''(x)|^2 + |g(x)|^2)dx.
$$

Define

$$
z_1(t) = w(\cdot,t),\ \ z_2(t) = \dot{w}(\cdot,t)
\tag{4.118}
$$

and

$$
A\begin{bmatrix} f(x) \\ g(x) \end{bmatrix} = \begin{bmatrix} g(x) \\ -f''''(x) \end{bmatrix},
\tag{4.119}
$$

with domain

$$
\begin{aligned}
D(A) = \big\{ &(f,g) \in H^4(0,1) \times H^2(0,1) \mid f(0) = f'(0) = 0 \\
& f''(1) = g(0) = g'(0) = 0,\ f'''(1) = kg'(1) \big\},
\end{aligned}
\tag{4.120}
$$

which is dense in $\mathcal{H}$. Then (4.117) can be written as

$$
\frac{d}{dt}\begin{bmatrix} z_1(t) \\ z_2(t) \end{bmatrix} = A\begin{bmatrix} z_1(t) \\ z_2(t) \end{bmatrix}.
\tag{4.121}
$$

It should be remarked that the operator $\mathcal{A}$ is not dissipative on $\mathcal{H}$; hence, the Lümer-Phillips theorem cannot be applied to show the existence of a C_0-semigroup associated with $\mathcal{A}$.

Lemma 4.23 $\mathcal{A}^{-1}$ *exists and is compact on* $\mathcal{H}$. *Therefore, the spectrum* $\sigma(\mathcal{A})$ *consists entirely of isolated eigenvalues.*

Proof. The proof is straightforward, and is thus omitted. $\square$

Lemma 4.24 $\lambda \in \sigma(\mathcal{A})$ *if and only if* λ *is a solution of* $G(\lambda) = 0$ *where* $G(\lambda)$ *is an entire function of* λ:

$$G(\lambda) = 4 + (1+k)(e^{\sqrt{2\lambda}} + e^{-\sqrt{2\lambda}}) + (1-k)(e^{i\sqrt{2\lambda}} + e^{-i\sqrt{2\lambda}}). \quad (4.122)$$

Proof. λ is an eigenvalue of $\mathcal{A}$ if and only if there exists a nonzero $\phi(x)$ such that

$$\begin{cases} \lambda^2 \phi(x) + \phi''''(x) = 0, \\ \phi(0) = \phi'(0) = \phi''(1) = 0, \phi'''(1) = k\lambda\phi'(1). \end{cases} \quad (4.123)$$

The solution of (4.123) is given by

$$\phi(x) = c_1 e^{\sqrt{\lambda i} x} + c_2 e^{-\sqrt{\lambda i} x} + c_3 e^{i\sqrt{\lambda i} x} + c_4 e^{-i\sqrt{\lambda i} x}$$

where c_i, $i = 1, 2, 3, 4$ should satisfy

$$\begin{bmatrix} 1 & 1 & 1 & 1 \\ 1 & -1 & i & -i \\ e^{\sqrt{\lambda i}} & e^{-\sqrt{\lambda i}} & -e^{i\sqrt{\lambda i}} & -e^{-i\sqrt{\lambda i}} \\ (i-k)e^{\sqrt{\lambda i}} & (k-i)e^{-\sqrt{\lambda i}} & (1-ki)e^{i\sqrt{\lambda i}} & (ki-1)e^{-i\sqrt{\lambda i}} \end{bmatrix} \begin{bmatrix} c_1 \\ c_2 \\ c_3 \\ c_4 \end{bmatrix} = 0.$$

$$\quad (4.124)$$

The boundary value problem (4.123) has a nontrivial solution if and only if the determinant of the coefficient matrix of c_i in (4.124) is equal to zero from which we obtain (4.122). $\square$

The following lemma provides an alternative characterization of the eigenvalues of $\mathcal{A}$.

Lemma 4.25 $\sigma(\mathcal{A})$ *consists of all zeros of the following meromorphic function*

$$F(\lambda) = 1 + \frac{k}{2}\lambda + 4k \sum_{n=1}^{\infty} \frac{\lambda^3}{\lambda^2 + \lambda_n} \gamma_n \beta_n^{-3} \|\phi_n\|^{-2} \quad (4.125)$$

where $\lambda_n, \phi_n(x), \beta_n$, *and* γ_n *are given in Lemma 4.6.*

Proof.   This can be similarly proved as we did in Lemma 4.17.   □

Both $F(\lambda)$ and $G(\lambda)$ will be used in the sequel.

Lemma 4.26 *Let λ be a zero of $F(\lambda)$. Then $\mathrm{Re}\lambda < 0$.*

Proof.   Let $\lambda = a + bi, a, b \in \mathbf{R}$, be a zero of $F(\lambda)$. Noting that $\lambda^3 = (a + bi)^3 = a^3 - 3ab^2 + (3a^2b - b^3)i$ and

$$
\begin{aligned}
\frac{\lambda^3}{\lambda^2 + \lambda_n} &= \frac{\lambda^3(\bar{\lambda}^2 + \lambda_n)}{|\lambda^2 + \lambda_n|^2} = \frac{|\lambda|^4\lambda + \lambda_n\lambda^3}{|\lambda^2 + \lambda_n|^2} \\
&= \frac{|\lambda|^4 a + \lambda_n(a^3 - 3ab^2) + i[|\lambda|^4 b + \lambda_n(3a^2b - b^3)]}{|\lambda^2 + \lambda_n|^2},
\end{aligned}
$$

we have

$$
\begin{cases}
1 + \dfrac{k}{2}a + 4k \displaystyle\sum_{n=1}^{\infty} \frac{|\lambda|^4 a + \lambda_n(a^3 - 3ab^2)}{|\lambda^2 + \lambda_n|^2}\gamma_n\beta_n^{-3}\|\phi_n\|^{-2} = 0, \\
\dfrac{k}{2}b + 4k \displaystyle\sum_{n=1}^{\infty} \frac{|\lambda|^4 b + \lambda_n(3a^2b - b^3)}{|\lambda^2 + \lambda_n|^2}\gamma_n\beta_n^{-3}\|\phi_n\|^{-2} = 0.
\end{cases} \tag{4.126}
$$

From the first equality above, it is seen that $a \neq 0$. Suppose $b \neq 0$. Then from the second equation in (4.126)

$$
\frac{k}{2} + 4k \sum_{n=1}^{\infty} \frac{|\lambda|^4 + \lambda_n(3a^2 - b^2)}{|\lambda^2 + \lambda_n|^2}\gamma_n\beta_n^{-3}\|\phi_n\|^{-2} = 0.
$$

Therefore,

$$
\begin{aligned}
&\frac{k}{2} + 4k \sum_{n=1}^{\infty} \frac{|\lambda|^4}{|\lambda^2 + \lambda_n|^2}\gamma_n\beta_n^{-3}\|\phi_n\|^{-2} \\
&= -4k \sum_{n=1}^{\infty} \frac{\lambda_n(3a^2 - b^2)}{|\lambda^2 + \lambda_n|^2}\gamma_n\beta_n^{-3}\|\phi_n\|^{-2}.
\end{aligned}
$$

By use of (4.126), we see that

$$
\begin{aligned}
0 &= \frac{1}{a} + \frac{k}{2} + 4k \sum_{n=1}^{\infty} \frac{|\lambda|^4 + \lambda_n(a^2 - 3b^2)}{|\lambda^2 + \lambda_n|^2}\gamma_n\beta_n^{-3}\|\phi_n\|^{-2} \\
&= \frac{1}{a} + 4k \sum_{n=1}^{\infty} \frac{\lambda_n(a^2 - 3b^2) - \lambda_n(3a^2 - b^2)}{|\lambda^2 + \lambda_n|^2}\gamma_n\beta_n^{-3}\|\phi_n\|^{-2} \\
&= \frac{1}{a} - 4k \sum_{n=1}^{\infty} \frac{2\lambda_n|\lambda|^2}{|\lambda^2 + \lambda_n|^2}\gamma_n\beta_n^{-3}\|\phi_n\|^{-2}.
\end{aligned}
$$

Consequently,

$$\frac{1}{a} = 8k|\lambda|^2 \sum_{n=1}^{\infty} \frac{\beta_n \gamma_n}{|\lambda^2 + \lambda_n|^2} \|\phi_n\|^{-2} < 0,$$

since $\gamma_n < 0$ as shown in Lemma 4.6. The proof is complete if one can show that $b \neq 0$ for any $a > 0$. In fact, if $a > 0$ and $b = 0$, then $\sqrt{2\lambda}$ is a positive real number. Therefore,

$$4 + (1+k)(e^{\sqrt{2\lambda}} + e^{-\sqrt{2\lambda}}) \geq 4 + 2(1+k) > 2|1-k| \geq |(1-k)(e^{i\sqrt{2\lambda}} + e^{-i\sqrt{2\lambda}})|,$$

which means $G(\lambda) \neq 0$, a contradiction. □

Lemma 4.27 *The resolvent operator of $\mathcal{A}$ has the following asymptotic property:*

$$\|R(\lambda, \mathcal{A})\| = O(|\lambda|^{-1/2}) \quad \text{for } \operatorname{Re}\lambda \geq 0. \tag{4.127}$$

Proof. For any $(f, g) \in \mathcal{H}$, we need to find $(\phi, \psi) \in D(\mathcal{A})$ such that

$$(\mathcal{A} - \lambda) \begin{bmatrix} \phi(x) \\ \psi(x) \end{bmatrix} = \begin{bmatrix} \psi(x) - \lambda\phi(x) \\ -\phi''''(x) - \lambda\psi(x) \end{bmatrix} = \begin{bmatrix} f(x) \\ g(x) \end{bmatrix},$$

from which we have

$$\psi(x) = \lambda\phi(x) + f(x), \tag{4.128}$$

with $\phi(x)$ satisfying

$$\begin{cases} \phi''''(x) + \lambda^2\phi(x) = -[\lambda f(x) + g(x)], \\ \phi(0) = \phi'(0) = \phi''(1) = 0, \\ \phi'''(1) = k[\lambda\phi'(1) + f'(1)]. \end{cases} \tag{4.129}$$

Since we only consider those λ with $\operatorname{Re}\lambda \geq 0$, we are able to write λ as

$$\lambda = |\lambda|e^{i\theta}, \quad -\frac{\pi}{2} \leq \theta \leq \frac{\pi}{2}.$$

Since the estimates for $-\frac{\pi}{2} \leq \theta \leq 0$ are similar, we consider only the case where $0 \leq \theta \leq \frac{\pi}{2}$. In this case, $\frac{\pi}{4} \leq \frac{\theta}{2} + \frac{\pi}{4} \leq \frac{\pi}{2}$ and the following equalities hold true:

$$\lambda i = |\lambda|e^{i(\theta + \frac{\pi}{2})}, \quad \sqrt{\lambda i} = \sqrt{|\lambda|}e^{i(\frac{\theta}{2} + \frac{\pi}{4})},$$

$$e^{-\sqrt{\lambda i}} = e^{\sqrt{|\lambda|}[-\cos(\frac{\theta}{2} + \frac{\pi}{4}) - i\sin(\frac{\theta}{2} + \frac{\pi}{4})]} = O(1),$$

$$e^{i\sqrt{\lambda i}} = e^{\sqrt{|\lambda|}[-\sin(\frac{\theta}{2} + \frac{\pi}{4}) - i\cos(\frac{\theta}{2} + \frac{\pi}{4})]} = O(1), \tag{4.130}$$

$$e^{i\sqrt{\lambda i}}e^{\sqrt{\lambda i}} = O(1). \tag{4.131}$$

These relations will be used frequently in the sequel. The solution $\phi(x)$ to (4.129) can be expressed as $\phi(x) = \tilde{\phi}(x) + \phi_p(x)$ where $\phi_p(x)$ satisfies

$$\begin{cases} \phi_p''''(x) + \lambda^2 \phi_p(x) = -[\lambda f(x) + g(x)], \\ \phi_p(0) = \phi_p'(0) = \phi_p'''(0) = 0, \end{cases} \tag{4.132}$$

and $\tilde{\phi}(x)$ satisfies

$$\begin{cases} \tilde{\phi}''''(x) + \lambda^2 \tilde{\phi}(x) = 0, \\ \tilde{\phi}(0) = \tilde{\phi}'(0) = 0, \\ \tilde{\phi}''(1) = A, \\ \tilde{\phi}'''(1) - k\lambda\tilde{\phi}'(1) = B \end{cases} \tag{4.133}$$

with

$$A = -\phi_p''(1), \; B = -[\phi_p'''(1) - k\lambda\phi_p'(1)] + kf'(1).$$

A particular solution of (4.132) is given by

$$\phi_p(x) = -\frac{1}{2}(\lambda i)^{-3/2} \int_0^x [\sinh\sqrt{\lambda i}(x-\xi) - \sin\sqrt{\lambda i}(x-\xi)][\lambda f(\xi) + g(\xi)]d\xi. \tag{4.134}$$

Clearly, $\|R(\lambda, \mathcal{A})\| = O(|\lambda|^{-1/2})$ holds if we can find a constant M, independent of λ, such that for all $\text{Re}\lambda \geq 0$,

$$\left(\int_0^1 [|\phi''(x)|^2 + |\psi(x)|^2]dx\right)^{1/2} \leq M|\lambda|^{-1/2}\left(\int_0^1 [|f''(x)|^2 + |g(x)|^2]dx\right)^{1/2}. \tag{4.135}$$

We derive this inequality in four steps.

• **1st Step: Estimation of $\phi_p''(x)$**: This is obtained by directly manipulating $\phi_p(x)$ in (4.134).

$$\begin{aligned} \phi_p''(x) &= \frac{1}{4}(\lambda i)^{-1/2}e^{\sqrt{\lambda i}x}\int_0^x e^{-\sqrt{\lambda i}\xi}[i\,f''(\xi) - g(\xi)]d\xi \\ &\quad -i\frac{1}{4}(\lambda i)^{-1/2}e^{-i\sqrt{\lambda i}x}\int_0^x e^{i\sqrt{\lambda i}\xi}[i\,f''(\xi) + g(\xi)]d\xi \\ &\quad + O(|\lambda|^{-1/2}[\|f''\| + \|g\|]). \end{aligned} \tag{4.136}$$

• **2nd Step: Estimations of A and B:**

In view of (4.136), we have

$$\begin{aligned} -\phi_p''(1) &= -\frac{1}{4}(\lambda i)^{-1/2}\int_0^1 e^{\sqrt{\lambda i}(1-\xi)}[i\,f''(\xi) - g(\xi)]d\xi \\ &\quad +i\frac{1}{4}(\lambda i)^{-1/2}\int_0^1 e^{-i\sqrt{\lambda i}(1-\xi)}[i\,f''(\xi) + g(\xi)]d\xi \\ &\quad + O(|\lambda|^{-1/2}[\|f''\| + \|g\|]). \end{aligned} \tag{4.137}$$

Also, it is easy to verify that

$$-\phi_p'''(1) = -\frac{1}{4}\int_0^1 e^{\sqrt{\lambda i}(1-\xi)}[i\,f''(\xi) - g(\xi)]d\xi$$

$$+\frac{1}{4}\int_0^1 e^{-i\sqrt{\lambda i}(1-\xi)}[i\,f''(\xi) + g(\xi)]d\xi$$

$$+O(\|f''\| + \|g\|), \tag{4.138}$$

and that

$$\phi_p'(1) = -\frac{1}{2}(\lambda i)^{-1}\int_0^1 [\cosh\sqrt{\lambda i}(1-\xi) - \cos\sqrt{\lambda i}(1-\xi)][\lambda f(\xi) + g(\xi)]d\xi$$

$$= -\frac{1}{2}(\lambda i)^{-1}\int_0^1 [\cosh\sqrt{\lambda i}(1-\xi) - \cos\sqrt{\lambda i}(1-\xi)]g(\xi)d\xi$$

$$+\frac{i}{2}(\lambda i)^{-1}\int_0^1 [\cosh\sqrt{\lambda i}(1-\xi) + \cos\sqrt{\lambda i}(1-\xi)]f''(\xi)d\xi$$

$$-i(\lambda i)^{-1}f'(1)$$

$$= -\frac{1}{4}(\lambda i)^{-1}\int_0^1 [e^{\sqrt{\lambda i}(1-\xi)} - e^{-i\sqrt{\lambda i}(1-\xi)}]g(\xi)d\xi$$

$$+\frac{i}{4}(\lambda i)^{-1}\int_0^1 [e^{\sqrt{\lambda i}(1-\xi)} + e^{-i\sqrt{\lambda i}(1-\xi)}]f''(\xi)d\xi$$

$$+O(|\lambda|^{-1}[\|f''\| + \|g\|]).$$

$$= \frac{1}{4}(\lambda i)^{-1}\int_0^1 e^{\sqrt{\lambda i}(1-\xi)}[i\,f''(\xi) - g(\xi)]d\xi$$

$$+\frac{1}{4}(\lambda i)^{-1}\int_0^1 e^{-i\sqrt{\lambda i}(1-\xi)}[i\,f''(\xi) + g(\xi)]d\xi + O(|\lambda|^{-1}[\|f''\| + \|g\|]).$$

Thus,

$$B = -[\phi_p'''(1) - k\lambda\phi_p'(1)] + kf'(1)$$

$$= -\frac{1}{4}\int_0^1 e^{\sqrt{\lambda i}(1-\xi)}[i\,f''(\xi) - g(\xi)]d\xi + \frac{1}{4}\int_0^1 e^{-i\sqrt{\lambda i}(1-\xi)}[i\,f''(\xi) + g(\xi)]d\xi$$

$$-\frac{k}{4}i\int_0^1 e^{\sqrt{\lambda i}(1-\xi)}[i\,f''(\xi) - g(\xi)]d\xi$$

$$-\frac{k}{4}i\int_0^1 e^{-i\sqrt{\lambda i}(1-\xi)}[i\,f''(\xi) + g(\xi)]d\xi + O(\|f''\| + \|g\|)$$

$$= -\frac{1}{4}(1 + ki)\int_0^1 e^{\sqrt{\lambda i}(1-\xi)}[i\,f''(\xi) - g(\xi)]d\xi$$

$$+\frac{1}{4}(1 - ki)\int_0^1 e^{-i\sqrt{\lambda i}(1-\xi)}[i\,f''(\xi) + g(\xi)]d\xi + O(\|f''\| + \|g\|).$$

• **3rd Step: Estimation of $\tilde{\phi}''(x)$:**

Solving (4.133) for $\tilde{\phi}''(x)$ we have

$$\tilde{\phi}(x) = A_1 e^{\sqrt{\lambda i}x} + A_2 e^{-\sqrt{\lambda i}x} + A_3 e^{i\sqrt{\lambda i}x} + A_4 e^{-i\sqrt{\lambda i}x}. \tag{4.139}$$

Here, $A_i, i = 1, 2, 3, 4$ are constants to be determined. By the boundary conditions $\tilde{\phi}(0) = \tilde{\phi}'(0) = 0$, we get

$$\begin{cases} A_1 = -\frac{1+i}{2}A_3 - \frac{1-i}{2}A_4, \\ A_2 = -\frac{1-i}{2}A_3 - \frac{1+i}{2}A_4. \end{cases} \tag{4.140}$$

On the other hand, from the boundary conditions $\tilde{\phi}''(1) = A, \tilde{\phi}'''(1) - k\lambda\tilde{\phi}'(1) = B$, we have

$$E\begin{bmatrix} A_3 \\ A_4 \end{bmatrix} = \begin{bmatrix} A/(\lambda i) \\ B/(\lambda i)^{3/2} \end{bmatrix}, \tag{4.141}$$

where $E = \{e_{ij}\}$ is a 2×2 matrix with elements

$$e_{11} = -\frac{1+i}{2}e^{\sqrt{\lambda i}} - \frac{1-i}{2}e^{-\sqrt{\lambda i}} - e^{i\sqrt{\lambda i}},$$

$$e_{12} = -\frac{1-i}{2}e^{\sqrt{\lambda i}} - \frac{1+i}{2}e^{-\sqrt{\lambda i}} - e^{-i\sqrt{\lambda i}},$$

$$e_{21} = -(1+ki)\frac{1+i}{2}e^{\sqrt{\lambda i}} + (1+ki)\frac{1-i}{2}e^{-\sqrt{\lambda i}} - (k+i)e^{i\sqrt{\lambda i}},$$

$$e_{22} = -(1+ki)\frac{1-i}{2}e^{\sqrt{\lambda i}} + (1+ki)\frac{1+i}{2}e^{-\sqrt{\lambda i}} + (k+i)e^{-i\sqrt{\lambda i}}.$$

Calculating the determinant of E (which is denoted by $\Delta = \det E$), we have

$$
\begin{aligned}
\Delta &= \left[\frac{1+i}{2}e^{\sqrt{\lambda i}} + \frac{1-i}{2}e^{-\sqrt{\lambda i}} + e^{i\sqrt{\lambda i}}\right] \\
&\quad \times \left[(1+ki)\frac{1-i}{2}e^{\sqrt{\lambda i}} - (1+ki)\frac{1+i}{2}e^{-\sqrt{\lambda i}} - (k+i)e^{-i\sqrt{\lambda i}}\right] \\
&\quad - \left[\frac{1-i}{2}e^{\sqrt{\lambda i}} + \frac{1+i}{2}e^{-\sqrt{\lambda i}} + e^{-i\sqrt{\lambda i}}\right] \\
&\quad \times \left[(1+ki)\frac{1+i}{2}e^{\sqrt{\lambda i}} - (1+ki)\frac{1-i}{2}e^{-\sqrt{\lambda i}} + (k+i)e^{i\sqrt{\lambda i}}\right] \\
&= 4(k-i) - 2(k+i) - i(1+k)\left(e^{(1-i)\sqrt{\lambda i}} + e^{(i-1)\sqrt{\lambda i}}\right) \\
&\quad -i(1-k)\left(e^{(1+i)\sqrt{\lambda i}} + e^{-(1+i)\sqrt{\lambda i}}\right) \\
&= -ie^{-i\sqrt{\lambda i}}\left[(1+k)e^{\sqrt{\lambda i}} + (1-k)e^{-\sqrt{\lambda i}} + O(|e^{i\sqrt{\lambda i}}|)\right].
\end{aligned}
$$

From (4.141), it is easy to see that

$$\Delta(\lambda i)^{3/2} A_3$$

$$= [-(1+ki)\frac{1-i}{2}e^{\sqrt{\lambda i}} + (1+ki)\frac{1+i}{2}e^{-\sqrt{\lambda i}} + (k+i)e^{-i\sqrt{\lambda i}}]\sqrt{\lambda i}A$$

$$+[\frac{1-i}{2}e^{\sqrt{\lambda i}} + \frac{1+i}{2}e^{-\sqrt{\lambda i}} + e^{-i\sqrt{\lambda i}}]B$$

$$= -\frac{1}{4}(1+k)(1+i)e^{-i\sqrt{\lambda i}}\int_0^1 e^{\sqrt{\lambda i}(1-\xi)}[i\ f''(\xi) - g(\xi)]d\xi$$

$$+\frac{i}{4}[-(1+k)e^{\sqrt{\lambda i}} + (1-k)e^{-\sqrt{\lambda i}}]\int_0^1 e^{-i\sqrt{\lambda i}(1-\xi)}[i\ f''(\xi) + g(\xi)]d\xi$$

$$+O(\|f''\| + \|g\|),$$

and that

$$\Delta(\lambda i)^{3/2} A_4$$

$$= [(1+ki)\frac{1+i}{2}e^{\sqrt{\lambda i}} - (1+ki)\frac{1-i}{2}e^{-\sqrt{\lambda i}} + (k+i)e^{i\sqrt{\lambda i}}]\sqrt{\lambda i}A$$

$$+[-\frac{1+i}{2}e^{\sqrt{\lambda i}} - \frac{1-i}{2}e^{-\sqrt{\lambda i}} - e^{i\sqrt{\lambda i}}]B$$

$$= \frac{1}{4}[(1-k)(1-i)e^{i\sqrt{\lambda i}}\int_0^1 e^{\sqrt{\lambda i}(1-\xi)}[i\ f''(\xi) - g(\xi)]d\xi$$

$$+\frac{1}{4}[-(1+k)e^{\sqrt{\lambda i}} + (k-1)e^{-\sqrt{\lambda i}}]\int_0^1 e^{-i\sqrt{\lambda i}(1-\xi)}[i\ f''(\xi) + g(\xi)]d\xi$$

$$+O(\|f''\| + \|g\|).$$

$$= \frac{1}{4}[-(1+k)e^{\sqrt{\lambda i}} + (k-1)e^{-\sqrt{\lambda i}}]\int_0^1 e^{-i\sqrt{\lambda i}(1-\xi)}[i\ f''(\xi) + g(\xi)]d\xi$$

$$+O(\|f''\| + \|g\|).$$

Therefore,

$$(\lambda i)^{3/2} A_3 = \frac{1}{4}\frac{(1+k)(1-i)}{(1+k) + (1-k)e^{-2\sqrt{\lambda i}}}\int_0^1 e^{-\sqrt{\lambda i}\xi}[i\ f''(\xi) - g(\xi)]d\xi$$

$$+\frac{1}{4}\frac{(1+k) - (1-k)e^{-2\sqrt{\lambda i}}}{(1+k) + (1-k)e^{-2\sqrt{\lambda i}}}\int_0^1 e^{i\sqrt{\lambda i}\xi}[i\ f''(\xi) + g(\xi)]d\xi$$

$$+O(|e^{i\sqrt{\lambda i}}|[\|f''\| + \|g\|]),$$

and

$$(\lambda i)^{3/2} A_4 = -\frac{i}{4}\int_0^1 e^{i\sqrt{\lambda i}\xi}[i\ f''(\xi) + g(\xi)]d\xi$$

$$+O(|e^{i\sqrt{\lambda i}}|[\|f''\| + \|g\|]).$$

The above equations, together with (4.140), yield

$$A_i = O(|\lambda|^{-3/2}[\|f''\| + \|g\|]), \quad i = 1, 2, 3, 4. \tag{4.142}$$

Based on these estimations, we are able to show that

$$\tilde{\phi}''(x)$$
$$= \lambda i A_1 e^{\sqrt{\lambda i}x} + \lambda i A_2 e^{-\sqrt{\lambda i}x} - \lambda i A_3 e^{i\sqrt{\lambda i}x} - \lambda i A_4 e^{-i\sqrt{\lambda i}x}$$
$$= \lambda i A_1 e^{\sqrt{\lambda i}x} - \lambda i A_4 e^{-i\sqrt{\lambda i}x} + O(|\lambda|^{-1/2}[\|f''\| + \|g\|])$$
$$= \lambda i[-\frac{1+i}{2}A_3 - \frac{1-i}{2}A_4]e^{\sqrt{\lambda i}x} - \lambda i A_4 e^{-i\sqrt{\lambda i}x} + O(|\lambda|^{-1/2}[\|f''\| + \|g\|])$$
$$= -\frac{1}{4}\frac{1+k}{(1+k) + (1-k)e^{-2\sqrt{\lambda i}}}e^{\sqrt{\lambda i}x}\int_0^1 e^{-\sqrt{\lambda i}\xi}[i\, f''(\xi) - g(\xi)]d\xi(\lambda i)^{-1/2}$$
$$+\frac{1+i}{8}\frac{1+k-(k-1)e^{-2\sqrt{\lambda i}}}{(1+k) + (1-k)e^{-2\sqrt{\lambda i}}}e^{\sqrt{\lambda i}x}\int_0^1 e^{i\sqrt{\lambda i}\xi}[i\, f''(\xi) + g(\xi)]d\xi(\lambda i)^{-1/2}$$
$$-\frac{1+i}{8}\frac{1+k-(1-k)e^{-2\sqrt{\lambda i}}}{(1+k) + (1-k)e^{-2\sqrt{\lambda i}}}e^{\sqrt{\lambda i}x}\int_0^1 e^{i\sqrt{\lambda i}\xi}[i\, f''(\xi) + g(\xi)]d\xi(\lambda i)^{-1/2}$$
$$+\frac{i}{4}(\lambda i)^{-1/2}e^{-i\sqrt{\lambda i}x}\int_0^1 e^{i\sqrt{\lambda i}\xi}[i\, f''(\xi) + g(\xi)]d\xi + O(|\lambda|^{-1/2}[\|f''\| + \|g\|])$$
$$= -\frac{1}{4}(\lambda i)^{-1/2}e^{\sqrt{\lambda i}x}\int_0^1 e^{-\sqrt{\lambda i}\xi}[i\, f''(\xi) - g(\xi)]d\xi$$
$$+\frac{i}{4}(\lambda i)^{-1/2}e^{-i\sqrt{\lambda i}x}\int_0^1 e^{i\sqrt{\lambda i}\xi}[i\, f''(\xi) + g(\xi)]d\xi + O(|\lambda|^{-1/2}[\|f''\| + \|g\|])$$
$$= -\phi_p''(x) + O(|\lambda|^{-1/2}[\|f''\| + \|g\|])$$

Thus,

$$\phi''(x) = \tilde{\phi}''(x) + \phi_p''(x) = O(|\lambda|^{-1/2}[\|f''\| + \|g\|]). \tag{4.143}$$

• Final Step: Estimation of $\psi(x)$

To complete the proof, we need to show that

$$\psi(x) = \lambda\phi(x) + f(x) = \lambda\tilde{\phi}(x) + \lambda\phi_p(x) + f(x) = O(|\lambda|^{-1/2}[\|f''\| + \|g\|]). \tag{4.144}$$

From the expression of $\phi_p(x)$, we have

$$\lambda\phi_p(x)$$
$$= \frac{i}{2}(\lambda i)^{-1/2}\int_0^x [\sinh\sqrt{\lambda i}(x - \xi) - \sin\sqrt{\lambda i}(x - \xi)][\lambda f(\xi) + g(\xi)]d\xi$$
$$= \frac{i}{4}(\lambda i)^{-1/2}\int_0^x [e^{\sqrt{\lambda i}(x-\xi)} - ie^{-i\sqrt{\lambda i}(x-\xi)}]g(\xi)d\xi + O(|\lambda|^{-1/2}[\|f''\| + \|g\|])$$

$$+ \frac{i}{2}(\lambda i)^{-1}\Big[-2\lambda f(x) + (\lambda i)^{-1/2}$$

$$\times \int_0^x [\sinh \sqrt{\lambda i}(x-\xi) + \sin \sqrt{\lambda i}(x-\xi)]\lambda f''(\xi)d\xi \Big]$$

$$= \frac{i}{4}(\lambda i)^{-1/2}\int_0^x [e^{\sqrt{\lambda i}(x-\xi)} - ie^{-i\sqrt{\lambda i}(x-\xi)}]g(\xi)d\xi + O(|\lambda|^{-1/2}[\|f''\| + \|g\|])$$

$$- \frac{i}{4}(\lambda i)^{-1/2}\int_0^x [e^{\sqrt{\lambda i}(x-\xi)} + ie^{-i\sqrt{\lambda i}(x-\xi)}]if''(\xi)d\xi - f(x)$$

$$= -\frac{i}{4}(\lambda i)^{-1/2}\int_0^x e^{\sqrt{\lambda i}(x-\xi)}[i\, f''(\xi) - g(\xi)]d\xi + O(|\lambda|^{-1/2}[\|f''\| + \|g\|])$$

$$+ \frac{1}{4}(\lambda i)^{-1/2}\int_0^x e^{-i\sqrt{\lambda i}(x-\xi)}[if''(\xi) + g(\xi)]d\xi - f(x).$$

That is,

$$\lambda\phi_p(x) + f(x) = -\frac{i}{4}(\lambda i)^{-1/2}\int_0^x e^{\sqrt{\lambda i}(x-\xi)}[i\, f''(\xi) - g(\xi)]d\xi$$

$$+ \frac{1}{4}(\lambda i)^{-1/2}\int_0^x e^{-i\sqrt{\lambda i}(x-\xi)}[if''(\xi) + g(\xi)]d\xi$$

$$+ O(|\lambda|^{-1/2}[\|f''\| + \|g\|]). \tag{4.145}$$

On the other hand,

$$\tilde{\phi}(x)$$

$$= A_1 e^{\sqrt{\lambda i}x} + A_2 e^{-\sqrt{\lambda i}x} + A_3 e^{i\sqrt{\lambda i}x} + A_4 e^{-i\sqrt{\lambda i}x}$$

$$= A_1 e^{\sqrt{\lambda i}x} + A_4 e^{-i\sqrt{\lambda i}x} + O(|\lambda|^{-1/2}[\|f''\| + \|g\|])$$

$$= \Big[-\frac{1+i}{2}A_3 - \frac{1-i}{2}A_4 \Big]e^{\sqrt{\lambda i}x} + A_4 e^{-i\sqrt{\lambda i}x} + O(|\lambda|^{-1/2}[\|f''\| + \|g\|])$$

$$= -\frac{1}{4}(\lambda i)^{-3/2}e^{\sqrt{\lambda i}x}\int_0^1 e^{-\sqrt{\lambda i}\xi}[i\, f''(\xi) - g(\xi)]d\xi$$

$$- \frac{1+i}{8}\frac{(1+k) - (1-k)e^{-2\sqrt{\lambda i}}}{(1+k) + (1-k)e^{-2\sqrt{\lambda i}}}(\lambda i)^{-3/2}e^{\sqrt{\lambda i}x}\int_0^1 e^{i\sqrt{\lambda i}\xi}[i\, f''(\xi) + g(\xi)]d\xi$$

$$+ \frac{1-i}{2}\frac{i}{4}(\lambda i)^{-3/2}e^{\sqrt{\lambda i}x}\int_0^1 e^{i\sqrt{\lambda i}\xi}[if''(\xi) + g(\xi)]d\xi$$

$$- \frac{i}{4}(\lambda i)^{-3/2}e^{-i\sqrt{\lambda i}x}\int_0^1 e^{i\sqrt{\lambda i}\xi}[if''(\xi) + g(\xi)]d\xi$$

$$+ O(|\lambda|^{-3/2}[\|f''\| + \|g\|])$$

$$= -\frac{1}{4}(\lambda i)^{-3/2}e^{\sqrt{\lambda i}x}\int_0^1 e^{-\sqrt{\lambda i}\xi}[if''(\xi) - g(\xi)]d\xi$$

$$- \frac{i}{4}(\lambda i)^{-3/2}e^{-i\sqrt{\lambda i}x}\int_0^1 e^{i\sqrt{\lambda i}\xi}[if''(\xi) + g(\xi)]d\xi$$

$$+O(|\lambda|^{-3/2}[\|f''\| + \|g\|])$$

Consequently,

$$
\begin{aligned}
\lambda\tilde{\phi}(x) &= \frac{i}{4}(\lambda i)^{-1/2}e^{\sqrt{\lambda i}x}\int_0^1 e^{-\sqrt{\lambda i}\xi}[if''(\xi) - g(\xi)]d\xi \\
&\quad -\frac{1}{4}(\lambda i)^{-1/2}e^{-i\sqrt{\lambda i}x}\int_0^1 e^{i\sqrt{\lambda i}\xi}[if''(\xi) + g(\xi)]d\xi \\
&\quad +O(|\lambda|^{-1/2}[\|f''\| + \|g\|]) \\
&= -\lambda\phi_p(x) - f(x) + O(|\lambda|^{-1/2}[\|f''\| + \|g\|])
\end{aligned}
$$

which is exactly (4.144). The lemma is proved.   □

We are ready to state the following theorem.

Theorem 4.28 *There exists a unique classical solution to (4.121) for each initial condition $W_0 \in D(\mathcal{A}^2)$.*

Proof. In view of (4.27), there exists an $M > 0$ such that

$$\|R(\lambda, \mathcal{A})\| \leq M|\lambda|^{-1/2} \leq M(1 + |\lambda|^{-1/2})$$

for all Re$\lambda \leq 0$. It follows from Theorem 2.91 that there exists a unique solution $W(t)$ for any $W_0 \in D(\mathcal{A}^2)$ and that the solution is 2-wellposed, i.e., there exist constant M_0 and ω such that $\|W(t)\| \leq M_0 e^{\omega t}\|W_0\|_{D(\mathcal{A}^2)}$. Moreover, by Corollary 2.93, $\mathcal{A}$ generates a 2−times integrated semigroup. □

We are interested in showing that the unique solution is actually exponentially stable, that is, the decay rate can be taken to be negative. The following theorem provides us with the answer to this problem. This result is important from a practical standpoint because it shows that shear force feedback can introduce damping for vibration of the rotating flexible arms.

Theorem 4.29 *Let $W(t)$ be the solution of (4.121) with initial condition W_0. Then, there exist constants $M > 0$ and $\omega > 0$, independent of W_0, such that*

$$\|W(t)\|_{\mathcal{H}} \leq Me^{-\omega t}\|W_0\|_{D(\mathcal{A})},$$

$$\text{for all } t \geq 1, \ W_0 \in D(\mathcal{A}^2) \subset D(\mathcal{A}).$$

Proof. The main idea is to first show that we can extend the domain, on which Lemma 4.27 holds, to the left-half complex plane, and then we can use the Paley-Wiener theorem to get the desired result. Let $\lambda = \sigma + i\tau$ with $\sigma, \tau \in \mathbf{R}$. From the resolvent equation

$$R(\sigma + i\tau, \mathcal{A}) - R(i\tau, \mathcal{A}) = -\sigma R(\sigma + i\tau, \mathcal{A})R(i\tau, \mathcal{A})$$

it is seen that $\|\sigma R(i\tau, \mathcal{A})\| \leq 1/2$, for $0 \geq \sigma \geq -\delta = -1/(2D)$, where $D = \sup_{\tau \in \mathbf{R}} \|R(i\tau, \mathcal{A})\| < \infty$. Hence, $[1 + \sigma R(i\tau, \mathcal{A})]^{-1}$ exists and $\|[1 + \sigma R(i\tau, \mathcal{A})]^{-1}\| \leq 2$. Thus, for $-\delta \leq \sigma \leq 0$,

$$\begin{aligned}
\|R(\sigma + i\tau, \mathcal{A})\| &\leq \|[1 + \sigma R(i\tau, \mathcal{A})]^{-1}\|\|R(i\tau, \mathcal{A})\| \\
&\leq 2\|R(i\tau, \mathcal{A})\| = O((\tau^2 + \sigma^2)^{-1/4}),
\end{aligned}$$

i.e.,

$$\|R(\lambda, \mathcal{A})\| = O(|\lambda|^{-1/2}), \text{ for all } \mathrm{Re}\lambda = \sigma \geq -\delta. \tag{4.146}$$

Now, let $\sigma \geq -\varepsilon > -\delta, \varepsilon > 0$ and $W_0 \in D(\mathcal{A})$. Then

$$\begin{aligned}
R(\sigma + i\tau, \mathcal{A})W_0 &= R(\sigma + i\tau, \mathcal{A})R(-\delta, \mathcal{A})(-\delta - \mathcal{A})W_0 \\
&= \frac{1}{\sigma + \delta + i\tau}[R(-\delta, \mathcal{A}) - R(\sigma + i\tau, \mathcal{A})](-\delta - \mathcal{A})W_0 \\
&= \frac{1}{\sigma + \delta + i\tau}W_0 - \frac{1}{\sigma + \delta + i\tau}R(\sigma + i\tau, \mathcal{A})(-\delta - \mathcal{A})W_0.
\end{aligned}$$

$$\tag{4.147}$$

Thus, there exists a constant $C > 0$ such that

$$\|R(\sigma + i\tau, \mathcal{A})W_0\|_{\mathcal{H}}^2 \leq \frac{C}{(\delta + \sigma)^2 + \tau^2} \leq \frac{C}{(\delta - \varepsilon)^2 + \tau^2},$$

which means

$$\sup_{\sigma \geq -\varepsilon} \int_{-\infty}^{\infty} \|R(\sigma + i\tau, \mathcal{A})W_0\|_{\mathcal{H}}^2 d\tau \leq \int_{-\infty}^{\infty} \frac{C}{(\delta - \varepsilon)^2 + \tau^2} d\tau < \infty.$$

According to the Paley-Wiener Theorem 3.37 in Chapter 3, there exists a $G(t) \in L^2(0, \infty; \mathcal{H})$ such that

$$R(\lambda, \mathcal{A})W_0 = \int_0^{\infty} e^{-(\lambda + \varepsilon)t}G(t)dt, \quad \text{for all } \mathrm{Re}\lambda \geq -\varepsilon. \tag{4.148}$$

On the other hand, if $W_0 \in D(\mathcal{A}^2)$, we already know from Theorem 4.28, that there is a solution $W(t)$, and that $\mathcal{A}$ generates a 2−times integrated semigroup $S(t)$. Since the solution $W(t)$ and $S(t)$ are related by the following equation

$$S(t)W_0 = \int_0^t (t - s)W(s)ds,$$

as is verified in (2.142), and since the resolvent $R(\lambda, \mathcal{A})$ and $S(t)$ are related by

$$R(\lambda, \mathcal{A})W_0 = \lambda^2 \int_0^\infty e^{-\lambda t} S(t)W_0 dt,$$

it is easy to show that

$$R(\lambda, \mathcal{A})W_0 = \int_0^\infty e^{-\lambda t} W(t) dt, \qquad (4.149)$$

for $\mathrm{Re}\lambda$ sufficiently large. By the uniqueness of the Fourier transform on $L^2(-\infty, \infty; \mathcal{H})$, (4.148) and (4.149) imply that

$$e^{\varepsilon t} W(t) = G(t) \in L^2(0, \infty; \mathcal{H}). \qquad (4.150)$$

Let ω be a constant satisfying $0 < \omega < \varepsilon$. Taking the derivative with respect to λ on both sides of (4.148), we have

$$R^2(\lambda, \mathcal{A})W_0 = \int_0^\infty e^{-\lambda t} t W(t) dt, \text{ for all } \mathrm{Re}\lambda \geq -\omega.$$

Since it can be verified that $e^{-\lambda t} t W(t) \in L^1(0, \infty; \mathcal{H})$ by (4.150), the above integration exists in the sense of the usual Bochner integral. Now considering a special case where $\lambda = -\omega + i\tau$, we have

$$R^2(-\omega + i\tau, \mathcal{A})W_0 = \int_0^\infty e^{-i\tau t} t e^{\omega t} W(t) dt.$$

By the inverse Fourier transform on $L^2(-\infty, \infty; \mathcal{H})$, it is easy to see that

$$
\begin{aligned}
t e^{\omega t} W(t) &= \frac{1}{2\pi} \int_{-\infty}^\infty e^{i\tau t} R^2(-\omega + i\tau, \mathcal{A})W_0 d\tau \\
&= \frac{1}{2\pi} \int_{-\infty}^\infty e^{i\tau t} R(-\omega + i\tau, \mathcal{A}) \Big[\frac{1}{\delta - \omega + i\tau} W_0 \\
&\quad - \frac{1}{\delta - \omega + i\tau} R(-\omega + i\tau, \mathcal{A})(-\delta - \mathcal{A})W_0 \Big] d\tau \\
&= \frac{1}{2\pi} \int_{-\infty}^\infty e^{i\tau t} \frac{1}{\delta - \omega + i\tau} R(-\omega + i\tau, \mathcal{A})W_0 d\tau \\
&\quad - \frac{1}{2\pi} \int_{-\infty}^\infty e^{i\tau t} \frac{1}{\delta - \omega + i\tau} R^2(-\omega + i\tau, \mathcal{A})(-\delta - \mathcal{A})W_0 d\tau.
\end{aligned}
$$

In view of (4.146), there exists a constant C_0 such that

$$\|R(-\omega + i\tau, \mathcal{A})\| \leq C_0(\omega^2 + \tau^2)^{-1/4}.$$

Therefore,

$$
\begin{aligned}
\|W(t)\|_{\mathcal{H}} \;\leq\; & \frac{C_0}{2\pi t} e^{-\omega t} \int_0^{\infty} [(\delta - \omega)^2 + \tau^2]^{-1/2} (\omega^2 + \tau^2)^{-1/4} d\tau \|W_0\|_{\mathcal{H}} \\
& + \frac{C_0^2}{2\pi t} e^{-\omega t} \int_0^{\infty} [(\delta - \omega)^2 + \tau^2]^{-1} (\omega^2 + \tau^2)^{-1/2} d\tau \\
& \times [\|\delta W_0\|_{\mathcal{H}} + \|\mathcal{A} W_0\|_{\mathcal{H}}] \\
\leq\; & M e^{-\omega t} \|W_0\|_{D(\mathcal{A})}, \quad \text{for all } t \geq 1, \; W_0 \in D(\mathcal{A}), \qquad (4.151)
\end{aligned}
$$

for some constant $M > 0$. This completes the proof. $\square$

Remark 4.30 *Theorem 4.29 implies that actually $\mathcal{A}$ generates a 1-time integrated semigroup.*

4.5.2 Asymptotic behavior of the spectrum

In this section, we shall estimate the spectrum with large moduli based on the characteristic equation $G(\lambda) = 0$ defined in Lemma 4.24. The following theorem establishes an explicit relationship between the feedback gain k and the eigenvalues λ_n, and indicates the interesting fact that the spectral distributions are totally different for $k = 1$ and $k \neq 1$.

Theorem 4.31 $G(\lambda) = 0$ *has solutions $\{\lambda_n\}$ and $\overline{\lambda}_n$ which satisfy*

$$
\begin{cases}
\lambda_n = -2(n\pi + \frac{\pi}{2})^2, \; \text{if } k = 1; \\[2mm]
\lambda_n = n\pi \log |\frac{k-1}{k+1}| + i\left[(n\pi)^2 - \frac{1}{4}\left(\log |\frac{k-1}{k+1}|\right)^2\right] + O(n^{-1}), \; \text{if } k > 1; \\[2mm]
\lambda_n = (n + \frac{1}{2})\pi \log |\frac{k-1}{k+1}| + i\left[(n\pi + \frac{\pi}{2})^2 - \frac{1}{4}\left(\log |\frac{k-1}{k+1}|\right)^2\right] + O(n^{-1}), k < 1,
\end{cases}
$$

where n is a large positive integer.

Proof. For $k = 1$, $G(\lambda) = 0$ reduces to

$$
2 + e^{\sqrt{2\lambda}} + e^{-\sqrt{2\lambda}} = 0
$$

the solutions of which are given by those λ_n satisfying $\sqrt{2\lambda_n} = (2n\pi + \pi)i$ for integer n, i.e.,

$$
\lambda_n = -2(n\pi + \frac{\pi}{2})^2.
$$

For the case $k \neq 1$, write $\lambda_n = |\lambda_n| e^{i\theta}$. θ must satisfy $\pi/2 < \theta < 3\pi/2$, since λ_n lie on the left-half plane as shown in Lemma 4.26.

Let $\delta > 0$ be a sufficiently small constant. Consider first the case $\frac{\pi}{2} < \theta \leq \pi - \delta$. Then

$$\sqrt{\lambda_n} = |\lambda_n|^{1/2} e^{i\theta/2} = |\lambda_n|^{1/2} [\cos \frac{\theta}{2} + i \, \sin \frac{\theta}{2}],$$

and

$$e^{-\sqrt{2\lambda_n}} = O(e^{-\gamma \sqrt{|\lambda_n|}}), \quad e^{i\sqrt{2\lambda_n}} = O(e^{-\gamma \sqrt{|\lambda_n|}})$$

for some $\gamma > 0$. From equation $G(\lambda_n) = 0$, we have

$$e^{\sqrt{2\lambda_n} + i\sqrt{2\lambda_n}} = \frac{k-1}{k+1} + O(e^{-\gamma \sqrt{|\lambda_n|}}) \tag{4.152}$$

which implies

$$e^{\sqrt{2|\lambda_n|}(\cos \frac{\theta}{2} - \sin \frac{\theta}{2})} = \frac{k-1}{k+1} e^{-i\sqrt{2|\lambda_n|}(\cos \frac{\theta}{2} + \sin \frac{\theta}{2})} + O(e^{-\gamma \sqrt{|\lambda_n|}}). \tag{4.153}$$

Since the left-hand side of (4.153) is a positive real number, we have

$$e^{\sqrt{2|\lambda_n|}(\cos \frac{\theta}{2} - \sin \frac{\theta}{2})} = |\frac{k-1}{k+1}| + O(e^{-\gamma \sqrt{|\lambda_n|}}), \tag{4.154}$$

$$e^{-i\sqrt{2|\lambda_n|}(\cos \frac{\theta}{2} + \sin \frac{\theta}{2})} = \text{sign}(k-1) + O(e^{-\gamma \sqrt{|\lambda_n|}}). \tag{4.155}$$

In view of (4.154), we see that

$$\theta \to \frac{\pi}{2}, \text{as } |\lambda_n| \to \infty$$

since otherwise the left-hand side of (4.154) would go to zero. Furthermore, from (4.154)

$$\sqrt{2|\lambda_n|}(\cos \frac{\theta}{2} - \sin \frac{\theta}{2}) = \log \left(|\frac{k-1}{k+1}| + O(e^{-\gamma \sqrt{|\lambda_n|}}) \right)$$

$$= \log |\frac{k-1}{k+1}| + O(e^{-\gamma \sqrt{|\lambda_n|}}).$$

Thus

$$2|\lambda_n|(1 - \sin \theta) = \left(\log |\frac{k-1}{k+1}| \right)^2 + O(e^{-\gamma \sqrt{|\lambda_n|}}),$$

or

$$\sin \theta = 1 - \frac{1}{2|\lambda_n|} \left(\log |\frac{k-1}{k+1}| \right)^2 + O(e^{-\gamma \sqrt{|\lambda_n|}}). \tag{4.156}$$

Hence,

$$\text{Re}\lambda_n = |\lambda_n| \cos \theta = -|\lambda_n| \sqrt{1 - \sin^2 \theta}$$

$$= -\sqrt{|\lambda_n|} \log |\frac{k-1}{k+1}| + O(|\lambda_n|^{-1/2}). \tag{4.157}$$

On the other hand, from (4.155) we get

$$\begin{cases} \sqrt{2|\lambda_n|}(\cos\frac{\theta}{2}+\sin\frac{\theta}{2}) = 2n\pi + O(e^{-\gamma\sqrt{|\lambda_n|}}), & \text{if } k > 1, \\ \sqrt{2|\lambda_n|}(\cos\frac{\theta}{2}+\sin\frac{\theta}{2}) = (2n+1)\pi + O(e^{-\gamma\sqrt{|\lambda_n|}}), & \text{if } k < 1 \end{cases} \tag{4.158}$$

for some large positive integer n. Thus, $|\lambda_n| = O(n^2)$ and

$$\begin{cases} 2|\lambda_n|(1+\sin\theta) = (2n\pi)^2 + O(ne^{-n}), & \text{if } k > 1, \\ 2|\lambda_n|(1+\sin\theta) = [(2n+1)\pi]^2 + O(ne^{-n}), & \text{if } k < 1. \end{cases} \tag{4.159}$$

In view of (4.156), we have

$$2|\lambda_n|(1+\sin\theta) = 4|\lambda_n| - \left(\log|\frac{k-1}{k+1}|\right)^2 + O(e^{-\gamma\sqrt{|\lambda_n|}}).$$

Hence,

$$\begin{cases} |\lambda_n| = (n\pi)^2 + \frac{1}{4}\left(\log|\frac{k-1}{k+1}|\right)^2 + O(ne^{-\gamma n}), & \text{if } k > 1 \\ |\lambda_n| = (n\pi + \frac{\pi}{2})^2 + \frac{1}{4}\left(\log|\frac{k-1}{k+1}|\right)^2 + O(ne^{-\gamma n}), & \text{if } k < 1 \end{cases} \tag{4.160}$$

and

$$\mathrm{Im}\lambda_n = |\lambda_n|\sin\theta = |\lambda_n| - \frac{1}{2}\left(\log|\frac{k-1}{k+1}|\right)^2 + O(e^{-\gamma\sqrt{|\lambda_n|}}). \tag{4.161}$$

Substituting (4.160) into (4.157) yields

$$\begin{cases} \mathrm{Re}\lambda_n = n\pi\log|\frac{k-1}{k+1}| + O(n^{-1}), & \text{if } k > 1, \\ \mathrm{Re}\lambda_n = (n+\frac{1}{2})\pi\log|\frac{k-1}{k+1}| + O(n^{-1}), & \text{if } k < 1. \end{cases} \tag{4.162}$$

Combining (4.160)-(4.162) yields the expressions for λ_n given in the Theorem.

We now show that, in the area $\pi - \delta < \theta \leq \pi$, there exist no zeros of $G(\lambda)$ for large $|\lambda|$. In this case, $\frac{\pi}{2} - \frac{\delta}{2} < \frac{\theta}{2} \leq \frac{\pi}{2}$ and

$$e^{-\sqrt{2\lambda}} = O(1), \ e^{i\sqrt{2\lambda}} = O(1).$$

If there exists a λ such that $G(\lambda) = 0$, then

$$e^{\sqrt{2\lambda}+i\sqrt{2\lambda}} = O(1)$$

or

$$(1+k)e^{|2\lambda|^{1/2}[\cos\frac{\theta}{2}+i\sin\frac{\theta}{2}]} = (1-k)e^{|2\lambda|^{1/2}[\sin\frac{\theta}{2}-i\cos\frac{\theta}{2}]} + O(1). \tag{4.163}$$

When $\theta \to \pi$, the modulus of the right-hand side of (4.163) is much larger than that of the left-hand side. Therefore, there exists no solution of $G(\lambda) = 0$ for $|\lambda|$ sufficiently large.

The existence of the zeros which satisfy the relation in this theorem remains to be shown. We only consider the case of $k > 1$ since the treatment for $k < 1$ is very similar. Returning to (4.152), we see that the solution of the equation

$$e^{(1+i)\mu} = \frac{k-1}{k+1}$$

can be expressed as

$$\mu_n = \frac{1}{2} \log \left| \frac{k-1}{k+1} \right| + n\pi + i \left[n\pi - \frac{1}{2} \log \left| \frac{k-1}{k+1} \right| \right]$$

with $|\mu_n| = O(n)$. Let O_n be the circle with center at μ_n and with radius $|\mu_n|^{-2}$:

$$O_n = \{ \lambda \mid |\lambda - \mu_n| \le |\mu_n|^{-2} \}.$$

Then for all λ located on the circumference of O_n, we see that

$$
\begin{aligned}
\left| e^{(1+i)\lambda} - \frac{k-1}{k+1} \right| &= \left| e^{(1+i)\lambda} - e^{(1+i)\mu_n} \right| \\
&= \left| \frac{k-1}{k+1} \right| \left| e^{(1+i)(\lambda - \mu_n)} - 1 \right| \\
&= \left| \frac{k-1}{k+1} \right| \left| e^{(1+i)|\mu_n|^{-2}e^{i\theta}} - 1 \right| \\
&= \left| \frac{k-1}{k+1} \right| \left| (1+i)|\mu_n|^{-2}e^{i\theta} + O(|\mu_n|^{-4}) \right| \\
&= \left| \frac{k-1}{k+1} \right| \left| 1+i \right| |\mu_n|^{-2} + O(|\mu_n|^{-4}) \\
&= \left| \frac{k-1}{k+1} \right| \left| 1+i \right| |\lambda|^{-2} + O(|\lambda|^{-4}) > O(e^{-\gamma\sqrt{|\lambda|}})
\end{aligned}
$$

holds for all sufficiently large n. By Rouché's theorem, there exists, inside O_n, a unique solution σ_n to equation

$$e^{(1+i)\lambda} = \frac{k-1}{k+1} + O(e^{-\gamma\sqrt{|\lambda|}})$$

such that

$$|\sigma_n - \mu_n| \le |\mu_n|^{-2}.$$

Hence, $|\sigma_n^2 - \mu_n^2| = |\sigma_n - \mu_n||\sigma_n + \mu_n| \le 2|\mu_n|^{-1} + |\mu_n|^{-4}$. In view of (4.160), we see that $\lambda_n = \frac{1}{2}\sigma_n^2$ is the unique solution of (4.152) such that

$$\left| \lambda_n - \frac{1}{2}\mu_n^2 \right| \le |\mu_n|^{-1} + \frac{1}{2}|\mu_n|^{-4}.$$

Since

$$\frac{1}{2}\mu_n^2 = n\pi \log |\frac{k-1}{k+1}| + i\left[(n\pi)^2 - \frac{1}{4}\left(\log |\frac{k-1}{k+1}|\right)^2\right],$$

taking n to be a positive integer, we have the desired result. □

4.6   Stability analysis of a hybrid system

4.6.1   Well-posedness and exponential stability

In the previous sections, we concentrated on the vibration suppression of flexible arms using the acceleration of the control motor as the control input. In this case, the motor angle is not controlled. In practice, it is usually desirable to control not only the vibration in flexible arms, but also the rotation angle of the motor. For this purpose, instead of (4.32), we can control the motor such that

$$\dot{\theta}(t) = -k_2(\theta(t) - \theta_d) - k_3\int_0^t (\theta(\tau) - \theta_d)d\tau + k_1 y_{xx}(0,t), \quad (4.164)$$
$$k_1 > 0, \; k_2 > 0, \; k_3 > 0$$

where θ_d is the desired motor angle and $k_i, i = 1, \cdots, 3$ are feedback gains. This is a proportional and integral (PI) + strain feedback control law. Taking the time derivative of both sides of (4.164), and substituting the resultant equation into (4.24), we have

$$\begin{cases} \dfrac{\partial^2 y(x,t)}{\partial t^2} + \dfrac{\partial^4 y(x,t)}{\partial x^4} + k_1 x \dfrac{\partial^3 y(0,t)}{\partial t \partial x^2} = k_2\dot{\theta}(t) + k_3(\theta(t) - \theta_d), \\ \qquad\qquad 0 < x < 1, t > 0, \\ y(0,t) = y_x(0,t) = y_{xx}(1,t) = y_{xxx}(1,t) = 0, \\ \ddot{\theta}(t) + k_2\dot{\theta}(t) + k_3(\theta(t) - \theta_d) = k_1 x y_{xxt}(0,t), \\ y(x,0) = y_0(x), y_t(x,0) = y_1(x), \theta(0) = \theta_0, \dot{\theta}(0) = \theta_{01}, \end{cases} \quad (4.165)$$

This is a *hybrid system* consisting of a coupled partial differential equation and an ordinary differential equation.

Similarly, for an XY robot with a flexible tip arm, as mentioned before, we can control the motor motion such that

$$\dot{s}(t) = -k_2(s(t) - s_d) - k_3\int_0^t (s(\tau) - s_d)d\tau - k_1 y_{xxx}(0,t), \quad (4.166)$$
$$k_1 > 0, \; k_2 > 0, \; k_3 > 0.$$

Substituting this equation into (4.24) yields the closed-loop system

$$
\begin{cases}
\dfrac{\partial^2 y(x,t)}{\partial t^2} + \dfrac{\partial^4 y(x,t)}{\partial x^4} - k_1 \dfrac{\partial^4 y(0,t)}{\partial t \partial x^3} = k_2 \dot{s}(t) + k_3(s(t) - s_d), \\[2mm]
\qquad\qquad 0 < x < 1, t > 0, \\[1mm]
y(0,t) = y_x(0,t) = y_{xx}(1,t) = y_{xxx}(1,t) = 0, \\[1mm]
\ddot{s}(t) + k_2 \dot{s}(t) + k_3(s(t) - s_d) = -k_1 y_{xxxt}(0,t), \\[1mm]
y(x,0) = y_0(x), y_t(x,0) = y_1(x), s(0) = s_0, \dot{s}(0) = s_{01}.
\end{cases}
\tag{4.167}
$$

We are interested in deriving conditions on PI feedback gains k_2, k_3 and the strain, or shear force feedback gain k_1 such that the closed-loop systems (4.165) and (4.167) are exponentially stable, which is a generalization of the results in the previous sections.

Since the stability analysis for the closed-loop equations of strain feedback control of rotating arms is very similar to that of shear force feedback control of translating arms, we only consider the stability of (4.167). For notational simplicity, let $z(t) = s(t) - s_d$. Adding a term $k_1 k_2 \dfrac{\partial^3 y(0,t)}{\partial x^3}$ to both sides of (4.167), we obtain

$$
\begin{cases}
\dfrac{\partial^2 y(x,t)}{\partial t^2} + \dfrac{\partial^4 y(x,t)}{\partial x^4} + k_1 k_2 \dfrac{\partial^3 y(0,t)}{\partial x^3} - k_1 \dfrac{\partial^4 y(0,t)}{\partial t \partial x^3} \\[2mm]
\qquad = k_2 \Big[\dot{z}(t) + k_1 \dfrac{\partial^3 y(0,t)}{\partial x^3}\Big] + k_3 z(t), \\[2mm]
y(0,t) = y_x(0,t) = y_{xx}(1,t) = y_{xxx}(1,t) = 0, \\[1mm]
\ddot{z}(t) + k_2 \dot{z}(t) + k_3 z(t) = -k_1 y_{xxxt}(0,t), \\[1mm]
y(x,0) = y_0(x), y_t(x,0) = y_1(x), z(0) = z_0, \dot{z}(0) = z_{01}.
\end{cases}
\tag{4.168}
$$

Let us first consider the dominant equation of this equation by setting the right-hand side of the first equation to zero. Namely, we consider

$$
\begin{cases}
\dfrac{\partial^2 \bar{y}(x,t)}{\partial t^2} + \dfrac{\partial^4 \bar{y}(x,t)}{\partial x^4} + k_1 k_2 \dfrac{\partial^3 \bar{y}(0,t)}{\partial x^3} - k_1 \dfrac{\partial^4 \bar{y}(0,t)}{\partial t \partial x^3} = 0, \\[2mm]
\bar{y}(0,t) = \bar{y}_x(0,t) = \bar{y}_{xx}(1,t) = \bar{y}_{xxx}(1,t) = 0, \\[1mm]
\ddot{z}_p(t) + k_2 \dot{z}_p(t) + k_3 z_p(t) = -k_1 \bar{y}_{xxxt}(0,t), \\[1mm]
\bar{y}(x,0) = y_0(x), \bar{y}_t(x,0) = y_1(x), z_p(0) = z_0, \dot{z}_p(0) = z_{01}.
\end{cases}
\tag{4.169}
$$

We use semigroup framework to explain the solution of (4.169). To this end, let

$$
\begin{cases}
\bar{y}_1(x,t) = \tfrac{1}{2}\{\bar{y}_{xx}(x,t) + [\bar{y}_t(x,t) - k_1 \bar{y}_{xxx}(0,t)]\}, \\[2mm]
\bar{y}_2(x,t) = \tfrac{1}{2}\{\bar{y}_{xx}(x,t) - [\bar{y}_t(x,t) - k_1 \bar{y}_{xxx}(0,t)]\}, \\[2mm]
z_1(t) = z_p(t), \; z_2(t) = \dot{z}_p(t) + k_1 \bar{y}_{xxx}(0,t),
\end{cases}
$$

and $\bar{Y}(t) = \left(\bar{y}_1(x,t), \bar{y}_2(x,t), z_1(t), z_2(t)\right)^T \in \mathcal{H} = L^2(0,1) \times L^2(0,1) \times R^2$.
Then equation (4.169) can be written as

$$\frac{d}{dt}\bar{Y}(t) = \mathcal{A}\bar{Y}(t),$$

where the operator $\mathcal{A} : D(\mathcal{A}) \subset \mathcal{H} \to \mathcal{H}$ is defined by

$$
\mathcal{A}\begin{bmatrix} \phi_1 \\ \phi_2 \\ z_1 \\ z_2 \end{bmatrix} = \begin{bmatrix} -\phi_2'' + k_2[\phi_1(0) - \phi_2(0)]/2 \\ \phi_1'' - k_2[\phi_1(0) - \phi_2(0)]/2 \\ z_2 + \phi_1(0) - \phi_2(0) \\ -k_3 z_1 - k_2 z_2 - k_2[\phi_1(0) - \phi_2(0)] \end{bmatrix} \tag{4.170}
$$

with domain

$$
D(\mathcal{A}) = \{(\phi_1, \phi_2, z_1, z_2)^T \in \mathcal{H} \,\big|
$$
$$
\phi_1(\ell) + \phi_2(\ell) = \phi_1'(\ell) + \phi_2'(\ell) = \phi_1'(0) - \phi_2'(0) = 0,
$$
$$
\phi_1(0) - \phi_2(0) = -k_1(\phi_1'(0) + \phi_2'(0))\}. \tag{4.171}
$$

Lemma 4.32 *There exists a positive constant $K > 0$, such that*

$$
Re\langle \mathcal{A}\phi, \phi \rangle_{\mathcal{H}} \leq K \langle \phi, \phi \rangle_{\mathcal{H}}, \text{ for all } \begin{bmatrix} \phi_1 \\ \phi_2 \\ z_1 \\ z_2 \end{bmatrix} \in D(\mathcal{A}). \tag{4.172}
$$

Proof. For any $\phi = (\phi_1, \phi_2, z_1, z_2)^T \in D(\mathcal{A})$, it is seen that

$$
Re\langle \mathcal{A}\phi, \phi \rangle_{\mathcal{H}} = \left\langle \begin{bmatrix} -\phi_2'' + \frac{k_2}{2}[\phi_1(0) - \phi_2(0)] \\ \phi_1'' - \frac{k_2}{2}[\phi_1(0) - \phi_2(0)] \\ z_2 + \phi_1(0) - \phi_2(0) \\ -k_3 z_1 - k_2 z_2 - k_2[\phi_1(0) - \phi_2(0)] \end{bmatrix}, \begin{bmatrix} \phi_1 \\ \phi_2 \\ z_1 \\ z_2 \end{bmatrix} \right\rangle_{\mathcal{H}}
$$

$$
= -\int_0^1 \phi_1(x)\overline{\phi_2''(x)}dx + \int_0^1 \phi_1''(x)\overline{\phi_2(x)}dx
$$

$$
+Re\frac{k_2}{2}[\phi_1(0) - \phi_2(0)][\int_0^1 \overline{\phi_1(x)}dx - \int_0^1 \overline{\phi_2(x)}dx]
$$

$$
+Re(1 - k_3)z_1\bar{z}_2 + Re[\phi_1(0) - \phi_2(0)]\bar{z}_1 - k_2|z_2|^2
$$
$$
-Re k_2 \bar{z}_2[\phi_1(0) - \phi_2(0)].
$$

Now, since

$$
\int_0^1 \phi_1''(x)\overline{\phi_2(x)}dx - \int_0^1 \phi_1(x)\overline{\phi_2''(x)}dx
$$

$$
\begin{aligned}
&= \phi_1'(1)\overline{\phi_2(1)} - \phi_1(1)\overline{\phi_2'(1)} + \phi_1(0)\overline{\phi_2'(0)} - \phi_1'(0)\overline{\phi_2(0)} \\
&= \phi_1'(1)[\overline{\phi_1(1)} + \overline{\phi_2(1)}] + \phi_1'(0)[\overline{\phi_1(0)} - \overline{\phi_2(0)}] \\
&= -k_1\phi_1'(0)[\overline{\phi_1'(0)} + \overline{\phi_2'(0)}] \\
&= -\frac{k_1}{2}|\phi_1'(0) + \phi_2'(0)|^2 \\
&= -\frac{1}{2k_1}|\phi_1(0) - \phi_2(0)|^2
\end{aligned}
$$

and

$$
\operatorname{Re}\frac{k_2}{2}[\phi_1(0) - \phi_2(0)]\int_0^1 \overline{\phi_1(x)}dx
$$

$$
\leq \frac{k_2}{4}\varepsilon|\phi_1(0) - \phi_2(0)|^2 + \frac{k_2}{4\varepsilon}\|\phi_1\|^2,
$$

$$
\operatorname{Re}\frac{k_2}{2}[\phi_1(0) - \phi_2(0)]\int_0^1 \overline{\phi_2(x)}dx
$$

$$
\leq \frac{k_2}{4}\varepsilon|\phi_1(0) - \phi_2(0)|^2 + \frac{k_2}{4\varepsilon}\|\phi_2\|^2,
$$

$$
\operatorname{Re}(1 - k_3)z_1\bar{z}_2 \leq \frac{1 - k_3}{2}|z_1|^2 + \frac{1 - k_3}{2}|z_2|^2,
$$

$$
\operatorname{Re}[\phi_1(0) - \phi_2(0)]\bar{z}_1 \leq \frac{1}{2}\varepsilon|\phi_1(0) - \phi_2(0)|^2 + \frac{1}{2\varepsilon}|z_1|^2,
$$

$$
-\operatorname{Re}k_2\bar{z}_2[\phi_1(0) - \phi_2(0)] \leq \frac{1}{2}k_2\varepsilon|\phi_1(0) - \phi_2(0)|^2 + \frac{1}{2\varepsilon}|z_2|^2,
$$

where we have used the inequality $2ab \leq \varepsilon a^2 + \frac{1}{\varepsilon}b^2$ for arbitrary $\varepsilon > 0$. Therefore,

$$
\begin{aligned}
\operatorname{Re}\langle \mathcal{A}\phi, \phi\rangle_{\mathcal{H}} \leq \ &-\Big(\frac{1}{2k_1} - \frac{k_2}{2}\varepsilon - \frac{1}{2}\varepsilon - \frac{1}{2}k_2\varepsilon\Big)|\phi_1(0) - \phi_2(0)|^2 \\
&+ \frac{k_2}{4\varepsilon}\|\phi_1\|^2 + \frac{k_2}{4\varepsilon}\|\phi_2\|^2 \\
&+ \Big(\frac{1 - k_3}{2} + \frac{1}{2\varepsilon}\Big)|z_1|^2 + \Big(\frac{1 - k_3}{2} - k_2 + \frac{k_2}{2\varepsilon}\Big)|z_2|^2.
\end{aligned}
$$

Taking ε to be small enough such that

$$
\frac{1}{2k_1} - \frac{k_2}{2}\varepsilon - \frac{1}{2}\varepsilon - \frac{1}{2}k_2\varepsilon > 0,
$$

and letting

$$
K = \max\{\frac{k_2}{4\varepsilon}, \frac{1 - k_3}{2} + \frac{1}{2\varepsilon}, \frac{1 - k_3}{2} - k_2 + \frac{k_2}{2\varepsilon}\},
$$

we arrive at (4.172). $\qquad\qquad\qquad\qquad\qquad\qquad\qquad\qquad\qquad$ □

Lemma 4.33 *Suppose $k_1 k_2 < 1$. Then the operator $\mathcal{A}$ is an infinitesimal generator of a C_0-semigroup $T(t)$ on $\mathcal{H}$ with exponential decay*

$$\|T(t)\| \leq M_d e^{-\omega_d t}, \text{ for some } M_d \geq 1, \ \omega_d > 0. \qquad (4.173)$$

Proof. It is obvious that $D(\mathcal{A}) \supset C_0^\infty(0,1) \times C_0^\infty(0,1) \times \mathbf{R}^2$ which is dense in $\mathcal{H}$. We claim that $\mathcal{A}^{-1}$ exists and is compact. In fact, solving equation

$$\mathcal{A} \begin{bmatrix} \phi_1 \\ \phi_2 \\ z_1 \\ z_2 \end{bmatrix} = \begin{bmatrix} g_1 \\ g_2 \\ c_1 \\ c_2 \end{bmatrix}, \quad \forall \begin{bmatrix} g_1 \\ g_2 \\ c_1 \\ c_2 \end{bmatrix} \in \mathcal{H},$$

we get

$$\begin{cases} \phi_1(0) - \phi_2(0) = k_1/(1 - k_1 k_2) \int_0^1 \big(g_2(x) - g_1(x)\big)dx, \\ (\phi_1(x) - \phi_2(x)) = \int_0^x (x - \tau)(g_2(\tau) + g_1(\tau))d\tau + (\phi_1(0) - \phi_2(0)), \\ (\phi_1(x) + \phi_2(x)) = -(\phi_1(0) - \phi_2(0)) \int_1^x (1/k_1 - k_2 s)ds \\ \qquad + \int_1^x \int_0^s (g_2(\tau) - g_1(\tau))d\tau ds, \\ z_1 = -(c_1 k_2 + c_2)/k_3, \ z_2 = c_1 - (\phi_1(0) - \phi_2(0)). \end{cases}$$

Thus, $\mathcal{A}^{-1}$ exists and is compact by the Sobolev imbedding theorem since from the expression above $\mathcal{A}^{-1}$ maps a bounded set of $\mathcal{H}$ into a bounded set of $H^1(0,1) \times H^1(0,1) \times \mathbf{R}^2$.

Now that $\mathcal{A}^{-1}$ is compact, $\sigma(\mathcal{A})$ consists only of isolated eigenvalues. Hence, there exists a $\lambda_0 > K$ such that $\lambda_0 + K \in \rho(\mathcal{A})$, or $\lambda_0 \in \rho(\mathcal{A} - K)$. Because $\mathcal{A} - K$ is dissipative, it follows from the Lümer-Phillips theorem that $\mathcal{A} - K$ generates a C_0-semigroup on H, and so does $\mathcal{A}$.

It remains to show that the semigroup generated by $\mathcal{A}$ is exponentially stable. This is accomplished if we can show

a) All eigenvalues of $\mathcal{A}$ lie on the left-half complex plane.

b) $\|R(i\omega, \mathcal{A})\|$ is uniformly bounded for all $\omega \in \mathbf{R}$.

We show a) first. It is easily seen that $\lambda \in \sigma_p(\mathcal{A})$ if and only if there exists a $\bar{\phi} \neq 0$ such that

$$\begin{cases} \lambda^2 \bar{\phi}(x) + \bar{\phi}''''(x) + (k_1 k_2 - k_1 \lambda)\bar{\phi}'''(0) = 0 \\ \bar{\phi}(0) = \bar{\phi}'(0) = \bar{\phi}''(1) = \bar{\phi}'''(1) = 0. \end{cases}$$

Let $\phi(x) = \bar{\phi}''(1 - x)$. Then $\phi \neq 0$ satisfies

$$\begin{cases} \lambda^2 \phi(x) + \phi''''(x) = 0, \\ \phi(0) = \phi'(0) = \phi'''(1) = 0, \phi''(1) = (-\lambda k_1 + k_1 k_2)\phi'(1). \end{cases}$$

By Lemma 4.6, $\phi'(1) \neq 0$. Taking inner product with $\phi(x)$ on both sides of the above equation yields

$$\lambda^2 \|\phi\|^2 - (-\lambda k_1 + k_1 k_2)|\phi'(1)|^2 + \|\phi''\|^2 = 0.$$

Writing $\lambda = a + bi$ in terms of its real part a and imaginary part b, we find from the above equation that

$$\begin{cases} (a^2 - b^2) \|\phi\|^2 + (ak_1 - k_1 k_2)|\phi'(1)|^2 + \|\phi''\|^2 = 0, \\ 2ab \|\phi\| + bk_1|\phi'(1)|^2 = 0. \end{cases}$$

If $b \neq 0$, then the second equality above gives

$$a = -\frac{k_1}{2} \frac{|\phi'(1)|^2}{\|\phi\|^2} < 0.$$

If $b = 0$, then clearly

$$a^2 \|\phi\|^2 + k_1 a|\phi'(1)|^2 + \|\phi''\|^2 - k_1 k_2|\phi'(1)|^2 = 0$$

holds. Since

$$\begin{aligned} \|\phi''\|^2 - k_1 k_2|\phi'(1)|^2 &= \|\phi''\|^2 - k_1 k_2 \left| \int_0^1 \phi''(x)dx \right|^2 \\ &\geq (1 - k_1 k_2) \|\phi''\|^2, \end{aligned}$$

we see that $a < 0$. Therefore, $\text{Re}\lambda < 0$ for all $\lambda \in \sigma(\mathcal{A})$. The verification of b) is straightforward but tedious. The interested reader is encouraged to complete this step following the proof procedure of Lemma 4.27. $\square$

Having made clear some properties of the dominant system (4.169), we now return to analyzing (4.168), which is our ultimate purpose. Again let

$$\begin{cases} y_1(x, t) = \frac{1}{2}\{y_{xx}(x, t) + [y_t(x, t) - k_1 y_{xxx}(0, t)]\}, \\ y_2(x, t) = \frac{1}{2}\{y_{xx}(x, t) - [y_t(x, t) - k_1 y_{xxx}(0, t)]\}, \qquad (4.174) \\ z_1(t) = z(t), \quad z_2(t) = \dot{z}(t) + k_1 y_{xxx}(0, t), \end{cases}$$

and $Y = \left(y_1(x, t), y_2(x, t), z_1(t), z_2(t)\right)^T \in \mathcal{H} = L^2(0, \ell) \times L^2(0, \ell) \times \mathbf{R}^2$. Then (4.168) can be written as

$$\frac{d}{dt}Y(t) = (\mathcal{A} + \mathcal{B})Y(t), \qquad (4.175)$$

where the operator $\mathcal{B}\colon D(\mathcal{B}) = \mathcal{H} \to \mathcal{H}$ is defined by

$$\mathcal{B}(\phi_1(x), \phi_2(x), z_1, z_2)^T = \left(\frac{k_3 z_1 + k_2 z_2}{2}, \frac{-k_3 z_1 - k_2 z_2}{2}, 0, 0\right)^T. \qquad (4.176)$$

Since $\mathcal{B}$ is compact, (4.175) admits a unique solution

$$Y(t) = S(t)Y(0),$$

where $S(t)$ is the C_0-semigroup on $\mathcal{H}$ generated by $\mathcal{A} + \mathcal{B}$. Moreover, by Corollary 3.54 the necessary and sufficient condition for $Y(t)$ to decay exponentially, i.e.,

$$\|S(t)\| \leq Me^{-\omega t}, \quad \text{for some } M \geq 1 \text{ and } \omega > 0$$

is that

$$S(\mathcal{A} + \mathcal{B}) < 0,$$

where $S(\mathcal{A} + \mathcal{B})$ denotes the spectral bound of $\mathcal{A} + \mathcal{B}$.

Corollary 4.34 *Assume the initial conditions of (4.168) satisfy*

$$y_0(0) = y_0'(0) = y_0''(1) = y_0'''(1) = y_1(0) = y_1'(0) = 0$$
$$y_0 \in H^4(0,1), \ y_1 \in H^2(0,1).$$

Then (4.168) admits a unique classical solution

$$\begin{cases} y_{xx}(x,t) \in C^1([0,\infty); H^2(0,1)), \\ y_t(x,t) - k_1 y_{xxx}(0,t) \in C^1([0,\infty); L^2(0,1)), \\ \dot{z}(t) + k_1 y_{xxx}(0,t) \in C^1([0,\infty); \mathbf{R}) \end{cases}$$

such that

$$\begin{cases} \dfrac{\partial}{\partial t}[y_t(x,t) - k_1 y_{xxx}(0,t)] + y_{xxxx}(x,t) + k_1 k_2 y_{xxx}(0,t) \\ \qquad = k_2[\dot{z}(t) + k_1 y_{xxx}(0,t)] + k_3 z(t) \\ y(0,t) = y_x(0,t) = y_{xx}(1,t) = y_{xxx}(1,t) = 0 \\ \dfrac{d}{dt}[\dot{z}(t) + k_1 y_{xxx}(0,t)] + k_2 \dot{z}(t) + k_3 z(t) = 0 \\ y(x,0) = y_0(x), \ y_t(x,0) = y_1(x). \end{cases}$$

4.6.2 Spectral analysis

In this section, we shall find the distribution of the spectrum of the operator $\mathcal{A} + \mathcal{B}$ on the complex plane. As usual, let A be the operator defined in (4.25). We are going to characterize the spectrum of $\mathcal{A} + \mathcal{B}$ by using the spectrum of A.

Lemma 4.35 *The spectrum set $\sigma(\mathcal{A}+\mathcal{B})$ of the operator $\mathcal{A}+\mathcal{B}$ consists only of its eigenvalues. $\lambda \in \sigma(\mathcal{A}+\mathcal{B})$ if and only if there exists a nonzero $\psi \in H$ such that*

$$\begin{cases} (\lambda^2 + A)\psi(x) = -\lambda^2, \\ \lambda^2 + \lambda k_2 + k_3 = -\lambda k_1 \psi'''(0), \end{cases} \tag{4.177}$$

or equivalently,

$$\begin{cases} \lambda^2 \phi(x) + \phi''''(x) = 0, \\ \phi(0) = \phi'(0) = \phi'''(1) = 0, \\ \phi''(1) = -\lambda^2, \\ \lambda^2 + \lambda k_2 + k_3 = \lambda k_1 \phi'(1), \end{cases} \tag{4.178}$$

hold for some nonzero $\phi \in H$.

Proof. The first conclusion follows from the fact that both $\mathcal{A}^{-1}$ and $\mathcal{B}$ are compact operators. Let $\phi = \psi''(1-x)$. (4.178) is deduced from (4.177). It is only necessary to notice the following relationship between the solution of (4.177) and the eigenvector $(\phi_1, \phi_2, z_1, z_2)^T$ of the operator $\mathcal{A}+\mathcal{B}$ corresponding to λ:

$$\begin{cases} \phi_1(x) = [\psi''(x) + (\lambda\psi(x) - k_1\psi'''(0))]/2, \\ \phi_2(x) = [\psi''(x) - (\lambda\psi(x) - k_1\psi'''(0))]/2, \\ z_1 = 1, \quad z_2 = \lambda + k_1\psi'''(0), \end{cases}$$

i.e., if ψ is a solution of (4.177), then $(\phi_1, \phi_2, z_1, z_2)^T$ defined above satisfies $(\mathcal{A}+\mathcal{B})(\phi_1, \phi_2, z_1, z_2)^T = \lambda(\phi_1, \phi_2, z_1, z_2)^T$. Conversely, if $(\mathcal{A}+\mathcal{B})(\phi_1, \phi_2, z_1, z_2)^T = \lambda(\phi_1, \phi_2, z_1, z_2)^T$, then $\lambda \neq 0$, $z_1 \neq 0$ and

$$\psi(x) = \frac{k_1}{z_1\lambda}[\phi_1'(0) + \phi_2'(0)] + \frac{1}{z_1\lambda}[\phi_1(x) - \phi_2(x)] \neq 0.$$

$\psi(x)$ satisfies (4.177). $\square$

The following theorem exploits some useful properties of the spectrum of $\mathcal{A}+\mathcal{B}$.

Theorem 4.36   *(i) $\sigma(\mathcal{A}+\mathcal{B})$ does not contain any purely imaginary complex number except $\pm i\sqrt{k_3}$.*

(ii) $\lambda \in \sigma(\mathcal{A}+\mathcal{B})$ if and only if λ is a zero of the following meromorphic function

$$F(\lambda) = \lambda^2 + \lambda k_2 + k_3 + \lambda^2 k_1 \sqrt{2\lambda} \frac{f(\lambda)}{g(\lambda)}, \tag{4.179}$$

where

$$
\begin{cases}
f(\lambda) = \exp(\sqrt{2\lambda}) - \exp(-\sqrt{2\lambda}) - i\left[\exp(i\sqrt{2\lambda}) - \exp(-i\sqrt{2\lambda})\right], \\
g(\lambda) = 4 + \exp(i\sqrt{2\lambda}) + \exp(-i\sqrt{2\lambda}) + \exp(\sqrt{2\lambda}) + \exp(-\sqrt{2\lambda}),
\end{cases}
$$

(4.180)

and for all zeros λ of $F(\lambda)$, there holds

$$
Re\lambda \to -\frac{1}{k_1} \quad as \quad |\lambda| \to \infty.
$$

(4.181)

(iii)

$$
F(\lambda) = \lambda^2 + \lambda k_2 + k_3 + k_1\lambda^3 \sum_{n=1}^{\infty} \frac{\lambda_n}{\lambda^2 + \lambda_n} c_n^2 \|\phi_n\|^2.
$$

(4.182)

(iv) For any $\lambda \in \sigma(\mathcal{A} + \mathcal{B})$ with $Re\ \lambda \geq 0$,

$$
|\lambda| \leq \sqrt{k_3}.
$$

(v) If $k_2 \geq 4k_1 k_3$, $k_3 < \lambda_1/2$, then

$$
Re\lambda < 0, \quad for\ any\ \lambda \in \sigma(\mathcal{A} + \mathcal{B}).
$$

(vi) For any $k_1 > 0, k_2 > 0$, there is a $k_3 \in (\lambda_n, \lambda_{n+1})$ such that $\lambda = i\sqrt{k_3} \in \sigma(\mathcal{A} + \mathcal{B})$.

(vii) $\lambda = i\sqrt{k_3} \in \sigma(\mathcal{A} + \mathcal{B})$ if and only if

$$
k_2 = k_1\sqrt{k_3}k_3^{1/4} \frac{K}{2 + \cos(k_3^{1/4})[\exp(k_3^{1/4}) + \exp(-k_3^{1/4})]},
$$

where

$$
\begin{aligned}
K = {}& \sin(k_3^{1/4})[\exp(k_3^{1/4}) - \exp(-k_3^{1/4})] \\
& + \cos(k_3^{1/4})[\exp(2k_3^{1/4}) - \exp(-2k_3^{1/4})].
\end{aligned}
$$

Proof. Let $\lambda \in \sigma(\mathcal{A} + \mathcal{B})$. Multiplying $\phi(x)$ on both sides of equation (4.178) and integrating by parts, we obtain

$$
\lambda^2\|\phi\|^2 + \alpha^2\|\phi''\|^2 + \lambda^2\bar{\phi}'(1) = 0.
$$

If $\lambda = i\omega, \omega \in \mathbf{R}$, then

$$
-\omega^2\|\phi\|^2 + \alpha^2\|\phi''\|^2 = \omega^2\bar{\phi}'(1).
$$

So $\bar{\phi}'(1) = \phi'(1) \in \mathbf{R}$. Substituting $\lambda = i\omega$ into the last equation in (4.178) yields

$$-\omega^2 + i\omega k_2 + k_3 = i\omega k_1 \phi'(1)$$

which implies $\omega^2 = k_3$. Thus (i) is proved. Solving (4.178), we get

$$\begin{aligned}
\phi(x) &= c_1 \exp(\sqrt{\lambda i}x) + c_2 \exp(-\sqrt{\lambda i}x) \\
&\quad + (c_2 - c_1)\sin(\sqrt{\lambda i}x) - (c_1 + c_2)\cos(\sqrt{\lambda i}x),
\end{aligned}$$

where c_1 and c_2 satisfy

$$\begin{cases}
c_1 \exp(\sqrt{\lambda i}) - c_2 \exp(-\sqrt{\lambda i}) + (c_1 - c_2)\cos(\sqrt{\lambda i}) - (c_1 + c_2)\sin(\sqrt{\lambda i}) = 0, \\
c_1 \exp(\sqrt{\lambda i}) + c_2 \exp(-\sqrt{\lambda i}) + (c_1 - c_2)\sin(\sqrt{\lambda i}) + (c_1 + c_2)\cos(\sqrt{\lambda i}) = \lambda i.
\end{cases}$$

From these, it is easy to see that

$$\phi'(1) = 2\sqrt{\lambda i}\left[c_1 \exp(\sqrt{\lambda i}) - c_2 \exp(-\sqrt{\lambda i})\right].$$

Let

$$q(\lambda) = 2 + \cos(\sqrt{\lambda i})\left[\exp(\sqrt{\lambda i}) + \exp(-\sqrt{\lambda i})\right].$$

Then the zeros of $q(\lambda)$ consist of $\{\pm i\sqrt{\lambda_n}\}$ (see, e.g., Lemma 4.18). We claim that

$$q(\lambda) \neq 0, \quad \text{for any } \lambda \in \sigma(\mathcal{A} + \mathcal{B}).$$

Since otherwise, from (i), we must have $\lambda = \pm i\sqrt{k_3}$ and

$$\begin{cases}
\left[\exp(ik_3^{1/4}) + \cos(ik_3^{1/4})\right]c_1 - \sin(ik_3^{1/4})c_2 = -\frac{\sqrt{k_3}}{2}, \\
k_2 = k_1\phi'(1) = 2k_1 k_3^{1/4}i\left[c_1 \exp(ik_3^{1/4}) - c_2 \exp(-ik_3^{1/4})\right].
\end{cases}$$

Comparing the imaginary part of the above equality leads to a contradiction. Thus

$$\begin{cases}
c_1 = \frac{\lambda i}{2}\left[\exp(-\sqrt{\lambda i}) + \sin(\sqrt{\lambda i}) + \cos(\sqrt{\lambda i})\right]/q(\lambda), \\
c_2 = \frac{\lambda i}{2}\left[\exp(\sqrt{\lambda i}) - \sin(\sqrt{\lambda i}) + \cos(\sqrt{\lambda i})\right]/q(\lambda),
\end{cases}$$

and

$$\begin{aligned}
F(\lambda) &= \lambda^2 + \lambda k_2 + k_3 - \lambda k_1 \phi'(1), \\
&= \lambda^2 + \lambda k_2 + k_3 - \lambda^2 k_1 \sqrt{\lambda i}\frac{p(\lambda)}{q(\lambda)}i,
\end{aligned}$$

where

$$p(\lambda) = \sin(\sqrt{\lambda i})[\exp(\sqrt{\lambda i}) + \exp(-\sqrt{\lambda i})] + \cos(\sqrt{\lambda i})[\exp(\sqrt{\lambda i}) - \exp(-\sqrt{\lambda i})].$$

In view of the following identities,

$$\frac{i+1}{\sqrt{i}} = \sqrt{2}, \ \frac{1-i}{\sqrt{i}} = -\sqrt{2}i, \ (1+i)\sqrt{\lambda i} = i\sqrt{2\lambda}, \ (1-i)\sqrt{\lambda i} = \sqrt{2\lambda},$$

we have (4.180). Observe that

$$F(\lambda) = \lambda^2 \left[1 + k_1 \sqrt{2\lambda} \frac{f(\lambda)}{g(\lambda)} + (\frac{k_2}{\lambda} + \frac{k_3}{\lambda^2})\right].$$

Then $F(\lambda) = 0$ iff

$$\hat{F}(\lambda) = 1 + k_1 \sqrt{2\lambda} \frac{f(\lambda)}{g(\lambda)} + (\frac{k_2}{\lambda} + \frac{k_3}{\lambda^2}) = 0.$$

This is exactly the same as the problem discussed in Section 4.4.2 for $k_2 = k_3 = 0$. Since the term $(k_2/\lambda + k_3/\lambda^2)$ does not affect the asymptotic behavior of the zero points of $\hat{F}(\lambda)$ following the proof of Theorem 4.19, we have immediately the estimation (4.181).

To prove (iii), let $\lambda \in \sigma(\mathcal{A} + \mathcal{B})$, and ϕ be the solution of (4.177). Then

$$\phi = -\lambda^2(\lambda^2 + A)^{-1} \cdot 1 = -\lambda^2 \sum_{n=1}^{\infty} \frac{1}{\lambda^2 + \lambda_n} c_n \phi_n(x).$$

Here, we have used the eigenfunction expansion of 1, stated in Lemma 4.6. Since

$$\langle 1, \ \phi \rangle = -\lambda^2 \sum_{n=1}^{\infty} \frac{1}{\lambda^2 + \lambda_n} c_n \langle 1, \ \phi_n \rangle = -\lambda^2 \sum_{n=1}^{\infty} \frac{1}{\lambda^2 + \lambda_n} c_n^2 \|\phi_n\|^2,$$

$$\lambda^2 \langle 1, \ \phi \rangle - \phi'''(0) = -\lambda^2,$$

we see that

$$\begin{aligned}
\lambda^2 + \lambda k_2 + k_3 &= -\lambda k_1 \phi'''(0) = -k_1 \lambda^3 (\langle 1, \ \phi \rangle + 1) \\
&= -k_1 \lambda^3 + k_1 \lambda^5 \sum_{n=1}^{\infty} \frac{1}{\lambda^2 + \lambda_n} c_n^2 \|\phi_n\|^2 \\
&= -k_1 \lambda^3 \sum_{n=1}^{\infty} \frac{\lambda_n}{\lambda^2 + \lambda_n} c_n^2 \|\phi_n\|^2,
\end{aligned}$$

i.e., $F(\lambda) = 0$. Let $\lambda = a + ib$, $b \neq 0$. Then

$$2a + k_2 + k_1 \sum_{n=1}^{\infty} \frac{[|\lambda|^4 + \lambda_n(3a^2 - b^2)]\lambda_n}{|\lambda^2 + \lambda_n|^2} c_n^2 \|\phi_n\|^2 = 0$$

and

$$a^2 - b^2 + ak_2 + k_3 + k_1 \sum_{n=1}^{\infty} \frac{[a|\lambda|^4 + \lambda_n(a^3 - 3ab^2)]\lambda_n}{|\lambda^2 + \lambda_n|^2} c_n^2 \|\phi_n\|^2 = 0,$$

from which we deduce that

$$k_3 = |\lambda|^2 + 2ak_1 \sum_{n=1}^{\infty} \frac{|\lambda|^2 \lambda_n^2}{|\lambda^2 + \lambda_n|^2} c_n^2 \|\phi_n\|^2.$$

Therefore, $|\lambda|^2 \leq k_3$ for $a \geq 0$, this proves (iv). To prove (v), notice that

$$2a < -k_2 + k_1 \sum_{n=1}^{\infty} \frac{\lambda_n^2 b^2}{|\lambda^2 + \lambda_n|^2} c_n^2 \|\phi_n\|^2.$$

Suppose $a \geq 0$. Then by the assumptions in (v) and by noticing that $\lambda_1 < \lambda_n$, we see that

$$|\lambda^2 + \lambda_n| \geq \lambda_n - |\lambda^2| \geq \lambda_n - k_3 \geq \lambda_n - \lambda_1/2 \geq \lambda_n/2,$$

and that

$$\begin{aligned} 2a &< -k_2 + 4k_1 b^2 \sum_{n=1}^{\infty} c_n^2 \|\phi_n\|^2 = -k_2 + 4k_1 b^2 \\ &\leq -k_2 + 4k_1 k_3 \leq 0, \end{aligned}$$

i.e., $a < 0$, a contradiction. Hence, $a = \mathrm{Re}\lambda < 0$ under the assumptions given in (v). (v) is thus proved.

Finally, consider the real function

$$\begin{aligned} f(k_3) &= -iF(i\sqrt{k_3})/\sqrt{k_3} = k_2 - k_1 k_3 \frac{\lambda_n}{\lambda_n - k_3} c_n^2 \|\phi_n\|^2 \\ &- k_1 k_3 \frac{\lambda_{n+1}}{\lambda_{n+1} - k_3} c_{n+1}^2 \|\phi_{n+1}\|^2 - k_1 k_3 \sum_{k \neq n, n+1}^{\infty} \frac{\lambda_k}{\lambda_k - k_3} c_k^2 \|\phi_k\|^2. \end{aligned}$$

(vi) follows from the fact that

$$f(k_3) \to \infty, \text{ as } k_3 \to \lambda_n, \quad f(k_3) \to -\infty, \text{ as } k_3 \to \lambda_{n+1}.$$

(vii) is calculated directly from the expression of $F(\lambda)$. $\square$

Remark 4.37 *(4.181) indicates that PI constants k_2 and k_3 do not affect the large behavior of the eigenfrequency.*

Corollary 4.38 *Suppose $k_1 k_2 < 1$, $k_2 \geq 4k_1 k_3$, and $k_3 < \lambda_1/2$, then the unique classical solution of (4.168) guaranteed by Corollary 4.34 has the following uniformly exponential decay:*

$$\int_0^1 \{y_{xx}^2(x,t) + [y_t(x,t) - k_1 y_{xxx}(0,t)]^2\} dx$$

$$+z^2(t) + [\dot{z}(t) + k_1 y_{xxx}(0,t)]^2$$

$$\leq Me^{-\omega t}\left[\int_0^1 \{y_0''(x)^2 + [y_1(x) - k_1 y_0'''(0)]^2\} dx + z_0^2 + [z_1 + k_1 y_0'''(0)]^2\right]$$

for some $M \geq 1, \omega > 0$.

4.7 Gain adaptive strain feedback control of Euler-Bernoulli beams

In Section 4.2, we introduced strain feedback control law with constant feedback gain to stabilize the rotating beam equation which is recapped as follows

$$\begin{cases} \dfrac{\partial^2 y(x,t)}{\partial t^2} + \dfrac{EI}{\rho}\dfrac{\partial^4 y(x,t)}{\partial x^4} = -x\ddot{\theta}(t), \\ y(0,t) = y_x(0,t) = y_{xx}(\ell,t) = y_{xxx}(\ell,t) = 0, \\ y(x,0) = y_0(x), \quad \dot{y}(x,0) = y_1(x). \end{cases} \tag{4.183}$$

Here, the beam's bending rigidity EI, the mass density per unit length ρ and the length of the beam ℓ are resumed because we are going to discuss control problems due to the variation of various physical parameters.

In this section, we are interested in a new kind of control law named *gain adaptive strain feedback control* in which the strain $y_{xx}(0,t)$ at the fixed end of the beam is measured, and the driving motor is controlled such that its angular velocity $\dot{\theta}(t)$ satisfies

$$\dot{\theta}(t) = k(t)y_{xx}(0,t). \tag{4.184}$$

What is different from before is that here the feedback gain is no longer a constant, but is time dependent. Since $y_{xx}(0,t)$ can be measured, we choose $k(t)$ according to

$$\begin{cases} \dot{k}(t) = \alpha[y_{xx}(0,t)]^2, \\ k(0) = k_0 \geq 0, \end{cases} \tag{4.185}$$

where α is a positive constant. Since $k_0 \geq 0$, we see that $k(t) \geq 0$, for all $t \geq 0$. Eq.(4.185) is called *gain adaptation law*.

Intuitively, this adaptation law implies that if the measured strain signal $y_{xx}(0, t)$ is large, then a large gain $k(t)$ should be taken. The great advantage of using this gain adaptation law is that good control performance (vibration suppression) can be automatically achieved even in the presence of tip load uncertainties or variations of the beam's physical parameters such as EI and ρ. This is because tip load uncertainties or parameter variations will be reflected in the measurement signal $y_{xx}(0, t)$, with large $y_{xx}(0, t)$ corresponding to massive tip load or stiffer beam in general.

Taking the time derivative on both sides of (4.184) yields

$$\ddot{\theta}(t) = \dot{k}(t)y_{xx}(0, t) + k(t)\dot{y}_{xx}(0, t).$$

Substituting this and (4.185) into (4.183), we obtain the following gain adaptive strain feedback controlled closed-loop system:

$$\begin{cases} \dfrac{\partial^2 y(x, t)}{\partial t^2} + k(t)x\dfrac{\partial^3 y(x, t)}{\partial x^2 \partial t} + \dfrac{EI}{\rho}\dfrac{\partial^4 y(x, t)}{\partial x^4} + \alpha x[y_{xx}(0, t)]^3 = 0, \\ y(0, t) = y_x(0, t) = y_{xx}(\ell, t) = y_{xxx}(\ell, t) = 0, \\ y(x, 0) = y_0(x), \ \dot{y}(x, 0) = y_1(x). \end{cases} \tag{4.186}$$

Eq.(4.186) represents a time-varying nonlinear initial-boundary value system whose stability is usually difficult to analyze. In the next section, we shall use the nonlinear semigroup approach presented in Chapter 2 to show the existence and uniqueness of the solution of (4.186), and the energy multiplier method to show the exponential stability of the unique solution.

To adopt the nonlinear semigroup approach, take $k(t)$ as a new state variable and rewrite equation (4.186) and (4.185) as

$$\begin{cases} \dfrac{d}{dt}[\dot{y}(x, t) + k(t)xy_{xx}(0, t)] + \dfrac{EI}{\rho}\dfrac{\partial^4 y(x, t)}{\partial x^4} = 0, \\ y(0, t) = y_x(0, t) = y_{xx}(\ell, t) = y_{xxx}(\ell, t) = 0, \\ \dot{k}(t) = \alpha[y_{xx}(0, t)]^2, \\ y(x, 0) = y_0(x), \ \dot{y}(x, 0) = y_1(x), \ k(0) = k_0 \ge 0. \end{cases} \tag{4.187}$$

Let $H = H_E^2(0, \ell) \times L^2(0, \ell) \times \mathbf{R}$ be the underlying real Hilbert state space with the inner product

$$\langle (u_1(x), v_1(x), r_1), (u_2(x), v_2(x), r_2) \rangle$$
$$= \int_0^\ell \left[u_1''(x)u_2''(x) + v_1(x)v_2(x)\right]dx + \frac{EI}{2\alpha\rho}r_1 r_2,$$

and the induced norm $\|\cdot\|$. Here, $H_E^2(0, \ell) = \{u \in H^2(0, \ell), u(0) = u'(0) = 0\}$. Let $H^+ = H_E^2(0, \ell) \times L^2(0, \ell) \times \mathbf{R}^+$, a closed convex subset of H. Define a nonlinear operator $\mathcal{A} : D(\mathcal{A}) \subset H^+ \to H$ by

$$\mathcal{A}(u(x), v(x), r) = (v(x) - rxu''(0), \ -\frac{EI}{\rho}u''''(x), \ \alpha[u''(0)]^2) \qquad (4.188)$$

with

$$D(\mathcal{A}) = \{(u(x), v(x), r) \in H^4(0, \ell) \times H^2(0, ,\ell) \times \mathbf{R}^+ \,\big|$$
$$u(0) = u'(0) = u''(\ell) = u'''(\ell) = 0, \ v(0) = 0, v'(0) = ru''(0)\}.$$

Then (4.187) can be written as a *nonlinear evolution equation* on H:

$$\begin{cases} \dot{Y}(t) = \mathcal{A}Y(t), \\ Y(0) = Y_0, \end{cases} \qquad (4.189)$$

where

$$\begin{aligned} Y(t) &= (y(x,t), \ \dot{y}(x,t) + k(t)xy_{xx}(0,t), \ k(t)), \\ Y_0 &= (y_0(x), \ y_1(x) + k_0 xy_0''(0), \ k_0) \in H^+. \end{aligned}$$

Let

$$E(t) = \int_0^\ell \Big\{ \frac{EI}{\rho}[y_{xx}(x,t)]^2 + [\dot{y}(x,t) + k(t)xy_{xx}(0,t)]^2 \Big\} dx. \qquad (4.190)$$

It is easily verified that

$$\|Y(t)\|^2 = E(t) + \frac{EI}{2\alpha\rho}k^2(t) \qquad (4.191)$$

and that

$$\frac{d}{dt}\|Y(t)\|^2 = -\frac{EI}{\rho}k(t)[y_{xx}(0,t)]^2 \le 0. \qquad (4.192)$$

Combining (4.187),(4.191), and (4.192), we obtain

$$\frac{d}{dt}\Big[\|Y(t)\|^2 + \frac{EI}{2\alpha\rho}k^2(t)\Big] = \frac{d}{dt}\Big[2\|Y(t)\|^2 - E(t)\Big] = 0. \qquad (4.193)$$

Lemma 4.39 *The operator $\mathcal{A}$ defined in (4.188) is closed and dissipative in H with the domain $D(\mathcal{A})$ dense in H^+ and*

$$H^+ \subset \mathcal{R}(I - \lambda\mathcal{A}) \quad \text{for all } \lambda > 0, \qquad (4.194)$$

where $\mathcal{R}(I - \lambda\mathcal{A})$ represents the range of $I - \lambda\mathcal{A}$.

Proof. The closedness of $\mathcal{A}$ is obvious. Let $U, V \in D(\mathcal{A})$ where

$$U = (u_1(x), v_1(x), r_1), \ V = (u_2(x), v_2(x), r_2).$$

Then a simple calculation leads to

$$\langle \mathcal{A}U - \mathcal{A}V, \ U - V \rangle = -\frac{1}{2}\frac{EI}{\rho}(r_1 + r_2)[u_1''(0) - u_2''(0)]^2 \leq 0,$$

which proves the dissipativity of $\mathcal{A}$. To prove (4.194), it is sufficient to show that for any $\lambda > 0$ and any $(f(x), g(x), r_0) \in H^+$ there exists $(u(x), v(x), r) \in D(\mathcal{A})$, such that

$$(I - \lambda\mathcal{A})(u(x), v(x), r) = (f(x), g(x), r_0),$$

i.e.,

$$\lambda v(x) = u(x) + rxu''(0) - f(x), \quad r = r_0 + \lambda\alpha[u''(0)]^2,$$

and u satisfies

$$\begin{cases} \lambda^2 \dfrac{EI}{\rho} u''''(x) + u(x) + \lambda rxu''(0) = f(x) + \lambda g(x), \\ u(0) = u'(0) = u''(\ell) = u'''(\ell) = 0, \\ r = r_0 + \lambda\alpha[u''(0)]^2. \end{cases} \quad (4.195)$$

(4.195) can be written in the operator form as

$$\begin{cases} (I + \lambda^2 A + \lambda rB)u = f + \lambda g, \\ r = r_0 + \lambda\alpha[u''(0)]^2, \end{cases} \quad (4.196)$$

where A, B and Q are operators defined by

$$\begin{cases} A\phi = \dfrac{EI}{\rho}\phi''''(x), \\ D(A) = \{\phi \in L^2(0, \ell) \big| \phi(0) = \phi'(0) = \phi''(\ell) = \phi'''(\ell) = 0\}, \\ B\phi = \phi''(0)x, \quad Q = BA^{-1}. \end{cases} \quad (4.197)$$

A is unbounded, self-adjoint, and positive definite in $L^2(0, \ell)$ and Q is bounded, self-adjoint, and positive semi-definite on $L^2(0, \ell)$.

Since $\langle(\lambda^2 + A^{-1} + \lambda rQ)\phi, \ \phi\rangle_{L^2} \geq \lambda^2\|\phi\|_{L^2}^2$ for all $\phi \in L^2(0, \ell)$ and $r \geq 0$, it is seen that $\lambda^2 + A^{-1} + \lambda rQ$ is invertible, bounded and self-adjoint on $L^2(0, \ell)$ and $\|(\lambda^2 + A^{-1} + \lambda rQ)^{-1}\| \leq 1/\lambda^2$. Thus, the equation

$$(\lambda^2 + A + \lambda rB)u = f + \lambda g$$

has a unique solution which is given by

$$u = u_r = A^{-1}(\lambda^2 + A^{-1} + \lambda rQ)^{-1}(f + \lambda g).$$

Clearly, u_r is continuous in $r \geq 0$ on $L^2(0, \ell)$ and

$$\|u_r\|_{L^2} \leq \|A^{-1}\| \|f + \lambda g\|_{L^2}/\lambda^2. \qquad (4.198)$$

On the other hand, multiplying x on both sides of the first equation of (4.195) and integrating from 0 to ℓ yields

$$u_r''(0) = [-\langle x, u_r \rangle_{L^2} + \langle x, f + \lambda g \rangle_{L^2}]/(\lambda^2 EI/\rho + \lambda r\ell^3/3) \qquad (4.199)$$

which together with (4.198) gives

$$|u_r''(0)| \leq \frac{\rho}{\lambda^2 EI} \|x\|_{L^2}(\|u_r\|_{L^2} + \|f + \lambda g\|_{L^2}) \qquad (4.200)$$

$$\leq \frac{\rho}{\lambda^4 EI} \|x\|_{L^2}(1 + \|A^{-1}\|)\|f + \lambda g\|_{L^2} \qquad (4.201)$$

$$\leq C_\lambda \|f + \lambda g\|_{L^2}, \quad \text{for all } r \geq 0 \qquad (4.202)$$

where C_λ is a fixed constant depending on λ. Define

$$p(r) = r_0 + \lambda\alpha[u_r''(0)]^2.$$

Then $p(r)$ is continuous on $\mathbf{R}^+$, and $p(\mathbf{R}^+) \subset [0, \ r_0 + \lambda\alpha C_\lambda^2 \|f + \lambda g\|_{L^2}^2]$. Hence, $p(r)$ has a fixed point r in $\mathbf{R}^+$, $p(r) = r$, $r \geq 0$ which implies that (4.195) admits a solution. So $H^+ \subset \mathcal{R}(I - \lambda)$. Finally, since

$$\{\phi \in C^\infty[0, \ell], \phi(0) = \phi'(0) = \phi^{(n)}(0) = \phi^{(n)}(\ell) = 0, \ n \geq 2\}$$
$$\times C_0^\infty[0, \ell] \times \mathbf{R}^+ \subset D(\mathcal{A})$$

we see that $D(\mathcal{A})$ is dense in H^+.   □

Recalling the Crandall-Liggett theorem in Chapter 2, we have the following existence and uniqueness result.

Theorem 4.40 *The operator $\mathcal{A}$ defined by (4.188) generates a unique nonlinear strong continuous semigroup of contractions $S(t)$ on H^+. Thus equation (4.189) admits a unique strong solution $Y(t)$ for each $Y_0 \in D(\mathcal{A})$ in the sense that, for any $y_0 \in D(A), y_1 \in D(A^{1/2}), k_0 \geq 0, y(t, x)$ satisfies*

$$\begin{cases} y \in L^\infty([0, \infty); H^4(0, \ell)), \dot{y} \in L^\infty([0, \infty); H^2(0, \ell)), \\ \dfrac{d}{dt}\left[\dot{y}(x, t) + k(t)xy_{xx}(0, t)\right] + \dfrac{EI}{\rho}\dfrac{\partial^4 y(x, t)}{\partial x^4} = 0, \ \dot{k}(t) = \alpha[y_{xx}(0, t)]^2, \\ \qquad \text{for } t \geq 0, a.e. \text{ in } H, \\ y(x, 0) = y_0(x), \ \dot{y}(x, 0) = y_1(x), \ k(0) = k_0. \end{cases}$$

Lemma 4.41 $0 \in \mathcal{R}(\mathcal{A})$ and $(I - \mathcal{A})^{-1}$ is compact.

Proof. $0 \in \mathcal{R}(\mathcal{A})$ is trivial. We need only to prove the second assertion. Let $\{V_n\} \subset H^+$, $\|V_n\| \leq K$ be a bounded sequence and $U_n = (u_n(x), v_n(x), r_n) \in H^+$ satisfy $(I - \mathcal{A})U_n = V_n$. Then, by the dissipativity of $\mathcal{A}$, $\|\mathcal{A}U_n\| \leq \|V_n\| \leq K$ and $\|U_n\| \leq \|V_n\| \leq K$. These imply that

$$\|u_n\|_{H^4} \leq C_1, \ \|v_n\|_{H^2} \leq C_1, \ r_n \leq C_1$$

for some uniform constant C_1. By the Sobolev imbedding theorem, there is a subsequence of U_n, still indexed by n for notational simplicity, and $U_0 = (u_0(x), v_0(x), r_0) \in H^+$ such that

$$U_n \to U_0 \quad \text{in the topology of } H.$$

Consequently, $u_0(0) = u_0'(0) = 0$. The proof is complete. $\qquad\qquad\square$

Lemma 4.42 *Let $S(t)$ be the semigroup generated by $\mathcal{A}$. If for some $Y_0 = (y_0(x), \ y_1(x) + k_0 x y_0''(0), \ k_0) \in D(\mathcal{A})$, there hold*

$$\frac{d}{dt}\|S(t)Y_0\| \equiv 0, \quad \forall t \geq 0,$$

then $Y_0 = (0, \ 0, \ k_0)$.

Proof. Let $Y(t) = (y(x,t), \dot{y}(x,t) + k(t)xy_{xx}(0,t), k(t))$ be the solution of (4.187) with initial value $Y(0) = Y_0$. In view of (4.192), $\frac{d}{dt}\|S(t)Y_0\| \equiv 0$ implies that $k(t)[y_{xx}(0,t)]^2 = 0$. If $k(t) \equiv 0$, i.e.,

$$k_0 + \int_0^t [y_{xx}(0,\tau)]^2 d\tau = 0, \quad \forall t \geq 0,$$

then $y_{xx}(0,t) = 0$ for all $t \geq 0$. Thus, (4.187) reduces to

$$\begin{cases} \dfrac{\partial^2 y(x,t)}{\partial t^2} + \dfrac{EI}{\rho}\dfrac{\partial^4 y(x,t)}{\partial x^4} = 0, \\ y(0,t) = y_x(0,t) = y_{xx}(\ell,t) = y_{xxx}(\ell,t) = 0, \ y_{xx}(0,t) = 0, \\ k(t) = k_0. \end{cases} \quad (4.203)$$

Multiplying $y_x(x,t)$ on both sides of (4.203), and integrating with respect to x and t from 0 to ℓ and 0 to T, respectively, we obtain

$$\int_0^T [\dot{y}(\ell,t)]^2 dt = 2 \int_0^\ell y_x(x,t)\dot{y}(x,t)\Big|_0^T dx. \quad (4.204)$$

Similarly, multiplying $xy_x(x,t)$ on both sides of (4.203), and integrating with respect to x and t from 0 to ℓ and 0 to T, respectively, we obtain

$$3\frac{EI}{\rho}\int_0^T\int_0^\ell [y_{xx}(x,t)]^2 dx dt + \int_0^T\int_0^\ell [\dot{y}(x,t)]^2 dx dt$$

$$= \ell \int_0^T [\dot{y}(\ell,t)]^2 dt - 2\int_0^\ell xy_x(x,t)\dot{y}(x,t)\Big|_0^T dx. \qquad (4.205)$$

Now noting that (4.203) is a conservative system, i.e., the energy

$$G(t) = \frac{EI}{\rho}\int_0^\ell [y_{xx}(x,t)]^2 dx + \int_0^\ell [\dot{y}(x,t)]^2 dx$$

stored in (4.203) is constant for any $t \geq 0$, it is deduced from (4.204) and (4.205) that there exists a constant K_0, independent of T, such that

$$\int_0^T G(t)dt \leq 3\frac{EI}{\rho}\int_0^T\int_0^\ell [y_{xx}(x,t)]^2 dx dt + \int_0^T\int_0^\ell [\dot{y}(x,t)]^2 dx dt$$

$$= 2\int_0^\ell (\ell-x)y_x(x,t)\dot{y}(x,t)dx\Big|_0^T$$

$$\leq K_0 G(0),$$

where the Cauchy-Schwartz inequality is used to obtain the last inequality. Therefore, $G(t) \equiv 0$, or equivalently $y(x,t) \equiv 0$ and $k(t) \equiv k_0$. □

Theorem 4.40, Lemmas 4.41 and 4.42, (4.192), and LaSalle's invariance principle [Theorem 3.64] imply the strong stability of (4.187) which is summarized as

Theorem 4.43 *Let y be the solution of (4.187). Then the energy function defined in (4.190) satisfies $\lim_{t\to\infty} E(t) = 0$ and*

$$\lim_{t\to\infty} k(t) = \sqrt{\frac{\alpha\rho}{EI}E(0) + k_0^2},$$

$$k(t) \leq \sqrt{\frac{\alpha\rho}{EI}E(0) + k_0^2}, \quad \forall t \geq 0. \qquad (4.206)$$

Proof. The first assertion $\lim_{t\to\infty} E(t) = 0$ is immediate. We only show (4.206). From (4.191) and (4.193), it is seen that

$$\dot{E}(t) = 2\frac{d}{dt}\|Y(t)\|^2.$$

Integrating both sides from 0 to ∞ yields

$$
\begin{aligned}
E(\infty) - E(0) &= 2\|Y(\infty)\|^2 - 2\|Y(0)\|^2 \\
&= \frac{EI}{\alpha\rho}k^2(\infty) - 2\left(E(0) + \frac{EI}{2\alpha\rho}k_0^2\right).
\end{aligned}
$$

Since $E(\infty) = 0$, it follows that

$$
k(\infty) = \sqrt{\frac{\alpha\rho}{EI}E(0) + k_0^2}.
$$

Since $\dot{k}(t) \geq 0$ as can be seen from (4.187), $k(t)$ is monotonically increasing, which shows (4.206). $\qquad\square$

Theorem 4.44 *Let $k_0 > 0$ and $E(t)$ be defined by (4.190). Then for every solution $y(x,t)$ of (4.187), there exist positive constants $M \geq 1$ and $\mu > 0$ such that*

$$
E(t) \leq Me^{-\mu t}, \tag{4.207}
$$

$$
|k^2(t) - k_0^2 - \frac{\alpha\rho}{EI}E(0)| \leq \frac{\alpha\rho}{EI}Me^{-\mu t}, \quad \forall t \geq 0. \tag{4.208}
$$

Proof. Since $D(\mathcal{A})$ is dense in H, we may assume $y_0 \in D(A), y_1 \in D(A^{1/2})$ without loss of generality. Then y is the strong solution of (4.187) by Theorem 4.40. Choose a multiplier

$$
\begin{aligned}
\beta(t) &= \int_0^\ell y(x,t)[\dot{y}(x,t) + k(t)xy_{xx}(0,t)]dx \\
&\quad +4\int_0^\ell xy_x(x,t)[\dot{y}(x,t) + k(t)xy_{xx}(0,t)]dx \\
&\quad -4\ell\int_0^\ell y_x(x,t)[\dot{y}(x,t) + k(t)xy_{xx}(0,t)]dx.
\end{aligned}
$$

Taking the derivative of β along the solution of (4.187) yields

$$
\begin{aligned}
\dot{\beta}(t) &= -E(t) - 6\frac{EI}{\rho}\int_0^\ell [y_{xx}(x,t)]^2 dx + 2\ell\frac{EI}{\rho}[y_{xx}(0,t)]^2 \\
&\quad -2\int_0^\ell [\dot{y}(x,t)]^2 dx \\
&\quad -5k(t)y_{xx}(0,t)\int_0^\ell x\dot{y}(x,t)dx \\
&\quad -4\int_0^\ell x\dot{y}(x,t)\dot{y}_x(x,t)dx
\end{aligned}
$$

$$+k^2(t)[y_{xx}(0,t)]^2 \int_0^\ell x^2 dx$$

$$+4\ell k(t)y_{xx}(0,t) \int_0^\ell \dot{y}(x,t)dx$$

$$+4\ell \int_0^\ell \dot{y}(x,t)\dot{y}_x(x,t)dx.$$

Since it is easy to verify that

$$-4\int_0^\ell x\dot{y}(x,t)\dot{y}_x(x,t)dx = -2\ell[\dot{y}(\ell,t)]^2 + 2\int_0^\ell [\dot{y}(x,t)]^2 dx,$$

$$4\ell \int_0^\ell \dot{y}(x,t)\dot{y}_x(x,t)dx = 2\ell[\dot{y}(\ell,t)]^2,$$

one has

$$\dot{\beta}(t) = -E(t) - 6\frac{EI}{\rho}\int_0^\ell [y_{xx}(x,t)]^2 dx + 2\ell\frac{EI}{\rho}[y_{xx}(0,t)]^2$$

$$-5k(t)y_{xx}(0,t)\int_0^\ell x[\dot{y}(x,t) + k(t)xy_{xx}(0,t)]dx$$

$$+4\ell k(t)y_{xx}(0,t)\int_0^\ell [\dot{y}(x,t) + k(t)xy_{xx}(0,t)]dx. \qquad (4.209)$$

For any real number a, b and $\delta > 0$, it is well known that the following inequality

$$2ab \le \delta a^2 + b^2/\delta$$

holds. This, together with the Cauchy-Schwartz inequality, can be used to infer the existence of positive constants K_1, K_2 and $\tilde{K}_3$ such that

$$|\beta(t)| \le K_1 E(t),$$

$$\dot{\beta}(t) \le -K_2 E(t) + \tilde{K}_3 k^2(t)[y_{xx}(0,t)]^2 + 2\ell\frac{EI}{\rho}[y_{xx}(0,t)]^2$$

$$= -K_2 E(t) + \left(\tilde{K}_3 + \frac{2\ell EI}{\rho k^2(t)}\right)k^2(t)[y_{xx}(0,t)]^2$$

$$\le -K_2 E(t) + \left(\tilde{K}_3 + \frac{2\ell EI}{\rho k_0^2}\right)k^2(t)[y_{xx}(0,t)]^2$$

$$= -K_2 E(t) + K_3 k^2(t)[y_{xx}(0,t)]^2,$$

where $K_3 = \tilde{K}_3 + \frac{2\ell EI}{\rho k_0^2}$. Let ε be a small constant such that

$$0 < \varepsilon < \min\left\{\frac{1}{K_1}, \frac{2EI}{\rho K_3\sqrt{\alpha\rho E(0)/EI + k_0^2}}\right\}. \qquad (4.210)$$

Define
$$E_\varepsilon(t) = E(t) + \varepsilon\beta(t).$$

Obviously, we have

$$(1 - \varepsilon K_1)E(t) \le E_\varepsilon(t) \le (1 + \varepsilon K_1)E(t).$$

In view of (4.206), it is easy to see that

$$
\begin{aligned}
\dot{E}_\varepsilon(t) &= \dot{E}(t) + \varepsilon\dot{\beta}(t) \\
&\le -2\frac{EI}{\rho}k(t)[y_{xx}(0,t)]^2 + \varepsilon\dot{\beta}(t) \\
&\le -k(t)[2\frac{EI}{\rho} - \varepsilon K_3 k(t)][y_{xx}(0,t)]^2 - \frac{\varepsilon K_2}{1 + \varepsilon K_1}E_\varepsilon(t) \\
&\le -\frac{\varepsilon K_2}{1 + \varepsilon K_1}E_\varepsilon(t).
\end{aligned}
$$

Therefore,

$$E(t) \le \frac{1 + \varepsilon K_1}{1 - \varepsilon K_1}E(0)\exp\left(-\frac{\varepsilon K_2}{1 + \varepsilon K_1}t\right) = Me^{-\mu t}, \quad \forall t \ge 0$$

where $M = \frac{1 + \varepsilon K_1}{1 - \varepsilon K_1}$ and $\mu = \frac{\varepsilon K_2}{1 + \varepsilon K_1}$. Eq.(4.208) follows from (4.207) and (4.191)-(4.193). $\square$

Remark 4.45 *The decay rate μ in Theorem 4.44 depends on the initial energy $E(0)$ of the system, so the exponential decay is not uniform, which is very common for nonlinear problems and is in sharp contrast with linear cases where exponentially asymptotically stable and exponentially stable mean the same thing. For practical problems, the initial energy is always finite. In this case, there always exists a positive constant μ which guarantees the exponential stability, although theoretically it might be interesting to further investigate whether the exponential stability is uniform.*

Remark 4.46 *The results in Theorem 4.43 indicate that for large initial energy we need a large strain feedback gain.*

4.8   Notes and references

Early research on vibration control of flexible continuum beams mainly focused on applications in connection with space structures [7], [107]. Book [20] perhaps was the first who foresaw the need to consider vibration control of multi-link light-weight robot arms. The literature on this respect has since become quite extensive, see, for instance, [22], [58], [144].

The model in Section 4.1 is obtained by Sakawa and Luo [143]. It is a generalization of those models developed in early published papers. The concept of the A-dependent operators in Section 4.1 is first introduced by Luo in [99] and is successfully used to argue the well-posedness and stability of the non-standard second order evolution equations in the subsequent sections. The direct strain and shear force feedback control laws are proposed in [99], [102]. The motivation for developing these kinds of sensor feedback control laws comes from the observation that feedback using higher order derivative information is more powerful for the stabilization of infinite dimensional systems [134]. It should be noted that, prior to the works [99] and [102], strain feedback was used in torque control to suppress vibration in flexible robot arms [57],[82]. The feedback law was derived by the Lyapunov method, but is easily shown to be a non-exponentially stabilizable feedback law. In contrast to this, the strain and shear force feedback in Section 4.4 can stabilize exponentially the rotating and translating beams, respectively. The spectral analysis for strain feedback controlled closed-loop system is discussed by Guo [62]. The fact that shear force feedback can also stabilize rotating beams, as discussed in Section 4.5, appeared in a recent paper by Luo and Guo [101] where the integrated semigroup introduced in Chapter 2 plays an important role. The stability analysis for the hybrid system in Section 4.6 can be found in [68] and the gain adaptive strain feedback control is addressed in [67].

There are actually too many articles which deal with vibration control of flexible beams (or arms) to be listed here. Most of them, however, are based on finite dimensional approximations, instead of the infinite dimensional models used in this book.

Chapter 5

Dynamic Boundary Control of Vibration Systems Based on Passivity

In this chapter, we consider feedback stabilization of a class of *passive* infinite dimensional systems by means of dynamic boundary control. The notion of passivity was developed in connection with circuit theory in the late '50s where the basic motivation was to investigate the behavior of circuits composed of passive circuit elements such as resistors, capacitors and inductors, see [61]. This concept was then introduced into control systems, see [2], [50], [156], [164]. To motivate the concept of passivity, let us consider the following situation: Let $\mathcal{S}$ be a dynamical system with an input vector $u = (u_1, \cdots, u_m)^T \in \mathbf{R}^m$ and an output vector $y = (y_1, \cdots, y_m)^T \in \mathbf{R}^m$. Let H be the Hilbert space in which the solutions of $\mathcal{S}$ evolves, and let $E(t) : H \to \mathbf{R}$ be a positive time function which depends on the solutions of $\mathcal{S}$. Assume that the time derivative of $E(t)$ along the solutions of $\mathcal{S}$ satisfies

$$\dot{E}(t) = u^T y = \sum_{i=1}^{m} u_i y_i. \tag{5.1}$$

In such systems, $E(t)$ can be thought of as the "internal energy" of the system, and (5.1) is no more than a description of the conservation of energy. The right-hand side of (5.1) represents the "external power" supplied to the system, and the left-hand side represents the rate of change of "internal energy". If we choose the control inputs u_i as

$$u_i = -\alpha_i y_i, \quad \alpha_i \geq 0, \tag{5.2}$$

and use (5.2) in (5.1), then

$$\dot{E} = -\sum_{i=1}^{m} \alpha_i (y_i)^2. \tag{5.3}$$

Hence, the control law given by (5.2) results in the dissipation of the internal energy of the system, and under appropriate assumptions one may conclude some stabilization results. For more details on the applications of passivity in the finite dimensional linear and nonlinear systems, see [21], [50], [76], [164], [165], and the related references therein.

In what follows, we first develop a general framework which characterizes the passivity for a class of infinite dimensional systems with boundary inputs and outputs. We show that some of the examples frequently encountered in the literature (e.g., the wave equation, the Euler-Bernoulli and the Timoshenko beam equations) belong to this class, and we present the stability results for such systems. In section 5.2, we extend the class of controllers given by (5.2) to a class of strictly positive real controllers. Here, the controller will represent a finite dimensional system whose input is y_i and whose output is $-u_i$ such that the associated transfer function is strictly positive real. We will prove that the results of Section 5.1 may be valid in this case as well. In Section 5.3, we apply the ideas presented in Sections 5.1 and 5.2 to the control of a rotating flexible beam. In Section 5.4, we discuss stability robustness problems with respect to small time delays in the feedback for damped wave equation with dynamic boundary control.

5.1   A general framework for system passivity

Let H be a Hilbert space and denote by $\langle \cdot, \cdot \rangle_H$ and $\|\cdot\|_H$ the inner product and the induced norm in H, respectively. Consider the following abstract equation in H:

$$w_{tt} + Aw = 0, \tag{5.4}$$

where A is a linear differential operator. For simplicity, assume that the spatial variable x belongs to $\mathbf{R}$ and takes values in $[0, 1]$. Suppose that associated with (5.4) are the following boundary conditions:

$$(B_i^1 w)(0) = f_i^1, \quad i = 1, 2, \ldots, k, \tag{5.5}$$

$$(B_i^2 w)(1) = f_i^2, \quad i = 1, 2, \ldots, l, \tag{5.6}$$

$$(B_i^3 w)(0) = 0, \quad i = 1, 2, \ldots, p, \tag{5.7}$$

$$(B_i^4 w)(1) = 0, \quad i = 1, 2, \ldots, r, \tag{5.8}$$

where B_i^j are various linear (not necessarily bounded) operators in H, k, l, p, r are some appropriate integers, and f_i^j are control inputs of our systems. In this chapter, we use the notation $(\cdot)_i^j$ where the indices take the values $i = 1, \ldots, k$ or l, $j = 1, 2$, whichever is appropriate, and in the sequel we will not state the range of indices, which should be obvious from the context. We note that here $(B_i^j w)(\cdot) : [0, 1] \to \mathbf{R}$ and $(B_i^j w)(c)$ denotes the value of $B_i^j w$ at $x = c$.

Let us define the following sets

$$\mathcal{S}_1 = \{w \in H \mid (B_i^1 w)(0) = 0, \quad (B_i^2 w)(1) = 0\}, \tag{5.9}$$

$$\mathcal{S}_2 = \{w \in H \mid (B_i^3 w)(0) = 0, \quad (B_i^4 w)(1) = 0\}. \tag{5.10}$$

Let $D(A) \subset H$ be the domain of A. For simplicity, we may take

$$D(A) = \{w \in H \mid Aw \in H \}. \tag{5.11}$$

Let A_{uc} denote the operator A with the following domain

$$D(A_{uc}) = D(A) \cap \mathcal{S}_1 \cap \mathcal{S}_2. \tag{5.12}$$

We make the following assumptions

Assumption 5.1 *$D(A)$ is dense in H.*

Assumption 5.2 *$D(A_{uc})$ is dense in H, A_{uc} is self-adjoint and coercive in H, i.e., the following holds for some $\alpha > 0$*

$$\langle w, A_{uc} w \rangle_H \geq \alpha \|w\|_H^2, \quad w \in D(A_{uc}). \tag{5.13}$$

From assumption 5.2 it follows that $A_{uc}^{1/2}$ exists, is self-adjoint and non-negative. Define the set V as

$$V = D(A_{uc}^{1/2}). \tag{5.14}$$

We make the following assumption for technical reasons.

Assumption 5.3 *The set $V \subset H$ satisfies the following*

$$V \cap \mathcal{S}_1 \neq V, \quad V \cap \mathcal{S}_2 = V. \tag{5.15}$$

We note that in most cases, the sets $\mathcal{S}_1$ and $\mathcal{S}_2$ impose certain conditions on $w \in H$ at the boundaries, and the set V could be redefined without changing the density arguments so that assumption 5.3 is satisfied.

5.1.1   Uncontrolled case

Let us consider the system given by (5.4)-(5.8) with $f_i^1 = f_i^2 = 0$ for $i = 1, \ldots, k, l$, whichever is appropriate. The resulting system is called *un-controlled* since the control inputs are set to zero. We can rewrite (5.4) as

$$\dot{z} = \mathcal{A}z, \quad z(0) \in X, \tag{5.16}$$

where $X = V \times H$, $z = (w, \ w_t)^T \in X$, and $\mathcal{A}$ is a linear operator defined on X as

$$\mathcal{A} = \begin{bmatrix} 0 & I \\ -A & 0 \end{bmatrix}, \tag{5.17}$$

with $D(\mathcal{A}) = D(A_{uc}) \times V$. For $z_1 = (u_1, \ v_1)^T, z_2 = (u_2, \ v_2)^T \in X$, the inner product and the norm on X is defined as

$$\langle z_1, \ z_2 \rangle_X = \langle A_{uc}^{1/2} u_1, \ A_{uc}^{1/2} u_2 \rangle_H + \langle v_1, \ v_2 \rangle_H, \tag{5.18}$$

$$\|z\|_X^2 = \left\| A_{uc}^{1/2} u \right\|_H^2 + \|v\|_H^2, \tag{5.19}$$

where $z = (u, \ v)^T \in X$. From Example 2.34, it follows that the operator $\mathcal{A}$ generates a C_0-semigroup of contractions on X. In the sequel, we will show that this property is preserved when the control inputs are chosen appropriately.

5.1.2   Controlled case

Consider the system given by (5.4)-(5.8). Our aim is to find control laws for f_i^j such that the resulting system possesses the following properties :

   i : There exists a solution to (5.4)-(5.8) in an appropriate Hilbert space and this solution is unique (well-posedness problem),

   ii : The solution of (5.4)-(5.8) decays to zero as $t \to \infty$ (asymptotic stability problem).

   In the sequel, we will propose a class of feedback control laws to solve the problems posed above. In such feedback schemes, the control inputs are appropriate functions of w and/or w_t, evaluated at the appropriate boundary. Such functions are naturally called the *outputs* of the system. The selection of appropriate outputs are necessary for the control schemes based on passivity; our next assumption clarifies this point.

Assumption 5.4 *Let $D_1 = D(A) \cap S_2$ and $D = D_1 \times V$. D_1 is dense in $D(A_{uc})$ and the following holds*

$$\langle z, \ \mathcal{A}z \rangle_X = \sum_{i=1}^{k} (\ B_i^1 u\ )(0)(\ O_i^1 v\ )(0) + \sum_{i=1}^{l} (\ B_i^2 u\ )(1)(\ O_i^2 v\ )(1), \tag{5.20}$$

where $z = (u,\ v)^T \in D$ and O_i^j, $i = 1, \ldots k$ or l, $j = 1, 2$, whichever appropriate, are linear (not necessarily bounded) operators in H. We call (5.20) the power form for the system given by (5.16). (cf. (5.1)).

Remark 5.5 *Assume that the operator $\mathcal{A}$ generates a C_0−semigroup of contractions on $X = V \times H$. Let $z(0) \in D(\mathcal{A})$, and let $z(t)$ be the solution of (5.16). We have $z(t) \in D(\mathcal{A})$ for $t \geq 0$ and $z(t)$ is differentiable, see Theorem 2.12. Let us define the energy $E(t)$ associated with (5.16) as*

$$E(t) = \frac{1}{2}\langle z(t),\ z(t)\rangle_X. \tag{5.21}$$

Differentiating (5.21) along the solution of (5.16) and using $f_i^j = 0$ and (5.20), we obtain $\dot{E} = 0$, i.e., the energy is conserved for the uncontrolled case. We will choose the control inputs appropriately by using the power form given by (5.20) so that the energy is dissipated and all solutions asymptotically decay to zero.

Let $z = (w,\ w_t)^T$ be the solution of (5.16). Keeping in mind (5.20), we define the outputs y_i^j of the system (5.16) as

$$y_i^1 = (\ O_i^1 w_t\ )(0), \quad i = 1, 2, \ldots k, \tag{5.22}$$

$$y_i^2 = (\ O_i^2 w_t\ )(1), \quad i = 1, 2, \ldots l. \tag{5.23}$$

In the sequel, we show, as an example, that the Timoshenko beam equation satisfies all the assumptions stated above.

Example 5.6 *Consider the following well-known Timoshenko beam equations:*

$$\rho y_{tt} - K(y_{xx} - \phi_x) = 0, \quad x \in (0, L) \tag{5.24}$$

$$I_\rho \phi_{tt} - EI\phi_{xx} + K(\phi - y_x) = 0, \tag{5.25}$$

where L is the length of the beam, x is the spatial coordinate along the beam, $y(x, t)$ is the displacement of the beam from its equilibrium position, and $\phi(x, t)$ is the angle of rotation of the beam cross-sections due to bending. The coefficients ρ, I_ρ and EI are the mass per unit length, the mass moment of inertia of the beam cross-sections, and the flexural rigidity of the beam, respectively. The coefficient K is equal to kGA where G is the shear modulus, A is the cross-sectional area, and k is a numerical factor depending on the beam shape of the beam cross-sections. All coefficients are assumed to be constant. For details of the model see [107],[153], and for its control see [85], [109], and [112].
     Equations (5.24) and (5.25) can be obtained through Hamilton's principle by using the natural energy of the beam given by :

$$E_B(t) = \frac{1}{2}\int_0^L \{\rho y_t^2 + I_\rho \phi_t^2 + K(\phi - y_x)^2 + EI\phi_x^2\}dx. \tag{5.26}$$

In (5.26), the first two terms in the integral represent the kinetic energy due to translation and rotation, and the last two terms represent the potential energy due to shearing deformation and bending, respectively, see [107].

The boundary conditions we have are :

$$y(0,t) = 0, \quad \phi(0,t) = 0, \tag{5.27}$$

$$-K(\phi(L,t) - y_x(L,t)) = f(t), \quad EI\phi_x(L,t) = g(t), \tag{5.28}$$

where $f(t)$ and $g(t)$ are the control force and torque applied at the boundary $x = L$.

For notational convenience, we will normalize all coefficients in (5.24)-(5.28) to unity. We note that all of the results we present for the normalized case also hold when all of these coefficients are positive, but are otherwise arbitrary. Hence, we consider the following equations

$$\begin{cases} y_{tt} - (y_{xx} - \phi_x) = 0, \quad 0 < x < 1, \quad t \geq 0, \\ \phi_{tt} - \phi_{xx} + (\phi - y_x) = 0, \\ y(0,t) = 0, \quad \phi(0,t) = 0, \\ y_x(1,t) - \phi(1,t) = f(t), \quad \phi_x(1,t) = g(t). \end{cases} \tag{5.29}$$

To analyze the system given by (5.29) using the passivity approach, we first define $H = L^2(0,1) \times L^2(0,1)$, and the operator A in H as

$$A \begin{bmatrix} u \\ \phi \end{bmatrix} = \begin{bmatrix} \phi_x - u_{xx} \\ -\phi_{xx} + \phi - u_x \end{bmatrix}, \quad \forall w = \begin{bmatrix} u \\ \phi \end{bmatrix} \in D(A), \tag{5.30}$$

with $D(A)$ chosen as

$$D(A) = \{(u, \phi)^T \in H \mid u', u'', \phi', \phi'' \in L^2(0,1)\}. \tag{5.31}$$

Obviously, $D(A)$ is dense in H; hence assumption 5.1 is satisfied. From the boundary conditions in (5.29), we observe that $k = r = 0$, $l = p = 2$, $B_1^2 w = -\phi + u_x$, $B_2^2 w = \phi_x$, $B_1^3 w = u$, $B_2^3 w = \phi$, $f_1^2(t) = f(t)$ and $f_2^2(t) = g(t)$. Accordingly, we have

$$S_1 = \{w \in H \mid \phi(1) - u_x(1) = 0, \quad \phi_x(1) = 0\}, \tag{5.32}$$

$$S_2 = \{w \in H \mid u(0) = \phi(0) = 0\}, \tag{5.33}$$

$$D(A_{uc}) = \{w \in H \mid w \in D(A), u(0) = \phi(0) = 0,$$
$$\phi(1) - u_x(1) = 0, \quad \phi_x(1) = 0\}. \tag{5.34}$$

To check assumption 5.2, let us define $w_1 = (u_1, \phi_1)^T, w_2 = (u_2, \phi_2)^T \in D(A_{uc})$. By using integration by parts, we obtain

$$
\begin{aligned}
\langle w_1, Aw_2 \rangle_H &= \int_0^1 u_1(\phi_{2x} - u_{2xx})dx + \int_0^1 \phi_1(-\phi_{2xx} + \phi_2 - u_{2x})dx \\
&= \int_0^1 \phi_{1x}\phi_{2x}dx + \int_0^1 (\phi_1 - u_{1x})(\phi_2 - u_{2x})dx \\
&= \langle Aw_1, w_2 \rangle_H,
\end{aligned}
\tag{5.35}
$$

hence, A_{uc} is symmetric in H. Note that for $w \in D(A_{uc})$ we have

$$
\langle w, A_{uc}w \rangle_H = \int_0^1 \phi_x^2 dx + \int_0^1 (\phi - u_x)^2 dx.
\tag{5.36}
$$

To check (5.13), note that since $u(0) = \phi(0) = 0$ for $w \in D(A_{uc})$, by using Cauchy-Schwartz inequality, we have

$$
\int_0^1 u^2 dx \leq \int_0^1 u_x^2 dx, \qquad \int_0^1 \phi^2 dx \leq \int_0^1 \phi_x^2 dx.
\tag{5.37}
$$

Applying the following simple inequality

$$
\mid ab \mid \leq 2 \mid ab \mid \leq \delta a^2 + \frac{b^2}{\delta}, \qquad a,b \in \mathbf{R}, \delta > 0,
\tag{5.38}
$$

and (5.37), we obtain

$$
\int_0^1 \phi_x^2 dx + \int_0^1 (\phi - u_x)^2 dx \geq (2 - \delta) \int_0^1 \phi^2 dx + (1 - \frac{1}{\delta}) \int_0^1 u^2 dx,
\tag{5.39}
$$

where $\delta > 0$ is an arbitrary constant. By choosing $1 < \delta < 2$, it follows that (5.13) is satisfied with $\alpha = \min\{1 - 1/\delta, 2 - \delta\}$. Hence, assumption 5.2 also is satisfied. Note that for $w \in D(A_{uc})$ we have

$$
\langle A_{uc}^{1/2}w, A_{uc}^{1/2}w \rangle_H = \langle w, A_{uc}w \rangle_H.
\tag{5.40}
$$

Therefore, we may choose V as

$$
V = \{w \in H \mid u', \phi' \in L_2(0,1), u(0) = \phi(0) = 0\};
\tag{5.41}
$$

hence, assumption 5.3 is satisfied.

Now consider the operator $\mathcal{A}$ given by (5.17) where A is given by (5.30) and $X = V \times H$, $D(\mathcal{A}) = D(A_{uc}) \times V$. Let $w = (u, \phi)^T, w_i = (u_i, \phi_i)^T \in V$, $\omega = (v, \xi)^T, \omega_i = (v_i, \xi_i)^T \in H$ for $i = 1, 2$ and set $z = (w, \omega)^T, z_i = (w_i, \omega_i)^T \in X$. Then by using (5.18), (5.19), (5.36), and (5.40) we obtain

$$
\begin{aligned}
\langle z_1, z_2 \rangle_X &= \int_0^1 \phi_{1x}\phi_{2x}dx + \int_0^1 (\phi_1 - u_{1x})(\phi_2 - u_{2x})dx \\
&\quad + \int_0^1 v_1 v_2 dx + \int_0^1 \xi_1 \xi_2 dx,
\end{aligned}
\tag{5.42}
$$

$$\|z\|_X^2 = \int_0^1 \phi_x^2 dx + \int_0^1 (\phi - u_x)^2 dx + \int_0^1 v^2 dx + \int_0^1 \xi^2 dx. \qquad (5.43)$$

Note that the natural energy of the Timoshenko beam given by (5.26) is the same as half of the norm given by (5.43) when $z = (u, \phi, u_t, \phi_t)^T$ is a solution of (5.29). To check assumption 5.4, first note that $D_1 = \{(u, \phi)^T \in D(A) \mid u(0) = \phi(0) = 0\}$, and $D = D_1 \times V$. For $w = (u, \phi)^T \in D_1, \omega = (v, \xi)^T \in V$ and $z = (w, \omega)^T \in D$, by using (5.17), (5.30), and (5.42), and by integrating by parts we obtain

$$
\begin{aligned}
\langle z, \mathcal{A}z \rangle_X &= \int_0^1 \phi_x \xi_x dx + \int_0^1 (\phi - u_x)(\xi - v_x) dx \\
&\quad + \int_0^1 v(u_{xx} - \phi_x) dx + \int_0^1 \xi(\phi_{xx} - \phi + u_x) dx \\
&= (u_x(1) - \phi_1(1))v(1) + \phi_x(1)\xi(1) \\
&= (B_1^2 w)(1)v(1) + (B_2^2 w)(1)\xi(1).
\end{aligned}
\qquad (5.44)
$$

Note that (5.44) has the same form as (5.20), cf. (5.1); hence assumption 5.4 is satisfied with $O_1^2 w = v$ and $O_2^2 w = \xi$. Therefore, according to our passivity formulation we should choose the outputs as

$$y_1^2(t) = y_t(1, t), \quad y_2^2(t) = \phi_t(1, t). \qquad (5.45)$$

Remark 5.7 *Similar analyzes also apply to wave equations and Euler-Bernoulli beam equations with boundary control.*

Let us consider the system given by (5.4)-(5.8) and assume that assumptions 5.1-5.4 hold. Here, f_i^j are the inputs, and the outputs are chosen as in (5.22), (5.23). We will denote the resulting system as $\mathcal{S}$. In this framework, the power form given by (5.20) takes the following form

$$\langle z, \mathcal{A}z \rangle_X = \sum_{i=1}^k f_i^1 y_i^1 + \sum_{i=1}^l f_i^2 y_i^2. \qquad (5.46)$$

For the system $\mathcal{S}$, the control problem we consider can be stated as follows : Find appropriate control laws for f_i^j by using the outputs y_i^j such that the resulting closed-loop system is well-posed and asymptotically stable. The block diagram of such a feedback configuration can be seen in Figure 5.1. Here, $\mathcal{C}$ represents the controller. While it is possible to use a general controller which relates the set of outputs to the set of inputs, here we will consider a simple choice in which f_i^j is related only to y_i^j as follows

$$f_i^j = -\alpha_i^j y_i^j, \qquad (5.47)$$

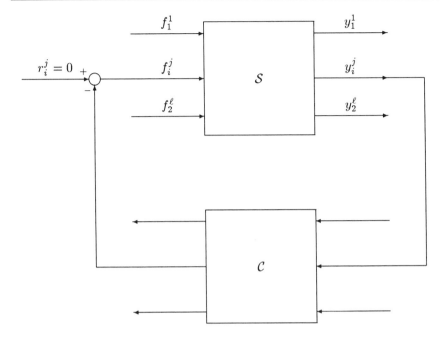

Figure 5.1: General Control System

where $\alpha_i^j \geq 0$, (cf. (5.2)). Such a selection is quite natural because the power form (5.46) becomes

$$\langle z, \ \mathcal{A}z\rangle_X = -\sum_{i=1}^{k} \alpha_i^1 (y_i^1)^2 - \sum_{i=1}^{l} \alpha_i^2 (y_i^2)^2. \tag{5.48}$$

Hence $\mathcal{A}$ becomes a dissipative operator with this controller. This property is of crucial importance in proving both the well-posedness of the closed-loop system and its asymptotic stability. For the asymptotic stability, in the sequel we will show that if we define the energy of the system $\mathcal{S}$ as $E(t) = \frac{1}{2} \|z(t)\|_X^2$, where $z(t)$ is a solution of the system, then the rate of energy is given by (5.48), cf. (5.3). If we can apply LaSalle's invariance principle, see theorem 3.64, then we can conclude that all solutions of the system $\mathcal{S}$ asymptotically tend to the maximal invariant set contained in

$$\mathcal{O} = \{z \in X \mid \langle z, \ \mathcal{A}z\rangle_X = 0\}. \tag{5.49}$$

Note that in the set $\mathcal{O}$, for any invariant solution we have $f_i^j(t) = 0$, and for any $\alpha_i^j > 0$ we have $y_i^j(t) = 0$ as well. If we can prove that under these conditions the only possible solution of the system $\mathcal{S}$ is the zero solution, then by LaSalle's invariance principle, we conclude that all solutions of the system

S asymptotically decay to zero. We note that in this case the inputs and the relevant outputs of the system S are zero, and the question of asymptotic stability is then related to the observability, see [41].

By using (5.22), (5.23) and (5.47) in (5.5), (5.6), we obtain

$$(B_i^1 w + \alpha_i^1 O_i^1 w_t)(0) = 0, \quad i = 1, \ldots, k, \tag{5.50}$$

$$(B_i^2 w + \alpha_i^2 O_i^2 w_t)(1) = 0, \quad i = 1, \ldots, l. \tag{5.51}$$

To account for the boundary conditions (5.50) and (5.51) of the closed-loop system, we define the following set

$$\begin{aligned} S_{1c} = \{(u, v)^T \in H \times H \mid (B_i^1 u + \alpha_i^1 O_i^1 v)(0) = 0, \quad i = 1, \ldots, k, \\ (B_i^2 u + \alpha_i^2 O_i^2 v)(1) = 0, \quad i = 1, \ldots, l\}, \end{aligned} \tag{5.52}$$

and define $D(A_c)$ as

$$D(A_c) = D(A) \cap S_2, \tag{5.53}$$

where S_2 is given by (5.10). By using the notation given above, the system S with the control law given by (5.47) can be rewritten as

$$\dot{z} = \mathcal{A}z, \quad z(0) \in X, \tag{5.54}$$

where $X = V \times H$, the operator $\mathcal{A}$ is given by (5.17) and

$$D(\mathcal{A}) = (D(A_c) \times V) \cap S_{1c}. \tag{5.55}$$

This system will be denoted by S_c.

Example 5.8 *Reconsider Example 5.6. By using (5.29), (5.45), and (5.47) we obtain the following equations :*

$$\begin{cases} y_{tt} - (y_{xx} - \phi_x) = 0, \\ \phi_{tt} - \phi_{xx} + \phi - y_x = 0, \\ y(0, t) = \phi(0, t) = 0, \\ y_x(1, t) - \phi(1, t) = -\alpha_1^2 y_t(1, t), \quad \phi_x(1, t) = -\alpha_2^2 \phi_t(1, t). \end{cases} \tag{5.56}$$

By using (5.52), (5.53) we obtain

$$\begin{aligned} S_{1c} = \{(u, \phi, v, \xi)^T \in H \times H \mid u'(1) - \phi(1) + \alpha_1^2 v(1) = 0, \\ \phi'(1) + \alpha_2^2 \xi(1) = 0\}, \end{aligned}$$

$$D(A_c) = \{(u, \phi)^T \in H \mid (u, \phi)^T \in D(A), u(0) = \phi(0) = 0\}, \tag{5.57}$$

where A, H, V and $D(A)$ are as given in Example 5.6. Hence, the system (5.56) can be put into the abstract form (5.54) with $X = V \times H$ and

$$\begin{aligned} D(\mathcal{A}) = \{(u, \phi, v, \xi)^T \in X \mid (u, \phi)^T \in D(A_c), (v, \xi)^T \in V, \\ u'(1) - \phi(1) + \alpha_1^2 v(1) = 0, \phi'(1) + \alpha_2^2 \xi(1) = 0\}. \end{aligned} \tag{5.58}$$

Now we consider the system $\mathcal{S}_c$ given by (5.54). For this system, we make the following assumption.

Assumption 5.9 *The operator* $\lambda I - \mathcal{A} : D(\mathcal{A}) \subset X \to X$ *is onto for all* $\lambda > 0$.

With these assumptions, we are ready to establish the following theorem.

Theorem 5.10 *Consider the system* $\mathcal{S}_c$ *given by (5.54) and let assumptions 5.1-5.4,5.9 hold. Then the operator* $\mathcal{A}$ *generates a* C_0-semigroup *of contractions* $T(t)$ *on* X. *If* $z(0) \in D(\mathcal{A})$, *then* $z(t) = T(t)z(0)$ *is the unique classical solution of (5.54) and* $z(t) \in D(\mathcal{A})$ *for* $t \geq 0$. *If* $z(0) \in X$, *then* $z(t) = T(t)z(0)$ *is the unique weak solution of (5.54).*

Proof. Since (5.48) holds, it follows that $\mathcal{A}$ is a dissipative operator on X. From assumption 5.9 and the Lümer-Phillips Theorem, see Theorem 2.27, it follows that $\mathcal{A}$ generates a C_0-semigroup of contractions on X. The rest follows from the standard properties of C_0-semigroups, see Chapter 2. □

To prove the asymptotic stability, we need the following assumptions.

Assumption 5.11 *The operator* $(\lambda I - \mathcal{A})^{-1} : X \to X$ *is compact for* $\lambda > 0$.

Assumption 5.12 *The only invariant solution of (5.54) in the set* $\mathcal{S}_1 \cap \mathcal{S}_2 \cap \mathcal{S}_3$ *is the zero solution, where* $\mathcal{S}_1$ *and* $\mathcal{S}_2$ *are given by (5.9), (5.10), and* $\mathcal{S}_3$ *is given by*

$$
\begin{aligned}
\mathcal{S}_3 = \{ (u,\ v)^T \in H \times H \mid (\ O_i^1 v\ )(0) = 0 \text{ for } \alpha_i^1 > 0, i = 1, \ldots, k, \\
(\ O_i^2 v\ )(1) = 0 \text{ for } \alpha_i^2 > 0, i = 1, \ldots, l \}.
\end{aligned} \tag{5.59}
$$

Remark 5.13 *We note that assumption 5.9 is quite natural in proving the existence and uniqueness of the solutions of (5.54) and is satisfied in most of the boundary control problems. The same statement also applies to assumption 5.11, and it can be said that most of the examples encountered in the literature satisfy this assumption, see, e.g., [84, p. 187]. To check assumption 5.12, we need to solve (5.4) with the boundary conditions given by* $\mathcal{S}_1, \mathcal{S}_2$, *and* $\mathcal{S}_3$. *Note that* $\mathcal{S}_3$ *gives extra boundary conditions, and due to these conditions in most of the examples, assumption 5.12 also is satisfied.*

Theorem 5.14 *Let assumptions 5.1-5.4, 5.9, 5.11, and 5.12 hold, consider the system* $\mathcal{S}_c$ *given by (5.54), and let* $T(t)$ *be the unique* C_0-semigroup *generated by* $\mathcal{A}$. *Then, the system* $\mathcal{S}_c$ *is globally asymptotically stable, that is for any* $z(0) \in X$, *the unique (classical or weak) solution* $z(t) = T(t)z(0)$ *of (5.54) asymptotically approaches to zero, i.e.,* $\lim_{t \to \infty} \|z(t)\|_X = 0$.

Proof. By Theorem 5.10, the operator $\mathcal{A}$ generates a unique C_0-semigroup of contractions $T(t)$ on X, and for any $z(0) \in X$, $z(t) = T(t)z(0)$ is the unique (classical or weak) solution of (5.54). Since $T(t)$ is a contraction, for any $z(0) \in X$, the solution $z(t)$ (classical or weak) is bounded in X. Then, by assumption 5.11 it follows that for any $z(0) \in X$, the trajectory of the solution $z(t)$, i.e., the set $\gamma(z(0)) = \{z(t) \in X, t \geq 0\}$ is precompact in X. It then follows that the ω-limit set $\omega(z(0))$ of the trajectory in question is nonempty, compact, and we have $z(t) \to \omega(z(0))$ as $t \to \infty$, see Theorem 3.61.

Next, we will show that $\omega(z(0))$ contains only the point zero. Since $T(t)$ is a contraction, it suffices to prove this statement for $z(0) \in D(\mathcal{A})$, the case $z(0) \in X$ will be treated then. Let $z(0) \in D(\mathcal{A})$; hence, we have $z(t) = T(t)z(0) \in D(\mathcal{A})$ for $t \geq 0$. Let us define the following energy function $E(t)$ on X

$$E(t) = \frac{1}{2}\langle z(t), z(t)\rangle_X = \frac{1}{2}\|z(t)\|_X^2. \tag{5.60}$$

Since $z(0) \in D(\mathcal{A})$, it follows from Theorem 2.12 that $z(t)$ is differentiable and that $\dot{z}(t) = \mathcal{A}z(t)$. Hence, $E(t)$ is differentiable, and by using (5.48) we obtain

$$\begin{aligned} \dot{E}(t) &= \langle z(t), \mathcal{A}z(t)\rangle_X \\ &= -\sum_{i=1}^{k} \alpha_i^1 \left(y_i^1(t)\right)^2 - \sum_{i=1}^{l} \alpha_i^2 \left(y_i^2(t)\right)^2. \end{aligned} \tag{5.61}$$

Now consider the set $\mathcal{O}$ given by (5.49) and let $\mathcal{M}$ be the largest invariant subset of $\mathcal{O}$. Since $\gamma(z(0))$ is precompact, it then follows from LaSalle's invariance principle that $z(t) \to \mathcal{M}$ as $t \to \infty$, see Theorem 3.64 in Chapter 3. We show that $\mathcal{M} = \{0\}$. Since $z(0) \in D(\mathcal{A})$, we have $\mathcal{M} \subset D(\mathcal{A})$, see [37], [42, Thm. 5]. Let $y \in \mathcal{M}$ and let $\hat{z}(t) = T(t)y$ be the corresponding solution. Since $\mathcal{M}$ is invariant, we have $\hat{z}(t) \in \mathcal{M}$ for $t \geq 0$. Hence, from (5.49) and (5.61) it follows that $\hat{z}(t)$ is the solution of (5.54) satisfying $\hat{z}(t) \in \mathcal{S}_1 \cap \mathcal{S}_2 \cap \mathcal{S}_3$ for $t \geq 0$. By assumption 5.12, we must have $\hat{z}(t) = 0$ for $t \geq 0$. It then follows that $y = 0$; hence, $\mathcal{M} = \{0\}$. Thus, $z(t) \to 0$ as $t \to \infty$ for any $z(0) \in D(\mathcal{A})$.

We prove that the same statement holds for $z(0) \in X$. Let $z(0) \in X$ and let $z(t) = T(t)z(0)$ be the corresponding (weak) solution of (5.54). Let $z_n \in D(\mathcal{A})$ be a sequence such that $\lim_{n\to\infty} z_n = z(0)$. Since $D(\mathcal{A})$ is dense in X, such a sequence exists. Then we have

$$\begin{aligned} \|T(t)z(0)\| &\leq \|T(t)(z(0) - z_n)\| + \|T(t)z_n\| \\ &\leq \|z(0) - z_n\| + \|T(t)z_n\|, \end{aligned} \tag{5.62}$$

where in the second step we used the fact that $T(t)$ is a contraction, i.e., $\|T(t)\| \leq 1$. Since $z_n \in D(\mathcal{A})$, we have $\lim_{t\to\infty} T(t)z_n = 0$. Consequently, $\lim_{t\to\infty} T(t)z(0) = 0$. $\quad\square$

We note that Theorem 5.14 only guarantees asymptotic stability, but not exponential (or uniform) stability. We now apply the energy multiplier method to deduce the latter.

Theorem 5.15 *Consider the system $\mathcal{S}_c$ given by (5.54) and let assumptions 5.1-5.4 and 5.9 hold. Let $T(t)$ be the C_0-semigroup of contractions generated by $\mathcal{A}$. Let $z = (u, v)^T \in X$ and let us define the projections $P_1 : X \to V$, $P_2 : X \to H$ as $P_1 z = u$, $P_2 z = v$. Let $z(0) \in D(\mathcal{A})$ and let $z(t)$ denote the solution of (5.54). Assume that there exists a linear map $Q : H \to H$ such that the following holds*

$$| \langle P_2 z(t), Q P_1 z(t) \rangle_H | \le CE(t), \tag{5.63}$$

$$\frac{d}{dt} \langle P_2 z(t), Q P_1 z(t) \rangle_H \le -E(t) + \sum_{i=1}^{k} a_i^1 (f_i^1)^2 + \sum_{i=1}^{l} a_i^2 (f_i^2)^2, \tag{5.64}$$

where $C > 0$ and a_i^j are arbitrary constants. Then $T(t)$ decays exponentially to zero, i.e., there are $M > 0$ and $\delta > 0$ such that

$$\|T(t)\| \le M e^{-\delta t}. \tag{5.65}$$

Proof. Let us define the following function

$$V(t) = tE(t) + \langle P_2 z(t), Q P_1 z(t) \rangle_H. \tag{5.66}$$

From (5.63) we obtain the following estimate for $t \ge 0$

$$(t - C)E(t) \le V(t) \le (t + C)E(t). \tag{5.67}$$

Let us first assume that $z(0) \in D(\mathcal{A})$. Hence, the unique (classical) solution of (5.54) is given by $z(t) = T(t)z(0)$ and is differentiable. Evaluating the time derivative of (5.66) and using (5.47), (5.61), (5.64) leads to

$$\begin{aligned}
\dot{V}(t) &= t\dot{E}(t) + E(t) + \frac{d}{dt} \langle P_2 z(t), Q P_1 z(t) \rangle_H \\
&\le -\sum_{i=1}^{k} \alpha_i^1 (t - \alpha_i^1 a_i^1)(y_i^1(t))^2 - \sum_{i=1}^{l} \alpha_i^2 (t - \alpha_i^2 a_i^2)(y_i^2(t))^2.
\end{aligned} \tag{5.68}$$

If we choose

$$T = \max\{\alpha_i^1 a_i^1, \ \alpha_i^2 a_i^2\}, \tag{5.69}$$

then

$$\dot{V}(t) \le 0, \quad t \ge T. \tag{5.70}$$

In view of (5.61), (5.67), and (5.70), we obtain

$$(t - C)E(t) \le V(t) \le V(T) \le (T + C)E(T) \le (T + C)E(0), \tag{5.71}$$

which implies the following

$$E(t) \leq \frac{T+C}{t-C} E(0), \quad t \geq \max\{T, C\}. \tag{5.72}$$

Substitution of (5.60) and $z(t) = T(t)z(0)$ into (5.72) gives

$$\|T(t)z(0)\| \leq \sqrt{\frac{T+C}{t-C}} \, \|z(0)\|, \quad t \geq \max\{T, C\}, \tag{5.73}$$

for $z(0) \in D(\mathcal{A})$. We show that (5.73) holds for $z(0) \in X$ as well. To see this, let $z(0) \in X$ be given and let $z_n \in D(\mathcal{A})$ be a sequence such that $z_n \to z(0)$ as $n \to \infty$. Then we have

$$
\begin{aligned}
\|T(t)z(0)\| &\leq \|T(t)(z(0) - z_n)\| + \|T(t)z_n\| \\
&\leq \|z(0) - z_n\| + \sqrt{\frac{T+C}{t-C}} \, \|z_n\| \\
&\leq (1 + \sqrt{\frac{T+C}{t-C}}) \|z(0) - z_n\| + \sqrt{\frac{T+C}{t-C}} \, \|z(0)\|, \quad (5.74)
\end{aligned}
$$

where we used the fact that $T(t)$ is a contraction. Passing to the limit $n \to \infty$ we obtain

$$\|T(t)z(0)\| \leq \sqrt{\frac{T+C}{t-C}} \, \|z(0)\|, \quad t \geq \max\{T, C\}. \tag{5.75}$$

Hence, for sufficiently large t we have $\|T(t)\| < 1$. It then follows from the Corollary 2.11 that (5.65) holds. $\qquad\square$

In the sequel we apply Theorems 5.10, 5.14 and 5.15 to Example 5.8.

Theorem 5.16 *Consider the system given by (5.56). Note that this system can be put into the form given by (5.54) where the operator $\mathcal{A}$ is given by (5.17). Let $\alpha_1^2 \geq 0, \alpha_2^2 \geq 0$. Then we have the following*

(i) *The operator $\mathcal{A}$ generates a C_0-semigroup of contractions $T(t)$ in X.*

(ii) *Let $\alpha_1^2 > 0, \alpha_2^2 > 0$. For any $z(0) \in X$, the (classical or weak) solutions of (5.54) asymptotically decay to zero.*

(iii) *For the case (ii), the semigroup $T(t)$ is exponentially stable, i.e., (5.65) holds.*

Remark 5.17 *Note that (iii) implies (ii). We included the latter since we need such a result in the following sections.*

Proof. (i) We will show that assumption 5.9 holds. Let $y = (f_1, f_2, h_1, h_2)^T \in X$ be given. We need to show that for some $z = (u_1, u_2, \phi_1, \phi_2)^T \in D(\mathcal{A})$, $(\lambda I - \mathcal{A})z = y$ holds for $\lambda > 0$. To prove this claim, let us write $\mathcal{A} = \mathcal{A}_u + \mathcal{A}_b$ where $\mathcal{A}_u$ is a linear and unbounded operator on X with $D(\mathcal{A}_u) = D(\mathcal{A})$ and

$$\mathcal{A}_u z = \begin{bmatrix} u_2 \\ u_1'' \\ \phi_2 \\ \phi_1'' \end{bmatrix}. \tag{5.76}$$

Note that $\mathcal{A}_b$ is a bounded operator on X. First we will show that assumption 5.9 holds for $\mathcal{A}_u$. To see this, consider $(\lambda I - \mathcal{A}_u)z = y$, which is equivalent to solving the following equations

$$\lambda u_1 - u_2 = f_1, \tag{5.77}$$

$$\lambda u_2 - u_1'' = f_2, \tag{5.78}$$

$$\lambda \phi_1 - \phi_2 = h_1, \tag{5.79}$$

$$\lambda \phi_2 - \phi_1'' = h_2, \tag{5.80}$$

$$u_1(0) = u_2(0) = 0, \quad \phi_1(0) = \phi_2(0) = 0, \tag{5.81}$$

$$u_1'(1) - \phi_1(1) + \alpha_1^2 u_2(1) = 0, \quad \phi_1'(1) + \alpha_1^2 \phi_2(1) = 0. \tag{5.82}$$

We show that (5.77)-(5.82) admits a unique solution in $D(\mathcal{A})$. In fact, u_1 and ϕ_1 are given by

$$u_1(x) = C_1 \sinh \lambda x - \frac{1}{\lambda} \int_0^x [f_2(s) + \lambda f_1(s)] \sinh \lambda(x - s) ds, \tag{5.83}$$

$$\phi_1(x) = C_2 \sinh \lambda x - \frac{1}{\lambda} \int_0^x [h_2(s) + \lambda h_1(s)] \sinh \lambda(x - s) ds, \tag{5.84}$$

where C_1 and C_2 are constants. Then u_2 and ϕ_2 can be found from (5.77), (5.79) and the constants C_1 and C_2 can be found from (5.82). After straightforward computations, it can easily be shown that C_1 and C_2 are uniquely determined if

$$\cosh \lambda + \alpha_1^2 \sinh \lambda \neq 0, \quad \cosh \lambda + \alpha_2^2 \sinh \lambda \neq 0, \tag{5.85}$$

which trivially holds since $\lambda > 0$. Hence $\lambda I - \mathcal{A}_u : X \to X$ is onto for $\lambda > 0$. Furthermore, we can show that $(\lambda I - \mathcal{A}_u)^{-1}$ is compact for $\lambda > 0$. Since $\mathcal{A}_b$ is bounded, it follows that $T = (\lambda I - \mathcal{A}_u)^{-1}\mathcal{A}_b$ also is compact. We claim that $I - T$ also is invertible. Since T is compact, it suffices to show that $\mu = 1$ is not an eigenvalue of T, i.e., $(I - T)y = 0$ does not have a nontrivial solution, see [84, p. 185]. If this holds, then by multiplying with $(\lambda I - \mathcal{A}_u)$, we obtain $y = \mathcal{A}y$. But since $\mathcal{A}$ is dissipative, we must have $y = 0$, which is a contradiction. Hence, $(I - T)^{-1}$ exists and we have

$$(\lambda I - \mathcal{A})^{-1} = (I - T)^{-1}(\lambda I - \mathcal{A}_u)^{-1}; \tag{5.86}$$

hence, $\lambda I - \mathcal{A}$ is onto for $\lambda > 0$. The result then follows from Theorem 5.4.

(ii) From (5.86) it follows that $(\lambda I - \mathcal{A})^{-1}$ is compact for $\lambda > 0$. To show that assumption 5.12 holds, we need to solve the following equations

$$y_{tt} - (y_{xx} - \phi_x) = 0, \tag{5.87}$$

$$\phi_{tt} - \phi_{xx} + \phi - y_x = 0, \tag{5.88}$$

$$y(0, t) = \phi(0, t) = 0, \quad y_x(1, t) - \phi(1, t) = \phi_x(1, t) = 0, \tag{5.89}$$

with the following extra conditions

$$y_t(1, t) = \phi_t(1, t) = 0. \tag{5.90}$$

Let us set $w = (y, \phi)^T$. By using separation of variables, we express the solutions in the form $w(x, t) = e^{\lambda t}\psi(x)$, where by a slight abuse of notation we set $\psi(x) = (u, \phi)^T$ as well. By using this form in (5.87), (5.88) and setting $\mu = -\lambda^2$, we obtain $A\psi = \mu\psi$, i.e., μ is an eigenvalue of A. Since A with the boundary conditions in (5.89), i.e.,

$$u(0) = \phi(0) = 0, \quad u'(1) = \phi'(1) = 0, \quad u(1) = \phi(1) = 0, \tag{5.91}$$

is coercive, see assumption 5.2, (5.13), it follows that $\mu > 0$. By multiplying $A\psi = \mu\psi$ with ψ' and after integration, we obtain

$$(u')^2 + (\phi')^2 + \mu u^2 + (\mu - 1)\phi^2 = C, \tag{5.92}$$

where C is an integration constant. Applying (5.91) in (5.92) yields $C = 0$. By using (5.36) and $A\psi = \mu\psi$, we obtain

$$\begin{aligned}
\langle \psi, A\psi \rangle_X &= \int_0^1 (\phi')^2 dx + \int_0^1 (\phi - u')^2 dx \\
&= \mu \int_0^1 u^2 dx + \mu \int_0^1 \phi^2 dx.
\end{aligned} \tag{5.93}$$

Using (5.92) with $C = 0$ in this equation yields

$$2\int_0^1 [(u')^2 + (\phi')^2]dx - 2\int_0^1 u'\phi dx = 0, \tag{5.94}$$

from which it follows that

$$\int_0^1 [(u')^2 + (\phi')^2]dx \leq 0. \tag{5.95}$$

That is, $\psi = 0$ or $w = 0$. The result then follows from Theorem 5.14.

(iii) Note that the energy given by (5.60) is

$$E(t) = \frac{1}{2} \int_0^1 [u_t^2 + \phi_t^2 + (u_x - \phi)^2 + \phi_x^2] dx, \qquad (5.96)$$

see (5.36), (5.19). We will use the following function $V(t)$,

$$V(t) = tE(t) + I_1(t) + I_2(t) \qquad (5.97)$$

where

$$I_1(t) = \int_0^1 u_t(\epsilon x u_x - \delta u) dx, \quad I_2(t) = \int_0^1 \phi_t(\epsilon x \phi_x + \delta \phi) dx, \qquad (5.98)$$

and $\epsilon > 0$ and $\delta > 0$ are constants to be determined later. Note that (5.97) has the same form as (5.66). Using (5.37) and (5.38) we can easily derive the following inequality.

$$\int_0^1 \phi_x^2 dx + \int_0^1 (\phi - u_x)^2 dx \geq (2 - \delta_0) \int_0^1 \phi^2 dx + (1 - \frac{1}{\delta_0}) \int_0^1 u_x^2 dx, \qquad (5.99)$$

where $1 < \delta_0 < 2$ is an arbitrary constant. In view of (5.38), (5.96), and (5.99), it can be verified that

$$|I_1(t)| \leq L_1 E(t), \qquad (5.100)$$

$$|I_2(t)| \leq L_2 E(t), \qquad (5.101)$$

where $L_1 > 0$ and $L_2 > 0$ are appropriate constants. This shows that (5.67) holds for $C = L_1 + L_2$.

To complete the proof, we need to calculate the time derivative of $V(t)$. To this end, differentiating (5.98), using (5.56), and integrating by parts we obtain

$$
\begin{aligned}
\dot{I}_1 + \dot{I}_2 = & -(\frac{\epsilon}{2} - \delta) \int_0^1 u_x^2 dx - (\frac{\epsilon}{2} + \delta) \int_0^1 u_t^2 dx - (\frac{\epsilon}{2} + \delta) \int_0^1 \phi_x^2 dx \\
& -(\frac{\epsilon}{2} - \delta) \int_0^1 \phi_t^2 dx + (\frac{\epsilon}{2} - \delta) \int_0^1 \phi^2 dx \\
& +\frac{\epsilon}{2} u_x^2(1,t) - \frac{\epsilon}{2} \phi(1,t)^2 - \delta(u_x(1,t) - \phi(1,t)) u(1,t) \\
& +\frac{\epsilon}{2} (u_t^2(1,t) + \phi_t^2(1,t)) + \delta \phi_x(1,t) \phi(1,t).
\end{aligned} \qquad (5.102)
$$

By using (5.37), (5.38), (5.56) and integrating by parts, we obtain the following inequalities

$$u_x^2(1,t) \leq \phi^2(1,t) + (1 + \frac{1}{\delta_1})(f_1^2(t))^2 + \delta_1 \int_0^1 \phi_x^2 dx, \qquad (5.103)$$

$$(u_x(1,t) - \phi(1,t))u(1,t) \le \frac{1}{\delta_1}(f_1^2(t))^2 + \delta_1 \int_0^1 u_x^2 dx, \qquad (5.104)$$

$$\phi_x(1,t)\phi(1,t) \le \frac{1}{\delta_1}(f_2^2(t))^2 + \delta_1 \int_0^1 \phi_x^2 dx, \qquad (5.105)$$

where $\delta_1 > 0$ is an arbitrary constant. Plugging (5.103)-(5.105) into (5.102) gives

$$\dot{I}_1 + \dot{I}_2 \le \frac{\epsilon}{2}[1 + \frac{1}{\delta_1} + \frac{\delta}{\epsilon\delta_1} + \frac{1}{(\alpha_1^2)^2}](f_1^2(t))^2 + [\frac{\epsilon}{2(\alpha_2^2)^2} + \frac{\delta}{\delta_1}](f_2^2(t))^2$$
$$-(\frac{\epsilon}{2} - \delta - \delta_1)\int_0^1 u_x^2 dx - (\frac{\epsilon}{2} - \delta)\int_0^1 \phi_t^2 dx$$
$$-(\frac{\epsilon}{2} + \delta)\int_0^1 u_t^2 dx - (2\delta - 2\delta_1)\int_0^1 \phi_x^2 dx. \qquad (5.106)$$

From (5.96) we obtain

$$E(t) \le \frac{1}{2}\int_0^1 [u_t^2 + \phi_t^2 + u_x^2 + 3\phi_x^2]dx. \qquad (5.107)$$

Hence, if the constants in (5.106) are chosen such that

$$\frac{\epsilon}{2} + \delta > \frac{1}{2} \;,\;\; \frac{\epsilon}{2} - \delta > \frac{1}{2} \;,\;\; 2\delta - 2\delta_1 > \frac{3}{2} \;,\;\; \frac{\epsilon}{2} - \delta - \delta_1 > 1, \qquad (5.108)$$

then it follows from (5.106) and (5.107) that (5.64) holds. By choosing $\delta_1 > 0$ arbitrarily small, these inequalities are satisfied if $4\delta > 3$ and $\epsilon > 2 + 2\delta$. It then follows from Theorem 5.15 that $T(t)$ is exponentially stable. $\qquad \square$

5.2   Dynamic boundary control using positive real controllers

In this section, we generalize the class of static controllers given by (5.47) to the class of positive real controllers. More precisely, we will replace the control law given by (5.47) with

$$\hat{f}_i^j(s) = -\alpha_i^j(s)\hat{y}_i^j(s), \qquad (5.109)$$

where $s \in \mathbf{C}$ is a complex variable, a hat denotes Laplace transform of the corresponding variable, and α_i^j denotes transfer function of the controller which is itself a dynamical system. In this sense, the control law in (5.109) is called *dynamic boundary control*, distinguishing it from the control law given in (5.47) which describes a static relation. We will assume that these transfer functions are positive real functions, see Definition 5.18. We will show that most of the results presented in the preceding section remain valid if the controller given by (5.109) is used instead of (5.47).

5.2.1 Positive real controllers and their characterizations

We note that (5.109) may be viewed as a linear, time-invariant, finite dimensional system whose input is $y_i^j(t)$ and whose output is $-f_i^j(t)$, and that $\alpha_i^j(s)$ is the transfer function of this system. An alternative characterization of such a system may be its state space representation. Let n_i^j be an integer and let $(A_i^j, b_i^j, c_i^j, d_i^j)$ represent a minimal realization of $\alpha_i^j(s)$ where $A_i^j \in \mathbf{R}^{n_i^j \times n_i^j}$ is a constant matrix, $b_i^j, c_i^j \in \mathbf{R}^{n_i^j}$ are constant vectors and $d_i^j \in \mathbf{R}$ is a constant real number. Hence, we have the following

$$\alpha_i^j(s) = c_i^{jT}(sI - A_i^j)^{-1} b_i^j + d_i^j. \tag{5.110}$$

To find such a realization, see, e.g., [25], [80]. The state space representation of the controller can be given as

$$\dot{z}_i^j(t) = A_i^j z_i^j(t) + b_i^j y_i^j(t), \tag{5.111}$$

$$-f_i^j(t) = c_i^{jT} z_i^j(t) + d_i^j y_i^j(t), \tag{5.112}$$

where $z_i^j \in \mathbf{R}^{n_i^j}$ is the state of the minimal realization. Note that when the controller is given by the transfer function $\alpha_i^j(s)$, then the corresponding minimal realization given by (5.111), (5.112) is not unique, but all such realizations are equivalent. More precisely, if $\hat{z}_i^j$ is the state of another minimal realization, then there exists an invertible matrix $T \in \mathbf{R}^{n_i^j \times n_i^j}$ such that $\hat{z}_i^j = T z_i^j$. Conversely, if the controller is given by (5.111)-(5.112), then by taking Laplace transform and using zero initial conditions we obtain (5.109) where $\alpha_i^j(s)$ is given by (5.110).

Definition 5.18 *Let $\alpha(s)$ be a rational function of a complex variable s. The function $\alpha(s)$ is called positive real (PR) if the following hold*

(i) $\alpha(\cdot)$ is analytic in $Re\{s\} > 0$.

(ii) $\alpha(p)$ is real for $p > 0$.

(iii)

$$Re\{\alpha(s)\} \geq 0 \; for \; Re\{s\} > 0. \tag{5.113}$$

The function $\alpha(s)$ is called strictly positive real (SPR) if $\alpha(s - \epsilon)$ is PR for some $\epsilon > 0$.

Condition (i) in Definition 5.18 implies that PR functions do not have any poles in the open right-half complex plane, i.e., any pole p_0 of $\alpha(s)$ necessarily satisfies $Re\{p_0\} \leq 0$. If $\alpha(s)$ is SPR and p_0 is a pole, then by the same argument we must have $\text{Re}\{p_0\} \leq -\epsilon$. Hence, while a PR function may have poles on the imaginary axis, all the poles of SPR functions are in the open left-half complex plane.

The properties of PR functions were originally investigated in the circuit theory literature, where a close connection between passive circuits and PR functions were well established. It has been shown that a function $\alpha(s)$ is PR if and only if it is an impedance (or admittance) function of a passive circuit containing resistors, capacitors and inductors, see ,e.g., [61]. The concept of PR functions was then extended to PR matrices, i.e., square matrices whose entries are rational functions of a complex variable, and it has been shown that such matrices are closely related with passive m-ports, see, e.g., [2], [125]. Around the same time, it was shown that PR functions could be used in the stabilization of some nonlinear feedback systems (e.g., in the context of Lur'e systems) and the work in this area led to the well-known *Kalman-Yakubovich lemma*, [81], [167]. This lemma was then extended to cover the PR matrices and is also known as the *positive real lemma* in the literature, see, e.g., [2], [156]. The PR functions are then studied in the systems and control literature in the context of stabilization of passive systems, see [50], [121], [148], [156].

Condition (5.113), given in Definition 5.18, usually is difficult to establish. The following theorem is useful in determination of PR and SPR functions.

Theorem 5.19 *A rational function $\alpha(s)$ of a complex variable s, with real coefficients, is PR if and only if the following conditions hold*

(i) *$\alpha(s)$ is a stable transfer function, i.e., any pole p_0 of $\alpha(s)$ satisfies $Re\{p_0\} \leq 0$.*

(ii) *The poles of $\alpha(s)$ on the imaginary axis are simple and the associated residues are real and nonnegative.*

(iii)

$$Re\{\alpha(i\omega)\} \geq 0, \tag{5.114}$$

for any $\omega \geq 0$ for which $i\omega$ is not a pole of $\alpha(s)$.

In addition, $\alpha(s)$ is SPR if and only if

(i) *$\alpha(s)$ is strictly stable, i.e., any pole p_0 of $\alpha(s)$ satisfies $Re\{p_0\} < 0$.*

(ii)

$$Re\{\alpha(i\omega)\} > 0, \quad \forall \omega \geq 0, \tag{5.115}$$

Proof.   See [2], [148].   ☐

The main advantage of PR or SPR functions is that they enable one
to use a Lyapunov function, and hence to apply Lyapunov stability theory
easily. The well-known Kalman-Yakubovich lemma is an important tool in
such stability analysis.

Lemma 5.20 (Kalman-Yakubovich) *Let $\alpha(s)$ be a rational function of
a complex variable s with real coefficients and let $\lim_{p\to\infty} \alpha(p) < \infty$. Let
(A, b, c, d) be a minimal realization of $\alpha(s)$, i.e., the triplet (A, b, c) is con-
trollable and observable (see, e.g., [80]) and the following holds*

$$\alpha(s) = c^T (sI - A)^{-1} b + d. \tag{5.116}$$

*If $\alpha(s)$ is SPR, then there exist symmetric and positive definite matrices
P and Q, and a vector q, all having appropriate dimensions, such that the
following hold*

$$A^T P + PA = -qq^T - Q, \tag{5.117}$$

$$Pb - c = \sqrt{2d}q. \tag{5.118}$$

Proof.   See, e.g., [148], [156].   ☐

We note that $\lim_{s\to\infty} c^T (sI - A)^{-1} b = 0$, hence from (5.113) it follows
that $d \geq 0$. This lemma also can be extended to PR functions for which Q
may become positive semi-definite, see [2].

5.2.2   Stability analysis of control systems with SPR controllers

Let us consider the system given by (5.4)-(5.8) with dynamic controllers
specified by (5.109), where α_i^j are SPR functions. Let assumptions 5.1-5.4
hold. We will show that under simple conditions, the feedback structure given
in Figure 5.1 with this class of controllers also has the same properties. Let
$(A_i^j, b_i^j, c_i^j, d_i^j)$ be a minimal realization of $\alpha_i^j(s)$, and consider the controller
given by (5.111), (5.112).

To analyze the closed-loop system with SPR controllers, we will first set
up some notation. As before, we denote by $\mathcal{S}$ the system given by (5.4)-(5.8),
(5.22) and (5.23). If the controller given by (5.47) is used, then the resulting
closed-loop system is denoted by $\mathcal{S}_c$. Note that $\mathcal{S}_c$ can be given by (5.54)
with X and $D(\mathcal{A})$ defined therein.

Now consider the system $\mathcal{S}$ with the controller given by (5.111), (5.112)
where the associated transfer function $\alpha_i^j(s)$ given by (5.110) is SPR. Denote

by $\mathcal{S}_{pr}$ the resulting closed-loop system. To obtain a state-space representation of $\mathcal{S}_{pr}$ similar to (5.54), first note that by using (5.22), (5.23), and (5.112) in (5.5) and (5.6), we obtain

$$(B_i^1 w + d_i^1 O_i^1 w_t)(0) + c_i^{1T} z_i^1 = 0, \quad i = 1, \ldots k, \qquad (5.119)$$

$$(B_i^2 w + d_i^2 O_i^2 w_t)(1) + c_i^{2T} z_i^2 = 0, \quad i = 1, \ldots l. \qquad (5.120)$$

To account for the boundary conditions (5.119) and (5.120), we define the following space

$$X_e = V \times H \times \prod_{i=1}^{k} \mathbf{R}^{n_i^1} \times \prod_{j=1}^{l} \mathbf{R}^{n_j^2}, \qquad (5.121)$$

and the following set $\mathcal{S}_{1pr} \subset X_e$

$$\begin{aligned}
\mathcal{S}_{1pr} = \ & \{ (u, \ v, \ z_1^1, \ldots, z_l^2)^T \in X_e \ | \\
& (B_i^1 u + d_i^1 O_i^1 v)(0) + c_i^{1T} z_i^1 = 0, \quad i = 1, \ldots k, \\
& (B_i^2 u + d_i^2 O_i^2 v)(1) + c_i^{2T} z_i^2 = 0, \quad i = 1, \ldots l. \}. \qquad (5.122)
\end{aligned}$$

In a way similar to that in Section 5.1, we can now formulate the system $\mathcal{S}_{pr}$ as the following state space form

$$\dot{z}_e = \mathcal{A}_e z_e, \quad z_e(0) \in X_e, \qquad (5.123)$$

where $z_e = (w, \ w_t, \ z_1^1, \ldots z_l^2)^T \in X_e$, $\mathcal{A}_e : X_e \to X_e$ is a linear operator defined as

$$\mathcal{A}_e z_e = \begin{bmatrix} v \\ -Au \\ A_1^1 z_1^1 + b_1^1 (O_1^1 v)(0) \\ \vdots \\ A_k^1 z_k^1 + b_k^1 (O_k^1 v)(0) \\ A_1^2 z_1^2 + b_1^2 (O_1^2 v)(1) \\ \vdots \\ A_l^2 z_l^2 + b_l^2 (O_l^2 v)(1) \end{bmatrix}, \qquad (5.124)$$

where $z_e = (u, \ v, \ z_1^1, \ldots z_l^2)^T \in D(\mathcal{A}_e)$ which is defined as

$$D(\mathcal{A}_e) = (D(A_c) \times V \times \prod_{i=1}^{k} \mathbf{R}^{n_i^1} \times \prod_{j=1}^{l} \mathbf{R}^{n_j^2}) \cap \mathcal{S}_{1pr}, \qquad (5.125)$$

where $D(A_c)$ is given by (5.53) and $\mathcal{S}_{1pr}$ is given by (5.122).

Since $\alpha_i^j(s)$ is SPR and $(A_i^j, b_i^j, c_i^j, d_i^j)$ is a minimal realization of it, by Lemma 5.20 the following holds

$$A_i^{jT} P_i^j + P_i^j A_i^j = -q_i^j q_i^{jT} - Q_i^j, \qquad (5.126)$$

$$P_i^j b_i^j - c_i^j = \sqrt{2 d_i^j} q_i^j \tag{5.127}$$

where $j = 1, 2, i = 1, \ldots \{k, l\}$, and P_i^j, Q_i^j are symmetric and positive definite matrices. The generalization of the inner product given by (5.18) for X to X_e is as follows

$$\langle z_e, \hat{z}_e \rangle_{X_e} = \langle z, \hat{z} \rangle_X + \frac{1}{2} \sum_{i=1}^{k} (z_i^{1T} P_i^1 \hat{z}_i^1 + \hat{z}_i^{1T} P_i^1 z_i^1)$$

$$+ \frac{1}{2} \sum_{i=1}^{l} (z_i^{2T} P_i^2 \hat{z}_i^2 + \hat{z}_i^{2T} P_i^2 z_i^2), \tag{5.128}$$

where we set $z_e = (z, z_1^1, \ldots, z_l^2)^T \in X_e$, $\hat{z}_e = (\hat{z}, \hat{z}_1^1, \ldots, \hat{z}_l^2)^T \in X_e$, $z, \hat{z} \in X$ and $\langle z, \hat{z} \rangle_X$ is given by (5.18). Since P_i^j are symmetric and positive definite, (5.128) defines an inner product in X_e. Accordingly, the norm in X_e is given as

$$\|z_e\|_{X_e}^2 = \|z\|_X^2 + \sum_{i=1}^{k} z_i^{1T} P_i^1 z_i^1 + \sum_{i=1}^{l} z_i^{2T} P_i^2 z_i^2. \tag{5.129}$$

Let us now consider the system $\mathcal{S}_c$ given by (5.54). Let $y = (f, g)^T \in X$ and $z = (u, v)^T \in D(\mathcal{A})$ such that $(\lambda I - \mathcal{A})z = y$ holds. By using (5.17), (5.50) and (5.51), this is equivalent to the following

$$\lambda u - v = f, \tag{5.130}$$

$$\lambda v + Au = g, \tag{5.131}$$

$$(B_i^3 u)(0) = 0, \quad (B_i^4 u)(1) = 0, \tag{5.132}$$

$$(B_i^1 u + \alpha_i^1 O_i^1 v)(0) = 0, \quad (B_i^2 u + \alpha_i^2 O_i^2 v)(1) = 0, \tag{5.133}$$

where $u \in D(A)$, $v \in V$ and the range of i should be determined according to (5.5)-(5.8). To guarantee the well-posedness of the system $\mathcal{S}_{pr}$, we need the following assumption.

Assumption 5.21 *Let $y = (f, g)^T \in X$, $\lambda > 0$, and $r_i^j \in \mathbf{R}$ be given. Instead of (5.133), consider the following*

$$(B_i^1 u + \alpha_i^1 O_i^1 v)(0) = r_i^1, \quad (B_i^2 u + \alpha_i^2 O_i^2 v)(1) = r_i^2. \tag{5.134}$$

Under these conditions, (5.130)-(5.132) and (5.134) has solution such that $u \in D(A)$ and $v \in V$.

Remark 5.22 *In most of the cases, the solvability of (5.130)-(5.133) is independent of the right-hand sides of these equations. This is usually the case when A is a differential operator. Hence, we can say that in most of the cases if assumption 5.9 holds, then assumption 5.21 holds as well.*

Theorem 5.23 *Consider the system S_{pr} given by (5.123). Let the transfer functions $\alpha_i^j(s)$ be SPR. Let assumptions 5.1-5.4, 5.9, and 5.21 hold for the associated system S_c. Then the operator A_e generates a C_0-semigroup of contractions $T_e(t)$ on X. If $z_e(0) \in D(A_e)$, then $z_e(t) = T_e(t)z_e(0)$ is the unique classical solution of (5.123) and $z(t) \in D(A)$ for $t \geq 0$. If $z_e(0) \in X_e$, then $z_e(t) = T_e(t)z_e(0)$ is the unique weak solution of (5.123).*

Proof. We will use the Lümer-Phillips theorem (see Theorem 2.27). Let $z_e \in D(A_e)$. By using (5.46), (5.124), (5.126)-(5.128), we obtain the following

$$
\begin{aligned}
\langle z_e, A_e z_e \rangle_{X_e} &= \langle z, Az \rangle_X + \frac{1}{2}\sum_{i=1}^{k}[z_i^{1T} P_i^1 (A_i^1 z_i^1 + b_i^1 y_i^1) \\
&\quad + (z_i^{1T} A_i^{1T} + y_i^1 b_i^{1T}) P_i^1 z_i^1] + \frac{1}{2}\sum_{i=1}^{l}[z_i^{2T} P_i^2 (A_i^2 z_i^2 + b_i^2 y_i^2) \\
&\quad + (z_i^{2T} A_i^{2T} + y_i^2 b_i^{2T}) P_i^2 z_i^2] \\
&= \sum_{i=1}^{k} f_i^1 y_i^1 + \sum_{i=1}^{l} f_i^2 y_i^2 + \frac{1}{2}\sum_{i=1}^{k}[z_i^{1T}(A_i^{1T} P_i^1 + P_i^1 A_i^1) z_i^1 \\
&\quad + 2z_i^{1T} P_i^1 b_i^1 y_i^1] + \frac{1}{2}\sum_{i=1}^{l}[z_i^{2T}(A_i^{2T} P_i^2 + P_i^2 A_i^2) z_i^2 \\
&\quad + 2z_i^{2T} P_i^2 b_i^2 y_i^2] \\
&= -\sum_{i=1}^{k} d_i^1 (y_i^1)^2 - \sum_{i=1}^{l} d_i^2 (y_i^2)^2 - \frac{1}{2}\sum_{i=1}^{k} z_i^{1T} Q_i^1 z_i^1 \\
&\quad - \frac{1}{2}\sum_{i=1}^{l} z_i^{2T} Q_i^2 z_i^2 + \sum_{i=1}^{k} z_i^{1T}(P_i^1 b_i^1 - c_i^1) y_i^1 \\
&\quad + \sum_{i=1}^{l} z_i^{2T}(P_i^2 b_i^2 - c_i^2) y_i^2 - \frac{1}{2}\sum_{i=1}^{k} z_i^{1T} q_i^1 q_i^{1T} z_i^1 \\
&\quad - \frac{1}{2}\sum_{i=1}^{l} z_i^{2T} q_i^2 q_i^{2T} z_i^2 \\
&= -\frac{1}{2}\sum_{i=1}^{k} z_i^{1T} Q_i^1 z_i^1 - \frac{1}{2}\sum_{i=1}^{k}[\sqrt{2d_i^1} y_i^1 - z_i^{1T} q_i^1]^2 \\
&\quad - \frac{1}{2}\sum_{i=1}^{l} z_i^{2T} Q_i^2 z_i^2 - \frac{1}{2}\sum_{i=1}^{l}[\sqrt{2d_i^2} y_i^2 - z_i^{2T} q_i^2]^2, \qquad (5.135)
\end{aligned}
$$

where in the first equality we used (5.124) and (5.128), in the second equality we used (5.46), in the third equality we used (5.112) and (5.126), and in the last equality we used (5.127). It follows from (5.135) that $\mathcal{A}_e$ is dissipative on X_e.

To prove that $\lambda I - \mathcal{A}_e$ is onto for $\lambda > 0$, let $y_e = (f,\ g,\ r_1^1, \dots, r_l^2)^T \in X_e$ be given. We need to find a $z_e = (u,\ v,\ z_1^1, \dots, z_l^2)^T \in D(\mathcal{A}_e)$ such that the following holds

$$(\lambda I - \mathcal{A}_e)z_e = y_e, \tag{5.136}$$

which is equivalent to the following set of equations

$$\lambda u - v = f, \tag{5.137}$$

$$\lambda v + Au = g, \tag{5.138}$$

$$\lambda z_i^1 - A_i^1 z_i^1 = b_i^1 (\ O_i^1 v\ )(0) + r_i^1, \quad i = 1, \dots, k, \tag{5.139}$$

$$\lambda z_i^2 - A_i^2 z_i^2 = b_i^2 (\ O_i^2 v\ )(1) + r_i^2, \quad i = 1, \dots, l, \tag{5.140}$$

$$(\ B_i^1 u\ )(0) + c_i^{1T} z_i^1 + d_i^1 (\ O_i^1 v\ )(0) = 0, \quad i = 1, \dots, k, \tag{5.141}$$

$$(\ B_i^2 u\ )(1) + c_i^{2T} z_i^2 + d_i^2 (\ O_i^2 v\ )(1) = 0, \quad i = 1, \dots, l, \tag{5.142}$$

such that $u \in D(A_c)$ and $v \in V$. Since $\alpha_i^j(s)$ are SPR, then from Definition 5.18 it follows that $(\lambda I - A_i^j)^{-1}$ exists for $\lambda > 0$. By using (5.139) and (5.140) we obtain

$$z_i^1 = (\lambda I - A_i^1)^{-1} b_i^1 (\ O_i^1 v\ )(0) + (\lambda I - A_i^1)^{-1} r_i^1, \quad i = 1, \dots, k, \tag{5.143}$$

$$z_i^2 = (\lambda I - A_i^2)^{-1} b_i^2 (\ O_i^2 v\ )(1) + (\lambda I - A_i^2)^{-1} r_i^2, \quad i = 1, \dots, l. \tag{5.144}$$

By using (5.143) and (5.144) in (5.141) and (5.142), respectively, we obtain

$$(\ B_i^1 u\ )(0) + \alpha_i^1(\lambda)(\ O_i^1 v\ )(0) = -c_i^{1T}(\lambda I - A_i^1)^{-1} r_i^1, \quad i = 1, \dots, k, \tag{5.145}$$

$$(\ B_i^2 u\ )(1) + \alpha_i^2(\lambda)(\ O_i^2 v\ )(1) = -c_i^{2T}(\lambda I - A_i^2)^{-1} r_i^2, \quad i = 1, \dots, l. \tag{5.146}$$

Note that since $\alpha_i^j(s)$ is SPR, from Definition 5.18 it follows that $\alpha_i^j(\lambda) \geq 0$ for $\lambda > 0$. Since assumptions 5.1-5.4, 5.9, and 5.21 hold, it follows that (5.137), (5.138), (5.145) and (5.146) has a solution $u \in D(A_c)$, $v \in V$. Then z_i^j could be found from (5.143) and (5.144). By construction $z_e \in D(\mathcal{A}_e)$. Hence, $\lambda I - \mathcal{A}_e$ is onto for $\lambda > 0$. The result then follows from the Lümer-Phillips theorem and the standard properties of C_0-semigroups, see Chapter 2. □

Next, we show that if assumption 5.11 holds for the associated system $\mathcal{S}_c$, then the same holds for the system $\mathcal{S}_{pr}$

Theorem 5.24 *Consider the system $\mathcal{S}_{pr}$ given by (5.123). Let the transfer functions $\alpha_i^j(s)$ be SPR. Let assumptions 5.1-5.4, 5.9, 5.11, and 5.21 hold for the associated system $\mathcal{S}_c$. Then $(\lambda I - \mathcal{A}_e)^{-1} X_e \to X_e$ is compact for $\lambda > 0$.*

Proof. Let $z_e = (z, \ z_r)^T \in X_e$ where $z = (u, \ v)^T \in X$ and $z_r = (z_1^1, \ldots, z_l^2)^T$. Let $\{(g_e)_n\}$ be a bounded sequence in X_e and let $\{(z_e)_n\} \in D(\mathcal{A}_e)$ such that $(z_e)_n = (\lambda I - \mathcal{A}_e)^{-1}(g_e)_n$. Since by Theorem 5.23, $\mathcal{A}_e$ generates a C_0-semigroup of contractions, by the Hille-Yosida theorem it follows that $\left\|(\lambda I - \mathcal{A}_e)^{-1}\right\| \leq \frac{1}{\lambda}$ for $\lambda > 0$. Hence, the sequence $\{(z_e)_n\}$ is bounded as well. From the proof of Theorem 5.23, it follows that to find z, we need to solve (5.137)-(5.146). As before, by setting $\alpha_i^j = \alpha_i^j(\lambda) \geq 0$, since $(\lambda I - A)^{-1}$ is compact, it follows that $(z)_n$ has a convergent subsequence, denoted by n_j. Since $(z_r)_n$ is bounded and belongs to a finite dimensional subspace of X_e, it follows that $(z_r)_n$ has another convergent subsequence on the subsequence n_j. Hence, $(z_e)_n$ has a convergent subsequence, and therefore $(\lambda I - \mathcal{A}_e)^{-1}$ is compact for $\lambda > 0$. □

Next, we state an asymptotic stability result similar to Theorem 5.14.

Theorem 5.25 *Consider the system $\mathcal{S}_{pr}$ given by (5.123). Let the transfer functions $\alpha_i^j(s)$ be SPR. If $\alpha_i^j(\lambda) > 0$ for some $\lambda > 0$ set $\alpha_i^j > 0$ in (5.59). Let the associated system $\mathcal{S}_c$ satisfy assumptions 5.1-5.4, 5.9, 5.11, 5.12, and 5.21. Then the system $\mathcal{S}_{pr}$ is globally asymptotically stable, i.e., for any $z_e(0) \in X_e$, the unique (classical or weak) solution $z_e(t)$ of (5.123) satisfy $\lim_{t \to \infty} z_e(t) = 0$.*

Proof. Since $T_e(t)$ is a contraction, it suffices to consider the case $z_e(0) \in D(\mathcal{A}_e)$, see Theorem 5.14; hence $z_e(t) = T_e(t)z_e(0) \in D(\mathcal{A}_e)$ for $t \geq 0$. Under this condition, let us define the energy $E_e(t)$ as

$$E_e(t) = \frac{1}{2}\langle z_e(t), \ z_e(t)\rangle_{X_e} = \frac{1}{2}\|z_e(t)\|_{X_e}^2, \tag{5.147}$$

cf. (5.60). Since $z_e(t)$ is differentiable and (5.123) holds, $E_e(t)$ is differentiable as well, see Theorem 2.12. By differentiating (5.147) and by using (5.135) we obtain

$$\begin{aligned}
\dot{E}_e(t) &= \langle z_e(t), \ \mathcal{A}_e z_e(t)\rangle_{X_e} \\
&= -\frac{1}{2}\sum_{i=1}^{k} z_i^{1T}Q_i^1 z_i^1 - \frac{1}{2}\sum_{i=1}^{k}[\sqrt{2d_i^1}y_i^1 - z_i^{1T}q_i^1]^2, \\
&\quad -\frac{1}{2}\sum_{i=1}^{l} z_i^{2T}Q_i^2 z_i^2 - \frac{1}{2}\sum_{i=1}^{l}[\sqrt{2d_i^2}y_i^2 - z_i^{2T}q_i^2]^2.
\end{aligned} \tag{5.148}$$

Since $(\lambda I - \mathcal{A}_e)^{-1}$ is compact and $T_e(t)$ is a contraction, it follows from LaSalle's invariance principle that all solutions of (5.123) asymptotically tend to the maximal invariant set $\mathcal{M}_e \subset \mathcal{O}_e$ where

$$\mathcal{O}_e = \{z_e \in X_e \mid \dot{E}_e(t) = 0\}. \tag{5.149}$$

From (5.148) and (5.149) it follows that $z_i^j(t) = 0$ in $\mathcal{O}_e$; hence, $\dot{z}_i^j(t) = 0$ as well. Then

a : if $b_i^j \neq 0$, then $y_i^j(t) = 0$ by (5.111) and $f_i^j(t) = 0$ by (5.112).

b : if $b_i^j = 0$, then $\alpha_i^j(s) = d_i^j$. For $\alpha_i^j > 0$, we have $d_i^j > 0$ as well, and by (5.148) we have $y_i^j(t) = 0$ and by (5.112) we have $f_i^j(t) = 0$.

Therefore from (5.123) and (5.124) it follows that the invariant solution of (5.123) in $\mathcal{O}_e$ reduces to the invariant solution of the associated system $\mathcal{S}_c$ in $\mathcal{S}_1 \cap \mathcal{S}_2 \cap \mathcal{S}_3$, see (5.59). Since assumption 5.12 holds, it follows that the only invariant solution of (5.123) in $\mathcal{O}_e$ is the zero solution. The result then follows from Theorem 5.14 $\qquad\square$

In the sequel, we will give a result on exponential stability of $\mathcal{S}_{pr}$. We first note that the strictly positive realness of $\alpha_i^j(s)$ alone may not be sufficient for exponential stability. To see this, let the associated system $\mathcal{S}_c$ be exponentially stable for a particular $\alpha_i^j > 0$, and consider the corresponding SPR function given by (5.110). By definition we have $d_i^j \geq 0$. If $d_i^j = 0$, then the corresponding energy expression (5.135) does not contain a term proportional to $\left(y_i^j\right)^2$, whereas for the associated system $\mathcal{S}_c$, the related expression does, see (5.48). Hence, by comparing (5.135) and (5.48), it seems that to guarantee exponential stability by using this approach, we need $d_i^j > 0$. For such technical reasons, we will make the following assumption for the SPR functions to guarantee exponential stability.

Assumption 5.26 *Let the transfer functions $\alpha_i^j(s)$ be SPR and let $d_i^j = \lim_{p \to \infty} \alpha_i^j(p) > 0$. For some $\gamma_i^j > 0$, such that $d_i^j \geq \gamma_i^j$, the following holds*

$$Re\{\alpha_i^j(i\omega)\} > \gamma_i^j, \quad \omega \geq 0. \tag{5.150}$$

Note that for a SPR function $\alpha(s)$, (5.115) holds and hence $Re\{\alpha(i\omega)\} > 0$ for $\omega \geq 0$. But it is possible to have $\lim_{\omega \to \infty} Re\{\alpha(i\omega)\} = 0$. In other words, the Nyquist plot of $\alpha(i\omega)$ is in the open right-half complex plane, but may not be bounded away from the imaginary axis. But if in addition $\alpha(s)$ satisfies assumption 5.26, then from (5.150) it follows that the Nyquist plot of $\alpha(i\omega)$ is bounded away from the imaginary axis.

Let us define a new function $h_i^j(s) = \alpha_i^j(s) - \gamma_i^j$, and assume that $\alpha_i^j(s)$ satisfies assumption 5.26, and let $(A_i^j, b_i^j, c_i^j, d_i^j)$ be a minimal representation of it. It follows easily that $h_i^j(s)$ is also SPR with a minimal realization $(A_i^j, b_i^j, c_i^j, d_i^j - \gamma_i^j)$. By using Kalman-Yakubovich lemma, the following equations also hold

$$A_i^{jT} P_i^j + P_i^j A_i^j = -q_i^j q_i^{jT} - Q_i^j, \tag{5.151}$$

$$P_i^j b_i^j - c_i^j = \sqrt{2(d_i^j - \gamma_i^j)} q_i^j, \tag{5.152}$$

where P_i^j, Q_i^j are symmetric and positive definite matrices. Without loss of generality, we will use these matrices in (5.128), (5.129) if $\alpha(s)$ satisfies assumption 5.26.

Lemma 5.27 *Consider the system S_{pr} given by (5.123) and let the transfer functions $\alpha_i^j(s)$ satisfy assumption 5.26. Let $z_e(0) \in D(\mathcal{A}_e)$ and let $z_e(t)$ be the corresponding solution of (5.123). Then the energy expression (5.135) takes the following form*

$$
\begin{aligned}
\langle z_e, \mathcal{A}_e z_e \rangle_{X_e} =\ & -\sum_{i=1}^{k} \gamma_i^1 (y_i^1)^2 - \sum_{i=1}^{l} \gamma_i^2 (y_i^2)^2 - \frac{1}{2} \sum_{i=1}^{k} z_i^{1T} Q_i^1 z_i^1 \\
& -\frac{1}{2} \sum_{i=1}^{k} \left[\sqrt{2(d_i^1 - \gamma_i^1)} y_i^1 - z_i^{1T} q_i^1 \right]^2 - \frac{1}{2} \sum_{i=1}^{l} z_i^{2T} Q_i^2 z_i^2 \\
& -\frac{1}{2} \sum_{i=1}^{l} \left[\sqrt{2(d_i^2 - \gamma_i^2)} y_i^2 - z_i^{2T} q_i^2 \right]^2 .
\end{aligned}
\tag{5.153}
$$

Proof. The proof easily follows from Theorem 5.23 (see (5.135)) when we use (5.151) and (5.152) instead of (5.126) and (5.127). □

We are now ready to prove the following exponential stability result.

Theorem 5.28 *Consider the system S_{pr} given by (5.123). Let the transfer functions $\alpha_i^j(s)$ be SPR. Let the associated system S_c satisfy the assumptions of Theorem 5.15. If $\alpha_i^j > 0$ in S_c, let the corresponding controller $\alpha_i^j(s)$ satisfy assumption 5.26. Then the semigroup $T_e(t)$ is exponentially stable.*

Proof. Let $z_e(0) \in D(\mathcal{A}_e)$ and let $z_e(t)$ be the corresponding solution of (5.123). Let us define $z_e(t) = (z(t),\ z_1^1(t), \ldots, z_l^2(t))^T$, $E(t) = \frac{1}{2}\langle z(t),\ z(t) \rangle_X$ and $E_e(t) = \frac{1}{2}\langle z_e(t),\ z_e(t) \rangle_{X_e}$. Let us define the projections $P_{1e} : X_e \to V$ and $P_{2e} : X_e \to H$ as $P_{1e} z_e = u$ and $P_{2e} z_e = v$ for $z = (u,\ v)^T$. As in Theorem 5.15, we define the following function

$$
V_e(t) = t E_e(t) + \langle P_{2e} z_e(t),\ Q P_{1e} z_e(t) \rangle_H.
\tag{5.154}
$$

Since $\|z(t)\|_X \leq \|z_e(t)\|_{X_e}$, from (5.63) it follows that

$$
\mid \langle P_{2e} z_e(t),\ Q P_{1e} z_e(t) \rangle_H \mid\ \leq C E_e(t).
\tag{5.155}
$$

Note that $P_{1e} z_e = P_1 z$, $P_{2e} z_e = P_2 z$, see (5.63). By using (5.64) and (5.112) we obtain

$$
\frac{d}{dt} \mid \langle P_{2e} z_e(t),\ Q P_{1e} z_e(t) \rangle_H \mid\ \leq\ -E(t) + \sum_{i=1}^{k} 2 a_i^1 (d_i^1)^2 (y_i^1(t))^2
$$

$$+ \sum_{i=1}^{l} 2a_i^2 (d_i^2)^2 (y_i^2(t))^2 + \sum_{i=1}^{k} 2a_i^1 (c_i^{1T} z_i^1(t))^2$$

$$+ \sum_{i=1}^{l} 2a_i^2 (c_i^{2T} z_i^2(t))^2, \tag{5.156}$$

where the constants a_i^j are the same as in (5.64). By using (5.155) in (5.154) it follows that

$$(t - C)E_e(t) \leq V_e(t) \leq (t + C)E_e(t). \tag{5.157}$$

Differentiating (5.154) and using (5.129), (5.147), (5.153) and (5.156), we obtain

$$\dot{V}_e(t) = t\dot{E}_e(t) + E_e(t) + \frac{d}{dt}\langle P_{2e} z_e(t), QP_{1e} z_e(t)\rangle_H$$

$$\leq -t \sum_{i=1}^{k} \gamma_i^1 (y_i^1)^2 - t \sum_{i=1}^{l} \gamma_i^2 (y_i^2)^2 - \frac{t}{2}\sum_{i=1}^{k} z_i^{1T} Q_i^1 z_i^1 - \frac{t}{2}\sum_{i=1}^{l} z_i^{2T} Q_i^2 z_i^2$$

$$+ \sum_{i=1}^{k} z_i^{1T} P_i^1 z_i^1 + \sum_{i=1}^{l} z_i^{2T} P_i^2 z_i^2 + \sum_{i=1}^{k} 2a_i^1 (d_i^1)^2 (y_i^1(t))^2$$

$$+ \sum_{i=1}^{l} 2a_i^2 (d_i^2)^2 (y_i^2(t))^2 + \sum_{i=1}^{k} 2a_i^1 (c_i^{1T} z_i^1(t))^2$$

$$+ \sum_{i=1}^{l} 2a_i^2 (c_i^{2T} z_i^2(t))^2. \tag{5.158}$$

Since P_i^j and Q_i^j are symmetric and positive definite, we have the following

$$\lambda_{min}(Q_i^j) \left\| z_i^j \right\|_2^2 \leq z_i^{jT} Q_i^j z_i^j \leq \lambda_{max}(Q_i^j) \left\| z_i^j \right\|_2^2, \tag{5.159}$$

$$\lambda_{min}(P_i^j) \left\| z_i^j \right\|_2^2 \leq z_i^{jT} P_i^j z_i^j \leq \lambda_{max}(P_i^j) \left\| z_i^j \right\|_2^2, \tag{5.160}$$

where $\lambda_{min}(\cdot)$ and $\lambda_{max}(\cdot)$ denote the minimum and maximum eigenvalue of the matrix in the argument, respectively, and $\|z\|_2^2 = z^T z$. We also have the following

$$(c_i^{jT} z_i^j(t))^2 \leq k_i^j \left\| z_i^j(t) \right\|_2^2, \tag{5.161}$$

where $k_i^j > 0$ are appropriate constants which depend only on c_i^j. By using (5.159)-(5.161) in (5.158) we obtain

$$\dot{V}_e(t) \leq -\sum_{i=1}^{k}[t\gamma_i^1 - 2a_i^1 (d_i^1)^2](y_i^1(t))^2 - \sum_{i=1}^{l}[t\gamma_i^2 - 2a_i^2 (d_i^2)^2](y_i^2(t))^2$$

$$-\sum_{i=1}^{k}(\frac{t}{2}\lambda_{min}(Q_i^1) - \lambda_{max}(P_i^1) - 2a_i^1 k_i^1) \left\| z_i^1(t) \right\|_2^2$$

$$-\sum_{i=1}^{l}(\frac{t}{2}\lambda_{min}(Q_i^2) - \lambda_{max}(P_i^2) - 2a_i^2 k_i^2) \left\| z_i^2(t) \right\|_2^2. \qquad (5.162)$$

Since the matrices Q_i^j are positive definite, it follows that $\lambda_{min}(Q_i^j) > 0$. Hence, for some $T > 0$, which is independent of $z_e(0)$, the following holds

$$\dot{V}_e(t) \le 0, \quad t \ge T. \qquad (5.163)$$

Following the proof of Theorem 5.15, we obtain the exponential stability result. $\qquad\qquad\qquad\qquad\qquad\qquad\qquad\qquad\qquad\qquad\qquad\qquad\qquad$ □

Remark 5.29 *Theorems 5.23-5.25, and 5.28 generalize the results to the corresponding system S_c for the system S_{pr}. More precisely, consider the system S_c and assume that the conditions in Theorems 5.10, 5.14, and 5.15 are satisfied. If we replace the controllers given by (5.47) with SPR controllers, then the closed-loop system is well-posed and the corresponding operator A_e generates a C_0-semigroup of contractions. Similarly, if for a set of controllers given by (5.47) the system S_c is asymptotically stable, then so is the system S_{pr} when we replace the controllers with SPR controllers. Note that if $\alpha_i^j = 0$ for S_c, then we may still use $\alpha_i^j(s) = 0$ for S_{pr} (i.e., do not apply a controller for the channel given by the indices (i, j)), or apply an SPR controller. If the system S_c satisfies the assumptions of Theorem 5.15, hence is exponentially stable, this result still holds if we replace the controllers corresponding to $\alpha_i^j > 0$ with SPR controllers which satisfy assumption 5.26. In this case, if $\alpha_i^j = 0$ for S_c, then we may still use $\alpha_i^j(s) = 0$ for S_{pr} (i.e., do not apply a controller for the channel given by the indices (i, j)), or apply an SPR controller, the resulting closed-loop system S_{pr} still remains exponentially stable.*

Example 5.30 *Consider the Timoshenko beam equation given by (5.56)-(5.58). For convenience, we will repeat the equations here.*

$$\begin{cases} y_{tt} - (y_{xx} - \phi_x) = 0, \\ \phi_{tt} - \phi_{xx} + \phi - y_x = 0, \\ y(0, t) = \phi(0, t) = 0, \\ y_x(1, t) - \phi(1, t) = f_1^2(t), \quad \phi_x(1, t) = f_2^2(t), \\ y_1^2(t) = y_t(1, t), \quad y_2^2(t) = \phi_t(1, t). \end{cases} \qquad (5.164)$$

For $i = 1, 2$, the SPR controllers are given by

$$\dot{z}_i^2(t) = A_i^2 z_i^2(t) + b_i^2 y_i^2(t), \qquad (5.165)$$

$$-f_i^2(t) = c_i^{2T} z_i^2(t) + d_i^2 y_i^2(t) \tag{5.166}$$

with the controller transfer functions being given by

$$\alpha_i^2(s) = c_i^{2T}(sI - A_i^2)^{-1}b_i^2 + d_i^2. \tag{5.167}$$

We set A, $D(A)$, $D(A_c)$, V, and H as in Example 5.6, and set $X_e = V \times H \times \mathbf{R}^{n_1^2} \times \mathbf{R}^{n_2^2}$, where V is given by (5.41),

$$\mathcal{A}_e \begin{bmatrix} u_1 \\ \phi_1 \\ u_2 \\ \phi_2 \\ z_1^2 \\ z_2^2 \end{bmatrix} = \begin{bmatrix} u_2 \\ \phi_2 \\ u_1'' - \phi_1' \\ \phi_1'' + \phi_1 - u_1' \\ A_1^2 z_1^2 + b_1^2 u_2(1) \\ A_2^2 z_2^2 + b_2^2 \phi_2(1) \end{bmatrix}, \tag{5.168}$$

$$\begin{aligned} D(\mathcal{A}_e) = \{ &(u_1,\ \phi_1,\ u_2,\ \phi_2,\ z_1^2,\ z_2^2)^T \in X_e \mid (u_1,\ \phi_1)^T \in D(A_c), \\ &(u_2,\ \phi_2)^T \in V, \\ &u_1'(1) - \phi_1(1) + c_1^{2T} z_1^2 + d_1^2 u_2(1) = 0, \\ &\phi_1'(1) + c_2^{2T} z_2^2 + d_2^2 \phi_2(1) = 0 \ \}, \end{aligned} \tag{5.169}$$

where $D(A_c)$ is given by (5.57).

Corollary 5.31 *Consider the system (5.123) where the operator $\mathcal{A}_e$ and $D(\mathcal{A}_e)$ are as given above. Let the transfer functions $\alpha_i^2(s)$, $i = 1,2$, given by (5.167) be SPR. Then*

(i) $\mathcal{A}_e$ given by (5.168) generates a C_0-semigroup of contractions $T_e(t)$ on X_e.

(ii) All solutions of (5.123) (classical or weak) asymptotically decay to zero.

(iii) If $\alpha_i^2(s)$, $i = 1,2$, satisfy assumption 5.26, then the semigroup $T_e(t)$ is exponentially stable.

Proof.   *Note that assumption 5.21 trivially holds in this case, see Theorem 5.16. The result then follows from Theorems 5.23-5.25, and 5.28.* $\square$

For a direct proof of the statements (i) and (iii) of Corollary 5.31, see [112].

5.3 Dynamic boundary control of a rotating flexible beam

In this section, we apply the passivity approach presented in Sections 5.1 and 5.2 to the stabilization of a rotating flexible beam equation. For convenience, we consider the normalized Euler-Bernoulli beam model which is derived in Chapter 4 and has the following form:

$$\begin{cases} y_{tt} + y_{xxxx} + x\ddot{\theta} = 0, & 0 < x < 1, \\ \ddot{\theta} = y_{xx}(0,t) + \tau(t), \\ y(0,t) = y_x(0,t) = 0, \\ -y_{xxx}(1,t) = f_1(t), & y_{xx}(1,t) = f_2(t), \end{cases} \tag{5.170}$$

where the first equation describes bending vibration of the beam, and the second equation describes motion of the control motor. Unlike Chapter 4, where $\ddot{\theta}(t)$ was taken as the control input and $f_1(t)$, $f_2(t)$ was set to zero, here we consider torque control $\tau(t)$ and two boundary control inputs $f_1(t)$, $f_2(t)$. We wish to discuss stabilization and orientation problems which are stated precisely as follows.

Stabilization Problem: Consider the system given by (5.170). Find appropriate control laws for $\tau(t)$, $f_1(t)$ and $f_2(t)$ such that the solutions $y(x,t)$, $y_t(x,t)$ and $\theta(t)$ satisfy the following asymptotic relations :

$$\lim_{t\to\infty} y(x,t) = 0, \qquad 0 \le x \le 1,$$
$$\lim_{t\to\infty} y_t(x,t) = 0, \qquad 0 \le x \le 1,$$
$$\lim_{t\to\infty} \dot{\theta}(t) = 0.$$

□

Orientation Problem: Consider the system (5.170). Let $\theta_0 \in [0, 2\pi)$, the desired angle, be given. Find appropriate control laws for $\tau(t)$, $f_1(t)$ and $f_2(t)$ such that the stabilization problem is solved, and moreover the following is satisfied

$$\lim_{t\to\infty} \theta(t) = \theta_0.$$

□

To motivate the analysis and to obtain an appropriate power form (see (5.20), (5.46)), we first define the following variable

$$w(x,t) = y(x,t) + x\theta(t). \tag{5.171}$$

By using (5.171) in (5.170), we obtain

$$\begin{cases} w_{tt} + w_{xxxx} = 0, \quad 0 < x < 1, \\ \ddot{\theta} = w_{xx}(0,t) + \tau(t), \\ w(0,t) = 0, \quad w_x(0,t) = \theta(t), \\ -w_{xxx}(1,t) = f_1(t), \quad w_{xx}(1,t) = f_2(t). \end{cases} \tag{5.172}$$

Define the following energy function

$$E_1(t) = \frac{1}{2} \int_0^1 w_t^2 dx + \frac{1}{2} \int_0^1 w_{xx}^2 dx + \frac{1}{2}\dot{\theta}^2, \tag{5.173}$$

which is suitable for the stabilization problem, since we are only interested in $\dot{\theta}$, not in θ. For the orientation problem, later we will modify (5.173). Evaluating the derivative along the solution of (5.172) leads to

$$\begin{aligned} \dot{E}_1(t) &= -w_{xxx}(1,t)w_t(1,t) + w_{xx}(1,t)w_{xt}(1,t) + \dot{\theta}(t)\tau(t) \\ &= f_1(t)w_t(1,t) + f_2(t)w_{xt}(1,t) + \dot{\theta}(t)\tau(t) \\ &= f_1(t)y_t(1,t) + f_2(t)y_{xt}(1,t) \\ &\quad + \dot{\theta}(t)(\tau(t) + f_1(t) + f_2(t)). \end{aligned} \tag{5.174}$$

If we define a new input term $f_3(t)$ as

$$f_3(t) = \tau(t) + f_1(t) + f_2(t), \tag{5.175}$$

then (5.174) becomes

$$\dot{E}_1(t) = f_1(t)y_t(1,t) + f_2(t)y_{xt}(1,t) + f_3(t)\dot{\theta}(t), \tag{5.176}$$

which suggests that we can take

$$y_1(t) = y_t(1,t), \quad y_2(t) = y_{xt}(1,t), \quad y_3(t) = \dot{\theta}(t) \tag{5.177}$$

as outputs. If we choose

$$f_1(t) = -\alpha_1 y_t(1,t), \quad f_2(t) = -\alpha_2 y_{xt}(1,t), \quad f_3(t) = -\alpha_3\dot{\theta}(t), \tag{5.178}$$

where $\alpha_i > 0$ are arbitrary constants, then it can be shown that the controller proposed in (5.178) solves the stabilization problem. If we use

$$f_3(t) = -\alpha_3\dot{\theta}(t) - \alpha_4(\theta(t) - \theta_0), \tag{5.179}$$

instead of the one used in (5.178), where $\alpha_4 > 0$, then the proposed controller solves the orientation problem as well. In the following section, we generalize the controllers given above to the class of SPR controllers.

5.3.1   Stabilization problem using SPR controllers

Let us consider the stabilization problem defined in the previous section for the system given by (5.170), (5.175) and (5.177). To generate the control inputs $f_i(t)$, we will use the following controllers

$$\dot{z}_i(t) = A_i z_i(t) + b_i y_i(t), \tag{5.180}$$

$$-f_i(t) = c_i^T z_i(t) + d_i y_i(t), \tag{5.181}$$

where, for $i = 1, 2, 3$, $z_i \in \mathbf{R}^{n_i}$ is the state of the controller, n_i is an integer, $A_i \in \mathbf{R}^{n_i \times n_i}$ is a constant matrix, $b_i, c_i \in \mathbf{R}^{n_i}$ are constant matrices and $d_i \in \mathbf{R}$ is a constant. We note that if the transfer function $\alpha_i(s)$ given as

$$\alpha_i(s) = c_i^T(sI - A_i)^{-1}b_i + d_i, \tag{5.182}$$

is SPR, then there exist symmetric and positive definite matrices P_i, Q_i and vector q_i such that

$$A_i^T P_i + P_i A_i = -q_i q_i^T - Q_i, \tag{5.183}$$

$$P_i b_i - c_i = \sqrt{2d_i} q_i, \tag{5.184}$$

and if $\alpha_i(s)$ satisfy assumption 5.26 for some $\gamma_i > 0$, then we may replace (5.184) with

$$P_i b_i - c_i = \sqrt{2d_i - \gamma_i} q_i, \tag{5.185}$$

see (5.152).

To analyze the system given by (5.170), (5.175), (5.177), (5.180), and (5.181), which is denoted by $\mathcal{S}_s$, we define $X_1 = V \times H \times \mathbf{R} \times \mathbf{R}^{n_1} \times \mathbf{R}^{n_2} \times \mathbf{R}^{n_3}$ where

$$H = L^2(0,1), \quad V = \{w \in H \mid w, w', w'' \in H, w(0) = w'(0) = 0\}.$$

Define the operator $\mathcal{A}_1 : X_1 \to X_1$ as

$$\mathcal{A}_1 \begin{bmatrix} u \\ v \\ \phi \\ z_1 \\ z_2 \\ z_3 \end{bmatrix} = \begin{bmatrix} v \\ -u'''' + x I_1 \\ -I_1 \\ A_1 z_1 + b_1 v(1) \\ A_2 z_2 + b_2 v'(1) \\ A_3 z_3 + b_3 \phi \end{bmatrix}, \tag{5.186}$$

where I_1 is defined as

$$I_1 = -\int_0^1 x u_{xxxx} dx + c_3^T z_3 + d_3 \phi. \tag{5.187}$$

Let the domain $D(\mathcal{A}_1)$ be defined as

$$D(\mathcal{A}_1) = \{(u, v, \phi, z_1, z_2, z_3)^T \in X_1 \mid u'''' \in H, u(0) = u'(0) = 0,\ v \in V,$$
$$u''(1) + c_1^T z_1 + d_1 v'(1) = 0,$$
$$-u'''(1) + c_2^T z_2 + d_2 v(1) = 0\}. \tag{5.188}$$

By using (5.186)-(5.188), the system $\mathcal{S}_s$ can be formulated as the following abstract equation:

$$\dot{z}(t) = \mathcal{A}_1 z(t), \quad z(0) \in X_1, \tag{5.189}$$

where $z = (y,\ y_t,\ \dot{\theta},\ z_1,\ z_2,\ z_3)^T \in X_1$.

Let $X = V \times H \times \mathbf{R}$ and $(u,\ v,\ \phi)^T \in X$. A natural norm in X is given by

$$\|(u,\ v,\ \phi)\|_*^2 = \int_0^1 v^2\,dx + \int_0^1 (u'')^2\,dx + \phi^2. \tag{5.190}$$

However, in deriving the power form given by (5.176) we used (5.173), see also (5.171), which can be rewritten as

$$\|(u,\ v,\ \phi)\|_X^2 = \int_0^1 (v + x\phi)^2\,dx + \int_0^1 (u'')^2\,dx + \phi^2. \tag{5.191}$$

We will use (5.191) as the norm in X since it yields the power form (5.176). It should be noted that the two norms defined above are actually equivalent.

Lemma 5.32 *The norms given by (5.190) and (5.191) are equivalent on X.*

Proof. By using Cauchy-Schwartz inequality, it can be easily shown that there exists a constant $K > 0$ such that the following holds :

$$\|(u,\ v,\ \phi)\|_X^2 \leq K \|(u,\ v,\ \phi)\|_*^2. \tag{5.192}$$

On the other hand, since

$$-2\int_0^1 x\phi v\,dx \leq \delta \int_0^1 v^2\,dx + M\phi^2/\delta \tag{5.193}$$

for an arbitrary constant $\delta > 0$ and $M = \int_0^1 x^2\,dx$, we have from (5.191) that

$$\|(u,\ v,\ \phi)\|_X^2 \geq (1 - \delta) \int_0^1 v^2\,dx + \int_0^1 u_{xx}^2\,dx + (1 - M/\delta + M)\phi^2. \tag{5.194}$$

By choosing $\delta > 0$ such that $M/(1 + M) < \delta < 1$, all coefficients in (5.194) can be made positive. Therefore, by comparing (5.190) and (5.194), we conclude that

$$\|(u,\ v,\ \phi)\|_X^2 \geq \min\{(1 - \delta),\ (1 - M/\delta + M)\} \|(u,\ v,\ \phi)\|_*^2. \tag{5.195}$$

The equivalence then follows from (5.192) and (5.195).   □

Consider the system S_s. Let the transfer functions $\alpha_i(s)$ given by (5.182) be SPR, and hence (5.183), (5.184) hold. Let $z = (u,\ v,\ \phi,\ z_1,\ z_2,\ z_3)^T$, $\hat{z} = (\hat{u},\ \hat{v},\ \hat{\phi},\ \hat{z}_1,\ \hat{z}_2,\ \hat{z}_3)^T \in X_1$. Similar to (5.128), we define the following inner product

$$\langle z,\ \hat{z}\rangle_{X_1} = \int_0^1 (v + x\phi)(\hat{v} + x\hat{\phi})dx + \int_0^1 (u'')(\hat{u}'')dx + \phi\hat{\phi}$$
$$+ \frac{1}{2}\sum_{i=1}^{3}(z_i^T P_i \hat{z}_i + \hat{z}_i^T P_i z_i), \tag{5.196}$$

and the corresponding norm

$$\|z\|_{X_1}^2 = \int_0^1 (v + x\phi)^2 dx + \int_0^1 (u'')^2 dx + \phi^2 + \sum_{i=1}^{3} z_i^T P_i z_i. \tag{5.197}$$

Theorem 5.33 *Consider the system S_s given by (5.189), where the operator $\mathcal{A}_1$ and $D(\mathcal{A}_1)$ are as given by (5.186), (5.188), respectively. Let the transfer functions $\alpha_i(s)$, $i = 1, 2, 3$, given by (5.182) be SPR. Then*

(i) *$\mathcal{A}_1$ generates a C_0-semigroup of contractions $T_1(t)$ on X_1.*

(ii) *All solutions of (5.189) (classical or weak) asymptotically decay to zero. This result also holds even if $\alpha_1(s) = 0$ and/or $\alpha_2(s) = 0$, i.e., even if any one or both of the boundary controllers are not used.*

(iii) *If $\alpha_1(s)$ and $\alpha_3(s)$ satisfy assumption 5.26, then the semigroup $T_1(t)$ is exponentially stable.*

Proof.   (i) Due to the power form (5.176) and the SPR controllers (5.180)-(5.181), the operator $\mathcal{A}_1$ is dissipative; moreover the following holds

$$\langle z,\ \mathcal{A}_1 z\rangle_{X_1} = -\frac{1}{2}\sum_{i=1}^{3} z_i^T Q_i z_i - \frac{1}{2}\sum_{i=1}^{3}\left(\sqrt{2d_i}y_i - z_i^T q_i\right)^2, \tag{5.198}$$

see Theorem 5.25.

To prove that $(\lambda I - \mathcal{A}_1)$ is onto for some $\lambda > 0$, we first decompose $\mathcal{A}_1$ as $\mathcal{A}_1 = \mathcal{A}_u + \mathcal{A}_d$ where

$$\mathcal{A}_u \begin{bmatrix} u \\ v \\ \phi \\ z_1 \\ z_2 \\ z_3 \end{bmatrix} = \begin{bmatrix} v \\ -u'''' \\ -c_3^T z_3 - d_3\phi \\ A_1 z_1 + b_1 v(1) \\ A_2 z_2 + b_2 v'(1) \\ A_3 z_3 + b_3\phi \end{bmatrix}, \tag{5.199}$$

with $D(\mathcal{A}_u) = D(\mathcal{A}_1)$, $\mathcal{A}_d = \mathcal{A}_1 - \mathcal{A}_a$ and $D(\mathcal{A}_1) \subset D(\mathcal{A}_d)$. By using the block diagonal form of $\mathcal{A}_u$, it can be easily shown that $\mathcal{A}_u$ generates a C_0-semigroup on X_1. Moreover, $(\lambda I - \mathcal{A}_u)^{-1}$ exists, is compact, and the following holds for some $M > 0$

$$\left\| (\lambda I - \mathcal{A}_u)^{-1} \right\|_{X_1} \le \frac{M}{\lambda}. \tag{5.200}$$

Note that to prove that $\mathcal{A}_u$ generates a semigroup, we may use the norm given by (5.190) to take advantage of the block diagonal form, whereas to prove the same for $\mathcal{A}_1$ we may use the norm given by (5.191) to take advantage of the power form.

Next, we note that the range of $\mathcal{A}_d$ is finite dimensional and the following holds

$$\|\mathcal{A}_d z\|_{X_1} \le a \|z\|_{X_1} + b \|\mathcal{A}_u z\|_{X_1}, \quad z \in D(\mathcal{A}_u), \tag{5.201}$$

for some constants $a > 0$ and $b > 0$. The estimate (5.201) can be easily obtained by direct computation. It then follows that $B = \mathcal{A}_d(\lambda I - \mathcal{A}_u)^{-1}$: $X_1 \to X_1$ is a compact operator, see, e.g., [84, p. 245]. We show that $I - B$ is invertible. Since B is compact, we need to show that $(I - B)y = 0$ implies $y = 0$, i.e., 1 is not an eigenvalue of B. Suppose that $By = y$ holds for some $y \neq 0 \in X_1$. By defining $x = (\lambda I - \mathcal{A}_u)^{-1} y$ and by multiplying this equation with $\lambda I - \mathcal{A}_1$ we obtain $(\lambda I - \mathcal{A}_1)x = 0$. But since $\mathcal{A}_1$ is dissipative, this implies that $x = 0$; hence $y = 0$. Therefore, $(I - B)^{-1}$ exists and is a bounded operator. Moreover we have

$$(\lambda I - \mathcal{A}_1)^{-1} = (\lambda I - \mathcal{A}_u)^{-1}(I - B)^{-1}; \tag{5.202}$$

hence, $(\lambda I - \mathcal{A}_1)$ is onto for $\lambda > 0$. The proof of (i) then follows from the Lümer-Phillips theorem.

(ii) From the arguments given in the case (i) and (5.202), it follows that $(\lambda I - \mathcal{A}_1)^{-1}$ is a compact operator for $\lambda > 0$. Let $z(0) \in D(\mathcal{A}_1)$, and let $z(t)$ be the corresponding solution of (5.189). Let us define the energy function as

$$E_1(t) = \frac{1}{2}\langle z(t),\ z(t)\rangle_{X_1}. \tag{5.203}$$

Since $z(t)$ is differentiable, so is $E_1(t)$. Differentiating (5.203) and using (5.189) and (5.198), we obtain

$$\dot{E}_1(t) = \langle z(t),\ \mathcal{A}_1 z(t)\rangle_{X_1}$$
$$= -\frac{1}{2}\sum_{i=1}^{3} z_i^T Q_i z_i - \frac{1}{2}\sum_{i=1}^{3} \left(\sqrt{2d_i}y_i - z_i^T q_i\right)^2, \tag{5.204}$$

where y_i are given by (5.177). By LaSalle's invariance principle, all solutions of (5.189) asymptotically approach to the maximal invariant subset in

$$\mathcal{O} = \{z \in X_1 \mid \dot{E}_1(t) = 0\}.$$

From (5.204), it follows that for the invariant solution in $\mathcal{O}$, we must have $z_i(t) = 0$; hence, $\dot{z}_i(t) = 0$, for $i = 1, 2, 3$. Therefore, from (5.180) and (5.181) it follows that $y_i(t) = 0$ and $f_i(t) = 0$ as well. Here, we will prove the asymptotic stability result *without* using $y_1(t) = y_2(t) = 0$. Hence, the asymptotic stability result will be valid even if $\alpha_1(s) = \alpha_2(s) = 0$, i.e., when the boundary controllers are not applied; hence, when the only control input is the torque applied to the rigid body. Note that under the conditions stated above, the equations of motion of the system $\mathcal{S}_s$ which is invariant in $\mathcal{O}$ must satisfy the following equations

$$
\begin{cases}
y_{tt} + y_{xxxx} = 0, & 0 < x < 1, \\
y(0, t) = y_x(0, t) = 0, \\
y_{xx}(1, t) = y_{xxx}(1, t) = 0, \\
y_{xx}(0, t) = 0.
\end{cases}
\tag{5.205}
$$

It can be easily shown that (5.205) has only the zero solution $y(x, t) = 0$. Thus, the asymptotic stability follows from LaSalle's invariance principle and Theorem 5.14.

(iii) Let $z(0) \in D(\mathcal{A}_1)$, $z(t)$ be the corresponding solution of (5.189), and let $E_1(t)$ be given by (5.203). We define the following function

$$
V_1(t) = tE_1(t) + 4 \int_0^1 xy_x(y_t + x\dot{\theta})dx.
\tag{5.206}
$$

Note that in this case, by using (5.185), similar to (5.153), instead of (5.204), we obtain

$$
\dot{E}_1 = -\sum_{i=1}^3 \gamma_i(y_i)^2 - \frac{1}{2}\sum_{i=1}^3 z_i^T Q_i z_i - \frac{1}{2}\sum_{i=1}^3 \left(\sqrt{2(d_i - \gamma_i)}y_i - z_i^T q_i\right)^2.
\tag{5.207}
$$

By using Cauchy-Schwartz inequality, it is straightforward to show that

$$
(t - C)E_1(t) \le V_1(t) \le (t + C)E_1(t)
\tag{5.208}
$$

holds for some $C > 0$. By differentiating the second term in (5.206) and using (5.170), we obtain

$$
\begin{aligned}
\frac{d}{dt}\int_0^1 xy_x(y_t + x\dot{\theta})dx \le &-\left(\frac{1}{2} - \delta\right)\int_0^1 y_t^2 dx - \left(\frac{3}{2} - 2\delta\right)\int_0^1 y_{xx}^2 dx \\
&+\left(\frac{3}{2} + \frac{2}{\delta}\right)(y_1)^2 + \left(\frac{1}{2} + \frac{2}{\delta}\right)(y_2)^2 \\
&+\frac{k_1}{\delta}\|z_1\|_2^2 + \left(\frac{1}{2} + \frac{2}{\delta}\right)k_2\|z_2\|_2^2 \\
&+\left[1 + \frac{1}{\delta}\int_0^1 (2x)^2 dx\right](y_3)^2,
\end{aligned}
\tag{5.209}
$$

where k_1 and k_2 are constants which depend only on c_1, c_2, and $\|z\|_2^2 = z^T z$ is the standard Euclidean norm in $\mathbf{R}^n$. By differentiating (5.206) and using (5.209) we obtain

$$\dot{V}_1(t) \le 0, \quad t \ge T,$$

for some $T \ge 0$, which does not depend on $z(0)$. The exponential stability result then follows from Theorem 5.15 and Theorem 5.28. $\square$

Remark 5.34 *We note that the asymptotic stability result holds even if* $\alpha_1(s) = \alpha_2(s) = 0$, *but* $\alpha_3(s)$ *is SPR. This corresponds to the case where the only control applied to the system is the torque control applied to the rigid hub. The exponential stability result holds even if* $\alpha_2(s) = 0$, *or is SPR, but* $\alpha_1(s)$ *and* $\alpha_3(s)$ *should satisfy assumption 5.26*

5.3.2   Orientation problem using positive real controllers

Consider the system given by (5.170). Following (5.179), we choose the following controller for $f_3(t)$, instead of (5.181),

$$-f_3(t) = c_3^T z_3(t) + d_3 \dot{\theta}(t) + k(\theta(t) - \theta_0), \tag{5.210}$$

where $k > 0$ is a constant. We assume that the controllers are SPR and given by (5.180), (5.181), with the exception for $f_3(t)$ given by (5.210). We denote this system by $\mathcal{S}_o$. Let us define the error angle

$$\theta_e(t) = \theta(t) - \theta_0. \tag{5.211}$$

Since θ_0 is a constant, we have $\dot{\theta}_e(t) = \dot{\theta}(t)$ and $\ddot{\theta}_e(t) = \ddot{\theta}(t)$. For this reason, we may use θ and θ_e interchangeably in this section. To analyze the system $\mathcal{S}_o$, we define

$$X_2 = V \times H \times \mathbf{R} \times \mathbf{R} \times \mathbf{R}^{n_1} \times \mathbf{R}^{n_2} \times \mathbf{R}^{n_3}, \tag{5.212}$$

and the operator $\mathcal{A}_2$ in X_2 as

$$\mathcal{A}_2 \begin{bmatrix} u \\ v \\ \theta \\ \phi \\ z_1 \\ z_2 \\ z_3 \end{bmatrix} = \begin{bmatrix} v \\ -u'''' + xI_2 \\ \phi \\ -I_2 \\ A_1 z_1 + b_1 v(1) \\ A_2 z_2 + b_2 v'(1) \\ A_3 z_3 + b_3 \phi \end{bmatrix}, \tag{5.213}$$

where I_2 is defined as

$$I_2 = -\int_0^1 x u_{xxxx} dx + c_3^T z_3 + d_3 \phi + k\theta, \tag{5.214}$$

with the following domain

$$D(\mathcal{A}_2) = \{(u,\ v,\ \theta,\ \phi,\ z_1,\ z_2,\ z_3)^T \in X_2 \ |$$
$$(u,\ v,\ \phi,\ z_1,\ z_2,\ z_3)^T \in D(\mathcal{A}_1), \theta \in \mathbf{R}\}, \quad (5.215)$$

and $D(\mathcal{A}_1)$ is given by (5.188). Then the abstract equation for the system S_o can be given as

$$\dot{z}(t) = \mathcal{A}_2 z(t), \quad z(0) \in X_2, \quad (5.216)$$

where $z = (y, y_t, \theta_e, \dot{\theta}_e, z_1, z_2, z_3)^T \in X_2$. Let us define $z = (u,\ v,\ \theta,\ \phi,\ z_1,\ z_2,\ z_3)^T \in X_2$, $z_r = (u,\ v,\ \phi,\ z_1,\ z_2,\ z_3)^T \in X_1$, and $\hat{z}, \hat{z}_r$ similarly. The inner product and the corresponding norm for X_2 can be given as

$$\langle z,\ \hat{z} \rangle_{X_2} = \langle z_r,\ \hat{z}_r \rangle_{X_1} + k\theta\hat{\theta}, \quad (5.217)$$

$$\|z\|_{X_2}^2 = \|z_r\|_{X_1}^2 + k\theta^2, \quad (5.218)$$

where the inner product and the norm for X_1 are given by (5.196), (5.197), respectively.

Theorem 5.35 *Consider the system S_o given by (5.216), where the operator $\mathcal{A}_2$ and $D(\mathcal{A}_2)$ are as given by (5.213), (5.215), respectively. Let the controllers be given by (5.180), (5.181), except for $f_3(t)$ which is given by (5.210). Let the transfer functions $\alpha_i(s)$, $i = 1, 2, 3$, given by (5.182) be SPR and $k > 0$. Then*

(i) $\mathcal{A}_2$ generates a C_0-semigroup of contractions $T_2(t)$ on X_2.

(ii) All solutions of (5.216) (classical or weak) asymptotically decay to zero. This result also holds even if $\alpha_1(s) = 0$ and/or $\alpha_2(s) = 0$, i.e., even if any one or both of the boundary controllers are not used, (cf. Theorem 5.33).

(iii) If $\alpha_1(s)$ and $\alpha_3(s)$ satisfy assumption 5.26, then the semigroup $T_2(t)$ is exponentially stable.

Remark 5.36 *The conclusion (iii), given above, holds even in $\alpha_2(s) = 0$, i.e., the corresponding controller is not used.*

Proof. (i) It is straightforward to show that

$$\langle z, \mathcal{A}_2 z \rangle_{X_2} = \langle z_r, \mathcal{A}_1 z_r \rangle_{X_1}. \quad (5.219)$$

Hence, $\mathcal{A}_2$ is dissipative. By using the block diagonal form of $\mathcal{A}_2$ and Theorem 5.33, it easily follows that $(\lambda I - \mathcal{A}_2)$ is onto for $\lambda > 0$. The operator $\mathcal{A}_2$ thus generates a C_0-semigroup of contractions $T_2(t)$ from the Lümer-Phillips theorem.

(ii) From Theorem 5.33 and the block diagonal form of $\mathcal{A}_2$, it easily follows that $(\lambda I - \mathcal{A}_2)^{-1}$ is compact for $\lambda > 0$. Let $z(0) \in D(\mathcal{A}_2)$, let $z(t)$ be the corresponding solution of (5.216), and define the "energy" $E_2(t)$ as

$$E_2(t) = \frac{1}{2}\langle z(t),\ z(t)\rangle_{X_2} = E_1(t) + \frac{1}{2}k\theta_e^2(t), \qquad (5.220)$$

where E_1 is given by (5.203). Since $z(t)$ is differentiable, so is $E_2(t)$; hence, by differentiating (5.220) and by using (5.198) and (5.219), we obtain

$$\dot{E}_2(t) = -\frac{1}{2}\sum_{i=1}^{3} z_i^T(t)Q_i z_i(t) - \frac{1}{2}\sum_{i=1}^{3}\left(\sqrt{2d_i}\,y_i(t) - z_i^T(t)q_i\right)^2. \qquad (5.221)$$

By LaSalle's invariance principle, all solutions of (5.216) asymptotically approach to the maximal invariant set in

$$\mathcal{O}_2 = \{z \in X_2 \mid \dot{E}_2(t) = 0\},$$

Clearly, the invariant solutions in $\mathcal{O}_2$ should also satisfy $z_i(t) = 0$; hence, $\dot{z}_i(t) = 0$ for $i = 1,2,3$, and by (5.180), (5.181) it follows that $y_i(t) = 0$, $f_1(t) = f_2(t) = 0$. Here, we will prove the asymptotic stability result *without* using $y_1(t) = y_2(t) = 0$. Hence, the asymptotic stability result will be valid even if $\alpha_1(s) = \alpha_2(s) = 0$, that is, when the boundary controllers are not applied. Note that in this case the only control input is the torque applied to the rigid body. Note that under the conditions stated above, the equations of motion of the system S_o which is invariant in $\mathcal{O}_2$ must satisfy

$$\begin{cases} y_{tt} + y_{xxxx} = 0, & 0 < x < 1, \\ y(0,t) = y_x(0,t) = 0, \\ y_{xx}(1,t) = y_{xxx}(1,t) = 0, \end{cases} \qquad (5.222)$$

and an additional condition

$$y_{xx}(0,t) - k\theta_e(t) = 0. \qquad (5.223)$$

Note that to obtain (5.223), we used $\dot{\theta}_e(t) = 0$, $\ddot{\theta}_e(t) = 0$ and (5.170), (5.175), (5.210). By differentiating (5.223), we obtain

$$y_{xxt}(0,t) = 0. \qquad (5.224)$$

By using Theorem 5.33, it can be easily shown that the only invariant solution of (5.222) and (5.224) in $\mathcal{O}_2$ is the zero solution. Hence, by LaSalle's invariance principle and Theorem 5.33, the asymptotic stability result follows.

(iii) Since $\mathcal{A}_2$ generates a C_0-semigroup of contractions, the solutions of (5.216) are bounded. By (ii), $(\lambda I - \mathcal{A}_2)^{-1}$ is compact for $\lambda > 0$; hence, the spectrum of $\mathcal{A}_2$ is discrete, see [84]. Since the solutions decay asymptotically to zero, there cannot be an eigenvalue of $\mathcal{A}_2$ on the imaginary axis. Hence, the imaginary axis belongs to the resolvent set of $\mathcal{A}_2$. To prove exponential decay, let $z = (u, \, v, \, \theta, \, \phi, \, z_1, \, z_2, \, z_3)^T \in D(\mathcal{A}_2)$ and let $y = (f, \, g, \, r_1, \, r_2, \, r_3, \, r_4, \, r_5)^T \in X_2$ be the solution of $(i\omega I - \mathcal{A}_2)z = y$, that is :

$$\begin{cases} i\omega u - v = f, \\ i\omega v + u_{xxxx} - xI_2 = g, \\ i\omega\theta - \phi = r_1, \\ i\omega\phi + I_2 = r_2, \\ (i\omega I - A_1)z_1 - b_1 v(1) = r_3, \\ (i\omega I - A_2)z_2 - b_2 v'(1) = r_4, \\ (i\omega I - A_3)z_3 - b_3 \phi = r_5, \end{cases} \tag{5.225}$$

where I_2 is given by (5.214).

Solving for θ from the third equation, we obtain

$$\theta = \frac{r_1 + \phi}{i\omega}. \tag{5.226}$$

Putting this in (5.225) and rearranging terms, we obtain

$$(i\omega I - \mathcal{A}_1) \begin{bmatrix} u \\ v \\ \phi \\ z_1 \\ z_2 \\ z_3 \end{bmatrix} = \begin{bmatrix} f \\ g + kx\theta \\ r_2 - k\theta \\ r_3 \\ r_4 \\ r_5 \end{bmatrix}, \tag{5.227}$$

where the operator $\mathcal{A}_1$ is given by (5.186). Let us define

$$z_r = (u, \, v, \, \phi, \, z_1, \, z_2, \, z_3)^T, \quad y_r = (f, \, g, \, r_2, \, r_3, \, r_4, \, r_5)^T,$$

$$y_f = (0, \, kx\theta, \, -k\theta, \, 0, \, 0, \, 0)^T.$$

From (5.227), we obtain

$$\|z_r\|_{X_1} \le \left\|(i\omega - \mathcal{A}_1)^{-1}\right\|_{X_1} \|y_r + y_f\|_{X_1}, \tag{5.228}$$

where the norm of X_1 is given by (5.197). Since $\mathcal{A}_1$ generates an exponentially decaying contraction semigroup, it follows from Corollary 3.36 that the following holds

$$\sup_{\omega \in \mathbf{R}} \left\|(i\omega - \mathcal{A}_1)^{-1}\right\|_{X_1} < \infty. \tag{5.229}$$

Combining (5.218), (5.226), (5.228) and (5.229), we obtain

$$\|z\|_{X_2} \le K_1 \|y\|_{X_2}, \tag{5.230}$$

for ω sufficiently large, where $K_1 > 0$ is an appropriate constant. This proves

$$\sup_{\omega \in \mathbf{R}} \left\| (i\omega - \mathcal{A}_2)^{-1} \right\|_{X_2} < \infty. \qquad (5.231)$$

Hence, it follows from Corollary 3.36 that $\mathcal{A}_2$ generates an exponentially decaying semigroup on X_2.   □

5.4   Stability robustness against small time delays

Most of the research in the area of boundary control of infinite dimensional flexible systems is concentrated on the problem of control and stabilization of conservative linear flexible systems, (e.g., strings or beams *without* damping). Such systems have infinitely many eigenvalues on the imaginary axis, and can be uniformly stabilized by using simple velocity feedback laws at their boundaries, see Sections 5.1, 5.2. However, it is well known now that these systems may become unstable when arbitrary small time delays were introduced into the feedback laws, see, e.g., [44], [45]. This lack of robustness and some other related results indicate that most of the conservative models in flexible structures may possess potential limitations for the feedback design, see [74]. Recently, in [43], it was argued that mathematical conservative models are never meant to represent physical systems for infinite time interval; hence, any control theory based on these models should attempt to justify its conclusions by using an appropriately damped version of the corresponding conservative model. This approach is then used in [115].

In this section, we consider a damped wave equation parameterized by a damping coefficient $a \geq 0$. When $a = 0$, this model reduces to the standard conservative wave equation. To stabilize this system, we apply the methodology introduced in Sections 5.1, 5.2. Following [43], we try to answer the following questions :

i: Does the proposed control law stabilize the conservative model and improve the stability of the damped model?

ii: Does the proposed control law robustly stabilize the damped model against small time delays in the feedback loop?

In the following subsection, we answer these questions for systems with static and dynamic feedback controllers separately. It should be emphasized that these controllers do not robustly stabilize the conservative wave equation against small time delays in the feedback loop.

To begin with, we first analyze, by using the frequency domain approach, the stability of the systems shown in Figure 5.2 when the feedback law is delayed, i.e., when we have $u(t) = -y(t - h)$ for some $h > 0$. Let $H(s)$

denote the transfer function of an SISO plant between its input u and its output y, see Figure 5.2. $H(s)$ is said to be well-posed if it is bounded on some right-half plane, and is said to be regular if it has a limit at $+\infty$ along the real axis. If we apply the unity feedback and set $u = r - y$, where r is the new input, then the closed-loop transfer function between r and y becomes $G^0(s) = H(s)(1 + H(s))^{-1}$. When there is a small time delay by ϵ in the feedback loop, the new transfer function $G^\epsilon(s)$ from r to y becomes $G^\epsilon(s) = H(s)(1 + e^{-\epsilon s}H(s))^{-1}$. We say that G^0 is robustly stable with respect to delays if there is an $\epsilon_0 > 0$ such that for any $\epsilon \in [0, \epsilon_0]$, G^ϵ is L_2-stable. If this property does not hold, then arbitrary small time delays destabilize G^0.

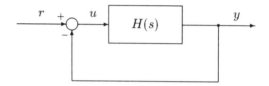

Figure 5.2: Unity Feedback System.

Let the transfer function $H(s)$ be meromorphic (i.e., analytic except at its poles) on the half plane $C_0 = \{s \in \mathbf{C} | Re\{s\} > 0\}$. Let $\mathcal{B}$ denote the (discrete) set of poles of H in C_0, and let γ^* be defined as :

$$\gamma^* = \lim_{|s| \to \infty} \sup_{s \in C_0 - \mathcal{B}} |H(s)|. \tag{5.232}$$

Theorem 5.37 *Let $H(s)$ be a regular transfer function and assume that $G^0 = H(1 + H)^{-1}$ is L_2-stable. Let γ^* be defined as in (5.232).*

(i) If $\gamma^* < 1$, then G^0 is robustly stable with respect to delays.

(ii) If $\gamma^* > 1$, then G^0 is not robustly stable with respect to delays.

Proof. See [97]. For a different version of this result, see [13]. □

Now let us turn to analyze the robustness with respect to small time delay of the following damped wave equation

$$\begin{cases} w_{tt} - w_{xx} + 2aw_t + a^2 w = 0, & 0 < x < 1, \ t \geq 0, \\ w(0, t) = 0, \quad w_x(1, t) = f(t), \end{cases} \tag{5.233}$$

where $a \geq 0$ is a damping constant, $f(t)$ is the boundary control input. For simplicity, some coefficients are chosen to be unity. The system given by (5.233) is first introduced in [45], and later investigated in [19], [43], [97], and [115]. For $a = 0$, the system given by (5.233) reduces to the standard conservative wave equation shown in Chapter 1.

It is advantageous to write (5.233) as an abstract equation. To this end, define $X = V \times H$, where $H = L^2(0, 1)$,

$$V = \{w \in H \mid w, w' \in H, w(0) = 0\},$$

and the operator $\mathcal{A}_d : X \to X$ is given by

$$\mathcal{A}_d \begin{bmatrix} u \\ v \end{bmatrix} = \begin{bmatrix} v \\ u'' - 2av - a^2 u \end{bmatrix}, \tag{5.234}$$

for $(u, v)^T \in D(\mathcal{A}_d)$. Before we specify the domain $D(\mathcal{A}_d)$, let us first define the following inner product and the norm on X:

$$\langle z, \hat{z} \rangle_1 = \int_0^1 v\hat{v}dx + \int_0^1 u'\hat{u}'dx + a^2 \int_0^1 u\hat{u}dx, \tag{5.235}$$

$$\|z\|_1^2 = \int_0^1 v^2 dx + \int_0^1 (u')^2 dx + a^2 \int_0^1 u^2 dx. \tag{5.236}$$

for $z = (u, v)^T$, $\hat{z} = (\hat{u}, \hat{v})^T \in X$.

To check whether the power form given by (5.20) holds, let

$$D_1 = \{u \in L^2 \mid u', u'' \in L^2, u(0) = 0\}$$

and let $z = (u, v)^T \in D_1 \times V$. Then we have

$$\langle z, \mathcal{A}_d z \rangle_1 = \int_0^1 v(u'' - 2av - a^2 u)dx + \int_0^1 u'v'dx + a^2 \int_0^1 u^2 dx$$

$$= u'(1)v(1) - 2a \int_0^1 v^2 dx. \tag{5.237}$$

Choose an output as

$$y(t) = w_t(1, t), \tag{5.238}$$

and the control input as

$$f(t) = -\alpha y(t), \tag{5.239}$$

where $\alpha \geq 0$ is a constant. Define

$$D(\mathcal{A}_d) = \{(u, v)^T \in X \mid u, u', u'' \in H, v, v' \in V,$$
$$u(0) = v(0) = 0, u'(1) + \alpha v(1) = 0\}.$$

The system specified by (5.233), (5.238), (5.239) can thus be written as

$$\dot{z} = \mathcal{A}_d z, \quad z(0) \in X. \tag{5.240}$$

We have the following theorem.

Theorem 5.38 *Consider the system given by (5.240). Let $\alpha \geq 0$ and $a \geq 0$. Then*

(i) *The operator $\mathcal{A}_d$ generates a C_0-semigroup of contractions on X.*

(ii) *For $a + \alpha > 0$, the solutions (classical or weak) of (5.240) asymptotically decay to zero.*

(iii) *For $a + \alpha > 0$, the semigroup is exponentially stable.*

Proof. By construction, $\mathcal{A}_d$ is dissipative. We can decompose $\mathcal{A}_d$ as $\mathcal{A}_d = \mathcal{A} + \mathcal{A}_b$ where $\mathcal{A}$ is the same as $\mathcal{A}_d$ with $a = 0$, and $\mathcal{A}_b = \mathcal{A}_d - \mathcal{A}$. Note that $\mathcal{A}_b$ is a bounded operator on X. It is easy to show that $\mathcal{A}$ generates a C_0-semigroup of contractions on X; hence, we have

$$\left\| (\lambda I - \mathcal{A})^{-1} \right\|_1 \leq \frac{M}{\lambda},$$

for $\lambda > 0$. Since

$$(\lambda I - \mathcal{A}_d)^{-1} = \left(I - (\lambda I - \mathcal{A})^{-1} \mathcal{A}_b \right)^{-1} (\lambda I - \mathcal{A})^{-1},$$

and $\mathcal{A}_b$ is bounded, it follows that for $\lambda > 0$ sufficiently large, $(\lambda I - \mathcal{A}_d)^{-1}$ exists, and since X is a Hilbert space, this property holds for all $\lambda > 0$. Since $(\lambda I - \mathcal{A})^{-1}$ is compact for $\lambda > 0$, so is $(\lambda I - \mathcal{A}_d)^{-1}$. We can easily show that

$$\sup_{\omega \in \mathbf{R}} \left\| (i\omega I - \mathcal{A}_d)^{-1} \right\|_1 < \infty.$$

Consequently, $\mathcal{A}_d$ generates an exponentially stable semigroup. □

Let us now consider the stability robustness of the above system with respect to small time delays. Here we assume that there is a small time delay in the feedback loop, that is, instead of (5.239), we let

$$f(t) = -\alpha y(t - h). \tag{5.241}$$

Denote by G^0 the system without delay, i.e., the system specified by (5.233), (5.238), and (5.239). We have the following corollary.

Corollary 5.39 *Let $a > 0$.*

(i) *If $\alpha < \frac{e^{2a}-1}{e^{2a}+1}$, then G^0 is robustly stable with respect to time delays,*

(ii) *If $\alpha > \frac{e^{2a}-1}{e^{2a}+1}$, then G^0 is not robustly stable with respect to time delays.*

Proof. First note that the related transfer function can be given as

$$H(s) = \frac{s\alpha}{(s+a)} \frac{1 - e^{-2(s+a)}}{1 + e^{-2(s+a)}}. \tag{5.242}$$

Since the system is exponentially stable, see Theorem 5.38, it follows that G^0 is L_2-stable; hence, Theorem 5.37 is applicable. We compute γ^* defined in (5.232). For this purpose, first note that $|1 - e^{-2(s+a)}| \leq 1 + e^{-2a}$, $|1 + e^{-2(s+a)}| \geq 1 - e^{-2a}$, and $|s/(s+a)| \leq 1$. This shows that $\gamma^* \leq \alpha \frac{e^{2a}+1}{e^{2a}-1}$. In fact the equality holds. To prove this, choose $s_n = 1/n + j(2n+1)\pi/2$ for $n \in \mathbf{N}$. It can be easily shown that $\lim_{n\to\infty} H(s_n) = \alpha \frac{e^{2a}+1}{e^{2a}-1}$. Consequently,

$$\gamma^* = \alpha \frac{e^{2a} + 1}{e^{2a} - 1}. \tag{5.243}$$

The result now follows from Theorem 5.37. □

According to Corollary 5.39, there does not exist a single $\alpha > 0$ such that G^0 is robustly stable with respect to small time delays for all $a > 0$. In other words, for a given $a > 0$, there always exists $\alpha > 0$ such that the closed-loop system becomes unstable when (5.241) is used instead of (5.239) for arbitrarily small $h > 0$.

Next, we will apply the SPR controllers to the stabilization of the model given by (5.233). Here, instead of the controller given by (5.239), we choose the following control law:

$$\dot{z} = A_1 z + b_1 y, \tag{5.244}$$

$$-f = c_1^T z + d_1 y, \tag{5.245}$$

where $A_1 \in \mathbf{R}^{n_1 \times n_1}$ is a constant matrix, $b_1, c_1 \in \mathbf{R}^{n_1}$ are constant vectors and $d_1 \in \mathbf{R}$ is a constant. We assume that the associated transfer function $\alpha_1(s)$ given by

$$\alpha_1(s) = c_1^T(sI - A_1)^{-1}b_1 + d_1 \tag{5.246}$$

is SPR. If we further assume that $\alpha_1(s)$ satisfies Assumption 5.26 for some $\gamma_1 > 0$, then there exist symmetric and positive matrices P, Q and a vector q, all having appropriate dimensions, such that the following holds.

$$A_1^T P + P A_1 = -qq^T - Q, \tag{5.247}$$

$$P b_1 - c_1 = \sqrt{2(d_1 - \gamma_1)}q, \tag{5.248}$$

and if $\alpha_1(s)$ is only SPR, we may take $\gamma_1 = 0$ in (5.248), see Lemma 5.20, (5.151), (5.152). Then the system (5.233), (5.238), (5.244), and (5.245) can be formulated as the following abstract equation:

$$\dot{z}_e = A_{de} z_e, \quad z_e(0) \in X_e, \tag{5.249}$$

where $X_e = X_1 \times \mathbf{R}^{n_1}$ and $\mathcal{A}_{de} : X_e \to X_e$ is given as

$$
\mathcal{A}_{de} \begin{bmatrix} u \\ v \\ z_1 \end{bmatrix} = \begin{bmatrix} v \\ u'' - 2av - a^2 u \\ A_1 z_1 + b_1 v(1) \end{bmatrix}, \tag{5.250}
$$

with

$$
\begin{aligned}
D(\mathcal{A}_{de}) \;=\; & \{(u,\ v,\ z_1)^T \in X_e \mid u, u', u'' \in H, v, v' \in H \\
& u(0) = v(0) = 0, \\
& u'(1) + c_1 z_1 + d_1 v(1) = 0\}.
\end{aligned}
$$

Let $z_e = (u,\ v,\ z_1)^T$, $\hat{z}_e = (\hat{u},\ \hat{v},\ \hat{z}_1)^T \in X_e$ and $z = (u,\ v)^T, \hat{z} = (\hat{u},\ \hat{v})^T \in X$. The inner product and the norm on X_e is defined as

$$
\langle z_e,\ \hat{z}_e \rangle_e = \langle z,\ \hat{z} \rangle_1 + \frac{1}{2}(z_1^T P \hat{z}_1 + \hat{z}_1^T P z_1), \tag{5.251}
$$

$$
\|z_e\|_e^2 = \|z\|_1^2 + z_1^T P z_1. \tag{5.252}
$$

Theorem 5.40 *Consider the system given by (5.249). Let $\alpha_1(s)$ be SPR and $a \geq 0$.*

(i) *The operator $\mathcal{A}_{de}$ generates a C_0-semigroup of contractions on X_e.*

(ii) *The solutions (classical or weak) of (5.249) asymptotically decay to zero.*

(iii) *Let $a > 0$, or $\alpha_1(s)$ satisfy assumption 5.26. Then the semigroup generated by $\mathcal{A}_{de}$ is exponentially stable.*

Proof. The proof easily follows from Theorem 5.38, and the results of Section 5.2. □

Now we state the following stability robustness result with respect to small time delays. Here, the system G^0 refers to the system given by (5.249).

Corollary 5.41 *Consider the system G^0 given by (5.249). Let $\alpha_1(s)$ be SPR and let $a > 0$.*

(i) *If $d_1 < \frac{e^{2a} - 1}{e^{2a} + 1}$, then G^0 is robustly stable with respect to time delays,*

(ii) *If $d_1 > \frac{e^{2a} - 1}{e^{2a} + 1}$, then G^0 is not robustly stable with respect to time delays.*

Proof. By using (5.233), (5.238), (5.244), and (5.245), it follows that the relevant transfer function $H(s)$ is given as

$$H(s) = \frac{s\alpha_1(s)}{(s+a)} \frac{1 - e^{-2(s+a)}}{1 + e^{-2(s+a)}}, \qquad (5.253)$$

where $\alpha_1(s)$ is given by (5.246), see [115]. Since by Theorem 5.40, the system is exponentially stable, it follows that G^0 is L_2-stable; hence, Theorem 5.37 is applicable. We compute γ^* given by (5.232). Note that $|g(s)|$ is bounded on C_0 and $\alpha_1(s) = d_1 + o(1/s)$ for large s. By using this, and following the proof of Corollary 5.39, it can be shown that

$$\gamma^* = d_1 \frac{e^{2a} + 1}{e^{2a} - 1}. \qquad (5.254)$$

The rest easily follows from Theorem 5.37. □

Remark 5.42 *Note that Corollary 5.41 is still valid when $d_1 = 0$, in which case the case (i) is trivially satisfied; hence, the corresponding G^0 is always robustly stable with respect to small time delays for all $a > 0$. Moreover, by Theorem 5.40, for the case $d_1 = 0$, the closed-loop system is exponentially stable for $a > 0$ and is asymptotically stable for $a = 0$. Hence, the controller given by (5.244) and (5.245) solves the problems stated in the introduction. Moreover, for the case $d_1 = 0$, both the corresponding controller transfer function $\alpha_1(s)$ and the open loop transfer function $H(s)$ are strictly proper, see (5.246) and (5.253). These points are important for actual implementation of $\alpha_1(s)$ and for the well-posedness of the model, see [74]. Also, for the application of the ideas presented in this section to some other damped models, see [115].*

5.5 Notes and references

The concept of passivity was introduced and developed in the late '50s for linear and finite dimensional systems in the context of passive electrical circuits. Most of the results developed in this context can be found in textbooks such as Guilliemin [61], Newcomb [125], and Anderson and Vongpanitlerd [2]. This idea was then extended to systems and control theory and was investigated in connection with feedback control systems; see Willems [164] and Desoer and Vidyasagar [50] for related results. Application of this concept to various nonlinear control systems can be found in [76] and [21].

In this chapter we developed a framework based on passivity for the analysis of certain boundary controlled conservative infinite dimensional systems. Boundary control of various conservative systems are extensively investigated in the literature; the wave equation was treated in Chen [26] and Lagnese [90]; the Euler-Bernoulli beam equation was studied in [28] and [30]; The Timoshenko beam equation was discussed in [85]. These results were extended to cover positive real controllers by Morgül in [114], [111], and [112]. The exponential stability results presented in this chapter utilize the energy multiplier method for which the textbook by Komornik [86] may be consulted for further reading. The lack of stability when small time delays are presented in the feedback for certain boundary control systems was first noted by Datko et al. in [45], and further in [43], [44], [97], and [115].

Chapter 6

Other Applications

In Chapter 4, we applied the energy multiplier method and frequency domain criteria to determine the exponential stability of linear dynamic systems which generate C_0-semigroups on Banach or Hilbert spaces. For systems which satisfy the spectrum-determined growth condition, the exponential stability can be examined by analyzing the spectrum distribution of the systems. One advantage of this method over others is that we are able to know not only whether the systems are exponentially stable, but also the exponential decay rate.

In this chapter, we shall consider two types of infinite dimensional systems — wave equations and thermoelastic equations with certain boundary stabilizers. Certainly, these systems can be viewed as linear Cauchy problems on appropriate Hilbert spaces. Therefore, the well-posedness of the problems can be treated in the framework of semigroups discussed in Chapter 2. Because the boundary stabilizers are usually designed to make the systems dissipative, it is routine to check the weak stability and asymptotic stability by using the theory developed in Chapter 3. Therefore, our major concern in this chapter is the exponential stability and the exponential decay rate for these systems. In Section 6.1, we discuss a general hyperbolic system consisting of a first order partial differential equation in time t and one-dimensional spatial variable x. This hyperbolic system, which originates from the work of Neves [124], is often encountered in the counter-flow heat exchanger process, gas absorber process, tubular reactor process, connected vibrating strings, and many other applications. In this general system, we allow only one boundary to possess dynamics, but the analysis method to be presented also can be applied to treat the cases with two dynamic boundaries. In order to exploit the solution properties that this general system possesses, we first consider three reduced systems which are simplified equations either by ignoring the couplings between state variables, or by replacing the dynamic boundary with a static boundary. For the most simple reduced system among the three, we show that the spectrum-determined growth condition holds. This, together with a number of results on the relationship among semigroups of each reduced system, demonstrates that the growth rate of the general system is determined completely by its own spectral bound and that of the most simple reduced system. Using this key result, we are able to show, in Sections 6.2-6.4, that the spectrum-determined growth condition is satisfied for wave equations describing the dynamics of serially connected vibrating strings with point stabilizers, as well as a vibrating cable with a tip mass. In Section 6.5 and 6.6, we consider a thermoelastic system with both Dirichlet-Dirichlet and Neumann-Dirichlet boundary conditions. It is shown that the spectrum-determined growth condition holds for these systems with the aid of a result due to Renardy [136] on spectrum-determined growth condition for a class of normal operators with bounded perturbations in Hilbert spaces. Finally, in Section 6.7, we present a counter-example given by Renardy [136] which shows that the spectrum alone cannot determine the growth rate for a two-dimensional hyperbolic system even with lower order perturbations. This negative result suggests that care must be taken when we treat practical infinite dimensional systems, and also demonstrates the importance of the theoretical developments presented in this chapter.

6.1 A General linear hyperbolic system

Consider the following one-dimensional linear homogeneous hyperbolic system

$$
\begin{cases}
\dfrac{\partial}{\partial t}\begin{bmatrix} u(x,t) \\ v(x,t) \end{bmatrix} + K(x)\dfrac{\partial}{\partial x}\begin{bmatrix} u(x,t) \\ v(x,t) \end{bmatrix} + C(x)\begin{bmatrix} u(x,t) \\ v(x,t) \end{bmatrix} = 0, 0 < x < 1, \\
\dfrac{d}{dt}[v(1,t) - Du(1,t)] = Fu(1,t) + Gv(1,t), \\
u(0,t) = Ev(0,t)
\end{cases}
\tag{6.1}
$$

for which the following assumptions are always imposed throughout the section:

(H1) $K(x)=\mathrm{diag}(\lambda_1(x),\ldots,\lambda_N(x),\mu_{N+1}(x),\ldots,\mu_n(x))$ is a diagonal $n \times n$ matrix with real entries $\lambda_i(x) \in C^1[0,1], \mu_j(x) \in C^1[0,1], \lambda_i(x) > 0, \mu_j(x) < 0$ for all $x \in [0,1]$ and $i = 1,2,\ldots,N, j = N+1, N+2,\ldots,n$.

(H2) $C(x)$ is an $n \times n$ matrix with entries $c_{ij}(x)$, which are continuous in $x \in [0,1]$.

(H3) if $i \neq j$ and $\lambda_i(x) = \lambda_j(x)$ (or $\mu_i(x) = \mu_j(x)$) somewhere in $[0,1]$, then $c_{ij}(x) \equiv 0$ on $[0,1]$.

(H4) $u(x) = (u_1(x), u_2(x),\ldots,u_N(x))^T$ is a column vector in $\mathbf{R}^N$ (or $\mathbf{C}^N$) and $v(x) = (v_{N+1}(x), v_{N+2}(x),\ldots,v_n(x))^T$ is a column vector in $\mathbf{R}^{n-N}$ (or $\mathbf{C}^{n-N}$).

(H5) D, E, F and G are real (or complex) constant matrices of appropriate sizes.

We shall find that it is convenient to denote a column vector u by $u = \text{col}[u_i]$ and matrices D, E, F, and G by $D = \{d_{ij}\}, E = \{e_{ij}\}, F = \{f_{ij}\}$, and $G = \{g_{ij}\}$, respectively.

For any $p, 1 \leq p < \infty$, let $X = (L^p(0,1))^n \times \mathbf{R}^{n-N}$ (or $(L^p(0,1))^n \times \mathbf{C}^{n-N}$), where $L^p(0,1)$ is the usual L^p function space with norm $\|\cdot\|_p$. Define a linear operator $A : D(A)(\subset X) \to X$ by

$$A(u, v, d) = \left[-K(x)\frac{d}{dx}\begin{bmatrix} u(x) \\ v(x) \end{bmatrix} - C(x)\begin{bmatrix} u(x) \\ v(x) \end{bmatrix}, Fu(1) + Gv(1)\right],$$

$$D(A) = \{(u, v, d) \in X \mid (u, v) \in (H^{1,p}(0,1))^N \times (H^{1,p}(0,1))^{n-N},$$
$$u(0) = Ev(0), d = v(1) - Du(1)\} \tag{6.2}$$

where $H^{1,p}(0,1)$ denotes the usual Sobolev space. With the operator A at hand, we can write (6.1) as an evolution equation on X:

$$\frac{dW(t)}{dt} = AW(t) \tag{6.3}$$

with $W(t) = (u(\cdot, t), v(\cdot, t), v(1, t) - Du(1, t))$.

Lemma 6.1 *The operator A defined by (6.2) has compact resolvent and hence $\sigma(A)$ consists of only isolated eigenvalues. Furthermore, $\sigma(A)$ consists of the zeros of an entire function.*

Proof. Given $(f, g, b) \in X$, we solve

$$(\lambda - A)(u, v, d) = (f, g, b),$$

that is,

$$\begin{cases} \dfrac{\partial}{\partial x}\begin{bmatrix} u(x) \\ v(x) \end{bmatrix} = -K^{-1}(x)[\lambda + C(x)]\begin{bmatrix} u(x) \\ v(x) \end{bmatrix} + K^{-1}(x)\begin{bmatrix} f(x) \\ g(x) \end{bmatrix}, \\ \lambda d = b + Fu(1) + Gv(1), u(0) = Ev(0), d = v(1) - Du(1). \end{cases} \tag{6.4}$$

Denote by $M(x, y, \lambda)$ the fundamental matrix of the system

$$\frac{d}{dx}\begin{bmatrix} u \\ v \end{bmatrix} = -K^{-1}(x)[\lambda + C(x)]\begin{bmatrix} u \\ v \end{bmatrix}.$$

It follows from (6.4) that

$$\begin{bmatrix} u(x) \\ v(x) \end{bmatrix} = M(x, 0, \lambda)\begin{bmatrix} E \\ I \end{bmatrix} v(0) + \int_0^x M(x, y, \lambda)K^{-1}(y)\begin{bmatrix} f(y) \\ g(y) \end{bmatrix} dy. \tag{6.5}$$

On the other hand, from the boundary conditions in (6.4), we see that

$$b = (-\lambda D - F, \lambda - G) \begin{bmatrix} u(1) \\ v(1) \end{bmatrix}$$

$$= (-\lambda D - F, \lambda - G) M(1, 0, \lambda) \begin{bmatrix} E \\ I \end{bmatrix} v(0)$$

$$+ (-\lambda D - F, \lambda - G) \int_0^1 M(1, y, \lambda) K^{-1}(y) \begin{bmatrix} f(y) \\ g(y) \end{bmatrix} dy.$$

Consequently,

$$H(\lambda) v(0) = b + \int_0^1 (\lambda D + F, G - \lambda) M(1, y, \lambda) K^{-1}(y) \begin{bmatrix} f(y) \\ g(y) \end{bmatrix} dy, \quad (6.6)$$

where

$$H(\lambda) = -(\lambda D + F, G - \lambda) M(1, 0, \lambda) \begin{bmatrix} E \\ I \end{bmatrix}. \quad (6.7)$$

Defining $h(\lambda)=\det H(\lambda)$, we see that $\lambda \in \sigma(A)$ if and only if λ is a zero of the entire function $h(\lambda)$. When $h(\lambda) \neq 0, \lambda \in \rho(A)$ and $R(\lambda, A)(f, g, b) = (u, v, d)$ where (u, v) is given by (6.5) with $v(0)$ determined by (6.6) and $d = v(1) - Du(1)$. It is seen from (6.5) that $R(\lambda, A)$ is compact for any $\lambda \in \rho(A)$. □

Theorem 6.2 *The operator A defined by (6.2) generates a C_0-semigroup $T(t)$ on X.*

Proof. We need only to prove the assertion for the case $C \equiv 0$, because C is a bounded operator by assumption (H2), and bounded perturbations do not affect C_0-semigroup generations. Assume, for simplicity, that X is real.

The trick is to define an equivalent norm on X by properly choosing some positive weighting functions $f_i(x), 1 \leq i \leq N$ and $g_i(x), N+1 \leq i \leq n$. Namely, define the norm on X as

$$\|(u, v, d)\|^p = \sum_{i=1}^N \int_0^1 f_i(x) |u_i(x)|^p dx + \sum_{j=N+1}^n \int_0^1 g_i(x) |v_i(x)|^p dx + \sum_{j=N+1}^n |d_i|^p.$$

$$(6.8)$$

It is easily verified that X^*, the dual space of X, consists of all elements (u^*, v^*, d^*) with

$$u_i^*(x) = \|(u, v, d)\|^{2-p} |u_i(x)|^{p/q} \operatorname{sign}(u_i(x)), \quad 1 \leq i \leq N,$$
$$v_j^*(x) = \|(u, v, d)\|^{2-p} |v_j(x)|^{p/q} \operatorname{sign}(v_j(x)), \quad N+1 \leq j \leq n,$$
$$d_j^* = \|(u, v, d)\|^{2-p} |d_j|^{p/q} \operatorname{sign}(d_j), \quad N+1 \leq j \leq n,$$

where q denotes the conjugate number of p, i.e., q satisfies $1/p + 1/q = 1$.

For any $(u, v, d) \in D(A), (u, v, d) \neq 0$ and any $(u^*, v^*, d^*) \in F((u, v, d)) \subset X^*$, where F denotes the duality set, a direct calculation shows that

$$\|(u, v, d)\|^{p-2}\langle(u^*, v^*, d^*), \ A(u, v, d)\rangle$$

$$= \sum_{i=1}^{N} \int_0^1 -\lambda_i(x) f_i(x) \frac{d}{dx} |u_i(x)|^p dx + \sum_{j=N+1}^{n} \int_0^1 -\mu_j(x) g_j(x) \frac{d}{dx} |v_j(x)|^p dx$$

$$+ \langle Fu(1) + Gv(1), \ [v(1) - Du(1)]^*\rangle$$

$$= -\sum_{i=1}^{N} \lambda_i(1) f_i(1) |u_i(1)|^p - \sum_{j=N+1}^{n} \mu_j(1) g_j(1) |v_j(1)|^p$$

$$+ \sum_{i=1}^{N} \lambda_i(0) f_i(0) |u_i(0)|^p + \sum_{j=N+1}^{n} \mu_j(0) g_j(0) |v_j(0)|^p$$

$$+ \sum_{i=1}^{N} \int_0^1 |u_i(x)|^p \frac{d}{dx} [\lambda_i(x) f_i(x)] dx + \sum_{j=N+1}^{n} \int_0^1 |v_j(x)|^p \frac{d}{dx} [\mu_j(x) g_j(x)] dx.$$

$$+ \langle Fu(1) + Gv(1), \ [v(1) - Du(1)]^*\rangle = I_1 + I_2 + I_3 + I_4.$$

We estimate I_i separately. It is clear from the expression of I_3 that

$$I_3 \leq C_0 \|(u, v, d)\|^p \tag{6.9}$$

where $C_0 = \max i, j \max_{x \in [0,1]} \{\frac{d}{dx}[\lambda_i(x) f_i(x)], \frac{d}{dx}[\mu_j(x) g_j(x)]\}$. Noting that $u_i(0) = \sum_{j=N+1}^{n} e_{ij} v_j(0)$, we see that

$$I_2 = \sum_{i=1}^{N} \lambda_i(0) f_i(0) |u_i(0)|^p + \sum_{j=N+1}^{n} \mu_j(0) g_j(0) |v_j(0)|^p$$

$$\leq \sum_{j=N+1}^{n} [\mu_j(0) g_j(0) + \sum_{i=1}^{N} \lambda_i(0) f_i(0) (\sum_{k=N+1}^{n} |e_{ik}|^q)^{p/q}] |v_j(0)|^p.$$

$$\tag{6.10}$$

Because $\lambda_i(0) > 0$ and $\mu_j(0) < 0$ from (H1), we can always find $g_j(0) > 0$ and $f_i(0) > 0$ such that

$$\mu_j(0) g_j(0) + \sum_{i=1}^{N} \lambda_i(0) f_i(0) (\sum_{k=N+1}^{n} |e_{ik}|^q)^{p/q} \leq 0, \quad N+1 \leq j \leq n \tag{6.11}$$

holds, which implies that $I_2 \leq 0$.

We now estimate I_4. Using the inequalities $(|a| + |b|)^p \leq 2^p(|a|^p + |b|^p)$ and $|a|^{1/p}|b|^{1/q} \leq |a|/p + |b|/q$ which hold for any real a and b, we have

$$I_4 \leq \sum_{j=N+1}^{n} |\sum_{i=1}^{N} f_{ji} u_i(1) + \sum_{i=N+1}^{n} g_{ji} v_i(1)| |v_j(1) - \sum_{i=1}^{N} d_{ji} u_i(1)|^{p/q}$$

$$\leq \frac{1}{p} \sum_{j=N+1}^{n} |\sum_{i=1}^{N} f_{ji}u_i(1) + \sum_{i=N+1}^{n} g_{ji}v_i(1)|^p$$

$$+\frac{1}{q} \sum_{j=N+1}^{n} |v_j(1) - \sum_{i=1}^{N} d_{ji}u_i(1)|^p$$

$$\leq \frac{2^p}{p} \sum_{j=N+1}^{n} \left[|\sum_{i=1}^{N} f_{ji}u_i(1)|^p + |\sum_{i=N+1}^{n} g_{ji}v_i(1)|^p \right] + \frac{1}{q} \|(u,v,d)\|^p$$

$$\leq \frac{2^p}{p} \sum_{j=N+1}^{n} \left(\sum_{i=1}^{N} |f_{ji}|^q \right)^{p/q} \sum_{i=1}^{N} |u_i(1)|^p$$

$$+\frac{2^p}{p} \sum_{j=N+1}^{n} \left(\sum_{i=N+1}^{n} |g_{ji}|^q \right)^{p/q} \sum_{i=N+1}^{n} |v_i(1)|^p + \frac{1}{q} \|(u,v,d)\|^p$$

$$= \sum_{i=1}^{N} \alpha_i |u_i(1)|^p + \sum_{j=N+1}^{n} \beta_j |v_j(1)|^p + \frac{1}{q} \|(u,v,d)\|^p$$

with α_i and β_j denoting the obvious constants.

Finally, it is verified that

$$I_1 + I_4 - \frac{1}{q} \|(u,v,d)\|^p$$

$$\leq \sum_{i=1}^{N} [-\lambda_i(1)f_i(1) + \alpha_i]|u_i(1)|^p + \sum_{j=N+1}^{n} [\beta_j - \mu_j(1)g_j(1)]|v_j(1)|^p$$

$$\leq \sum_{i=1}^{N} [-\lambda_i(1)f_i(1) + \alpha_i]|u_i(1)|^p$$

$$+2^p \sum_{j=N+1}^{n} |\beta_j - \mu_j(1)g_j(1)||v_j(1) - \sum_{i=1}^{N} d_{ji}u_i(1)|^p$$

$$+2^p \sum_{j=N+1}^{n} |\beta_j - \mu_j(1)g_j(1)||\sum_{i=1}^{N} d_{ji}u_i(1)|^p$$

$$\leq \sum_{i=1}^{N} [-\lambda_i(1)f_i(1) + \alpha_i]|u_i(1)|^p$$

$$+ \sum_{j=N+1}^{n} 2^p |\beta_j - \mu_j(1)g_j(1)||v_j(1) - \sum_{i=1}^{N} d_{ji}u_i(1)|^p$$

$$+2^p \sum_{j=N+1}^{n} |\beta_j - \mu_j(1)g_j(1)|(\sum_{i=1}^{N} |d_{ji}|^q)^{p/q} \sum_{i=1}^{N} |u_i(1)|^p$$

$$= \sum_{i=1}^{N} \left[-\lambda_i(1)f_i(1) + \alpha_i + 2^p \sum_{j=N+1}^{n} |\beta_j - \mu_j(1)g_j(1)|(\sum_{i=1}^{N} |d_{ji}|^q)^{p/q} \right] |u_i(1)|^p$$

$$+ \sum_{j=N+1}^{n} 2^p |\beta_j - \mu_j(1)g_j(1)||v_j(1) - \sum_{i=1}^{N} d_{ji}u_i(1)|^p.$$

If we choose $f_i(1) > 0, g_j(1) > 0$ such that

$$\begin{cases} -\lambda_i(1)f_i(1) + \alpha_i + 2^p \sum_{j=N+1}^{n} |\beta_j - \mu_j(1)g_j(1)|(\sum_{i=1}^{N} |d_{ji}|^q)^{p/q} \leq 0, \\ 2^p |\beta_j - \mu_j(1)g_j(1)| \leq C \end{cases}$$
$$(6.12)$$

for any $1 \leq i \leq N$ and $N + 1 \leq j \leq n$, then

$$I_1 + I_4 \leq (C + \frac{1}{q}) \, \|(u, v, d)\|^p.$$

The estimations of I_i above show that there exists a constant M such that

$$\langle (u^*, v^*, d^*), \, A(u, v, d) \rangle \leq M \, \|(u, v, d)\|^2 \qquad (6.13)$$

if we choose the weighting functions $f_i(x)$ and $g_i(x)$ such that they satisfy (6.11) and (6.12) and define the norm in X according to (6.8).

Because $A - M$ is dissipative and A has the properties stated in Lemma 6.1, by the standard argument, we conclude that A generates a C_0-semigroup on X. □

We now consider three reduced systems associated with (6.1):

$$\begin{cases} \dfrac{\partial}{\partial t} \begin{bmatrix} u(x,t) \\ v(x,t) \end{bmatrix} + K(x)\dfrac{\partial}{\partial x} \begin{bmatrix} u(x,t) \\ v(x,t) \end{bmatrix} + C_0(x) \begin{bmatrix} u(x,t) \\ v(x,t) \end{bmatrix} = 0, \\ \dfrac{d}{dt}[v(1,t) - Du(1,t)] = Fu(1,t) + Gv(1,t), \\ u(0,t) = Ev(0,t). \end{cases}$$
$$(6.14)$$

$$\begin{cases} \dfrac{\partial}{\partial t} \begin{bmatrix} u(x,t) \\ v(x,t) \end{bmatrix} + K(x)\dfrac{\partial}{\partial x} \begin{bmatrix} u(x,t) \\ v(x,t) \end{bmatrix} + C_0(x) \begin{bmatrix} u(x,t) \\ v(x,t) \end{bmatrix} = 0, \\ \dfrac{d}{dt}[v(1,t) - Du(1,t)] = 0, \\ u(0,t) = Ev(0,t). \end{cases}$$
$$(6.15)$$

$$\begin{cases} \dfrac{\partial}{\partial t} \begin{bmatrix} u(x,t) \\ v(x,t) \end{bmatrix} + K(x)\dfrac{\partial}{\partial x} \begin{bmatrix} u(x,t) \\ v(x,t) \end{bmatrix} + C_0(x) \begin{bmatrix} u(x,t) \\ v(x,t) \end{bmatrix} = 0, \\ v(1,t) = Du(1,t), u(0,t) = Ev(0,t), \end{cases}$$
$$(6.16)$$

where $C_0(x) = \mathrm{diag}(c_{ii}(x))$.

Each system above can be viewed as an evolution equation $\dot{W}(t) = A_i W(t), i = 2, 3, 4$ where A_i and $D(A_i)$ are defined similarly as A and $D(A)$ in (6.3). X still serves as the state spaces for (6.14) and (6.15) as for (6.1). But for (6.16), a closed subspace $X^0 = \{(u, v, 0) \in X\}$ of X should be taken as the state space. X^0 will be identified with $(L^p(0,1))^n$ when there is no confusion from context.

Similar to Lemma 6.1, we can show that each A_i generates a C_0-semigroup $T_i(t)$, respectively, on their state spaces.

The main result of this section can be stated as

Theorem 6.3 *For any $(u, v, d) \in X$, denote by $\Pi : X \to X^0$; $\Pi(u, v, d) = (u, v, 0)$ the projection from X to X^0. Under assumptions (H1)-(H5), $T(t) - T_4(t)\Pi$ is compact on X for each $t \geq 0$. Therefore, $\omega_{ess}(A) = \omega_{ess}(A_4)$.*

The proof of this theorem is completed by using the results of the following lemmas which show that $T(t) - T_2(t), T_2(t) - T_3(t)$ and $T_3(t) - T_4(t)\Pi$ are compact for each $t \geq 0$, respectively. First of all, the compactness of $T_3(t) - T_4(t)\Pi$ follows from

$$T_3(t) - T_4(t)\Pi = T_3(t) - T_3(t)\Pi = T_3(t)(I - \Pi),$$

where the first equality follows from the fact that the restriction of $T_3(t)$ on X^0 is equal to $T_4(t)$. Since for each $t \geq 0$ and $(u, v, d) \in X$, $T_3(t)(I - \Pi)(u, v, d) = dT_3(t)(0, 0, 1)$ is a rank one bounded operator, $T_3(t) - T_4(t)\Pi$ is compact for every $t \geq 0$.

Before proceeding, we state a theorem on the compactness of an integral operator on L^p space, which is essential for the proof in the sequel.

Lemma 6.4 *Let $\mathbf{I} \subset \mathbf{R}$ be a compact interval. If*

- *$\alpha(s, x) : \mathbf{I} \times [a, b] \to [a, b]$ is continuous and C^1 in s with $(\partial \alpha / \partial s)(s, x) \neq 0$ everywhere,*

- *$r_1(x), r_2(x) : [0, 1] \to \mathbf{I}$ are continuous, and*

- *$K : [a, b] \times \mathbf{I} \to \mathbf{R}(or\ \mathbf{C})$ is continuous,*

then $\mathcal{K} : L^p(a, b) \to L^p(a, b)$ defined by

$$(\mathcal{K}f)(x) = \int_{r_1(x)}^{r_2(x)} K(x, s) f(\alpha(s, x)) ds$$

is compact.

Proof. Making change of variable $\alpha(s, x) = \tau$ and denoting by $\beta(\tau, x)$ its inverse $s = \beta(\tau, x)$, we have

$$(\mathcal{K}f)(x) = \int_{\alpha(r_1(x),x)}^{\alpha(r_2(x),x)} K(x, \beta(\tau, x))\beta_\tau'(\tau, x)f(\tau)d\tau,$$

which is obviously a compact operator on $L^p(a, b)$. □

The case where $\alpha(s, x) = x$ provides an example that condition $(\partial\alpha/\partial s) \neq 0$ cannot be removed in Lemma 6.4.

Lemma 6.5 $T_2(t) - T_3(t)$ *is compact for every* $t \geq 0$.

Proof. We first note that if $T_2(t) - T_3(t)$ is compact in $t \in [0, t_0]$ for some $t_0 > 0$ then so is it for every $t \geq 0$. Indeed, for any $\epsilon > 0, \epsilon \leq t_0$,

$$T_2(t_0 + \epsilon) - T_3(t_0 + \epsilon)$$
$$= T_2(\epsilon)T_2(t_0) - T_3(\epsilon)T_3(t_0)$$
$$= T_2(\epsilon)[T_2(t_0) - T_3(t_0)] + [T_2(\epsilon) - T_3(\epsilon)]T_3(t_0),$$

and so $T_2(t_0 + \epsilon) - T_3(t_0 + \epsilon)$ is compact. Iterating this process, we see that $T_2(t) - T_3(t)$ is compact for every $t \geq 0$. We show that if we take $t_0 = c$ with

$$c = \min\{\frac{1}{\lambda_i(x)}, \frac{1}{-\mu_j(x)} \mid x \in [0, 1], 1 \leq i \leq N, N + 1 \leq j \leq n\}, \quad (6.17)$$

then $T_2(t) - T_3(t)$ is compact for every $t \in [0, t_0]$. We do this by directly solving (6.14) and (6.15) using the method of characteristics.

Let $T_2(t)(u_0, v_0, d_0) = (u, v, d)$ and write

$$u(x, t) = \text{col } [u_i(x, t)], \quad v(x, t) = \text{col } [v_j(x, t)],$$
$$u_0(x) = \text{col } [u_{i0}(x)], \quad v_0(x) = \text{col } [v_{j0}(x)].$$

Integrating along the characteristic line, we find the general solution of (6.14) which is given by

$$\begin{cases} u_i(x, t) = \phi_i(t - \int_0^x \frac{1}{\lambda_i(\tau)}d\tau)e^{-\int_0^x c_{ii}(\rho)/\lambda_i(\rho)d\rho}, & 1 \leq i \leq N, \\ v_j(x, t) = \psi_j(t + \int_x^1 \frac{1}{\mu_j(\tau)}d\tau)e^{-\int_0^x c_{jj}(\rho)/\mu_j(\rho)d\rho}, & N + 1 \leq j \leq n, \end{cases} \quad (6.18)$$

where ϕ_i and ψ_j, defined on $[-\int_0^1 \frac{1}{\lambda_i(\tau)}d\tau, \infty)$ and $[\int_0^1 \frac{1}{\mu_j(\tau)}d\tau, \infty)$, respectively, are arbitrary functions to be determined by the initial and boundary conditions. By the initial conditions, we have

$$\phi_i\left(-\int_0^x \frac{1}{\lambda_i(\tau)}d\tau\right) = u_{i0}(x)e^{\int_0^x c_{ii}(\rho)/\lambda_i(\rho)d\rho}, \quad 0 \leq x \leq 1,$$

$$\psi_j\left(\int_x^1 \frac{1}{\mu_j(\tau)}d\tau\right) = v_{j0}(x)e^{\int_0^x c_{jj}(\rho)/\mu_j(\rho)d\rho}, \quad 0 \le x \le 1,$$

or

$$\begin{cases} \phi_i(\theta) = u_{i0}(x_i(\theta))e^{\int_0^{x_i(\theta)} c_{ii}(\rho)/\lambda_i(\rho)d\rho}, & \theta \in [-\int_0^1 \frac{1}{\lambda_i(\tau)}d\tau, 0], \\ \psi_j(\theta) = v_{j0}(x_j(\theta))e^{\int_0^{x_j(\theta)} c_{jj}(\rho)/\mu_j(\rho)d\rho}, & \theta \in [\int_0^1 \frac{1}{\mu_j(\tau)}d\tau, 0], \end{cases} \quad (6.19)$$

where $x_i(\theta)$ and $x_j(\theta)$ are continuously differentiable functions which are determined by

$$-\int_0^{x_i(\theta)} \frac{1}{\lambda_i(\tau)}d\tau = \theta, \int_{x_j(\theta)}^1 \frac{1}{\mu_j(\tau)}d\tau = \theta, x_i'(\theta) \ne 0, x_j'(\theta) \ne 0. \quad (6.20)$$

Substituting (6.19) into (6.18), one has

$$\begin{cases} u_i(x,t) = u_{i0}(x_i(\theta_i(x,t)))e^{\int_x^{x_i(\theta_i(x,t))} c_{ii}(\rho)/\lambda_i(\rho)d\rho}, & t \le \int_0^x \frac{1}{\lambda_i(\tau)}d\tau, \\ v_j(x,t) = v_{j0}(x_j(\theta_j(x,t)))e^{\int_x^{x_j(\theta_j(x,t))} c_{jj}(\rho)/\mu_j(\rho)d\rho}, & t \le -\int_x^1 \frac{1}{\mu_j(\tau)}d\tau, \end{cases} \quad (6.21)$$

where $\theta_i(x,t) = t - \int_0^x \frac{1}{\lambda_i(\tau)}d\tau$ and $\theta_j(x,t) = t + \int_x^1 \frac{1}{\mu_j(\tau)}d\tau$.

Next, note that $t - \int_0^x \frac{1}{\lambda_i(\tau)}d\tau + \int_0^1 \frac{1}{\mu_j(\tau)}d\tau \le 0$ for any $1 \le i \le N$ and $N+1 \le j \le n$ when $t \le c$. From the boundary condition $u(0,t) = Ev(0,t)$, which is rewritten as

$$\text{col }[\phi_i(t)] = E \text{ col}\left[\psi_j(t + \int_0^1 \frac{d\tau}{\mu_j(\tau)})\right], \quad t \ge 0$$

in terms of ϕ_i and ψ_j, we have, for $t \le c, 0 \le x \le 1$, and $t \ge \int_0^x \frac{1}{\lambda_i(\tau)}d\tau$, that

$$u_i(x,t)$$
$$= \phi_i(t - \int_0^x \frac{1}{\lambda_i(\tau)}d\tau)e^{-\int_0^x c_{ii}(\rho)/\lambda_i(\rho)d\rho}$$
$$= e^{-\int_0^x c_{ii}(\rho)/\lambda_i(\rho)d\rho} \sum_{j=N+1}^n e_{ij}\psi_j(t - \int_0^x \frac{1}{\lambda_i(\tau)}d\tau + \int_0^1 \frac{d\tau}{\mu_j(\tau)})$$
$$= e^{-\int_0^x c_{ii}(\rho)/\lambda_i(\rho)d\rho} \sum_{j=N+1}^n e_{ij}v_{j0}(x_j(\theta_{ij}(x,t)))e^{\int_0^{x_j(\theta_{ij}(x,t))} c_{jj}(\rho)/\mu_j(\rho)d\rho},$$

$$(6.22)$$

where $\theta_{ij}(x,t) = t - \int_0^x \frac{1}{\lambda_i(\tau)}d\tau + \int_0^1 \frac{d\tau}{\mu_j(\tau)}$ and $E = \{e_{ij}\}$.

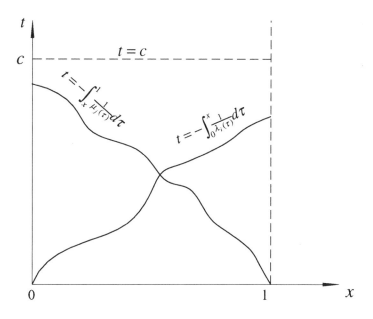

Figure 6.1: Characteristic Lines

(6.21) and (6.22) clearly show that for all $x \in [0,1], t \in [0,c], u_i(x,t)$ is determined completely by the initial conditions and the matrix E; in other words, it is independent of F and G.

Finally, from

$$
\begin{aligned}
d(t) &= v(1,t) - Du(1,t) \\
&= d_0 + \int_0^t [Fu(1,\tau) + Gv(1,\tau)]d\tau \\
&= d_0 + \int_0^t [(GD + F)u(1,\tau) + Gd(\tau)]d\tau \\
&= d_0 + G\int_0^t d(\tau)d\tau + \int_0^t (GD + F)u(1,\tau)d\tau,
\end{aligned}
$$

we have

$$
d(t) = v(1,t) - Du(1,t) = e^{Gt}d_0 + \int_0^t e^{G(t-s)}[GD + F]u(1,s)ds. \quad (6.23)
$$

Therefore,

$$v(1,t) = Du(1,t) + e^{Gt}d_0 + \int_0^t e^{G(t-s)}[GD + F]u(1,s)ds, t \geq 0. \quad (6.24)$$

It follows that

$$\mathrm{col}\left[\psi_j(t)e^{-\int_0^1 c_{jj}(\rho)/\mu_j(\rho)d\rho}\right]$$

$$= D\,\mathrm{col}\left[\phi_i(t - \int_0^1 \frac{d\tau}{\lambda_i(\tau)})e^{-\int_0^1 c_{ii}(\rho)/\lambda_i(\rho)d\rho}\right] + e^{Gt}d_0$$

$$+ \int_0^t e^{G(t-s)}[GD + F]\,\mathrm{col}\left[\phi_i(s - \int_0^1 \frac{d\tau}{\lambda_i(\tau)})e^{-\int_0^1 c_{ii}(\rho)/\lambda_i(\rho)d\rho}\right]ds.$$

If $t \leq c$, then $t - \int_0^1 \frac{1}{\lambda_i(\tau)}d\tau \leq 0$ for any $1 \leq i \leq N$. Therefore, for $c \geq t \geq -\int_x^1 \frac{1}{\mu_j(\tau)}d\tau$,

$$v_j(x,t) = \psi_j(t + \int_x^1 \frac{1}{\mu_j(\tau)}d\tau)e^{-\int_0^x c_{jj}(\rho)/\mu_j(\rho)d\rho}$$

$$= e^{\int_x^1 c_{jj}(\rho)/\mu_j(\rho)d\rho}\left[\Delta_j + \sum_{i=1}^N d_{ij}u_{i0}(x_i(\tilde{\theta}_{ij}(x,t)))e^{-\int_{x_i(\tilde{\theta}_{ij}(x,t))}^1 c_{ii}(\rho)/\lambda_i(\rho)d\rho}\right.$$

$$+ \int_0^{t+\int_x^1 \frac{d\tau}{\mu_j(\tau)}} \sum_{i=1}^N f_{ij}(t + \int_x^1 \frac{1}{\mu_j(\tau)}d\tau - s)u_{i0}(x_i(\theta_i(1,s)))$$

$$\left. \times e^{-\int_{x_i(\theta_i(1,s))}^1 c_{ii}(\rho)/\lambda_i(\rho)d\rho}ds\right]$$

$$N+1 \leq j \leq n,$$

$$(6.25)$$

where Δ_j represents the j-th row of $e^{Gt}d_0$, $e^{G(t-s)}[GD + F] = \{f_{ij}(t - s)\}$, $d = \{d_{ij}\}$, and $\tilde{\theta}_{ij}(x,t) = t + \int_x^1 \frac{1}{\mu_j(\tau)}d\tau - \int_0^1 \frac{1}{\lambda_i(\tau)}d\tau$.

Up to now, we have found the explicit solution of (6.14) in $(t,x) \in [0,c] \times [0,1]$ which is expressed by (6.21), (6.22) and (6.25). Let $T_3(t)(u_0,v_0,d_0) = (\tilde{u},\tilde{v},\tilde{d})$. Note that (6.15) is a special case of (6.14) for $F = G = 0$, we have, by (6.21) and (6.22),

$$u(x,t) - \tilde{u}(x,t) = 0, \text{ for } 0 \leq t \leq c, 0 \leq x \leq 1, \quad (6.26)$$

and from (6.23)

$$d(t) - \tilde{d}(t)$$

$$= (e^{Gt} - 1)d_0 + \int_0^t e^{G(t-s)}(GD + F)u(1,s)ds$$

$$= (e^{Gt} - 1)d_0 + \int_0^t \sum_{i=1}^N f_{ij}(t-s)u_{i0}(x_i(\theta_i(1,s)))e^{\int_x^{x_i(\theta_i(1,s))} c_{ii}(\rho)/\lambda_i(\rho)d\rho}ds.$$

$$(6.27)$$

Moreover, for $0 \le t \le -\int_x^1 \frac{1}{\mu_j(\tau)}d\tau$ and $0 \le x \le 1$, by (6.21) we have

$$v_j(x,t) - \tilde{v}_j(x,t) = 0,$$

and for $-\int_x^1 \frac{1}{\mu_j(\tau)}d\tau \le t \le c$, by (6.25) we have

$$v_j(x,t) - \tilde{v}_j(x,t)$$
$$= e^{\int_x^1 c_{jj}(\rho)/\mu_j(\rho)d\rho}\Big[\tilde{\Delta}_j$$
$$+ \int_0^{t+\int_x^1 \frac{1}{\mu_j(\tau)}d\tau} \sum_{i=1}^N f_{ij}(t + \int_x^1 \frac{1}{\mu_j(\tau)}d\tau - s)u_{i0}(x_i(\theta_i(1,s)))$$
$$\times e^{-\int_{x_i(\theta_i(1,s))}^1 c_{ii}(\rho)/\lambda_i(\rho)d\rho}ds\Big],$$

$$t \ge -\int_x^1 \frac{1}{\mu_j(\tau)}d\tau, \ N+1 \le j \le n,$$

$$(6.28)$$

where $\tilde{\Delta}_j$ denotes the j−th row of $(e^{Gt}-1)d_0$. From (6.20), we already know that $x_i(\theta_i(1,s))$ is a continuous function of s, and it satisfies $(\partial x_i(\theta_i(1,s))/\partial s) \ne 0$ everywhere. According to Lemma 6.4 and (6.26)-(6.28), we see that $T_2(t) - T_3(t)$ is compact for all $0 \le t \le c$. □

In order to show the compactness of $T(t) - T_2(t)$, the following lemma is needed.

Lemma 6.6 *Let $S(t)$ be a C_0-semigroup and L a bounded linear operator on a Banach space X. If $\int_0^t S(t-s)LS(s)ds$ is compact for each fixed $t \in [0, t_0]$, then*

$$\{\int_0^t S(t-s)LS(s)xds, x \in X, \|x\| \le 1, t \in [0, t_0]\}$$

is precompact. Moreover, $\int_0^t S(t-s)LS(s)ds$ is compact in $C(0, t_0; X)$.

Proof. Given $\epsilon > 0$ and let $n_0 = n_0(\epsilon)$ be an integer such that

$$\left\|\int_t^{t+r} S(t+r-s)LS(s)ds\right\| < \epsilon$$

for all $t \in [0, t_0]$ and all $r \geq 0$ satisfying $r \leq t_0/n_0$. Let t_n be defined by $t_n = nt_0/n_0, n = 0, 1, \ldots n_0 - 1$. For $t \in [t_n, t_{n+1}]$ and $r = t - t_n$, we have

$$
\int_0^t S(t-s)LS(s)ds
$$

$$
= \int_0^{t_n+r} S(t_n+r-s)LS(s)ds
$$

$$
= S(r)\int_0^{t_n+r} S(t_n-s)LS(s)ds
$$

$$
= S(t-t_n)\int_0^{t_n} S(t_n-s)LS(s)ds + \int_{t_n}^{t_n+r} S(r+t_n-s)LS(s)ds.
$$

Since $\{\int_0^{t_n} S(t_n-s)LS(s)xds, \|x\| \leq 1\}$ is precompact, so is

$$
K_n = \{S(t-t_n)\int_0^{t_n} S(t_n-s)LS(s)xds, \|x\| \leq 1, t \in [t_n, t_{n+1}]\}
$$

(see the proof of Theorem 3.53). Hence $K_\epsilon = K_0 \cup \ldots \cup K_{n_0-1}$ is an ϵ net of $\{\int_0^t S(t-s)LS(s)xds, \|x\| \leq 1, t \in [0, t_0]\}$. This proves the first part of the lemma. Secondly, for any $h > 0$

$$
\left\| \int_0^{t+h} S(t+h-s)LS(s)xds - \int_0^t S(t-s)LS(s)xds \right\|
$$

$$
\leq \left\| \int_t^{t+h} S(t+h-s)LS(s)xds \right\|
$$

$$
+ \left\| [S(h)-I]\int_0^t S(t-s)LS(s)xds \right\| \to 0, \quad \text{as } h \to 0,
$$

uniformly for all $t \in [0, t_0]$ and $\|x\| \leq 1$. The second part thus follows from the Arzela-Ascoli theorem. $\qquad \square$

Lemma 6.7 $T(t) - T_2(t)$ *is compact for all* $t \geq 0$.

Proof. Let $t_0 = c$ be as defined in (6.17). For the same reason as seen in the proof of Lemma 6.5, if we can show that $T(t) - T_2(t)$ is compact for each $t \leq t_0$, then $T(t) - T_2(t)$ is compact for every $t \geq 0$. Let

$$
B(u, v, d) = (\hat{C}(x)\begin{bmatrix} u(x) \\ v(x) \end{bmatrix}, 0)
$$

where $\hat{C}(x) = \{\hat{c}_{ij}(x)\} = -C(x) + \mathrm{diag}(c_{ii}(x))$. Then

$$T(t) - T_2(t) = \int_0^t T_2(t-s)BT_2(s)ds + \int_0^t T_2(t-s)B[T(s) - T_2(s)]ds.$$

That is, $T(t) - T_2(t)$ is the solution of the Volterra integral equation and so

$$T(t) - T_2(t) = \mathcal{K} \int_0^t T_2(t-s)BT_2(s)ds,$$

where $\mathcal{K} : C(0, t_0; X) \rightarrow C(0, t_0; X)$ is a continuous operator. If for any $t \geq 0$, $\int_0^t T_2(t-s)BT_2(s)ds$ is compact, then by Lemma 6.6, $\int_0^t T_2(t-s)BT_2(s)ds$ is compact in $C(0, t_0; X)$. Therefore, $T(t) - T_2(t)$ is compact in $C(0, t_0; X)$. In particular, $T(t) - T_2(t)$ is compact for any $t \geq 0$. Consequently, in order to show that $T(t) - T_2(t)$ is compact, it is sufficient to show that $\int_0^t T_2(t-s)BT_2(s)ds$ is compact. This is equivalent to showing that $\int_0^t T_3(t-s)BT_3(s)ds$ is compact for each $t \leq t_0$ by using the following equality

$$\begin{aligned} T_2(t-s)BT_2(s) &= T_3(t-s)BT_3(s) + [T_2(t-s) - T_3(t-s)]BT_3(s) \\ &\quad + T_2(t-s)B[T_2(s) - T_3(s)] \end{aligned}$$

and the property shown in Lemma 6.5. Similarly, since $T_3(t) - T_4(t)\Pi$ is compact for any $t \geq 0$, from

$$\begin{aligned} &T_3(t-s)BT_3(s) \\ &= T_4(t-s)\Pi BT_4(s)\Pi + [T_3(t-s) - T_4(t-s)\Pi]BT_4(s)\Pi \\ &\quad + T_3(t-s)B[T_3(s) - T_4(s)\Pi], \end{aligned}$$

it suffices to show that

$$\int_0^t T_4(t-s)\Pi BT_4(s)\Pi ds$$

is compact on X for $t \leq t_0$. Identifying B with

$$B(u, v) = \hat{C}(x) \begin{bmatrix} u(x) \\ v(x) \end{bmatrix}$$

on $(L^p(0,1))^{2N+2}$ and identifying $T_4(t)$ with the C_0-semigroup on $(L^p(0,1))^{2N+2}$, we need only to show that

$$\int_0^t T_4(t-s)BT_4(s)ds$$

is compact on $(L^p(0,1))^{2N+2}$ for any $t \leq t_0$.

Let $(u(x,s), v(x,s)) = T_4(s)(u_0(x), v_0(x))$. Then by letting $F = G = d_0 = 0$ in (6.21), (6.22), and (6.25), we have

$$
\begin{cases}
u_i(x,s) = u_{i0}(x_i(\theta_i(x,s)))e^{\int_x^{x_i(\theta_i(x,s))} c_{ii}(\rho)/\lambda_i(\rho)d\rho}, \quad s \le \int_0^x \frac{1}{\lambda_i(\tau)}d\tau, \\[2mm]
u_i(x,s) = e^{-\int_0^x c_{ii}(\rho)/\lambda_i(\rho)d\rho} \sum_{j=N+1}^n e_{ij}v_{j0}(x_j(\theta_{ij}(x,s))) \\[2mm]
\qquad\qquad \times e^{\int_0^{x_j(\theta_{ij}(x,s))} c_{jj}(\rho)/\mu_j(\rho)d\rho}, \quad c \ge s \ge \int_0^x \frac{1}{\lambda_i(\tau)}d\tau,\ 1 \le i \le N,
\end{cases}
\tag{6.29}
$$

and

$$
\begin{cases}
v_j(x,s) = v_{j0}(x_j(\theta_j(x,s)))e^{\int_x^{x_j(\theta_j(x,s))} c_{jj}(\rho)/\mu_j(\rho)d\rho}, \quad s \le -\int_x^1 \frac{1}{\mu_j(\tau)}d\tau, \\[2mm]
v_j(x,s) = \sum_{i=1}^N d_{ij}u_{i0}(x_i(\tilde\theta_{ij}(x,s)))e^{-\int_{x_i(\tilde\theta_{ij}(x,s))}^1 c_{ii}(\rho)/\lambda_i(\rho)d\rho} \\[2mm]
\qquad\qquad \times e^{\int_x^1 c_{jj}(\rho)/\mu_j(\rho)d\rho} \\[2mm]
\qquad c \ge s \ge -\int_x^1 \frac{1}{\mu_j(\tau)}d\tau,\ N+1 \le j \le n,
\end{cases}
\tag{6.30}
$$

where $x_i(\theta)$ and $x_j(\theta)$ are functions satisfying (6.20) and

$$
\begin{cases}
\theta_i(x,s) = s - \int_0^x \frac{1}{\lambda_i(\tau)}d\tau, \quad \theta_j(x,s) = s + \int_x^1 \frac{1}{\mu_j(\tau)}d\tau, \\[2mm]
\tilde\theta_{ij}(x,s) = s + \int_x^1 \frac{1}{\mu_j(\tau)}d\tau - \int_0^1 \frac{1}{\lambda_i(\tau)}d\tau, \\[2mm]
D = \{d_{ij}\},\ 1 \le i \le N,\ N+1 \le j \le n.
\end{cases}
$$

Let

$$
(\tilde u(x,t-s), \tilde v(x,t-s)) = T_4(t-s)\hat C(x)(u(x,s), v(x,s)) = T_4(t-s)(\hat u(x,s), \hat v(x,s)),
$$

where

$$
\begin{cases}
\hat u_i(x,s) = \sum_{k=1}^N \hat c_{ik}(x)u_k(x,s) + \sum_{k=N+1}^n \hat c_{ik}(x)v_k(x,s), \\[2mm]
\qquad\qquad 1 \le i \le N, \\[2mm]
\hat v_j(x,s) = \sum_{k=1}^N \hat c_{jk}(x)u_k(x,s) + \sum_{k=N+1}^n \hat c_{jk}(x)v_k(x,s), \\[2mm]
\qquad\qquad N+1 \le j \le n.
\end{cases}
\tag{6.31}
$$

Then

$$
\begin{cases}
\tilde u_i(x,t-s) = \hat u_i(x_i(\theta_i(x,t-s)),s)e^{\int_x^{x_i(\theta_i(x,t-s))} c_{ii}(\rho)/\lambda_i(\rho)d\rho}, \\[2mm]
\qquad\qquad t-s \le \int_0^x \frac{1}{\lambda_i(\tau)}d\tau, \\[2mm]
\tilde u_i(x,t-s) = e^{-\int_0^x c_{ii}(\rho)/\lambda_i(\rho)d\rho} \sum_{j=N+1}^n e_{ij}\hat v_j(x_j(\theta_{ij}(x,t-s)),s) \\[2mm]
\qquad\qquad \times e^{\int_0^{x_j(\theta_{ij}(x,t-s))} c_{jj}(\rho)/\mu_j(\rho)d\rho}, \\[2mm]
\qquad c \ge t-s \ge \int_0^x \frac{1}{\lambda_i(\tau)}d\tau,\ 1 \le i \le N,
\end{cases}
\tag{6.32}
$$

and

$$
\begin{cases}
\tilde{v}_j(x, t - s) = \hat{v}_j(x_j(\theta_j(x, t - s)), s)e^{\int_x^{x_j(\theta_j(x,t-s))} c_{jj}(\rho)/\mu_j(\rho)d\rho}, \\
\qquad\qquad t - s \leq -\int_x^1 \frac{1}{\mu_j(\tau)}d\tau, \\
\\
\tilde{v}_j(x, t - s) = \sum_{i=1}^N d_{ij}\hat{u}_i(x_i(\tilde{\theta}_{ij}(x, t - s)), s)e^{-\int_{x_i(\tilde{\theta}_{ij}(x,s))}^1 c_{ii}(\rho)/\lambda_i(\rho)d\rho} \\
\qquad\qquad \times e^{\int_x^1 c_{jj}(\rho)/\mu_j(\rho)d\rho}, \; c \geq t - s \geq -\int_x^1 \frac{1}{\mu_j(\tau)}d\tau, \; N + 1 \leq j \leq n.
\end{cases}
\tag{6.33}
$$

It is seen that $\int_0^t T_4(t - s)BT_4(s)(u_0, v_0, d_0)ds, t \leq c$, equals the sum of the terms of the form $\int_{r_1(x,t)}^{r_2(x,t)} f(x, t, s)p(x, t, s)ds$, where f, r_1, r_2 are continuous functions of their variables, and $p(x, t, s)$ is one of $\hat{u}_i(x_i(\theta_i(x, t - s)), s)$, $\hat{u}_i(x_i(\tilde{\theta}_{ij}(x, t - s)), s)$, $\hat{v}_j(x_j(\theta_{ij}(x, t - s)), s)$ and $\hat{v}_j(x_j(\theta_j(x, t - s)), s)$. Let us consider, for instance, the following term

$$
\int_{r_1(x,t)}^{r_2(x,t)} f(x, t, s)\hat{u}_i(x_i(\theta_i(x, t - s)), s)ds
\tag{6.34}
$$

to see what properties it has. The discussions of the other three cases are similar. Let $p_i(x, t, s) = x_i(\theta_i(x, t - s))$. Then from (6.31), $\hat{u}_i(p_i, s)$ are the sum of the terms $\hat{c}_{ik}(p_i)u_k(p_i, s), 1 \leq k \leq N$ and $\hat{c}_{ik}(p_i)v_k(p_i, s), N \leq k \leq n$, where $u_k(p_i, s)$ and $v_k(p_i, s)$ can be obtained through initial conditions by (6.29) and (6.30). By these relations, we can express the integral (6.34) as the sum of the terms of the form

$$
\int_{r_{ik1}(x,t)}^{r_{ik2}(x,t)} g_{ik}(x, t, s)u_{k0}(\alpha_{ik}(x, t, s))ds, \; 1 \leq i, k \leq N,
$$

$$
\int_{r_{jk1}(x,t)}^{r_{jk2}(x,t)} g_{jk}(x, t, s)v_{k0}(\alpha_{jk}(x, t, s))ds, \; N + 1 \leq j, k \leq n,
$$

where $r_{ik1}, r_{jk2}, g_{ik}, g_{jk}\alpha_{ik}, \alpha_{jk}$ are continuous functions of their variables. It can be checked that all $\partial\alpha_{ik}(x, t, s)/\partial s \neq 0, \partial\alpha_{jk}(x, t, s)/\partial s \neq 0$ under the condition that $c_{ij} = 0$ if either $\lambda_i(x) = \lambda_j(x)$ or $\mu_i(x) = \mu_j(x)$ somewhere for $i \neq j$. Actually, the possibility of $\partial\alpha_i(x, t, s)/\partial s = 0$ happens when, for example, such kind of integrals include the integrand such as

$$
g_{ik}(x, t, s)\hat{c}_{ik}(p_i)u_{k0}(x_k(\vartheta_k(p_i(x, s), s))),
$$

where $\alpha_{ik}(x, t, s) = x_k(\vartheta_k(p_i(x, s), s))$ satisfies $\partial\alpha_{ik}(x, t, s)/\partial s = 0$ if $\lambda_k(p_i) = \lambda_i(p_i)$. But this is saved by our assumption and $\hat{c}_{ii} = 0$. The proof is complete. $\qquad\qquad\qquad\qquad\qquad\qquad\qquad\qquad\qquad\qquad\qquad\qquad\qquad\qquad\square$

Combining Lemmas 6.5 and 6.7 and noticing the compactness of $T_3(t) - T_4(t)\Pi$, we conclude Theorem 6.3.

Example 6.8 *It is remarked that assumption (H3) in the beginning of this section cannot be removed for our problem. To see this, consider*

$$\begin{cases} \dfrac{\partial}{\partial t}\begin{bmatrix} u_1(x,t) \\ u_2(x,t) \\ v(x,t) \end{bmatrix} + K\dfrac{\partial}{\partial x}\begin{bmatrix} u_1(x,t) \\ u_2(x,t) \\ v(x,t) \end{bmatrix} + C\begin{bmatrix} u_1(x,t) \\ u_2(x,t) \\ v(x,t) \end{bmatrix} = 0, \\ v(1,t)=0,\ (u_1(0,t),u_2(0,t))=Ev(0,t), \end{cases} \tag{6.35}$$

where $0 < x < 1$ and

$$K = \begin{bmatrix} 1 & 0 & 0 \\ 0 & 1 & 0 \\ 0 & 0 & -1 \end{bmatrix}, \quad E=(1,0), \quad C = \begin{bmatrix} 0 & 0 & 0 \\ 1 & 0 & 0 \\ 0 & 0 & 0 \end{bmatrix}.$$

This is a special case of system (6.16) where both K and C are constant matrices. Let $T_c(t)$ be the associated semigroup of (6.35) and $T(t)$ the semigroup reduced by $T_c(t)$ with $C=0$ in (6.35). Then $T_c(t) - T(t)$ is not compact for $t < 2$.

Indeed, a simple calculation shows that

$$\{[T_c(t) - T(t)](u_{01},u_{02},v_0)\}(x) = \left(0, \begin{cases} tu_{01}(x-t),\ 0 \le x - t \le 1; \\ -xv_0(t-x),\ 0 \le t - x \le 1; \\ 0,\quad otherwise. \end{cases} ,0\right),$$

which is not compact for $t < 2$

Before ending this section, we study the system (6.16) in which there are no dynamics on the boundaries. The purpose for doing this is to show that the spectrum-determined growth condition holds for T_4, i.e., $\omega_{ess}(A_4) = S(A_4)$. If this is true, then we see that

$$\begin{aligned} \omega_0(A) &= \max\{S(A), \omega_{ess}(A)\} \\ &= \max\{S(A), \omega_{ess}(A_4)\} \\ &= \max\{S(A), S(A_4)\}, \end{aligned} \tag{6.36}$$

where the second equality follows from Theorem 6.3. This result clearly shows that the growth rate of (6.1) is completely determined by the spectral bound of A and A_4, which serves as a basic tool for us to prove the spectrum-determined growth condition for various wave equations to be discussed in the next three sections.

Lemma 6.9 A_4 *has compact resolvent and $\sigma(A_4) = \{\lambda \mid h_4(\lambda) = 0\}$, where $h_4(\lambda)$ is an exponential polynomial of the form:*

$$h_4(\lambda) = \sum_{k=1}^{m} b_k e^{\lambda \omega_k}$$

for some integer m, some constants b_k and some real $\omega_k, k = 1, 2, \dots, m$.

Proof. Given $(f, g) \in (L^p(0,1))^n$, we solve

$$(\lambda - A_4)(u, v) = (f, g),$$

that is,

$$\begin{cases} \dfrac{d}{dx}\begin{bmatrix} u(x) \\ v(x) \end{bmatrix} = -K^{-1}(x)[\lambda + C_0(x)]\begin{bmatrix} u(x) \\ v(x) \end{bmatrix} + K^{-1}(x)\begin{bmatrix} f(x) \\ g(x) \end{bmatrix}, \\ u(0) = Ev(0), v(1) = Du(1). \end{cases}$$

Denote by $Y(x, y, \lambda)$ the fundamental matrix of the system

$$\frac{d}{dx}\begin{bmatrix} u \\ v \end{bmatrix} = -K^{-1}(x)[\lambda + C_0(x)]\begin{bmatrix} u \\ v \end{bmatrix}.$$

Due to the special forms of $K(x)$ and $C_0(x)$, we are able to obtain the explicit expression of $Y(x, y, \lambda)$ as follows:

$$Y(x, y, \lambda) = \begin{bmatrix} Y_1(x, y, \lambda) & 0 \\ 0 & Y_2(x, y, \lambda) \end{bmatrix},$$

$$Y_1(x, y, \lambda) = \mathrm{diag}(\exp(-\lambda \int_y^x \frac{ds}{\lambda_i(s)} - \int_y^x \frac{c_{ii}(s)}{\lambda_i(s)}ds)),$$

$$Y_2(x, y, \lambda) = \mathrm{diag}(\exp(-\lambda \int_y^x \frac{ds}{\mu_j(s)} - \int_y^x \frac{c_{jj}(s)}{\mu_j(s)}ds)).$$

Then

$$\begin{bmatrix} u(x) \\ v(x) \end{bmatrix} = Y(x, 0, \lambda)\begin{bmatrix} E \\ I \end{bmatrix} v(0) + \int_0^x Y(x, y, \lambda)K^{-1}(y)\begin{bmatrix} f(y) \\ g(y) \end{bmatrix} dy.$$

Moreover, $v(1) = Du(1)$ implies that

$$H_0(\lambda)v(0) = \int_0^1 (D, -I)Y(1, y, \lambda)K^{-1}(y)\begin{bmatrix} f(y) \\ g(y) \end{bmatrix} dy,$$

where $H_0(\lambda) = (-I, D)Y(1, 0, \lambda)\begin{bmatrix} E \\ I \end{bmatrix}$. Therefore, $\lambda \in \sigma(A_4)$ if and only if

$$h_4(\lambda) = \det H_0(\lambda) = 0. \tag{6.37}$$

Because the elements of $H_0(\lambda)$ are the linear combinations of the exponential functions of the form $e^{-\lambda \int_0^1 1/\lambda_i(s)ds}$ and $e^{-\lambda \int_0^1 1/\mu_j(s)ds}$, $h_4(\lambda)$ is therefore an exponential polynomial. For any $\lambda \in \rho(A)$,

$$R(\lambda, A_4)\begin{bmatrix} f \\ g \end{bmatrix}(x)$$

$$= Y(x, 0, \lambda)\begin{bmatrix} Ev(0) \\ v(0) \end{bmatrix} + \int_0^x Y(x, y, \lambda)K^{-1}(y)\begin{bmatrix} f(y) \\ g(y) \end{bmatrix} dy \tag{6.38}$$

with

$$v(0) = H_0^{-1}(\lambda) \int_0^1 (D, -I)Y(1, y, \lambda)K^{-1}(y) \begin{bmatrix} f(y) \\ g(y) \end{bmatrix} dy. \qquad (6.39)$$

It is readily seen in (6.38) that $R(\lambda, A_4)$ is compact for any $\lambda \in \rho(A_4)$. This proves the lemma. $\square$

Before proving the spectrum-determined growth condition for system (6.16), we need several properties of almost periodic functions.

Recall that a continuous function $g(x)$ $(-\infty < x < \infty)$ is said to be *almost periodic* if for every $\varepsilon > 0$, there is a quantity $\ell > 0$ such that every interval of the x-axis of length ℓ contains at least one number τ such that

$$|g(x + \tau) - g(x)| < \varepsilon \quad \text{for all } x \in (-\infty, \infty).$$

Consider an exponential polynomial

$$f(z) = \sum_{n=1}^N b_n e^{\omega_n z}, \qquad (\omega_n \text{ are real}) \qquad (6.40)$$

where b_n, ω_n are constants. Then $g(x) = f(\alpha + ix)$ is an almost periodic function for any fixed $\alpha \in \mathbf{R}$ (see [91]).

Lemma 6.10   *(i) Let $f(z)$ be defined by (6.40). Let $\Omega = \{z \mid f(z) = 0\}$. Then for all z satisfying $\text{dist}(z, \Omega) \geq \delta > 0$, $\alpha \leq \text{Re} z \leq \beta$ and $|\text{Im} z| \geq M$, there exists a constant $m(\delta, \alpha, \beta, M) > 0$ such that $|f(z)| > m(\delta, \alpha, \beta, M)$, where α, β, δ, and M are constants.*

(ii) For an exponential polynomial $f(z)$ such as (6.40), if $f(\lambda) \neq 0$ for all $\lambda = \sigma + i\tau$ with fixed σ, then

$$\frac{1}{f(\lambda)} = \sum_{k=1}^\infty a_k e^{\gamma_k \lambda}$$

with $\sum_{k=1}^\infty |a_k| e^{\sigma \gamma_k} < \infty$, where γ_k are real, and a_k are complex constants.

Proof.   (ii) is a generalization of the well-known Wiener's theorem on Fourier series; its proof can be found in [130]. Here, we give a proof of (i). Suppose otherwise, then there are an infinite number of z_m with $\alpha \leq \text{Re} z_m \leq \beta$ such that $f(z_m) \to 0$ as $m \to \infty$. Let $z_m = \alpha_m + i\beta_m$. Then $|\beta_m| \to \infty$ as $m \to \infty$. We may assume, without loss of generality, that all β_m are positive. Consider functions

$$\phi_m(z) = f(z + i\beta_m) = \sum_{n=1}^N a_n e^{\omega_n z} e^{i\omega_n \beta_m}.$$

It is obvious that $\{\phi_m(z)\}$ are uniformly bounded in $\{z | \alpha - \delta/2 < \text{Re}z < \beta + \delta/2\}$. By the Montel's theorem [38], there is a subseries of $\phi_m(z)$, still denoted by $\phi_m(z)$, such that $\phi_m(z)$ converges uniformly to an analytic function $\phi(z)$ on $\{z | \alpha - \delta/2 < \text{Re}z < \beta + \delta/2\}$. Suppose, without loss of generality, that $\alpha_m \to \alpha_0$ as $m \to \infty, \alpha \leq \alpha_0 \leq \beta$. Therefore,

$$\phi(\alpha_0) = \lim_{m \to \infty} \phi_m(\alpha_m) = \lim_{m \to \infty} f(\alpha_m + i\beta_m) = 0.$$

On the other hand, since $\phi_m(z)$ does not equal zero for z satisfying $\alpha - \delta/2 < \text{Re}z < \beta + \delta/2$ and $\text{Im}z \geq -\delta$ for sufficient large m by assumption, ϕ must be identical to zero by the Hurwitz theorem [38]. It follows that $f(x) = \phi_m(x - i\beta_m) \to 0$ for all $\alpha - \delta/2 < x < \beta + \delta/2$ and so $f \equiv 0$ by the analyticity of f, which is impossible since the distance between the zeros of f and z with $\alpha \leq \text{Re}z \leq \beta$ and $|\text{Im}z| \geq M$ is positive by assumption. This proves the theorem. $\qquad \square$

Theorem 6.11 Let $S(A_4), \omega_{ess}(A_4)$ and $\omega_0(A_4)$ be the spectral bound, essential growth rate and growth rate of the C_0-semigroup $T_4(t)$, respectively. Then

$$S(A_4) = \omega_{ess}(A_4) = \omega_0(A_4).$$

Proof. We first show that $S(A_4) = \omega_0(A_4)$. Let $S(A_4) = \alpha_0$. It suffices to show that for any $\alpha > \alpha_0$, there is an $M > 0$ such that

$$\|T_4(t)W\| \leq Me^{\alpha t} \|W\|$$

for any W in a dense subset of X^0. We shall use the following inverse Laplace transform

$$T_4(t)W = \frac{1}{2\pi i} \int_{\beta - i\infty}^{\beta + i\infty} e^{\lambda t} R(\lambda, A_4) W d\lambda, \tag{6.41}$$

which holds for some $\beta > 0$ and all $W \in D(A_4^2)$ (see Theorem 2.40). We choose those W which are sufficiently smooth such that they have compact supports in $(0,1)$.

For any $\alpha > \alpha_0$, we claim that the path of integration in (6.41) can be shifted to $\text{Re}\lambda = \alpha$ by showing that

$$\int_{\Omega(R)} e^{\lambda t} R(\lambda, A_4) W d\lambda \to 0, \text{ as } R \to \infty,$$

where $\Omega(R) = \{\lambda \mid \alpha \leq \text{Re}\lambda \leq \beta, |\text{Im}\lambda| = R\}$. To this end, we have to estimate the resolvent which is given in (6.38). Since the resolvent $R(\lambda, A_4)$ consists of two parts, we estimate only the first part. The estimation of the second part is simple. Write $H_0^{-1}(\lambda) = \frac{L(\lambda)}{h_4(\lambda)}$, where $L(\lambda)$ is the matrix whose elements are composed of the algebraic cofactors of $H_0(\lambda)$. Then the first part of $R(\lambda, A_4)$ in (6.38) can be written as

$$\frac{1}{h_4(\lambda)} Y(x,0,\lambda) \begin{bmatrix} E \\ I \end{bmatrix} L(\lambda)(D,-I) \int_0^1 Y(1,y,\lambda) K^{-1}(y) W(y) dy. \tag{6.42}$$

By the expression of $H_0(\lambda)$, the elements of $\begin{bmatrix} E \\ I \end{bmatrix} L(\lambda)(D,-I)$ takes the form $ae^{b\lambda}$, where a and b are real. Therefore, the component of (6.42) takes the form

$$\frac{1}{h_4(\lambda)} b_1(x) e^{-\lambda(a+c(x))} \int_0^1 b_2(y) e^{-\lambda d(y)} w(y) dy, \tag{6.43}$$

where b_1, b_2 are C^1 functions and a is real, w is a component of W, $c(x) = \int_0^x p(s)ds, d(y) = \int_y^1 q(s)ds, p,q$ are one of $1/\lambda_i$ and $1/\mu_j$. Since dist$(\lambda, \sigma(A)) > \epsilon > 0$ for some $\epsilon > 0$ and all $\lambda \in \Omega(R)$, it follows from Lemma 6.10 that there is a positive constant $\delta > 0$ such that

$$\mid h_4(\lambda) \mid \geq \delta, \text{ for all } \lambda \in \Omega(R). \tag{6.44}$$

By (6.43) and (6.44), we can write (6.43) in a compact form

$$b(x,\lambda) \int_0^1 b_2(y) e^{-\lambda d(y)} w(y) dy,$$

where $b(x,\lambda)$ is bounded for $0 \leq x \leq 1$ and $\alpha \leq \text{Re}\lambda \leq \beta$. Integrating by parts yields

$$\int_0^1 b_2(y) e^{-\lambda d(y)} w(y) dy = \frac{1}{\lambda} \int_0^1 e^{-\lambda d(y)} \frac{d}{dy}[b_2(y)w(y)/d'(y)]dy,$$

which tends to zero as $\mid \lambda \mid \to \infty, \alpha \leq \text{Re}\lambda \leq \beta$. Thus we conclude that

$$T_4(t)W = \frac{1}{2\pi i} \int_{\alpha-i\infty}^{\alpha+i\infty} e^{\lambda t} R(\lambda, A_4) W d\lambda. \tag{6.45}$$

Now, for $\lambda_0 = \alpha + i\tau$, by Lemma 6.10, we can write

$$\frac{1}{h_4(\lambda_0)} = \sum_{k=1}^{\infty} a_k e^{\lambda_0 \gamma_k}, \tag{6.46}$$

where γ_k is real and $\sum_{k=1}^{\infty}|a_k|e^{\alpha\gamma_k} < \infty$. Then for $\lambda = \lambda_0$, (6.43) is reduced to

$$\sum_{k=1}^{\infty} a_k e^{\alpha\gamma_k} b_1(x) e^{-\alpha(a+c(x))} \int_0^1 b_2(y) e^{-\alpha d(y)} e^{i\tau(\gamma_k - a - c(x) - d(y))} w(y) dy$$

$$= \sum_{k=1}^{\infty} a_k e^{\alpha\gamma_k} d_1(x) \int_0^1 d_2(y) e^{i\tau(\gamma_k - a - c(x) - d(y))} w(y) dy, \qquad (6.47)$$

where $d_1(x) = b_1(x) e^{-\alpha(a+c(x))}, d_2(y) = b_2(y) e^{-\alpha d(y)}$ are C^1 functions. Therefore, to estimate the integral of the right-hand side of (6.45), we have to estimate

$$\lim_{R\to\infty} \int_{-R}^{R} e^{\alpha t} \sum_{k=1}^{\infty} a_k e^{\alpha\gamma_k} d_1(x) \int_0^1 d_2(y) e^{i\tau(t+\gamma_k - a - c(x) - d(y))} w(y) dy$$

$$= \lim_{R\to\infty} e^{\alpha t} \sum_{k=1}^{\infty} a_k e^{\alpha\gamma_k} d_1(x) \int_0^1 d_2(y) w(y) dy \int_{-R}^{R} e^{i\tau(t+\gamma_k - a - c(x) - d(y))} d\tau$$

$$= \lim_{R\to\infty} 2e^{\alpha t} \sum_{k=1}^{\infty} a_k e^{\alpha\gamma_k} d_1(x)$$

$$\times \int_0^1 d_2(y) w(y) \frac{\sin R[t + \gamma_k - a - c(x) - d(y)]}{t + \gamma_k - a - c(x) - d(y)} dy$$

$$= -\lim_{R\to\infty} 2e^{\alpha t} \sum_{k=1}^{\infty} a_k e^{\alpha\gamma_k} d_1(x)$$

$$\times \int_0^{d(0)} d_3(z) w(\eta(z)) \frac{\sin R[t + \gamma_k - a - c(x) - z]}{t + \gamma_k - a - c(x) - z} dz$$

$$= \lim_{R\to\infty} 2e^{\alpha t} \sum_{k=1}^{\infty} a_k e^{\alpha\gamma_k} d_1(x) \int_{t+\gamma_k - a - c(x)}^{t+\gamma_k - a - c(x) - d(0)} L_{k,t,x}(z) \frac{\sin Rz}{z} dz,$$

$$(6.48)$$

where $L_{k,t,x}(z) = d_3(t + \gamma_k - a - c(x) - z) w(\eta(t + \gamma_k - a - c(x) - z))$ is a C^1 function of z for any fixed (k, t, x) and $d_3(z) = d_2(\eta(z))\eta'(z)$ for $z = d(y), y = \eta(z)$ is also C^1 with respect to z. Now, if $t + \gamma_k - a - c(x)$ and $t + \gamma_k - a - c(x) - d(0)$ have the same sign, then by integrating by parts, we have

$$\lim_{R\to\infty} \int_{t+\gamma_k - a - c(x)}^{t+\gamma_k - a - c(x) - d(0)} L_{k,t,x}(z) \frac{\sin Rz}{z} dz = 0.$$

If $t + \gamma_k - a - c(x) = 0$, then by using the fact that $\lim_{R\to\infty} \int_0^{\delta} \frac{\sin Rz}{z} dz = \pi/2$ for any positive δ, we have

$$\lim_{R\to\infty} \int_0^{-d(0)} L_{k,t,x}(z) \frac{\sin Rz}{z} dz$$

$$= \lim_{R \to \infty} \int_0^{-d(0)} [L_{k,t,x}(z) - L_{k,t,x}(0)] \frac{\sin Rz}{z} dz + L_{k,t,x}(0) \int_0^{-d(0)} \frac{\sin Rz}{z} dz$$

$$= \lim_{R \to \infty} L_{k,t,x}(0) \int_0^{-d(0)} \frac{\sin Rz}{z} dz = \pm \frac{\pi}{2} L_{k,t,x}(0).$$

If $t + \gamma_k - a - c(x)$ and $t + \gamma_k - a - c(x) - d(0)$ have the opposite sign, then

$$\lim_{R \to \infty} \int_{t+\gamma_k - a - c(x)}^{t+\gamma_k - a - c(x) - d(0)} L_{k,t,x}(z) \frac{\sin Rz}{z} dz = \pm \pi L_{k,t,x}(0).$$

Therefore, the limit of the integral of the right-hand side of (6.48) is bounded by $\pi \mid d_3(t + \gamma_k - a - c(x)) w(\eta(t + \gamma_k - a - c(x))) \mid$ as R tends to ∞. Here we set $w(\eta) = 0$ for η outside $[0, 1]$. Since

$$|d_1(x) d_3(t + \gamma_k - a - c(x)) w(\eta(t + \gamma_k - a - c(x)))| \le M |w(\eta(t + \gamma_k - a - c(x)))|$$

for some M which is independent of t and k, we have

$$\|d_1(x) d_3(t + \gamma_k - a - c(x)) w(\eta(t + \gamma_k - a - c(x)))\|_{L^p(0,1)}$$

$$\le M \left(\int_0^1 \mid w(\eta(t + \gamma_k - a - c(x))) \mid^p dx \right)^{1/p}$$

$$= M \left(\int_0^{t+\gamma_k - a - c(1)} \frac{1}{-\eta'(d(z)) c'(x)} \mid w(z) \mid^p dz \right)^{1/p}$$

$$\le \hat{M} \|w\|_{L^p(0,1)}$$

where $\hat{M}$ is a constant independent of t and k. Finally, it is easily seen that the integrals on the right-hand side of (6.48) are uniformly bounded with respect to k; hence, the limit in R and the summation in k are commutable. Therefore,

$$\lim_{R \to \infty} \int_{-R}^R e^{\alpha t} \sum_{k=1}^\infty a_k e^{\alpha \gamma_k} d_1(x) \int_0^1 d_2(y) e^{i\tau(t + \gamma_k - a - c(x) - d(y))} w(y) dy$$

$$= e^{\alpha t} \sum_{k=1}^\infty a_k e^{\alpha \gamma_k} p_{k,t}(x),$$

where $\|p_{k,t}(x)\|_{L^p}$ is bounded uniformly by $\|w\|_{L^p(0,1)}$ for all t and k. Summarizing, we have proven that for any $\alpha > \alpha_0$, there exists a constant $\tilde{M} > 0$ such that $\|T_4(t)\| \le \tilde{M} e^{\alpha t}$. Thus, $\omega_0(A_4) = \alpha_0$.

Furthermore, we claim that there is a sequence λ_n such that $h_4(\lambda_n) = 0$, $\mathrm{Re} \lambda_n \to \alpha_0$, and $|\lambda_n| \to \infty$. Suppose it is not true. Then by the definition of α_0, there exists a λ_1 such that $h(\lambda_1) = 0$, $\mathrm{Re} \lambda_1 = \alpha_0$. On the other hand, the assumption implies that there is a $\epsilon > 0$ such that for all λ with $|\lambda|$

sufficiently large on the vertical line $\mathrm{Re}\lambda = \alpha_0, |h_4(\lambda)| \geq \epsilon$ by (i) of Lemma 6.10. This, together with $h(\lambda_1) = 0$, contradicts that $h_4(\alpha + i\tau)$ is almost periodic with respect to τ. Therefore, the assertion is true.

Suppose, without loss of generality, that $\mathrm{Re}\lambda_i \neq \mathrm{Re}\lambda_j$ for $i \neq j$. From the spectral mapping theorem and Lemma 6.9, it follows that $\{e^{\lambda_n t}\} \subset \sigma(T_4(t))$. However, $e^{\lambda_n t}$ are distinct for different n and $e^{\lambda_n t} \to e^{\alpha_0 t}$. Hence $e^{\alpha_0 t} \in \sigma_{ess}(T_4(t))$. Consequently $e^{\alpha_0 t} \leq \gamma_{ess}(T_4(t))$ and so $\alpha_0 \leq w_{ess}(A_4)$. This together with $\alpha_0 = w_0(A_4)$ completes the proof. $\qquad \Box$

6.2 Stabilization of serially connected vibrating strings

In this section, we consider $(N + 1)$-serially connected vibrating strings with the following dynamic model [27]

$$y_{tt}(x,t) - c_i^2 y_{xx}(x,t) = 0, \; i - 1 < x < i, i = 1, 2, \ldots, N, N + 1, \quad (6.49)$$

where $y(x,t)$ denotes vibration magnitude of the connected string at time t and position x, and c_i represents the wave speed of the i-th string. Notice that, for notational simplicity, we have assumed that the length of each segment of the connected string is 1. At the left end, $x = 0$, and the right end, $x = N + 1$, the following conservative conditions

$$y(0,t) = 0 \text{ or } y_x(0,t) = 0; \; y(N+1,t) = 0 \text{ or } y_x(N+1,t) = 0 \quad (6.50)$$

or dissipative boundary conditions

$$y_x(0,t) = k_0^2 y_t(0,t), \quad y_x(N+1,t) = -k_{N+1}^2 y_t(N+1,t), \quad (6.51)$$
$$k_0^2, k_{N+1}^2 \geq 0$$

are imposed.

At the i-th intermediate node $x = i, i = 1, 2, \ldots, N$, either force feedback dissipative conditions

$$\begin{cases} y_t(i^-,t) - y_t(i^+,t) = -k_i^2 c_i^2 y_x(i^-,t), \\ c_i^2 y_x(i^-,t) - c_{i+1}^2 y_x(i^+,t) = 0 \end{cases} \quad (6.52)$$

or velocity feedback conditions

$$\begin{cases} y(i^-,t) - y(i^+,t) = 0, \\ c_i^2 y_x(i^-,t) - c_{i+1}^2 y_x(i^+,t) = -k_i^2 y_t(i^+,t) \end{cases} \quad (6.53)$$

are assumed.

The energy of the system is defined by

$$E(t) = \frac{1}{2} \sum_{i=1}^{N+1} \int_{i-1}^{i} [c_i^2 y_x^2(x,t) + y_t^2(x,t)]dx \qquad (6.54)$$

and formally, it is easy to show that

$$\frac{dE(t)}{dt} = c_{N+1}^2 y_x(N+1,t)y_t(N+1,t) - c_1^2 y_x(0,t)y_t(0,t)$$

$$+ \sum_{i=1}^{N} [c_i^2 y_x(i^-,t)y_t(i^-,t) - c_{i+1}^2 y_x(i^+,t)y_t(i^+,t)] \le 0.$$

$$(6.55)$$

Since we are concerned with the energy dissipation of system (6.49) - (6.53), we wish to define a suitable state space in which the norm of the state variable (y, y_t) is equivalent to the system energy $E(t)$ defined in (6.54). Denote X_0 by $X_0 = \text{span}\{\phi_0(x)\}$, where $\phi_0(x)$ is the solution of the equilibrium of system (6.49)-(6.53), i.e., it satisfies

$$\begin{cases} \phi_0''(x) = 0, \quad i-1 < x < i, \\ \phi_0(0) = 0, \text{ or } \phi_0'(0) = 0; \ \phi_0(N+1) = 0 \text{ or } \phi_0'(N+1) = 0, \\ \qquad \text{if (6.50) is imposed,} \\ \phi_0'(0) = 0, \ \phi_0'(N+1) = 0 \quad \text{if (6.51) is imposed,} \\ \phi_0'(i^-) = \phi_0'(i^+) = 0 \quad \text{if (6.52) is imposed,} \\ \begin{cases} \phi_0(i^-) = \phi_0(i^+), \\ c_i^2 \phi_0'(i^-) = c_{i+1}^2 \phi_0'(i^+), \quad \text{if (6.53) is imposed.} \end{cases} \end{cases}$$

Let

$$X = \{(\phi,\psi) \big| \ (\phi(x), \psi(x)) \in H^1(i-1,i) \times L^2(i-1,i),$$
$$\text{for } x \in (i-1,i), \ i = 1, 2, \cdots, N+1,$$
$$\phi(0) = 0, \text{ or } \phi(N+1) = 0, \quad \text{if (6.50) is assumed.}$$
$$\phi(i^-) = \phi(i^+), \quad \text{if (6.53) is assumed.}\}$$

In the state space X/X_0 which denotes the quotient space, system (6.49)-(6.53) can be written as an evolution equation:

$$\frac{d}{dt} \begin{bmatrix} y \\ y_t \end{bmatrix} = B \begin{bmatrix} y \\ y_t \end{bmatrix},$$

where

$$B \begin{bmatrix} \phi(x) \\ \psi(x) \end{bmatrix} = \begin{bmatrix} \psi(x) \\ c_i^2 \phi''(x) \end{bmatrix}, \ i-1 < x < i$$

with the domain $D(B)$ being defined in an obvious way from the boundary conditions (6.50)-(6.53). Since $\frac{dE(t)}{dt} \leq 0$, B is dissipative. Moreover, B has no zero eigenvalues in X/X_0. It can be shown easily that B^{-1} exists and hence B generates a C_0-semigroup of contractions on X/X_0 by virtue of the Lümer-Phillips theorem.

Now make a linear bounded transformation $P : X/X_0 \to (L^2(0,1))^{2N+2}$ by

$$\begin{bmatrix} u \\ v \end{bmatrix} = P \begin{bmatrix} y_x \\ y_t \end{bmatrix},$$

where

$$u(x,t) = \mathrm{col}[u_1(x,t), u_2(x,t), \ldots, u_{N+1}(x,t)],$$
$$v(x,t) = \mathrm{col}[v_1(x,t), v_2(x,t), \ldots, v_{N+1}(x,t)],$$

and

$$\begin{bmatrix} u_i(x,t) \\ v_i(x,t) \end{bmatrix} = \frac{1}{2} \begin{bmatrix} (-1)^i c_i y_x(p_i(x),t) + y_t(p_i(x),t) \\ -(-1)^i c_i y_x(p_i(x),t) + y_t(p_i(x),t) \end{bmatrix} \tag{6.56}$$

with $p_i(x) = (-1)^i(1/2 - x) + i - 1/2$. Then P is a one-to-one mapping and the norm of $\begin{bmatrix} u \\ v \end{bmatrix}$ in H is equal to the energy $E(t)$ of system (6.49)-(6.53).

We shall show that $\begin{bmatrix} u \\ v \end{bmatrix}$ satisfies the evolution equation (6.16), and hence there exists a C_0-semigroup $T(t)$ such that

$$\begin{bmatrix} u \\ v \end{bmatrix} = T(t) \begin{bmatrix} u_0 \\ v_0 \end{bmatrix} = T(t)P \begin{bmatrix} y_0 \\ y_1 \end{bmatrix},$$

where $y_0 = y_x(x,0), y_1 = y_t(x,0)$ are the initial values of (6.49) - (6.53). Thus,

$$E(t) = \frac{1}{2} \left\| \begin{bmatrix} u \\ v \end{bmatrix} \right\| = \frac{1}{2} \left\| T(t)P \begin{bmatrix} y_0 \\ y_1 \end{bmatrix} \right\| \leq \|T(t)\| \, \|P\| \left\| \begin{bmatrix} y_0 \\ y_1 \end{bmatrix} \right\|,$$

which means that the energy $E(t)$ decays asymptotically (or exponentially) to zero when $T(t)$ is asymptotically (or exponentially) stable. In particular, when P maps X/X_0 onto H, then P^{-1} is bounded,

$$\begin{bmatrix} y_x \\ y_t \end{bmatrix} = P^{-1}T(t)P \begin{bmatrix} y_0 \\ y_1 \end{bmatrix} = S(t) \begin{bmatrix} y_0 \\ y_1 \end{bmatrix}$$

and $S(t)$ is a C_0-semigroup on X/X_0. In this case, the stability of $T(t)$ is equivalent to the stability of $S(t)$. Unfortunately, P^{-1} may not always be bounded. The reason for this is that, in making the transformation, we actually used the boundary condition $y_t(i^-, t) = y_t(i^+, t)$, instead of the original condition $y(i^-, t) = y(i^+, t)$. Although the latter guarantees the former, the converse is not always true. This causes the trouble that P is not always onto. In the next section, we shall explain how to deal with the case where P is not onto. In general, we have to reduce the state space H in order to assure that P is onto.

We now continue to write (6.49)-(6.53) as the form (6.16) in the previous section, by using the transformation (6.56).

First, by using (6.56), (6.49) can be written as the following first order system on H:

$$\frac{\partial}{\partial t} \begin{bmatrix} u(x,t) \\ v(x,t) \end{bmatrix} + K \frac{\partial}{\partial x} \begin{bmatrix} u(x,t) \\ v(x,t) \end{bmatrix} = 0, \;\; 0 < x < 1 \tag{6.57}$$

where $K = \mathrm{diag}(c_1, \ldots, c_{N+1}, -c_1, \ldots, -c_{N+1})$ which is a special case of (6.16) in Section 6.1. In order to arrange the boundary conditions in the form of (6.16), we need to express $u_i(0,t)$ in terms of $v_i(0,t)$, and $v_i(1,t)$ in terms of $u_i(1,t)$. From condition (6.50), we have

$$\begin{cases} u_1(0,t) = -v_1(0,t), \\ u_{N+1}(1,t) = -v_{N+1}(1,t), & N+1 \text{ is odd}, \\ u_{N+1}(0,t) = -v_{N+1}(0,t), & N+1 \text{ is even}, \end{cases} \tag{6.58}$$

or

$$\begin{cases} v_1(0,t) = u_1(0,t), \\ v_{N+1}(1,t) = u_{N+1}(1,t), & N+1 \text{ is odd}, \\ u_{N+1}(0,t) = v_{N+1}(0,t), & N+1 \text{ is even}, \end{cases} \tag{6.59}$$

and from (6.51) we have

$$\begin{cases} u_1(0,t) = \frac{1-k_0^2/c_1}{1+k_0^2/c_1} v_1(0,t), \\ v_{N+1}(1,t) = \frac{1-k_{N+1}^2/c_{N+1}}{1+k_{N+1}^2/c_{N+1}} u_{N+1}(1,t), & N+1 \text{ is odd}, \\ u_{N+1}(0,t) = \frac{1-k_{N+1}^2/c_{N+1}}{1+k_{N+1}^2/c_{N+1}} v_{N+1}(0,t), & N+1 \text{ is even}. \end{cases} \tag{6.60}$$

When i is odd, condition (6.52) implies that

$$\begin{cases} [v_i(1,t) + u_i(1,t)] - [v_{i+1}(1,t) + u_{i+1}(1,t)] = -k_i^2 c_i[v_i(1,t) - u_i(1,t)], \\ c_i[v_i(1,t) - u_i(1,t)] = c_{i+1}[u_{i+1}(1,t) - v_{i+1}(1,t)], \end{cases}$$

or in matrix form

$$\begin{bmatrix} -1 - k_i^2 c_i & 1 \\ -c_i & -c_{i+1} \end{bmatrix} \begin{bmatrix} v_i(1,t) \\ v_{i+1}(1,t) \end{bmatrix} = \begin{bmatrix} 1 - k_i^2 c_i & -1 \\ -c_i & -c_{i+1} \end{bmatrix} \begin{bmatrix} u_i(1,t) \\ u_{i+1}(1,t) \end{bmatrix}.$$

Likewise, condition (6.53) implies that

$$\begin{cases} v_i(1,t) + u_i(1,t) = v_{i+1}(1,t) + u_{i+1}(1,t) \\ c_i[v_i(1,t) - u_i(1,t)] - c_{i+1}[u_{i+1}(1,t) - v_{i+1}(1,t)] = -k_i^2[v_i(1,t) + u_i(1,t)], \end{cases}$$

or in matrix form

$$\begin{bmatrix} -1 & 1 \\ -k_i^2 - c_i & -c_{i+1} \end{bmatrix} \begin{bmatrix} v_i(1,t) \\ v_{i+1}(1,t) \end{bmatrix} = \begin{bmatrix} 1 & -1 \\ k_i^2 - c_i & -c_{i+1} \end{bmatrix} \begin{bmatrix} u_i(1,t) \\ u_{i+1}(1,t) \end{bmatrix}.$$

Similarly, when i is even, conditions (6.52) and (6.53) lead, respectively, to

$$\begin{bmatrix} 1 + k_i^2 c_i & -1 \\ c_i & c_{i+1} \end{bmatrix} \begin{bmatrix} u_i(0,t) \\ u_{i+1}(0,t) \end{bmatrix} = \begin{bmatrix} k_i^2 c_i - 1 & 1 \\ c_i & c_{i+1} \end{bmatrix} \begin{bmatrix} v_i(0,t) \\ v_{i+1}(0,t) \end{bmatrix},$$

and

$$\begin{bmatrix} 1 & -1 \\ k_i^2 + c_i & c_{i+1} \end{bmatrix} \begin{bmatrix} u_i(0,t) \\ u_{i+1}(0,t) \end{bmatrix} = \begin{bmatrix} -1 & 1 \\ -k_i^2 + c_i & c_{i+1} \end{bmatrix} \begin{bmatrix} v_i(0,t) \\ v_{i+1}(0,t) \end{bmatrix}.$$

Now, introduce matrices E_i and D_i as

$$\begin{cases} E_i = \begin{bmatrix} 1 + k_i^2 c_i & -1 \\ c_i & c_{i+1} \end{bmatrix}^{-1} \begin{bmatrix} k_i^2 c_i - 1 & 1 \\ c_i & c_{i+1} \end{bmatrix}, \\ D_i = \begin{bmatrix} 1 & -1 \\ c_i + k_i^2 & c_{i+1} \end{bmatrix}^{-1} \begin{bmatrix} -1 & 1 \\ c_i - k_i^2 & c_{i+1} \end{bmatrix}. \end{cases}$$

Then we can rewrite (6.52) and (6.53) in a compact form as

$$\begin{cases} \begin{bmatrix} v_i(1,t) \\ v_{i+1}(1,t) \end{bmatrix} = E_i \begin{bmatrix} u_i(1,t) \\ u_{i+1}(1,t) \end{bmatrix}, \ i \text{ is odd}, \\ \begin{bmatrix} u_i(0,t) \\ u_{i+1}(0,t) \end{bmatrix} = E_i \begin{bmatrix} v_i(0,t) \\ v_{i+1}(0,t) \end{bmatrix}, \ i \text{ is even}, \ 1 \le i \le N, \end{cases} \tag{6.61}$$

and

$$\begin{cases} \begin{bmatrix} v_i(1,t) \\ v_{i+1}(1,t) \end{bmatrix} = D_i \begin{bmatrix} u_i(1,t) \\ u_{i+1}(1,t) \end{bmatrix}, \ i \text{ is odd}, \\ \begin{bmatrix} u_i(0,t) \\ u_{i+1}(0,t) \end{bmatrix} = D_i \begin{bmatrix} v_i(0,t) \\ v_{i+1}(0,t) \end{bmatrix}, \ i \text{ is even}, \ 1 \le i \le N, \end{cases} \tag{6.62}$$

respectively. Thus the boundary conditions can be written as

$$u(0,t) = Ev(0,t), \quad v(1,t) = Du(1,t), \tag{6.63}$$

where E and D are constant matrices constructed, in an obvious way, from E_i and D_i as well as (6.58)-(6.60).

Summarizing, we have successfully transformed the equation describing the dynamics of the serially connected strings and its boundary conditions into the form of the hyperbolic system (6.16) in Section 6.1, as is evident from (6.57) and (6.63). Let us put them together for the convenience of references.

$$\begin{cases} \dfrac{\partial}{\partial t}\begin{bmatrix} u(x,t) \\ v(x,t) \end{bmatrix} + K\dfrac{\partial}{\partial x}\begin{bmatrix} u(x,t) \\ v(x,t) \end{bmatrix} = 0, \;\; 0 < x < 1, \\[2mm] u(0,t) = Ev(0,t), \\ v(1,t) = Du(1,t). \end{cases} \tag{6.64}$$

Now, define the operator $A : D(A) \subset H \to H$ by

$$\begin{cases} A\begin{bmatrix} u \\ v \end{bmatrix} = -K\dfrac{\partial}{\partial x}\begin{bmatrix} u \\ v \end{bmatrix}, \\[2mm] D(A) = \left\{ \begin{bmatrix} u \\ v \end{bmatrix} \in (H^2(0,1))^{2N+2} \mid u(0) = Ev(0), v(1) = Du(1) \right\}. \end{cases} \tag{6.65}$$

With this setting, system (6.64) can be rewritten as an abstract equation on the state space $H = (L^2(0,1))^{2N+2}$:

$$\frac{\partial}{\partial t}\begin{bmatrix} u(x,t) \\ v(x,t) \end{bmatrix} = A\begin{bmatrix} u(x,t) \\ v(x,t) \end{bmatrix}. \tag{6.66}$$

The operator A defined above has the following properties.

Theorem 6.12   *(i) A is an infinitesimal generator of a C_0-semigroup of contractions on H.*

(ii) The resolvent operator $R(\lambda, A)$ is compact (whenever it is well defined), and is expressed as

$$R(\lambda, A)\begin{bmatrix} f \\ g \end{bmatrix} = e^{-\lambda K^{-1}x}\begin{bmatrix} E \\ I \end{bmatrix}v(0) \tag{6.67}$$

$$+ \int_0^x e^{-\lambda K^{-1}(x-s)}K^{-1}\begin{bmatrix} f(s) \\ g(s) \end{bmatrix}ds, \tag{6.68}$$

where

$$v(0) = H^{-1}(\lambda)(D, -I)\int_0^1 e^{-\lambda K^{-1}(1-s)}K^{-1}\begin{bmatrix} f(s) \\ g(s) \end{bmatrix}ds,$$

and $H(\lambda) = (-D, I)e^{-\lambda K^{-1}}\begin{bmatrix} E \\ I \end{bmatrix}$, *for any* $\begin{bmatrix} f \\ g \end{bmatrix} \in H$.

(iii) The spectrum-determined growth condition holds:

$$\omega_0(A) = \omega_{ess}(A) = S(A).$$

(iv) $\sigma(A) = \{\lambda | h(\lambda) = 0\}$, *where* $h(\lambda)$ *is an exponential polynomial* $h(\lambda) = det H(\lambda)$. *Moreover, corresponding to any* $\lambda \in \sigma(A)$, *there exists an independent eigenfunction*

$$\Phi_\lambda = e^{-\lambda K^{-1}x} \begin{bmatrix} E \\ I \end{bmatrix} y,$$

where y *is any vector satisfying* $H(\lambda)y = 0$. *In particular, if* $0 \in \sigma(A)$, *then* $\Phi_0 = \begin{bmatrix} E \\ I \end{bmatrix} y$ *for any* y *satisfying* $(D, -I) \begin{bmatrix} E \\ I \end{bmatrix} y = 0$.

By employing the result in the previous section, we can prove very simply the following theorem which is obtained in [94] in a different way.

Theorem 6.13 *Let* $h(\lambda)$ *be defined as in Theorem 6.12. Then system (6.66) is exponentially stable if and only if*

$$\inf\{|h(i\omega)|, \omega \in \mathbf{R}\} > 0. \tag{6.69}$$

Proof. If system (6.66) is exponentially stable, then it is necessary that $S(A) < 0$. This implies that the distance between the set $\{i\omega, \omega \in \mathbf{R}\}$ and the set $\{\lambda \mid h(\lambda) = 0\}$ does not equal zero. In view of (i) of Lemma 6.10, $|h(i\omega)|$ is bounded below from zero, i.e., (6.69) holds. Conversely, if (6.69) holds, then from (6.68), $\|R(i\omega, A)\|$ is uniformly bounded for $\omega \in \mathbf{R}$. Since $\|e^{At}\| \leq 1$, by the frequency domain criteria of exponential stability discussed in Chapter 3 (Corollary 3.36), it follows that e^{At} is exponentially stable. The theorem is thus proved. □

So far, we have discussed the exponential stability. We now prove an asymptotic stability result and, furthermore, we establish a relationship between the exponential stability and asymptotic stability.

Theorem 6.14 *The semigroup* e^{At} *generated by* A *is asymptotically stable if and only if* $Re\lambda < 0$ *for all* $\lambda \in \sigma(A)$, *or equivalently,*

(a). $0 \notin \sigma(A)$;

(b). $Re\lambda < 0$ *for all* $(\lambda, \phi), \lambda \neq 0, \phi \neq 0$ *satisfying*

$$\begin{cases} \lambda^2\phi(x) - c_i^2\phi''(x) = 0, & i-1 < x < i, \ i = 1,2,\ldots,N, N+1, \\ \phi(0) = 0 \ or \ \phi'(0) = 0; \ \phi(N+1) = 0 \ or \ \phi'(N+1) = 0, \\ \begin{cases} \lambda\phi(i^-) - \lambda\phi(i^+) = -k_i^2 c_i^2\phi'(i^-), \\ c_i^2\phi'(i^-) - c_{i+1}^2\phi'(i^+) = 0; \end{cases} & if \ (6.52) \ is \ imposed, \\ \begin{cases} \phi(i^-) - \phi(i^+) = 0, \\ c_i^2\phi'(i^-) - c_{i+1}^2\phi'(i^+) = -k_i^2\lambda\phi(i^+), \end{cases} & if \ (6.53) \ is \ imposed. \end{cases} \tag{6.70}$$

Proof. Since A has compact resolvent, it follows from Theorem 3.26 that e^{At} is asymptotically stable if and only if $\text{Re}\lambda < 0$ for all $\lambda \in \sigma(A)$. If $(\lambda, \phi), \lambda \neq 0, \phi \neq 0$ solves the eigenvalue problem (6.70), then let

$$\begin{bmatrix} u_i(x) \\ v_i(x) \end{bmatrix} = \frac{1}{2} \begin{bmatrix} (-1)^i c_i \phi'(p_i(x)) + \lambda\phi(p_i(x)) \\ -(-1)^i c_i \phi'(p_i(x)) + \lambda\phi(p_i(x)) \end{bmatrix}, \quad \forall x \in [0, 1], \quad (6.71)$$

where $p_i(x) = (-1)^i(1/2 - x) + i - 1/2$ is defined before. Since $\phi(p_i(x)) = \frac{1}{\lambda}[u_i(x) + v_i(x)]$, so $u_i(x), v_i(x), i = 1, 2, \ldots, N+1$ cannot be identically zero and

$$(u, v) = [u_1(x), \ldots, u_{N+1}, v_1(x), \ldots, v_{N+1}(x)]$$

is the eigenvector of A corresponding to λ. Conversely, if (u, v) is the eigenvector of A corresponding to $\lambda, \lambda \neq 0$, then let

$$\phi(p_i(x)) = \frac{1}{\lambda}[u_i(x) + v_i(x)], \qquad 0 \leq x \leq 1.$$

Since $u_i(x) = u_i(0)e^{\lambda/c_i x}, v_i(x) = v_i(0)e^{-\lambda/c_i x}, i = 1, 2, \ldots N+1$ cannot be zero identically, $\phi(x), x \in [0, N+1]$ defined above cannot be zero, and it can be verified easily that (λ, ϕ) solves (6.70). $\qquad\square$

Theorem 6.15 *If the semigroup generated by A is asymptotically stable, and in addition, $c_i, i = 1, 2, \ldots N+1$ are commensurable, i.e., there is a constant c such that $c_i = c \cdot n_i$ with n_i being some integers, or equivalently, $c_1/c_i =$rational for $i = 1, 2, \ldots N+1$. Then the semigroup is exponentially stable.*

Proof. We note from Theorem 6.12 that the exponential polynomial $h(\lambda) = \sum_{k=1}^{m} b_k e^{\lambda \omega_k}$ where $\omega_k = \sum_{i=1}^{n_k} \pm 1/c_i$. Since $c_i, i = 1, 2, \ldots N+1$ are commensurable, so does $1/c_i$; hence, there is a constant $c \neq 0$ such that $1/c_i = c n_i$ for some integer n_i. In this case, $\omega_k = \sum_{i=1}^{n_k} \pm 1/c_i = cN_k$, where N_k is an integer. Therefore, $h(i\omega) = \sum_{k=1}^{m} b_k e^{i\omega c N_k}$ for $\omega \in \mathbf{R}$. $h(i\omega)$ is then periodic with respect to ω with period $2\pi/c$. Consequently,

$$\inf\{|h(i\omega)|, \omega \in \mathbf{R}\} = \inf\{|h(i\omega)|, \omega \in [0, 2\pi/c]\}.$$

However, the asymptotic stability just proved implies that the right-hand side above is positive, thus e^{At} is exponentially stable by Theorem 6.13. $\qquad\square$

6.3   Two coupled vibrating strings

In this section, we consider two coupled vibrating strings with a force stabilizer in the middle point which is a special case of the problem in the previous section.

For notational convenience, suppose that each segment of the coupled string has unit length. Let $y(x,t)$ denote the vibrating magnitude of the string for $x \in [0,2]$. Then the dynamics of the vibrating string can be written as

$$
\begin{cases}
m_1 y_{tt}(x,t) - T_1 y_{xx}(x,t) = 0, x \in (0,1), t > 0, \\
m_2 y_{tt}(x,t) - T_2 y_{xx}(x,t) = 0, x \in (1,2), t > 0, \\
y(0,t) = y(2,t) = 0, \\
y(1^-,t) = y(1^+,t), \\
T_1 y_x(1^-,t) - T_2 y_x(1^+,t) = -k y_t(1,t), \quad k > 0, \\
y(x,0) = y_0(x), y_t(x,0) = y_1(x),
\end{cases}
\tag{6.72}
$$

where T_i and $m_i, i = 1,2$, denote the tension constants and mass densities per unit length on the i-th string, respectively. The constants $c_i = \sqrt{T_i/m_i}$ correspond to the wave speeds. Here, we consider the case where both ends of the string are fixed, i.e., $y(0,t) = y(2,t) = 0$. We call this *symmetric boundary conditions*. The cases such as $y(0,t) = y_x(2,t) = 0$ will be called *asymmetric boundary conditions*.

We show that the solution is asymptotically stable if and only if c_1/c_2 is *irrational*, but the solution is not exponentially stable regardless of the value of c_1/c_2. We prove these main results by introducing the linear bounded transformation P, as mentioned before, to transform (6.72) into the general linear hyperbolic system in Section 6.1. However, even for this simple problem, the variable transformation P is not an onto mapping. Thus, the original system and the transformed system is not equivalent. In fact, we will see that zero is an eigenvalue of the transformed system, but the original system does not contain a zero eigenvalue. By excluding the eigenspace corresponding to the zero eigenvalue from the state space, then P becomes a one-to-one mapping.

We first observe that the energy in system (6.72) is dissipative. That is, if we denote by

$$
\begin{aligned}
E(t) &= \frac{1}{2} \int_0^1 [m_1 y_t^2(x,t) + T_1 y_x^2(x,t)] dx \\
&\quad + \frac{1}{2} \int_1^2 [m_2 y_t^2(x,t) + T_2 y_x^2(x,t)] dx
\end{aligned}
\tag{6.73}
$$

the total energy of the system, then formally

$$\frac{dE(t)}{dt} \le -ky_t^2(1,t)$$

along the solution of (6.72). Furthermore, we consider (6.72) in the state space $X = H_0^1(0,2) \times L^2(0,2)$, which is a Hilbert space when equipped with the inner product induced norm

$$\|(\phi,\psi)\|^2 = \int_0^1 [m_1|\psi(x)|^2 + T_1|\phi(x)|^2]dx + \int_1^2 [m_2|\psi(x)|^2 + T_2|\phi(x)|^2]dx.$$

It is easy to see that

$$E(t) = \frac{1}{2} \left\| \begin{bmatrix} y \\ y_t \end{bmatrix} \right\|^2.$$

Now define a transformation $P : X \to (L^2(0,1))^4$ by

$$
\begin{cases}
u_1(x,t) = \frac{1}{2}\sqrt{m_1}[-c_1 y_x(x,t) + y_t(x,t)], \\
u_2(x,t) = \frac{1}{2}\sqrt{m_2}[c_2 y_x(2-x,t) + y_t(2-x,t)], \\
v_1(x,t) = \frac{1}{2}\sqrt{m_1}[c_1 y_x(x,t) + y_t(x,t)], \\
v_2(x,t) = \frac{1}{2}\sqrt{m_2}[-c_2 y_x(2-x,t) + y_t(2-x,t)].
\end{cases}
\tag{6.74}
$$

Using this, we can write (6.72) as the following abstract equation on $H = (L^2(0,1))^4$:

$$\frac{\partial}{\partial t}\begin{bmatrix} u(x,t) \\ v(x,t) \end{bmatrix} = A \begin{bmatrix} u(x,t) \\ v(x,t) \end{bmatrix}, \tag{6.75}$$

where

$$A\begin{bmatrix} u \\ v \end{bmatrix} = -K\frac{\partial}{\partial x}\begin{bmatrix} u \\ v \end{bmatrix}, \tag{6.76}$$

$$K = \operatorname{diag}(\sqrt{T_1}, \sqrt{T_2}, -\sqrt{T_1}, -\sqrt{T_2}),$$

$$D(A) = \left\{ \begin{bmatrix} u \\ v \end{bmatrix} \in (H^1(0,1))^4 \mid u(0) = Ev(0), v(1) = Du(1) \right\}$$

with

$$E = \begin{bmatrix} -1 & 0 \\ 0 & -1 \end{bmatrix}$$

and

$$D = \frac{1}{\sqrt{m_1 T_1} + \sqrt{m_2 T_2} + k} \begin{bmatrix} \sqrt{m_1 T_1} - \sqrt{m_2 T_2} - k & 2\sqrt{m_1 T_2} \\ 2\sqrt{m_2 T_1} & \sqrt{m_2 T_2} - \sqrt{m_1 T_1} - k \end{bmatrix}.$$

The energy $E(t)$ can be alternatively expressed as

$$E(t) = \frac{1}{2} \left\| \begin{bmatrix} u \\ v \end{bmatrix} \right\|_H^2.$$

Lemma 6.16 *Let A be defined as in (6.76). Then $0 \in \sigma(A)$. But 0 is not an eigenvalue of system (6.72).*

Proof. It is seen from Theorem 6.12 that any eigenfunction of A corresponding to 0 is a constant vector Φ_0, which is found to be

$$\Phi_0 = (\sqrt{T_2}, -\sqrt{T_1}, -\sqrt{T_2}, \sqrt{T_1})^T. \tag{6.77}$$

This shows that 0 is an eigenvalue of A. However, if there is a nonzero ϕ satisfying the following eigenproblem of system (6.72):

$$\begin{cases} m_1 \lambda^2 \phi(x) - T_1 \phi''(x) = 0, & x \in (0, 1), \\ m_2 \lambda^2 \phi(x) - T_2 \phi''(x) = 0, & x \in (1, 2), \\ \phi(0) = \phi(2) = 0, & \\ \phi(1^-) = \phi(1^+), & \\ T_1 \phi'(1^-) - T_2 \phi'(1^+) = -k\lambda\phi(1), & k > 0, \end{cases} \tag{6.78}$$

then λ satisfies the characteristic equation

$$(a + b + k)e^{(1+d)\mu} + (b - a - k)e^{\mu} + (a - b - k)e^{d\mu} - (a + b - k) = 0, \tag{6.79}$$

where

$$a = \frac{T_1}{c_1}, \ b = \frac{T_2}{c_2}, \ d = \frac{c_1}{c_2}, \ \mu = 2\frac{\lambda}{c_1}.$$

Since $\lambda = 0$ does not satisfy this equation, 0 is not an eigenvalue of system (6.72). $\square$

System (6.72) provides an example that the original system (6.72) is not equivalent to the transformed system (6.75). The connection between system (6.72) and (6.75) is, however, obvious in the sense that they have the same nonzero eigenvalues. This motivates us to consider the following modified state space

$$H = \{W \in (L^2(0, 1))^4 \mid \langle W, \Phi_0 \rangle = 0\}. \tag{6.80}$$

H is a Hilbert space with the inner product of $(L^2(0, 1))^4$. Denote by $A_H = A|_H$ the restriction of A on H. It is observed that when $W \in D(A)$, then $\langle AW, \Phi_0 \rangle = 0$, i.e., $AD(A) \subset H$. Thus,

$$A_H W = AW, \ \forall W \in D(A_H), \ D(A_H) = D(A) \cap H. \tag{6.81}$$

Hence, A is completely reducible by H and span$\{\Phi_0\}$, and A_H is a semigroup generator on H. It is seen that A and A_H have the same nonzero spectrum. Moreover, it can be readily shown that the transformation (6.74) is a one-to-one transformation from $H_0^1(0,2) \times L^2(0,2)$, the original state space for system (6.72), to H. Indeed, for any $(\phi, \psi) \in H_0^1(0,2) \times L^2(0,2)$, let

$$(u_1, u_2, v_1, v_2)^T = P(\phi, \psi)^T. \tag{6.82}$$

Then it is easy to show that

$$\langle (u_1, u_2, v_1, v_2)^T, \ \Phi_0 \rangle = 0, \tag{6.83}$$

That is, $(u_1, u_2, v_1, v_2) \in H$. Conversely, for any $(u_1, u_2, v_1, v_2) \in H$, there exists a $(\phi, \psi) \in H_0^1(0,2) \times L^2(0,2)$ such that (6.82) is satisfied.

Based on this observation, we see that the stability of

$$\frac{\partial}{\partial t} \begin{bmatrix} u(x,t) \\ v(x,t) \end{bmatrix} = A_H \begin{bmatrix} u(x,t) \\ v(x,t) \end{bmatrix} \tag{6.84}$$

on H is equivalent to the stability of (6.72) on X.

We now concentrate on analyzing the stability of (6.84).

Theorem 6.17 $\omega_{ess}(A_H) = \omega_0(A_H) = S(A_H) = 0$. *Therefore, system (6.84) is not exponentially stable.*

Proof. For the operator A defined in (6.76) which shares all the properties stated in Theorem 6.1, since $0 \in \sigma(A)$, we have $\omega_{ess}(A) = \omega_0(A) = S(A) = 0$. By the proof of Theorem 6.11, we know that there are $\lambda_n \in \sigma(A)$ such that $\mathrm{Re}\lambda_n \to S(A) = 0$ and $|\lambda_n| \to \infty$ as $n \to \infty$. The conclusion follows. □

The asymptotic stability of (6.84) depends on the ratio of the two wave speeds.

Theorem 6.18 *All eigenvalues of A_H satisfy $\mathrm{Re}\lambda < 0$ if and only if $d = c_1/c_2$ is irrational. Consequently, system (6.72) is asymptotically stable if and only if c_1/c_2 is irrational.*

Proof. Since the system is dissipative, $\mathrm{Re}\lambda \leq 0$. As we have seen in Theorem (6.14), $\mathrm{Re}\lambda < 0$ if and only if there does not exist a nonzero solution of (6.78) for any $i\omega, \omega \in \mathbf{R}$. Suppose there is a $\omega \in \mathbf{R}$ such that $\lambda = i\omega$ satisfies (6.78) with some $\phi \neq 0$. We show that c_1/c_2 must be rational. Multiplying the first two equations of (6.78) by $\overline{\phi}(x)$, where the overbar denotes complex conjugate, integrating the resultant equations over $[0,1]$ and $[1,2]$, respectively, adding the two expressions and using the boundary conditions in (6.78), yield

$$k\lambda|\phi(1)|^2 + \int_0^1 [T_1|\phi'(x)|^2 + m_1\lambda^2|\phi(x)|^2]dx$$

$$+ \int_1^2 [T_2|\phi'(x)|^2 + m_2\lambda^2|\phi(x)|^2]dx = 0, \qquad (6.85)$$

where $\lambda = i\omega$. Letting the imaginary part of (6.85) be equal to zero gives $\phi(1) = 0$. When $\lambda = i\omega$ and $\phi(1) = 0$, we can solve the differential equations in (6.78) directly to get

$$\phi(x) = \begin{cases} A_1 \sin \omega x/c_1, & x \in (0,1), \\ A_2 \sin \omega(x-2)/c_2, & x \in (1,2) \end{cases} \qquad (6.86)$$

for some constants A_1 and A_2. Again, by using the fact that $\phi(1) = 0$, we see that

$$A_1 \sin \omega/c_1 = A_2 \sin \omega/c_2 = 0. \qquad (6.87)$$

We claim that both A_1 and A_2 are not zeros. Indeed, if $A_1 = 0$, then $\phi(x) = 0$ for all $x \in [0,1]$. Since we have assumed $i\omega$ is an eigenvalue, $\phi(x)$ cannot be identically zero in $x \in [1,2]$, so $A_2 \neq 0$. By (6.87), $\sin \omega/c_2 = 0$ and so $\cos \omega/c_2 \neq 0$. However, $T_2\phi'(1^+) = \omega/c_2 T_2 A_2 \cos \omega/c_2 = T_1\phi'(1^-) = 0$ leads to a contradiction that $A_2 = 0$. Hence, $A_1 \neq 0$. Similarly, we can show that $A_2 \neq 0$. Then (6.87) implies that $\omega/c_1 = q\pi$ and $\omega/c_2 = p\pi$ for some integers p and q. Therefore, $d = c_1/c_2 = p/q$ is rational.

Conversely, if $d = c_1/c_2 = p/q$ for some coprime positive integers p and q, let $\omega = c_1 q = c_2 p$. It can be shown easily that

$$\lambda_n = in\omega\pi = ic_1 qn\pi, \quad n = \pm 1, \pm 2, \ldots$$

satisfy (6.79). The proof is complete.

$\square$

When zero is not an eigenvalue of the operator A, a different method is required to discuss the stability. The following example demonstrates how to treat such problems using a general approach. Consider the following equation with asymmetric conditions:

$$\begin{cases} m_1 y_{tt}(x,t) - T_1 y_{xx}(x,t) = 0, & x \in (0,1), t > 0, \\ m_2 y_{tt}(x,t) - T_2 y_{xx}(x,t) = 0, & x \in (1,2), t > 0, \\ y(0,t) = y_x(2,t) = 0, \\ y(1^-,t) = y(1^+,t), \\ T_1 y_x(1^-,t) - T_2 y_x(1^+,t) = -ky_t(1,t), & k > 0. \end{cases} \qquad (6.88)$$

The characteristic equation of (6.88) is

$$(a + b + k)e^{(1+d)\mu} + (a - b + k)e^{\mu} + (a - b - k)e^{d\mu} + (a + b - k) = 0$$

$$(6.89)$$

where a, b and μ are the same as those defined in (6.79).

Parallel to Theorem 6.18, we have the following proposition.

Proposition 6.19 *System (6.88) is equivalent to its transformed system under the transformation P defined in (6.74). Moreover, system (6.88) is asymptotically stable if and only if $d = c_1/c_2$ is irrational.*

To show that system (6.88) is not exponentially stable, we need the following lemma.

Lemma 6.20 *For any irrational d, there is an infinity of fractions p_n/q_n satisfying*

$$\left| d - \frac{p_n}{q_n} \right| < \frac{12}{q_n^2},$$

where p_n is odd, q_n is even, and p_n and q_n are coprime.

Proof. We consider only the case when $d > 0$. The proof for $d < 0$ is similar. For irrational $d > 0$, $2d$ also is irrational. From the well-known results of irrational number approximations by rational numbers [73], we know that there is an infinity of fractions p_n/q_n which are coprime and satisfy

$$\left| 2d - \frac{p_n}{q_n} \right| < \frac{1}{q_n^2}.$$

If p_n is odd, then $|d - p_n/(2q_n)| < 2/(2q_n)^2$. If p_n is even, since p_n and q_n are coprime, there exist integers f_n, g_n such that $|p_n f_n - q_n g_n| = 1$. We may assume, without loss of generality, that $0 < |f_n| < |q_n|$, since if $|f_n| > q_n$, we can write $|f_n| = a_n q_n + b_n$ for some integers a_n, b_n with $0 < b_n < q_n$. Hence, $|p_n b_n - q_n c_n| = 1$ for some integer c_n. As p_n is even, g_n must be odd. Noting that $q_n + f_n \le 2q_n$ implies $1/q_n \le 2/(q_n + f_n)$, we have

$$\left| 2d - \frac{p_n + g_n}{q_n + f_n} \right| \le \left| 2d - \frac{p_n}{q_n} \right| + \left| \frac{p_n + g_n}{q_n + f_n} - \frac{p_n}{q_n} \right|$$

$$\le \frac{1}{q_n^2} + \left| \frac{p_n f_n - q_n g_n}{q_n(q_n + f_n)} \right|$$

$$\le \frac{1}{q_n^2} + \frac{1}{q_n(q_n + f_n)} \le \frac{6}{(q_n + f_n)^2}. \qquad (6.90)$$

Hence,

$$\left| d - \frac{p_n + g_n}{2(q_n + f_n)} \right| \le \frac{12}{[2(q_n + f_n)]^2}.$$

Because $|(p_n + g_n)f_n - (q_n + f_n)g_n| = 1$, $p_n + g_n$ and $q_n + f_n$ is coprime. Since $p_n + g_n$ is also odd, $p_n + g_n$ and $2(q_n + f_n)$ are coprime. Because q_n are infinite in number, so are $2(q_n + f_n)$. Hence, by (6.90) $p_n + g_n$ are also infinite in number. Renaming $p_n + g_n$ by p_n and $2(q_n + f_n)$ by q_n in (6.90) yields the inequality in the Lemma.   □

Using Lemma 6.20, we are in a position to prove the following lemma.

Lemma 6.21 *For any irrational* $d = c_1/c_2$, *there exist* $w_n \in \mathbf{R}$ *such that* $|w_n| \to \infty$, $e^{iw_n} \to 1$ *and* $e^{idw_n} \to -1$ *as* $n \to \infty$.

Proof. By Lemma 6.20, $d = \frac{p_n}{q_n} + \frac{c_n}{q_n^2}$ with $|c_n| \leq 12$, q_n being even and p_n being odd. Take $w_n = \pi q_n$. Then $e^{iw_n} = 1$ and

$$e^{idw_n} = e^{ip_n \pi} \cdot e^{ic_n \pi/q_n} = -e^{ic_n \pi/q_n} \to -1.$$

This is the result required.   □

Theorem 6.22 *For any irrational* $d = c_1/c_2 > 0$, *there exists an infinite number of solutions to (6.89) with* $\mathrm{Re}\lambda_n \to 0$ *as* $n \to \infty$. *Hence, system (6.88) is not exponentially stable.*

Proof. Suppose this is not true. Then there exist an $M > 0$ and $\varepsilon > 0$ such that $|f(i\omega)| > \varepsilon$ for all $|\omega| \geq M$ by Lemma 6.10 (i), where $f(\mu)$ denotes the left-hand side of (6.89). But for ω_n in Lemma 6.21, a direct computation shows that

$$|f(i\omega_n)| \to 0 \quad \text{as } n \to \infty,$$

which is a contradiction.   □

6.4   A vibration cable with a tip mass

In this section, we consider a pinched vibration cable with a tip mass whose motion is described by the following partial differential equation:

$$\begin{cases} y_{tt}(x, t) - y_{xx}(x, t) = 0, & x \in (0, 1), \\ y(0, t) = 0, \\ y_x(1, t) + m y_{tt}(1, t) = -u(t), \\ y(x, 0) = y_0(x), y_t(x, 0) = y_1(x), \end{cases} \quad (6.91)$$

where $m > 0$ is the tip mass and $u(t)$ is the boundary control force applied at the free end and (y_0, y_1) is the initial condition. Our purpose is to find a feedback control law $u(t)$ so that the energy $E(t)$ of the resulting closed-loop system decays uniformly to zero, both for the cable deflection and the motion of the tip mass.

One of the natural control is the linear feedback control law:

$$u(t) = \alpha y_t(1, t), \alpha > 0, \tag{6.92}$$

by which the closed-loop system will be

$$\begin{cases} y_{tt}(x, t) - y_{xx}(x, t) = 0, \\ y(0, t) = 0, \\ y_x(1, t) + m y_{tt}(1, t) + \alpha y_t(1, t) = 0. \end{cases} \tag{6.93}$$

Let $V = \{u \mid u \in H^1(0, 1), u(0) = 0\}$. Consider system (6.93) in the state Hilbert space $H = V \times L^2(0, 1) \times \mathbf{R}$ with the inner product

$$\langle (u_1, v_1, d_1), (u_2, v_2, d_2) \rangle = \int_0^1 [u_1'(x)\overline{u_2'(x)} + v_1(x)\overline{v_2(x)}]dx. + m d_1 \bar{d_2}$$

Introduce a linear transformation P:

$$P(u(x), v(x), d)^T = (\frac{1}{2}(v(x) - u'(x)), \frac{1}{2}(v(x) + u'(x)), d)^T. \tag{6.94}$$

It is easy to see that P is a unitary mapping from H to $(L^2(0, 1))^2 \times \mathbf{R}$. Set

$$\begin{bmatrix} u(x, t) \\ v(x, t) \\ d \end{bmatrix} = P \begin{bmatrix} y(x, t) \\ y_t(x, t) \\ d \end{bmatrix}.$$

Then system (6.93) is transformed into an equivalent form under P:

$$\begin{cases} \frac{\partial}{\partial t} \begin{bmatrix} u(x, t) \\ v(x, t) \end{bmatrix} + K \frac{\partial}{\partial x} \begin{bmatrix} u(x, t) \\ v(x, t) \end{bmatrix} = 0, \\ \frac{d}{dt}[v(1, t) + u(1, t)] = \frac{1-\alpha}{m}u(1, t) - \frac{1+\alpha}{m}v(1, t), \\ u(0, t) = -v(0, t), \end{cases} \tag{6.95}$$

which is of the form of (6.1) with $N = 1, K(x) = \text{diag}(1, -1), C(x) = 0, D = -1, F = (1 - \alpha)/m, G = -(1 + \alpha)/m$ and $E = -1$. Now the reduced system (6.16) becomes

$$\begin{cases} \frac{\partial}{\partial t} \begin{bmatrix} u(x, t) \\ v(x, t) \end{bmatrix} + K \frac{\partial}{\partial x} \begin{bmatrix} u(x, t) \\ v(x, t) \end{bmatrix} = 0, \\ v(1, t) = -u(1, t), u(0, t) = -v(0, t). \end{cases} \tag{6.96}$$

Under P^{-1} defined by (6.94), system (6.96) is equivalent to the undamped vibration system

$$\begin{cases} y_{tt}(x,t) - y_{xx}(x,t) = 0, \ x \in (0,1), \\ y(0,t) = 0, y(1,t) = 0. \end{cases} \tag{6.97}$$

It is well known that all the eigenvalues of this system lie on the imaginary axis; hence, system (6.96) is not asymptotically stable. Therefore, there is at least one essential spectrum z, with $|z| = 1$, of the associated semigroup. From results in Section 6.1, system (6.95) is a compact perturbation of system (6.96); hence, it is only asymptotically stable (all its eigenvalues locate strictly on the left-half complex plane), but not exponentially stable. Actually, let ω_{ess} denote the essential growth rate of system (6.95). Then

$$\omega_{ess} = 0. \tag{6.98}$$

To achieve exponential stability, we propose another linear feedback control law:

$$u(t) = ay_{xt}(1,t) + \alpha y_t(1,t), \quad a > 0, \alpha > 0. \tag{6.99}$$

Substituting $u(t)$ into (6.91) yields the closed-loop system equation:

$$\begin{cases} y_{tt}(x,t) - y_{xx}(x,t) = 0, x \in (0,1), \\ y(0,t) = 0, \\ y_x(1,t) + my_{tt}(1,t) + ay_{xt}(1,t) + \alpha y_t(1,t) = 0. \end{cases} \tag{6.100}$$

Define the energy function $E(t)$ by

$$E(t) = \frac{1}{2} \int_0^1 [y_t^2(x,t) + y_x^2(x,t)]dx + \frac{1}{2} \frac{1}{m+\alpha a}[ay_x(1,t) + my_t(1,t)]^2. \tag{6.101}$$

It is easy to verify that

$$\frac{dE(t)}{dt} \le 0$$

along the solutions of (6.100). On the other hand, using the transformation P defined by (6.94) again, (6.100) is equivalent to the following system

$$\begin{cases} \dfrac{\partial}{\partial t}\begin{bmatrix} u(x,t) \\ v(x,t) \end{bmatrix} + K\dfrac{\partial}{\partial x}\begin{bmatrix} u(x,t) \\ v(x,t) \end{bmatrix} = 0, \\ \dfrac{d}{dt}[v(1,t) + \frac{m-a}{m+a}u(1,t)] = \frac{1-\alpha}{m+a}u(1,t) - \frac{1+\alpha}{m+a}v(1,t), \\ u(0,t) = -v(0,t) \end{cases} \tag{6.102}$$

on the state space $H = (L^2(0,1))^2 \times \mathbf{R}$ which is a Hilbert space with the inner product induced norm:

$$\|(u,v,d)\|^2 = \int_0^1 |u(x)|^2 dx + \int_0^1 |v(x)|^2 dx + \frac{1}{m+a}|d|^2.$$

Let

$$W(t) = \begin{bmatrix} u(\cdot,t) \\ v(\cdot,t) \\ v(1,t) + \frac{m-a}{m+a}u(1,t) \end{bmatrix} = \begin{bmatrix} (y_t(\cdot,t) - y_x(\cdot,t))/2 \\ y_t(\cdot,t) + y_x(\cdot,t))/2 \\ \frac{1}{m+a}(my_t(1,t) + ay_x(1,t)) \end{bmatrix}.$$

Then (6.102) can be written as

$$\begin{cases} \dot{W}(t) = AW(t) \\ W(0) = W_0 \end{cases} \tag{6.103}$$

with A being defined in an obvious way. Clearly, $E(t) = \frac{1}{2}\|W(t)\|^2$. Hence, $E(t)$ decays exponentially if and only if $\|W(t)\|$ decays exponentially. We show that $\|W(t)\|$ decays exponentially by using the spectrum-determined growth condition. For this purpose, we need to consider the following reduced system of (6.102)

$$\begin{cases} \dfrac{\partial}{\partial t}\begin{bmatrix} u(x,t) \\ v(x,t) \end{bmatrix} + K\dfrac{\partial}{\partial x}\begin{bmatrix} u(x,t) \\ v(x,t) \end{bmatrix} = 0, \\ v(1,t) = -\frac{m-a}{m+a}u(1,t), \ u(0,t) = -v(0,t). \end{cases} \tag{6.104}$$

Let B denote the associated generator of (6.104). Then Theorems 6.3 and 6.11 imply that

$$\omega_{ess}(A) = \omega_{ess}(B) = \omega_0(B) = S(B), \tag{6.105}$$

and $e^{At} - e^{Bt}$ is compact for any $t \geq 0$. Our first conclusion is the following theorem.

Theorem 6.23 *The spectrum-determined growth condition holds for system (6.103), i.e.,*

$$\omega_0(A) = S(A).$$

Proof. If we can show that $S(B) \leq S(A)$, then from (6.36), we see that

$$S(A) \leq \omega_0(A) = \max\{S(A), S(B)\} = S(A);$$

hence, $S(A) = \omega_0(A)$.

We know from Section 6.1 that both operators A and B have compact resolvents, so the spectrum of these two operators contains only isolated eigenvalues with finite algebraic multiplicities.

Let $\lambda \in \mathbf{C}$ be an eigenvalue of A and let $\Phi = (u, v, d)$ be an associated eigenvector with λ. Then $(\lambda - A)\Phi = 0$. This is equivalent to finding a $\phi \neq 0$, satisfying the following equation:

$$\begin{cases} \lambda^2 \phi(x) - \phi''(x) = 0, & x \in (0, 1), \\ \phi(0) = 0, \\ (1 + a\lambda)\phi'(1) + (m\lambda^2 + \alpha\lambda)\phi(1) = 0. \end{cases} \tag{6.106}$$

Solving (6.106), we see that $f(\lambda) = 0$ in order that $\phi(x) \neq 0$, where

$$f(\lambda) = ((1 - \alpha) + (a - m)\lambda)e^{-2\lambda} + (1 + \alpha) + (a + m)\lambda. \tag{6.107}$$

Similarly, $\sigma(B)$ consists of roots of $g(\lambda) = 0$, where

$$g(\lambda) = (m + a)e^{2\lambda} - m + a. \tag{6.108}$$

Two cases should be distinguished:

Case 1: $m \neq a$. In this case, all solutions of $g(\lambda) = 0$ lie on the vertical line:

$$\mathrm{Re}\lambda = \frac{1}{2}\log\left|\frac{m - a}{m + a}\right|. \tag{6.109}$$

On the other hand, $f(\lambda) = 0$ can be written asymptotically as

$$e^{2\lambda} = \frac{\alpha - 1 + (m - a)\lambda}{1 + \alpha + (m + a)\lambda} = \frac{m - a}{m + a} + \mathcal{O}(|\lambda|^{-1}). \tag{6.110}$$

Using the technique in Chapter 4, we can easily show, by Rouché's theorem, that there exist an infinite number of solutions to (6.110). Obviously, the solutions of $f(\lambda) = 0$ asymptotically satisfy

$$\mathrm{Re}\lambda \to \frac{1}{2}\log\left|\frac{m - a}{m + a}\right| \text{ as } |\lambda| \to \infty,$$

which together with (6.109) shows that $S(B) \leq S(A)$.

Case 2: $m = a$. In this case, there is no finite solution to $g(\lambda) = 0$, i.e., $S(B) = -\infty$. Therefore, $S(B) \leq S(A)$. A special situation in this case is when $m = a$ and $\alpha = 1$ in which $f(\lambda) = 0$ has only one solution $\lambda = -1/m$. □

From the proof of case 1 in Theorem 6.23 and the relationship between (6.100) and (6.102), we have

Corollary 6.24 *If $m \neq a$, then system (6.100) is exponentially stable with decay rate*

$$\omega_0 = \frac{1}{2} \log \left| \frac{m-a}{m+a} \right| < 0.$$

That is, for any $\epsilon > 0$ such that $\omega_0 + \varepsilon < 0$, there exists an $M_\epsilon > 0$ such that

$$E(t) \leq M_\epsilon e^{(\omega_0+\epsilon)t} E(0)$$

where $E(t)$ is defined in (6.101). When $m = a$ and $\alpha = 1$, then $\omega_0 = -1/m$.

This corollary clearly shows that the solution of (6.103) is exponentially stable. But this alone does not guarantee that the motion of the tip body decays exponentially because we have only shown that $[ay_x(1,t)+my_t(1,t)]^2$ decays exponentially. For smooth initial conditions, e.g., $W_0 \in D(A)$, however, we can further show that

$$
\begin{aligned}
\|W(t)\|_{D(A)} &= \left\|e^{At}W_0\right\|_{D(A)} = \left\|e^{At}W_0\right\| + \left\|Ae^{At}W_0\right\| \\
&= \left\|e^{At}W_0\right\| + \left\|e^{At}AW_0\right\| \\
&\leq M_\epsilon e^{(\omega_0+\epsilon)t} \|W_0\| + M_\epsilon e^{(\omega_0+\epsilon)t} \|AW_0\| \\
&= M_\epsilon e^{(\omega_0+\epsilon)t} \|W_0\|_{D(A)}.
\end{aligned}
\tag{6.111}
$$

On the other hand, since

$$\|W(t)\|_{D(A)} = \|W(t)\|_H + \|AW(t)\|_H$$

and $\|AW(t)\|$ contains the term $\int_0^1 y_{xt}^2(x,t)dx$, which decays exponentially, it turns out that $y_t^2(1,t)$ decays exponentially with the same decay rate ω_0 as $E(t)$ because

$$y_t^2(1,t) \leq \int_0^1 y_{xt}^2(x,t)dx.$$

Similarly, since

$$y^2(1,t) \leq \int_0^1 y_x^2(x,t)dx,$$

it follows that $y^2(1,t)$ also decays exponentially because the right-hand side in this inequality is a part of $E(t)$, which is already shown to be exponentially stable in Corollary 6.24. Consequently, we have shown that the motion of the tip body decays exponentially with the decay rate ω_0.

Since we have only considered the energy decay for the case $m \neq a$ and the case $m = a, \alpha = 1$, we now turn to consider the energy decay for the remaining case $m = a, \alpha \neq 1$. In this case, the reduced system (6.104) becomes

$$
\begin{cases}
\dfrac{\partial}{\partial t} \begin{bmatrix} u(x,t) \\ v(x,t) \end{bmatrix} + K \dfrac{\partial}{\partial x} \begin{bmatrix} u(x,t) \\ v(x,t) \end{bmatrix} = 0, \\
v(1,t) = 0, u(0,t) = -v(0,t).
\end{cases}
\tag{6.112}
$$

By using the method of characteristics, the solution of (6.112) can be readily found to be

$$u(x,t) = \begin{cases} u_0(x-t), & x \geq t, \\ -v_0(t-x), & x \leq t \leq 1+x, \\ 0, & t > 1+x, \end{cases} \tag{6.113}$$

$$v(x,t) = \begin{cases} v_0(x+t), & x+t \geq 1, \\ 0, & x+t > 1. \end{cases} \tag{6.114}$$

for any initial condition $u(x,0) = u_0(x), v(x,0) = v_0(x)$. By the compactness of $e^{At} - e^{Bt}$ for any $t \geq 0$, we have the following interesting results.

Theorem 6.25 *When $m = a$, the semigroup e^{At} associated with system (6.102) is compact for $t \geq 2$, but is not compact for $t < 2$.*

We wish to characterize the decay rate for the case $m = a, \alpha \neq 1$ which is summarized in the following theorem.

Theorem 6.26 *Suppose $m = a, \alpha \neq 1$. Let $\lambda \in \sigma(A)$. Then*

(i) There are no other points in $\sigma(A)$, except λ and $\bar{\lambda}$, which lie on the vertical line $\{z \mid Rez = Re\lambda\}$.

(ii) If $\alpha > 1$, then there is only one real negative eigenvalue λ_0 which is the dominant eigenvalue of A; that is, for all $\lambda \in \sigma(A), \lambda \neq \lambda_0$, it holds that $Re\lambda < \lambda_0$. Furthermore,

$$-\frac{1}{a+\alpha-1} < \lambda_0 < -\frac{1}{2}\log\left(\frac{\alpha+1+a}{\alpha-1+a}\right).$$

(iii) If $\alpha < 1$, then there are at most two negative real eigenvalues $\{\lambda_{01}, \lambda_{02}\}$. Moreover, (a) if $a \leq 1-\alpha$, then there is no real eigenvalue, and all eigenvalues satisfy

$$Re\lambda < -\frac{1}{2}\log\frac{1}{1-\alpha}.$$

(b) if there is no real eigenvalue when $a > 1-\alpha$, then all eigenvalues $\lambda \in \sigma(A)$ satisfy

$$Re\lambda < -\frac{1}{2}\log\frac{a}{1-\alpha}.$$

(c) if there is at least one real eigenvalue, then $a > 1-\alpha$ and all eigenvalues satisfy

$$Re\lambda < -\frac{1}{a-(1-\alpha)}.$$

Besides, the maximal real eigenvalue in case (c) is the dominant eigenvalue as in case (ii).

Proof. (i). Let $\lambda = \sigma + i\tau \in \sigma(A)$. Then $f(\lambda) = 0$ reduces to

$$\begin{cases} (\alpha - 1)e^{-2\sigma}\cos 2\tau = 1 + \alpha + 2a\sigma, \\ (1 - \alpha)e^{-2\sigma}\sin 2\tau = 2a\tau. \end{cases} \tag{6.115}$$

Suppose that $\sigma + i\xi \in \sigma(A)$. Then from the first equality of (6.115), we deduce that $\cos 2\tau = \cos 2\xi$, which implies $\tau + \xi = n\pi$ or $\tau - \xi = n\pi$ for some integer n. By the second equality of (6.115), it follows that $n = 0$. Therefore, $\xi = \tau$ or $\xi = -\tau$. This proves (i). To prove (ii), note that $f'(\sigma) = 2(\alpha - 1)e^{-2\sigma} + 2a > 0$ for any real σ by $\alpha > 1$, $f(0) = 2$ and $f(-\infty) = -\infty$. So, $f(\lambda) = 0$ has only one real negative solution λ_0. For any other $\lambda = \sigma + i\tau \in \sigma(A)$, by (6.115)

$$1 + \alpha + 2a\sigma = (\alpha - 1)e^{-2\sigma}\cos 2\tau \le (\alpha - 1)e^{-2\sigma},$$

that is, $f(\sigma) \le 0$ which implies that $\sigma \le \lambda_0$. This, together with (i), implies the dominance of λ_0. Furthermore, by

$$\begin{aligned} 0 = f(\lambda_0) &= (1 - \alpha)e^{-2\lambda_0} + 1 + \alpha + 2a\lambda_0 \\ &< (1 - \alpha)(1 - 2\lambda_0) + 1 + \alpha + 2a\lambda_0, \end{aligned}$$

we have $\lambda_0 > -\dfrac{1}{a + \alpha - 1}$. On the other hand, since λ_0 is negative,

$$\begin{aligned} 0 = f(\lambda_0) &= (1 - \alpha)e^{-2\lambda_0} + 1 + \alpha + 2a\lambda_0 \\ &> (1 - \alpha)e^{-2\lambda_0} + 1 + \alpha + a(1 - e^{-2\lambda_0}), \end{aligned}$$

which implies that

$$\lambda_0 < -\frac{1}{2}\log\left(\frac{\alpha + 1 + a}{\alpha - 1 + a}\right).$$

Finally, let $\alpha < 1$. Then there are three cases:
(a). Note that $f''(\sigma) > 0$ for any $\sigma \in \mathbf{R}$, the function $f'(\sigma) = -2((1 - \alpha)e^{-2\sigma} - a)$ has a unique real zero $\lambda^* = -\frac{1}{2}\log\frac{a}{1-\alpha}$, $f(0) = 2$ and $f(-\infty) = \infty$. If $a \le 1 - \alpha$, $f(\sigma) > f(0) = 2$ for all $\sigma < 0$. Therefore, for any $\lambda = \sigma + i\tau, \sigma < 0$, by the first equality of (6.115)

$$(\alpha - 1)e^{-2\sigma}cos2\tau = 1 + \alpha + 2a\sigma \ge 2 - (1 - \alpha)e^{-2\sigma},$$

and so

$$2(1 - \alpha)e^{-2\sigma} \ge (1 - \alpha)e^{-2\sigma}(1 - cos2\tau) \ge 2,$$

which implies that $Re\lambda = \sigma \le -\frac{1}{2}\log\frac{1}{1-\alpha}$.
(b). If $f(\lambda)$ has no real zero with $a > 1 - \alpha$, it is implied from the above that $f(\lambda^*) > 0$. For any complex eigenvalue $\lambda = \sigma + i\tau$, it follows from the second

quality of (6.115) that $\sin 2\tau = \frac{2\tau a}{1-\alpha}e^{2\sigma}$, which implies that $\frac{a}{1-\alpha}e^{2\sigma} \leq 1$ and so $\mathrm{Re}\lambda = \sigma \leq \lambda^*$. Therefore, in this case all $\lambda \in \sigma(A)$ satisfy

$$\mathrm{Re}\lambda \leq -\frac{1}{2}\log\frac{a}{1-\alpha}.$$

(c). If there is at least one real eigenvalue, it is implied that $a > 1 - \alpha$ and $f(\lambda^*) \leq 0$. There is only one real eigenvalue when $f(\lambda^*) = 0$ and two real eigenvalues while $f(\lambda^*) < 0$. It is obvious that in both cases, the maximal real eigenvalue is greater than or equal to λ^*. For any complex eigenvalue $\lambda = \sigma + i\tau$, it follows from the second equality of (6.115) that $\sin 2\tau = \frac{a}{1-\alpha}e^{2\sigma}2\tau$ which implies that $\frac{a}{1-\alpha}e^{2\sigma} \leq 1$ and so $\mathrm{Re}\lambda = \sigma \leq \lambda^*$. Furthermore, for a real zero λ of $f(\lambda) = 0$, we have

$$-2a\lambda = (1-\alpha)e^{-2\lambda} + 1 + \alpha > (1-\alpha)(1-2\lambda) + 1 + \alpha = 2 - 2(1-\alpha)\lambda,$$

which implies $\lambda < -\frac{1}{a-(1-\alpha)}$. (c) is thus proved. $\qquad\square$

The spectrum-determined growth condition proved in Theorem 6.23 enables us to determine the decay rate rigorously in most cases by finding the upper bound of the real parts of eigenvalues of the system. When $m \neq a$, Corollary 6.24 shows that the decay rate has nothing to do with the velocity feedback gain $\alpha > 0$. However, for the case of $m = a$, things are completely different. For example, for given m and $\alpha > 1$, it follows from (ii) of Theorem 6.26 that the decay rate tends to zero as α goes to infinity. On the other hand, for the case of $\alpha > 1$, from (ii) of Theorem 6.26, we may make the decay rate as small as possible by letting a $\to 0$ and $\alpha \to 1$; for the case of $\alpha < 1$, the decay rate also could be made as small as possible by taking a $= \epsilon, 1 - \alpha = \epsilon^2$ and $\epsilon \to 0$ (see (iii) of Theorem 6.26).

6.5 Thermoelastic system with Dirichlet - Dirichlet boundary conditions

In this section, we study the stability of a normalized one-dimensional linear model describing longitudinal vibrations within a thermoelastic rod which is assumed to be a homogeneous and isotropic cylindrical body positioned along the x-axis. Let $u(x,t)$ and $\vartheta(x,t)$ denote, respectively, the displacement and temperature of the rod at position x and time t. Then the normalized equations, which govern the coupled thermoelastic motions, are given by

$$\begin{cases} u_{tt}(x,t) - u_{xx}(x,t) + \gamma\vartheta_x(x,t) = 0, & x \in (0,1), \\ \vartheta_t(x,t) + \gamma u_{xt}(x,t) - k\vartheta_{xx}(x,t) = 0. \end{cases} \tag{6.116}$$

For the detailed derivation and the physical meaning of this model, the reader is referred to Day [47] and Carlson [23]. See also Liu and Zheng [95]. In

(6.116), the coupling constant γ is generally less than 1 and is a measure of the mechanical-thermal coupling presenting in the system. The constant k represents the heat conductivity.

Introduce the heat flux q, the stress σ, and the velocity v as

$$q = -\vartheta_x, \quad \sigma = u_x - \gamma\vartheta, \quad v = u_t. \tag{6.117}$$

Then the balancing of power flow at each end of the rod leads to the following boundary conditions

$$v(i,t)\sigma(i,t) - q(i,t)\vartheta(i,t) = 0, \quad i = 0, 1, \tag{6.118}$$

which will be called *natural boundary conditions*. (6.118) can be alternatively written as

$$\det \begin{bmatrix} v(i,t) & q(i,t) \\ \vartheta(i,t) & \sigma(i,t) \end{bmatrix} = 0, \quad i = 0, 1.$$

That is, the two-row vectors, or the two-column vectors, are linearly dependent, so there exist constants α and β with $\alpha^2 + \beta^2 = 1$ such that

$$\alpha v(i,t) + \beta q(i,t) = 0, \quad \alpha\vartheta(i,t) + \beta\sigma(i,t) = 0, \quad i = 0, 1 \tag{6.119}$$

or

$$\alpha v(i,t) + \beta\vartheta(i,t) = 0, \quad \alpha q(i,t) + \beta\sigma(i,t) = 0, \quad i = 0, 1. \tag{6.120}$$

Up to equivalence, (6.119) and (6.120) cover all natural boundary conditions.

For brevity of exposition, we shall consider only a special natural boundary condition: Dirichlet-Dirichlet boundary condition, as specified in the following

$$\begin{cases} u_{tt}(x,t) - u_{xx}(x,t) + \gamma\vartheta_x(x,t) = 0, & 0 < x < 1, \\ \vartheta_t(x,t) + \gamma u_{xt}(x,t) - k\vartheta_{xx}(x,t) = 0, \\ u(i,t) = \vartheta(i,t) = 0, & i = 0, 1. \end{cases} \tag{6.121}$$

The system energy associated with (6.121) is given by

$$E(t) = \frac{1}{2}\int_0^1 [u_x^2(x,t) + u_t^2(x,t) + \vartheta^2(x,t)]dx. \tag{6.122}$$

Formally, it is easily verified that

$$\frac{d}{dt}E(t) = -k\int_0^1 \vartheta_x^2(x,t)dx \leq 0 \tag{6.123}$$

along the solutions of (6.121). That is, the system is dissipative.

To write (6.121) as an abstract equation, let us introduce a Hilbert space $H = H_0^1(0,1) \times L^2(0,1) \times L^2(0,1)$ with the inner product

$$\langle (u_1, v_1, \vartheta_1), (u_2, v_2, \vartheta_2) \rangle = \int_0^1 [u_{1x} \overline{u}_{2x} + v_1 \overline{v}_2 + \vartheta_1 \overline{\vartheta}_2] dx.$$

Define the operator $A : D(A) \to H$ by

$$\begin{cases} A(u,v,\vartheta) = (v, u_{xx} - \gamma \vartheta_x, k\vartheta_{xx} - \gamma v_x), \\ D(A) = (H^2(0,1) \cap H_0^1(0,1)) \times H_0^1(0,1) \times (H^2(0,1) \cap H_0^1(0,1)). \end{cases} \quad (6.124)$$

Then (6.121) can be written as

$$\frac{dw(t)}{dt} = Aw(t) \quad (6.125)$$

with $w(t) = (u(\cdot, t), u_t(\cdot, t), \vartheta(\cdot, t))$. Since the energy is dissipative, the following lemma is readily shown.

Lemma 6.27 *The operator A defined by (6.124) is dissipative; A^{-1} exists and is compact. Therefore, by the Lümer-Phillips theorem, A generates a C_0-semigroup of contractions on H.*

We are interested in the problem of whether this semigroup is exponentially stable. Using frequency domain criteria, Liu and Zheng [95] proved that system (6.125) is exponentially stable. Here, we first show that system (6.125) satisfies the spectrum-determined growth condition by using a result due to Renardy [136], and then we show the system is exponentially stable by analyzing the spectrum distribution of the operator A.

Theorem 6.28 *[136] Let A be the infinitesimal generator of a C_0-semigroup on a Hilbert space H. Assume that A is normal, that is, A and A^* commute, and there exist a constant M and an integer n such that the following hold:*

(i) For each $\lambda \in \sigma(A)$ with $|\lambda| > M - 1, \lambda$ is an isolated eigenvalue of A with finite algebraic multiplicity.

(ii) For each $z \in \mathbf{C}, |z| > M$, the number of eigenvalues of A in the unit disk centered at z (counted by algebraic multiplicity) does not exceed n.

Then for any bounded operator $B \in \mathcal{L}(H)$, the spectrum-determined growth condition holds:

$$S(A + B) = \omega_0(A + B).$$

Proof. From Corollary 3.40, it suffices to show that

$$\sup_{\tau \in \mathbf{R}} \| R(\omega + i\tau, A + B) \| < \infty \text{ for any } \omega > S(A + B). \qquad (6.126)$$

Clearly, for any fixed $\omega > S(A+B)$, $\| R(\omega + i\tau, A + B) \|$ is uniformly bounded for all τ on any compact segment of a real line, so it is sufficient to prove (6.126) for those $\lambda \in \Gamma = \{ \lambda \mid \mathrm{Re}\lambda = \omega, |\lambda| > M + K \}$, where K is a large number to be chosen. For any given $h \in H$ and $\lambda \in \Gamma$, since $\lambda \in \rho(A + B)$, there is a unique u satisfying the following equation

$$Au - \lambda u + Bu = h. \qquad (6.127)$$

Let $P(\lambda)$ be the orthogonal projection onto the span of all eigenvectors of A for which the associated eigenvalues lie in a disk of radius K centered at λ. Let $Q(\lambda) = I - P(\lambda)$. Then H can be decomposed as the direct sum of $X = P(\lambda)H$ and $Y = Q(\lambda)H$, i.e.,

$$H = X \oplus Y.$$

Set $P(\lambda)u = x, Q(\lambda)u = y, P(\lambda)h = f, Q(\lambda)h = g$ and rewrite (6.127) as

$$Ax - \lambda x + P(\lambda)Bx + P(\lambda)By = f, \qquad (6.128)$$
$$Ay - \lambda y + Q(\lambda)By + Q(\lambda)Bx = g. \qquad (6.129)$$

We see that $A - \lambda + Q(\lambda)B$, when restricted to Y, is invertible for $K > \|B\|$ and

$$\left\| (A - \lambda + Q(\lambda)B) \left|_Y^{-1} \right\| \le \frac{1}{K - \|B\|}, \quad \lambda \in \Gamma. \right.$$

Indeed, by assumption, $A \left|_Y - \lambda \right.$ is normal, so it follows that

$$\left\| (A - \lambda)^{-1} \left|_Y \right\| = \frac{1}{d(\lambda, \sigma(A \left|_Y))} \le \frac{1}{K}, \right. \right.$$

where $d(\lambda, \sigma(A \left|_Y))$ denotes the distance between λ and $\sigma(A \left|_Y)$ (see [84, p.277]). Hence $\left\| (A - \lambda)^{-1}Q(\lambda)B \right\| \le \|B\| / K < 1$ as $K > \|B\|$. It follows that $I + (A - \lambda)^{-1}Q(\lambda)B$ is invertible in Y and

$$\begin{aligned}
\left\| (A - \lambda + Q(\lambda)B)^{-1} \left|_Y \right\| \right. &= \left\| (I + (A - \lambda)^{-1}Q(\lambda)B)^{-1}(A - \lambda)^{-1} \right\| \\
&\le \frac{1}{K - \|B\|},
\end{aligned}$$

which holds for all $\lambda \in \Gamma$. Let $\lambda \in \Gamma$. Solving y from (6.129) and plugging the result into (6.128), we obtain

$$\Delta(\lambda)x = f - P(\lambda)B[A - \lambda + Q(\lambda)B]^{-1}g, \qquad (6.130)$$

where $\Delta(\lambda) = A - \lambda + P(\lambda)B - P(\lambda)B[A - \lambda + Q(\lambda)B]^{-1}Q(\lambda)B$. Since $\lambda \in \rho(A+B)$, $\Delta(\lambda)$ is invertible and the uniform boundedness of these inverse will certainly induce the uniform boundedness of $\|R(\omega + i\tau, A + B)\|$, since the operator of the right-hand side of (6.130) is uniformly bounded.

Now in the space X, we can write

$$A - \lambda = \sum_{i=1}^{m}(\lambda_i - \lambda)P_{\lambda_i},$$

where P_{λ_i} is the eigenprojection of A corresponding to the eigenvalue λ_i and m is an integer which depends on λ but has a uniform upper bound N for all $\lambda \in \Gamma$ by assumption. Therefore,

$$\|(A - \lambda)|_X\| \le NK.$$

Using this, it is clear that

$$\|\Delta(\lambda)\| \le NK + \|B\| + \frac{\|B\|^2}{K - \|B\|}. \tag{6.131}$$

This indicates that $\Delta(\lambda)$ is uniformly bounded on X for all $\lambda \in \Gamma$. Since X is a finite dimensional space, the operator $\Delta(\lambda)$ has an $n \times n$ matrix representation, still denoted by $\Delta(\lambda)$, in terms of the orthonormal eigenvectors of A. (6.131) implies that the elements of this matrix are uniformly bounded for all $\lambda \in \Gamma$. Define

$$f_\lambda(\mu) = \det[A - \lambda - \mu + P(\lambda)B - P(\lambda)B[A - \lambda - \mu + Q(\lambda)B]^{-1}Q(\lambda)B].$$

Then $f_\lambda(\mu)$ is uniformly bounded for all $\lambda \in \Gamma$ and μ in any compact set.

Since $\mathrm{Re}\lambda = \omega$ is away from $\sigma(A+B)$ with a positive distance and hence there is a small $\delta > 0$ such that $\lambda + \mu \in \rho(A + B)$ for all $\lambda \in \Gamma$ and $|\mu| \le \delta$. It is seen from previous arguments that $\lambda + \mu \in \rho(A + B)$ if and only if $f_\lambda(\mu) \ne 0$. Therefore,

$$f_\lambda(\mu) \ne 0 \text{ for all } \lambda \in \Gamma, |\mu| \le \delta.$$

Because a lower bound on the determinant trivially yields an upper bound on the norm of the inverse matrix, our proof will be complete if we can show that $|f_\lambda(0)|$ is bounded from below for $\lambda \in \Gamma$.

Suppose that there exists a sequence $\lambda_n \in \Gamma$ such that $f_{\lambda_n}(0) \to 0$ as $n \to \infty$. Since $f_{\lambda_n}(\mu)$ are uniformly bounded in the disk $|\mu| \le \delta$, it follows that there is a subsequence of $\{f_{\lambda_n}\}$, still denoted by f_{λ_n}, and an analytic function $f(\mu)$ such that

$$f_{\lambda_n} \to f$$

uniformly in $|\mu| \le \delta/2$. This conclusion comes from Mentel's theorem which says that a sequence of locally uniformly bounded analytic functions contains a subsequence which converges to an analytic function [38]. Since $f_{\lambda_n}(\mu) \neq 0$ in $|\mu| \leq \delta/2$ and $f(0) = 0$, we have $f \equiv 0$ on $\{\mu \mid |\mu| < \delta\}$ by Hurwitz's theorem (see, e.g., [38]).

On the other hand, since $f_\lambda(\mu)$ is well defined for $\lambda \in \Gamma$ and $|\mu| < \delta$, by analytic extension, $f_\lambda(\mu)$ are defined on all $S_\delta = \{\mu \in C \mid \mathrm{Re}\mu \geq 0, -\delta < Im\mu < \delta\} \subset \rho(A+B)$, and they are uniformly bounded for all $\lambda \in \Gamma$ and μ in any compact subset of S_δ. Using Mentel's theorem again, we conclude that $f_{\lambda_n} \to 0$ for each $\mu \in S_\delta$. But this is clearly not the case for large μ by the Hille-Yosida theorem, and the theorem follows from this contradiction. □

The following theorem also is useful.

Theorem 6.29 *Let A be the infinitesimal generator of a C_0-semigroup on a Hilbert space H. Assume that there is a Hilbert space K, a normal operator A_0 in K, and a family of bounded operators $C(\lambda)$ on K with the following properties:*

(i) $C(\lambda)$ is defined for $\mathrm{Re}\lambda > \eta_0$ and its norm is uniformly bounded in any closed half-plane $\mathrm{Re}\lambda \geq \lambda_0 > \eta_0$.

(ii) A_0 satisfies all those hypotheses for the operator A in Theorem 6.28.

(iii) For $\mathrm{Re}\lambda > \eta_0, A - \lambda$ has a bounded, everywhere defined inverse if and only if $A_0 - \lambda - C(\lambda)$ has a bounded, everywhere defined inverse. Moreover, there is a constant C_0 such that

$$\|R(\lambda, A)\| \leq C_0 \|R(\lambda, A_0 + C(\lambda))\|. \tag{6.132}$$

Then

$$\omega_0(A) \leq \max\{S(A), \eta_0\}.$$

Proof. This is done easily by viewing $C(\lambda)$ as the operator B in Theorem 6.28 and by repeating the proof process of Theorem 6.28.   □

Let us now turn back to analyze the stability of (6.125) by using the results stated in the above theorems.

Choose $K = H_0^1(0,1) \times L^2(0,1)$ and define

$$A_0(u,v) = (v, u_{xx}) \tag{6.133}$$

with domain $(H^2(0,1) \cap H_0^1(0,1)) \times H_0^1(0,1)$. Then $A_0^* = -A_0$ is skew-adjoint; hence, it is normal. It is also clearly seen that A_0^{-1} is compact. Thus, the requirements on A_0 in Theorem 6.29 are satisfied. Let

$$C(\lambda)(u,v) = (0, -\gamma^2 D(\lambda - kD^2)^{-1}Dv) \tag{6.134}$$

where D is used to denote $\frac{\partial}{\partial x}$ and D^2 represents the Laplacian with Dirichlet boundary conditions. Now we verify that the operator $C(\lambda)$ defined by (6.134) does satisfy the conditions of Theorem 6.29 with $\eta_0 = -k\pi^2$. To this end, we first notice that

$$
\begin{aligned}
-\gamma^2 D(\lambda - kD^2)^{-1}D\phi &= -\gamma^2(\lambda - kD^2)^{-1}D(D\phi) \\
&= \frac{\gamma^2}{k}(\lambda - kD^2)^{-1}(-kD^2\phi) \\
&= \frac{\gamma^2}{k}(\lambda - kD^2)^{-1}(\lambda - kD^2 - \lambda)\phi \\
&= -\frac{\gamma^2}{k}(\lambda(\lambda - kD^2)^{-1}\phi - \phi) \qquad (6.135)
\end{aligned}
$$

for any $\phi \in C_0^\infty(0,1)$. Since D^2 is the Laplacian with Dirichlet boundary conditions, it is well known that it has a set of eigenvalues $\{-kn^2\pi^2, n \geq 1\}$ and associated eigenvectors $\{\sin n\pi x, n \geq 1\}$ which form an orthogonal basis of $L^2(0,1)$. In terms of this basis, we can expand $\phi(x)$ as $\phi(x) = \sum_{n=1}^\infty c_n \sin n\pi x$. This yields

$$
\lambda(\lambda - kD^2)^{-1}\phi(x) = \sum_{n=1}^\infty \frac{\lambda}{\lambda + kn^2\pi^2} c_n \sin n\pi x.
$$

Therefore, there is a constant $M > 0$ depending only on λ_0 such that for all $\mathrm{Re}\lambda \geq \lambda_0 > \eta_0 = -k\pi^2$, $\left\|\lambda(\lambda - kD^2)^{-1}\phi\right\| \leq M\left\|\phi\right\|$ in $L^2(0,1)$. Since $C_0^\infty(0,1)$ is dense in $L^2(0,1)$, we see that

$$
\left\|-\gamma^2 D(\lambda - kD^2)^{-1}D\right\| \leq \gamma^2(1+M)/k \qquad (6.136)
$$

holds uniformly for all $\mathrm{Re}\lambda \geq \lambda_0$. This verifies the condition (i) in Theorem 6.29. We show that the condition (iii) in Theorem 6.29 also is satisfied. For this purpose, taking any $(f,g,h) \in H$ and solving the following equation

$$
\begin{aligned}
(\lambda - A)(u,v,\vartheta) &= (\lambda u - v, \lambda v - u_{xx} + \gamma\vartheta_x, \lambda\vartheta - kv_{xx} + \gamma v_x) \\
&= (f,g,h),
\end{aligned}
$$

we have

$$
\begin{aligned}
(\lambda u - v, \lambda v - u_{xx} + \gamma\vartheta_x) &= (f,g), \\
\lambda\vartheta - kv_{xx} + \gamma v_x &= h.
\end{aligned}
$$

Replacing the spatial derivative by the operator D yields

$$
\begin{aligned}
\vartheta &= (\lambda - kD^2)^{-1}(h - \gamma Dv), \\
\vartheta_x &= D(\lambda - kD^2)^{-1}(h - \gamma Dv), \\
\lambda - (A_0 + C(\lambda))(u,v) &= (f,g) + (0, -\gamma D(\lambda - kD^2)^{-1}h).
\end{aligned}
$$

This clearly shows that for any $\mathrm{Re}\lambda > \eta_0, \lambda \in \rho(A)$ if and only if $\lambda \in \rho(A_0 + C(\lambda))$. Moreover,

$$R(\lambda, A)(f, g, h) = (u, v, (\lambda - kD^2)^{-1}(h - \gamma Dv)),$$
$$(u, v) = R(\lambda, A_0 + C(\lambda))[(f, g) + (0, -\gamma D(\lambda - kD^2)^{-1}h)].$$

$$(6.137)$$

Consequently, all the conditions of Theorem 6.29 are satisfied and it follows from the result of Theorem 6.29 that the growth rate of the thermoelastic system (6.125) satisfies

$$\omega_0(A) \leq \max\{S(A), -k\pi^2\}. \qquad (6.138)$$

In view of this relation, it remains to analyze the spectrum distribution of A in order to prove the exponential stability of (6.125). The rest of this section is devoted to the spectral analysis for the operator A.

Recall that A^{-1} is compact, as stated in Lemma 6.27. Thus, $\sigma(A)$ consists only of the eigenvalues of A. It is easily seen that $\lambda \in \sigma(A)$ if and only if there exists $(\phi, \psi) \neq 0$ such that

$$\begin{cases} \lambda^2\phi(x) - \phi''(x) + \gamma\psi'(x) = 0, \\ \lambda\psi(x) + \lambda\gamma\phi'(x) - k\psi''(x) = 0, \\ \phi(i) = \psi(i) = 0, \quad i = 0, 1. \end{cases} \qquad (6.139)$$

To eliminate ψ, we differentiate the first equation of (6.139) and substitute ψ'' into the second equation to obtain

$$\begin{cases} \lambda^2\phi(x) - \phi''(x) + \gamma\psi'(x) = 0, \\ \gamma\lambda\psi(x) + \lambda(\lambda k + \gamma^2)\phi'(x) - k\phi'''(x) = 0, \\ \phi(i) = \lambda(\lambda k + \gamma^2)\phi'(i) - k\phi'''(i) = 0, \quad i = 0, 1. \end{cases} \qquad (6.140)$$

Differentiating the second equation of (6.140) and substituting ψ' into the first equation of (6.140), we get

$$\begin{cases} k\phi''''(x) - \lambda(k\lambda + \gamma^2 + 1)\phi''(x) + \lambda^3\phi(x) = 0, \\ \phi(i) = \lambda(\lambda k + \gamma^2)\phi'(i) - k\phi'''(i) = 0, \quad i = 0, 1 \end{cases} \qquad (6.141)$$

with

$$\gamma\lambda\psi(x) = k\phi'''(x) - \lambda(\lambda k + \gamma^2)\phi'(x). \qquad (6.142)$$

Thus, solving the eigenvalue problem (6.139) is equivalent to finding a pair of $(\lambda, \phi) \in \mathbf{C} \times H^4(0, 1)$ such that $\phi \neq 0$ and equation (6.141) holds.

Lemma 6.30 *The general solution of (6.141) is*

$$\phi(x) = c_1 e^{a_1 x} + c_2 e^{-a_1 x} + c_3 e^{a_2 x} + c_4 e^{-a_2 x}, \tag{6.143}$$

where $c_i, i = 1, 2, 3, 4$ are arbitrary constants and

$$\begin{cases} a_1 = \sqrt{\frac{\lambda}{2k}[k\lambda + \gamma^2 + 1 + \sqrt{(k\lambda + \gamma^2 + 1)^2 - 4k\lambda}]}, \\ a_2 = \sqrt{\frac{\lambda}{2k}[k\lambda + \gamma^2 + 1 - \sqrt{(k\lambda + \gamma^2 + 1)^2 - 4k\lambda}]}. \end{cases} \tag{6.144}$$

Proof. The proof is almost obvious by noting that the characteristic equation of the first equation of (6.141) is

$$ka^4 - \lambda(k\lambda + \gamma^2 + 1)a^2 + \lambda^3 = 0 \tag{6.145}$$

which has four roots specified by

$$a^2 = \frac{\lambda}{2k}[k\lambda + \gamma^2 + 1 \pm \sqrt{(k\lambda + \gamma^2 + 1)^2 - 4k\lambda}]. \tag{6.146}$$

$\square$

We now derive the characteristic equation which an eigenvalue λ should satisfy. First, it is remarked that $\text{Re}\lambda < 0$ for any $\lambda \in \sigma(A)$, because we have already seen that A is dissipative in Lemma 6.27. Suppose $(k\lambda + \gamma^2 + 1)^2 - 4k\lambda = 0$, then we have

$$\lambda = \frac{1 - \gamma^2}{k} \pm \frac{2\gamma}{k}i, \tag{6.147}$$

which is impossible since γ is a constant less than 1, as we stated in the beginning of this section. Therefore, $(k\lambda + \gamma^2 + 1)^2 - 4k\lambda \neq 0$. The boundary conditions at $x = 0$ in (6.141) imply that

$$\begin{bmatrix} c_1 \\ c_2 \end{bmatrix} = -\frac{1}{2g_1} \begin{bmatrix} (g_1 + g_2)c_3 + (g_1 - g_2)c_4 \\ (g_1 - g_2)c_3 + (g_1 + g_2)c_4 \end{bmatrix}, \tag{6.148}$$

where

$$g_i = a_i(k\lambda^2 + \gamma^2\lambda - ka_i^2) \neq 0, \quad i = 1, 2. \tag{6.149}$$

Similarly, the boundary conditions at $x = 1$ yield

$$\begin{bmatrix} c_1 \\ c_2 \end{bmatrix} = -\frac{1}{2g_1} \begin{bmatrix} \exp(a_2 - a_1)(g_1 + g_2)c_3 + \exp(-a_2 - a_1)(g_1 - g_2)c_4 \\ \exp(a_2 + a_1)(g_1 - g_2)c_3 + \exp(a_1 - a_2)(g_1 + g_2)c_4 \end{bmatrix}. \tag{6.150}$$

Thus, from (6.148) and (6.150), the necessary and sufficient condition for (6.141) to possess a nonzero solution ϕ is

$$\det \begin{bmatrix} [1 - \exp(a_2 - a_1)](g_1 + g_2) & [1 - \exp(-a_2 - a_1)](g_1 - g_2) \\ [1 - \exp(a_2 + a_1)](g_1 - g_2) & [1 - \exp(a_1 - a_2)](g_1 + g_2) \end{bmatrix} = 0,$$

$$(6.151)$$

which is equivalent to

$$8g_1g_2 - [\exp(a_1 - a_2) + \exp(a_2 - a_1)](g_1 + g_2)^2$$
$$+[\exp(-a_1 - a_2) + \exp(a_2 + a_1)](g_1 - g_2)^2 = 0. \qquad (6.152)$$

An alternative expression of (6.152) is given below.

Lemma 6.31 *The characteristic equation which an eigenvalue λ of A should satisfy is given by*

$$8\gamma^2 \sqrt{k}\lambda \exp(a_1 + a_2)$$
$$+[\exp(2a_1) + \exp(2a_2)](k\lambda + \gamma^2 + 1 + 2\sqrt{k\lambda})(1 - \sqrt{k\lambda})^2$$
$$-[1 + \exp(2a_1 + 2a_2)](k\lambda + \gamma^2 + 1 - 2\sqrt{k\lambda})(1 + \sqrt{k\lambda})^2 = 0.$$

$$(6.153)$$

Proof. From (6.149) and (6.144), it is routine to check that

$$\begin{aligned} g_1g_2 &= a_1a_2[\lambda^2(k\lambda + \gamma^2)^2 - \lambda k(k\lambda + \gamma^2)(a_1^2 + a_2^2) + k^2a_1^2a_2^2], \\ g_1 - g_2 &= (a_1 - a_2)(-\lambda - \lambda\sqrt{k\lambda}), \\ g_1 + g_2 &= (a_1 + a_2)(-\lambda + \lambda\sqrt{k\lambda}), \\ a_1a_2 &= \lambda\sqrt{\lambda}/\sqrt{k}, \\ a_1^2 + a_2^2 &= \lambda(k\lambda + \gamma^2 + 1)/k, \end{aligned}$$

and so

$$\begin{aligned} g_1g_2 &= -\frac{\gamma^2}{\sqrt{k}}\sqrt{\lambda}\lambda^3, \\ (g_1 - g_2)^2 &= \frac{\lambda^3}{k}(k\lambda + \gamma^2 + 1 - 2\sqrt{k\lambda})(1 + \sqrt{k\lambda})^2, \\ (g_1 + g_2)^2 &= \frac{\lambda^3}{k}(k\lambda + \gamma^2 + 1 + 2\sqrt{k\lambda})(1 - \sqrt{k\lambda})^2. \end{aligned}$$

Consequently, we deduce (6.153) from (6.152). $\square$

Proposition 6.32 $\lambda \in \sigma(A)$ *if and only if $\lambda \neq 0$ is a root of equation (6.153).*

Proof. This result easily follows from Lemma 6.31. □

In the sequel, we will give an asymptotic analysis for the roots of equation (6.153). To simplify our presentation, we will use λ in replace of $k\lambda$ throughout our analysis, and restore it when we draw our final conclusions. Write (6.153) as

$$
\begin{aligned}
& 8\gamma^2\sqrt{\lambda}\exp(b_1+b_2) \\
& +[\exp(2b_1)+\exp(2b_2)](\lambda+\gamma^2+1+2\sqrt{\lambda})(1-\sqrt{\lambda})^2 \\
& -[1+\exp(2b_1+2b_2)](\lambda+\gamma^2+1-2\sqrt{\lambda})(1+\sqrt{\lambda})^2 = 0,
\end{aligned}
$$
(6.154)

where

$$
b_1 = \frac{\lambda}{\sqrt{2k}}f(\frac{1}{\lambda}), \quad b_2 = \frac{\sqrt{2\lambda}}{k}\frac{1}{f(\frac{1}{\lambda})},
$$
(6.155)

and

$$
f(\frac{1}{\lambda}) = \sqrt{1+\frac{\gamma^2+1}{\lambda}+\sqrt{\left(1+\frac{\gamma^2+1}{\lambda}\right)^2-\frac{4}{\lambda}}}.
$$

The Taylor expansions of the complex functions $f(1/\lambda)$ and $1/f(1/\lambda)$ at $\lambda = \infty$ are given by

$$
\begin{cases}
f(\frac{1}{\lambda}) = \sqrt{2}+\frac{\gamma^2}{\sqrt{2}}\frac{1}{\lambda}+\frac{\gamma^2}{\sqrt{2}}\left(1-\frac{\gamma^2}{4}\right)\frac{1}{\lambda^2}+O(|\lambda|^{-3}), \\
\frac{1}{f(\frac{1}{\lambda})} = \frac{1}{\sqrt{2}}-\frac{\gamma^2}{2\sqrt{2}}\frac{1}{\lambda}-\frac{\gamma^2}{2\sqrt{2}}\left(1-\frac{3\gamma^2}{4}\right)\frac{1}{\lambda^2}+O(|\lambda|^{-3}).
\end{cases}
$$
(6.156)

And hence as $|\lambda| \to \infty$,

$$
\begin{cases}
b_1 = \frac{\lambda}{k}+\frac{\gamma^2}{2k}+\frac{\gamma^2}{2k}\left(1-\frac{\gamma^2}{4}\right)\frac{1}{\lambda}+O(|\lambda|^{-2}), \\
b_2 = \frac{\sqrt{\lambda}}{k}-\frac{\gamma^2}{2k}\frac{1}{\sqrt{\lambda}}-\frac{\gamma^2}{2k}\left(1-\frac{3\gamma^2}{4}\right)\frac{1}{\lambda^{3/2}}+O(|\lambda|^{-5/2}).
\end{cases}
$$
(6.157)

Using the polar representation of the complex number, we can write

$$
\lambda = |\lambda|e^{i\vartheta}.
$$

As the closed right-half plane does not contain any eigenvalues and since λ is symmetric with respect to the real axis, we need only consider those λ where $\pi/2 < \vartheta \le \pi$. Furthermore, we divide our analysis into two parts and formulate the results into two lemmas, and then draw our concluding theorem from them.

Lemma 6.33 *Let $\delta > 0$ be sufficiently small. Then for all $\pi - \delta < \vartheta \leq \pi$*

$$\lambda = -(kn\pi)^2 + \gamma^2 + O(|\lambda|^{-1})$$

where n denotes some positive integer.

Proof. We need to carry out an asymptotic analysis on (6.154). To begin with, let us note that

$$
\begin{aligned}
\exp(b_1) &= \exp(\frac{\lambda}{k}) + \frac{\gamma^2}{2k}\exp(O(|\lambda|^{-1})) \\
&= \exp(\frac{|\lambda|}{k}\cos\vartheta + \frac{\gamma^2}{2k})\exp(\frac{i|\lambda|}{k}\sin\vartheta)\exp(O(|\lambda|^{-1})) \\
&= O(\exp(-\gamma_1|\lambda|)), \text{ for some } \gamma_1 > 0, \\
\exp(-b_2) &= \exp(-\frac{\sqrt{\lambda}}{k})\exp(O(|\lambda|^{-1/2})) \\
&= \exp(-\frac{|\lambda|^{1/2}}{k}\cos\frac{\vartheta}{2})\exp(O(|\lambda|^{-1/2})) = O(1).
\end{aligned}
$$

Multiplying $\exp(-2b_2)$ on both sides of (6.154) gives

$$
\begin{aligned}
&(\lambda + \gamma^2 + 1 + 2\sqrt{\lambda})(1 - \sqrt{\lambda})^2 - \exp(-2b_2)(\lambda + \gamma^2 + 1 - 2\sqrt{\lambda})(1 + \sqrt{\lambda})^2 \\
&= O(|\lambda|^2 \exp(-\gamma_1|\lambda|)).
\end{aligned}
$$

A simple calculation indicates that

$$
\begin{aligned}
\exp(-2b_2) &= \frac{(\lambda + \gamma^2 + 1 + 2\sqrt{\lambda})(1 - \sqrt{\lambda})^2}{(\lambda + \gamma^2 + 1 - 2\sqrt{\lambda})(1 + \sqrt{\lambda})^2} + O(\exp(-\gamma_1|\lambda|)) \\
&= 1 - \frac{4\gamma^2}{\lambda^{3/2}} + O(|\lambda|^{-5/2}), \tag{6.158}
\end{aligned}
$$

which will be our main building block to prove our assertion. Rewrite (6.158) as

$$e^{\mathrm{Re}(-2b_2)} = e^{i\mathrm{Im}(2b_2)} + \mathcal{O}(|\lambda|^{-3/2}). \tag{6.159}$$

From (6.158), we see that

$$1 - O(|\lambda|^{-3/2}) \leq e^{\mathrm{Re}(-2b_2)} \leq 1 + O(|\lambda|^{-3/2}).$$

Hence,

$$
\begin{aligned}
e^{i\mathrm{Im}(2b_2)} &= 1 + O(|\lambda|^{-3/2}), \\
e^{\mathrm{Re}(-2b_2)} &= 1 + O(|\lambda|^{-3/2}).
\end{aligned}
$$

On the other hand, the second equality above, together with (6.157), implies that

$$\vartheta \to \pi, \quad \text{as} \quad |\lambda| \to \infty. \tag{6.160}$$

Indeed, we may even show that $\vartheta = \pi$ for large enough $|\lambda|$. To see this, we analyze (6.159) again. Since $e^{i\text{Im}(2b_2)} = 1 + O(|\lambda|^{-3/2})$,

$$\sin(\text{Im}(2b_2)) = O(|\lambda|^{-3/2})$$

and

$$\cos(\text{Im}(2b_2)) = \sqrt{1 + \sin^2(\text{Im}(2b_2))} = \sqrt{1 + O(|\lambda|^{-3})} = 1 + O(|\lambda|^{-3}).$$

Plugging these into (6.158), we obtain

$$\begin{aligned}
e^{\text{Re}(-2b_2)} &= [1 - 4\gamma^2|\lambda|^{-3/2}e^{-i3/2\vartheta}]e^{i\text{Im}(2b_2)} + O(|\lambda|^{-5/2}) \\
&= [1 - 4\gamma^2|\lambda|^{-3/2}e^{-i3/2\vartheta}][\cos(\text{Im}(2b_2)) + i\sin(\text{Im}(2b_2))] \\
&\quad + O(|\lambda|^{-5/2}) \\
&= 1 - 4\gamma^2|\lambda|^{-3/2}\cos\frac{3\vartheta}{2} + O(|\lambda|^{-5/2}).
\end{aligned}$$

Taking logarithm on both sides of the above equation yields

$$\text{Re}(-2b_2) = -4\gamma^2|\lambda|^{-3/2}\cos\frac{3\vartheta}{2} + O(|\lambda|^{-5/2}).$$

Putting this into (6.157), we obtain

$$\begin{aligned}
&\left(-\frac{2}{k}|\lambda|^{1/2} + \frac{\gamma^2}{k}|\lambda|^{-1/2}\right)\cos\frac{\vartheta}{2} - \frac{\gamma^2}{2k}\left(1 - \frac{3\gamma^2}{4}\right)|\lambda|^{-3/2}\cos\frac{3\vartheta}{2} \\
&= -4\gamma^2|\lambda|^{-3/2}\cos\frac{3\vartheta}{2} + O(|\lambda|^{-5/2}), \tag{6.161}
\end{aligned}$$

which implies that $\cos\frac{\vartheta}{2} = O(|\lambda|^{-2})$. If $\vartheta \neq \pi$ for large $|\lambda|$, then by (6.160)

$$\cos\frac{3\vartheta}{2} \Big/ \cos\frac{\vartheta}{2} \to -1, \quad \text{as} \quad |\lambda| \to \infty.$$

But (6.161) then leads to a contradiction that $|\lambda|^{1/2} = O(1)$. Thus

$$\vartheta = \pi, \quad \text{for large } |\lambda|. \tag{6.162}$$

Finally, from (6.157) and (6.162),

$$\text{Im}(2b_2) = \frac{2}{k}|\lambda|^{1/2} + \frac{\gamma^2}{k}|\lambda|^{-1/2} + O(|\lambda|^{-3/2}).$$

Since $\sin(\text{Im}(2b_2)) = O(|\lambda|^{-3/2})$, we have

$$\frac{2}{k}|\lambda|^{1/2} + \frac{\gamma^2}{k}|\lambda|^{-1/2} = 2n\pi + O(|\lambda|^{-3/2})$$

or

$$\lambda = -(kn\pi)^2 + \gamma^2 + O(|\lambda|^{-1}), \tag{6.163}$$

where n is a sufficiently large integer. $\square$

Lemma 6.34 *Let $\delta > 0$ be sufficiently small, then for $\pi/2 < \vartheta \le \pi/2 + \delta$,*

$$\text{Re}\lambda = -\frac{\gamma^2}{2} + O(|\lambda|^{-3/2}), \quad \text{and}$$

$$|\lambda| = kn\pi + O(|\lambda|^{-1})$$

where n is a sufficiently large integer.

Proof. In this case, $\pi/4 < \vartheta/2 \le \pi/4 + \delta/2$. Thus, there is a $\gamma_2 > 0$ such that

$$\begin{aligned}
\exp(-b_2) &= \exp(-\frac{\sqrt{\lambda}}{k})\exp(O(|\lambda|^{-1/2})) \\
&= \exp(-\frac{|\lambda|^{1/2}}{k}\cos\frac{\vartheta}{2})\exp(O(|\lambda|^{-1/2})) \\
&= O(\exp(-\gamma_2|\lambda|^{1/2})),
\end{aligned}$$

and

$$\begin{aligned}
\exp(b_1) &= \exp(\frac{\lambda}{k} + \frac{\gamma^2}{2k})\exp(O(|\lambda|^{-1})) \\
&= \exp(\frac{|\lambda|}{k}\cos\vartheta + \frac{\gamma^2}{2k})\exp(i\frac{|\lambda|}{k}\sin\vartheta)\exp(O(|\lambda|^{-1})) \\
&= O(1).
\end{aligned}$$

The rest of the proof is very similar to that of Lemma 6.33.

Substituting the two equations above into (6.154) we deduce that

$$\begin{aligned}
&(\lambda + \gamma^2 + 1 + 2\sqrt{\lambda})(1 - \sqrt{\lambda})^2 - \exp(2b_1)(\lambda + \gamma^2 + 1 - 2\sqrt{\lambda})(1 + \sqrt{\lambda})^2 \\
&= O(|\lambda|^2\exp(-\gamma_2|\lambda|^{1/2})),
\end{aligned}$$

and

$$\begin{aligned}
\exp(2b_1) &= \frac{(\lambda + \gamma^2 + 1 + 2\sqrt{\lambda})(1 - \sqrt{\lambda})^2}{(\lambda + \gamma^2 + 1 - 2\sqrt{\lambda})(1 + \sqrt{\lambda})^2} + O(\exp(-\gamma_2|\lambda|^{1/2})) \\
&= 1 + O(|\lambda|^{-3/2}). \tag{6.164}
\end{aligned}$$

Write (6.164) as

$$e^{\text{Re}(2b_1)} = e^{i\text{Im}(-2b_1)} + O(|\lambda|^{-3/2}).$$

Comparing this with (6.164) gives

$$e^{\text{Re}(2b_1)} = 1 + O(|\lambda|^{-3/2}), \tag{6.165}$$
$$e^{i\text{Im}(-2b_1)} = 1 + O(|\lambda|^{-3/2}).$$

By noting (6.157), we have

$$\text{Re}(2b_1) = \frac{2}{k}|\lambda|\cos\vartheta + \frac{\gamma^2}{k} + \frac{\gamma^2}{k}(1 - \frac{\gamma^2}{4})|\lambda|^{-1}\cos\vartheta + O(|\lambda|^{-2}),$$

$$\tag{6.166}$$

$$\text{Im}(2b_1) = \frac{2}{k}|\lambda|\sin\vartheta + O(|\lambda|^{-1}). \tag{6.167}$$

From (6.165) and (6.166), we immediately conclude that

$$\vartheta \to \pi/2, \text{ as } |\lambda| \to \infty. \tag{6.168}$$

Furthermore, from (6.165) we have

$$\text{Re}(2b_1) = O(|\lambda|^{-3/2}).$$

Putting this into (6.166) yields

$$\frac{2}{k}|\lambda|\cos\vartheta + \frac{\gamma^2}{k} + \frac{\gamma^2}{k}(1 - \frac{\gamma^2}{4})\frac{1}{|\lambda|}\cos\vartheta = O(|\lambda|^{-3/2});$$

hence,

$$\frac{2}{k}|\lambda|\cos\vartheta + \frac{\gamma^2}{k} = O(|\lambda|^{-3/2}),$$

or

$$\cos\vartheta = -\frac{\gamma^2}{2|\lambda|} + O(|\lambda|^{-5/2}). \tag{6.169}$$

Since $\cos\vartheta = \text{Re}\lambda/|\lambda|$, (6.169) then implies that

$$\text{Re}\lambda = -\frac{\gamma^2}{2} + O(|\lambda|^{-3/2}). \tag{6.170}$$

Finally, from (6.169) and (6.167) we see that

$$\sin\vartheta = \sqrt{1 - \cos^2\vartheta} = \sqrt{1 + O(|\lambda|^{-2})} = 1 + O(|\lambda|^{-2}),$$
$$\text{Im}(2b_1) = \frac{2}{k}|\lambda|\sin\vartheta + O(|\lambda|^{-1}) = \frac{2}{k}|\lambda| + O(|\lambda|^{-1}).$$

These, together with (6.165), yield

$$|\lambda| = kn\pi + O(|\lambda|^{-1}), \tag{6.171}$$

where n is a sufficiently large positive integer. $\square$

Combining Lemma 6.33 and Lemma 6.34, we can prove

Theorem 6.35 *The eigenvalues of the operator A consist of a real sequence $\{\lambda_n\}$ and a sequence of conjugate pairs $\{\sigma_n, \bar{\sigma}_n\}$ with*

$$\begin{cases} \lambda_n = -k(n\pi)^2 + \frac{\gamma^2}{k} + O(n^{-2}), \\ \sigma_n = -\frac{\gamma^2}{2k} + in\pi + O(n^{-1}), \quad Re\sigma_n = -\frac{\gamma^2}{2k} + O(n^{-3/2}), \end{cases} \tag{6.172}$$

where n is a sufficiently large positive integer.

Proof. We shall establish the existence of the eigenvalues of the operator A in a similar way as we did in Chapter 4.
Case 1. First, we assume that $\lambda = |\lambda|e^{i\vartheta}, \pi - \delta < \vartheta \leq \pi$, with $\delta > 0$ sufficiently small. Then by (6.158)

$$\exp(-2b_2) = 1 + O(|\lambda|^{-3/2}). \tag{6.173}$$

From (6.157), $-2b_2 = -2\frac{\sqrt{\lambda}}{k} + \frac{\gamma^2}{k}\frac{1}{\sqrt{\lambda}} + O(|\lambda|^{-3/2})$, so (6.173) implies that

$$\exp(-2\frac{\sqrt{\lambda}}{k} + \frac{\gamma^2}{k}\frac{1}{\sqrt{\lambda}}) = 1 + O(|\lambda|^{-3/2}). \tag{6.174}$$

It is easy to see that $\exp(-2\frac{\sqrt{\lambda}}{k} + \frac{\gamma^2}{k}\frac{1}{\sqrt{\lambda}}) = 1$ has solutions

$$\lambda_n = -(kn\pi)^2 + \gamma^2 + O(|\lambda_n|^{-1}), \quad n = 0, \pm 1, \ldots \tag{6.175}$$

Let $\sqrt{\lambda}$ denote the positive branch of the square root of a complex λ, and so the mapping: $\lambda \to \sqrt{\lambda}$ maps the negative real axis of the complex plane to the positive imaginary axis. Instead of (6.174), we let $\mu = \sqrt{\lambda}$ and consider

$$\exp(-2\frac{\mu}{k} + \frac{\gamma^2}{k}\frac{1}{\mu}) = 1 + O(|\mu|^{-3}). \tag{6.176}$$

Let

$$d_2 = -2\frac{\mu}{k} + \frac{\gamma^2}{k}\frac{1}{\mu}.$$

Let $\mathcal{O}_n$ be a circle with radius $\alpha\lambda_n^{-1/2}, \alpha > 0$ and centered at $i\sqrt{\lambda_n}$. Then the circumference of this circle can be described by

$$\Gamma\mathcal{O}_n : \mu = i\sqrt{\lambda_n} + \alpha\lambda_n^{-1/2}e^{i\vartheta}, \quad 0 \leq \vartheta \leq 2\pi.$$

For $\mu \in \Gamma\mathcal{O}_n$,

$$|\mu| = \sqrt{\lambda_n}[1 + O(\lambda_n^{-1})]. \tag{6.177}$$

In view of (6.176) and (6.175), we conclude that, for all $\mu \in \Gamma\mathcal{O}_n$,

$$\mathrm{Re}d_2 = -\frac{2\alpha}{k}\lambda_n^{-1/2}\cos\vartheta + O(\lambda_n^{-3/2}).$$

$$\mathrm{Im}d_2 = -\frac{2}{k}\sqrt{\lambda_n} - \frac{2\alpha}{k}\lambda_n^{-1/2}\sin\vartheta - \frac{\gamma^2}{k}\lambda_n^{-1/2} + O(\lambda_n^{-3/2})$$

$$= -2n\pi - \frac{2\alpha}{k}\lambda_n^{-1/2}\sin\vartheta + O(\lambda_n^{-3/2}).$$

Consequently,

$$|1 - \exp(d_2)|^2 = [1 - \exp(\mathrm{Re}d_2)]^2 + 2\exp(\mathrm{Re}d_2)[1 - \cos(\mathrm{Im}d_2)]$$

$$= \frac{4\alpha^2}{k^2}\lambda_n^{-1} + O(\lambda_n^{-3/2}). \tag{6.178}$$

Since $i\sqrt{\lambda_n}$ is the unique root of $\exp(d_2) = 1$ inside $\mathcal{O}_n$, we can apply Rouché's theorem to functions $\exp(d_2) - 1$ and $\exp(d_2) - 1 - O(|\mu|^{-3})$, and conclude that there exists a unique zero μ_n to the equation $\exp(d_2) - 1 - O(|\mu|^{-3}) = 0$ insides $\mathcal{O}_n$ for all $n \geq N$ when N is sufficiently large.

Let $\hat{\lambda}_n = \mu_n^2$. Since μ_n is inside $\mathcal{O}_n$, we know that

$$|\mu_n^2 + \lambda_n| \leq 2\alpha + O(\lambda_n^{-1}), \quad \text{for all } n \geq N. \tag{6.179}$$

Thus, $\hat{\lambda}_n$ is the unique root of (6.174) and hence of (6.173) in the order of $O(n^2)$. Furthermore, by the arbitrariness of $\alpha > 0$, we see that

$$\lim_{n\to\infty}|\hat{\lambda}_n + \lambda_n| = 0. \tag{6.180}$$

Case 2. As for now, we assume that $\lambda = |\lambda|e^{i\vartheta}, \pi/2 < \vartheta \leq \pi/2 + \delta$, with $\delta > 0$ sufficiently small. From (6.164)

$$\exp(2b_1) = 1 + O(|\lambda|^{-3/2}). \tag{6.181}$$

Since $b_1 = \frac{\lambda}{k} + \frac{\gamma^2}{2k} + O(|\lambda|^{-1})$, so (6.181) implies that

$$\exp(2\frac{\lambda}{k} + \frac{\gamma^2}{k}) = 1 + O(|\lambda|^{-1}). \tag{6.182}$$

Obviously,

$$\sigma_n = -\frac{\gamma^2}{2} + ikn\pi, \quad n = 0, \pm1, \pm2, \cdots \tag{6.183}$$

are the solutions of $\exp(2\frac{\lambda}{k} + \frac{\gamma^2}{k}) = 1$.

Let $\hat{\mathcal{O}}_n$ be the circle with radius $|\sigma_n|^{-1/2}$ and centered at σ_n. Then

$$\Gamma\hat{\mathcal{O}}_n = \{\lambda = |\lambda|e^{i\vartheta} \mid |\lambda - \sigma_n| \leq |\sigma_n|^{-1/2}\}.$$

For any $\lambda \in \Gamma\hat{\mathcal{O}}_n$, we have

$$|\lambda| = |\sigma_n|[1 + O(|\sigma_n|^{-3/2})].$$

Let $c_1 = 2\frac{\lambda}{k} + \frac{\gamma^2}{k}$. Then for all $\lambda \in \Gamma\hat{\mathcal{O}}_n$,

$$\mathrm{Re}(c_1) = \frac{2}{k}|\sigma_n|^{-1/2}\cos\vartheta,$$

$$\mathrm{Im}(c_1) = 2n\pi + \frac{2}{k}|\sigma_n|^{-1/2}\sin\vartheta,$$

and so

$$\begin{aligned}|1 - \exp(c_1)|^2 &= [1 - \exp(\mathrm{Re}(c_1))]^2 + 2\exp(\mathrm{Re}(c_1))[1 - \cos(\mathrm{Im}(c_1))]\\ &= \frac{4}{k^2}|\sigma_n|^{-1} + O(|\sigma_n|^{-3/2}) = O(|\lambda|^{-1}).\end{aligned}$$

Applying Rouché's theorem to functions $\exp(2\frac{\lambda}{k} + \frac{\gamma^2}{k}) - 1$ and $\exp(2\frac{\lambda}{k} + \frac{\gamma^2}{k}) - 1 - O(|\lambda|^{-1})$, we conclude that there exists a unique zero $\hat{\sigma}_n$ for $\exp(2\frac{\lambda}{k} + \frac{\gamma^2}{k}) - 1 - O(|\lambda|^{-1}) = 0$ inside $\hat{\mathcal{O}}_n$ for all $n \geq \hat{N}$ when $\hat{N}$ is sufficiently large.

Since $\hat{\sigma}_n$ lies inside $\hat{\mathcal{O}}_n$, by definition,

$$|\hat{\sigma}_n - \sigma_n| \leq |\sigma_n|^{-1/2}, \text{ for all } n \geq \hat{N}. \tag{6.184}$$

This together with (6.171) implies that this $\hat{\sigma}_n$ is the unique root of (6.181) with order $O(n^{-1})$ for positive n. $\qquad\square$

The above spectral analysis results clearly indicate that there exist a set of eigenvalues on the real axis which corresponds to the spectra of the heat equation, and a set of complex conjugate eigenvalues which corresponds to the vibration modes of the wave equation. Turning back to (6.138), we see that, in order for the spectrum determined growth condition to hold for (6.125), a sufficient condition is that $S(A) \geq -k\pi^2$. Physically, this is possible because $-k\pi^2$, the first eigenvalue of the "pure" (without vibration coupling, i.e., $\gamma = 0$ in (6.116)) heat equation, determines the energy decay rate of the heat equation, and we cannot expect the energy decays more rapidly when there are coupled vibrations which produce additional heat. To prove this, it is sufficient to show that (6.153) has a real root in $(-k\pi^2, 0)$. We do this below.

Since we are only concerned with the negative real eigenvalues of A which are the negative real roots of the characteristic equation (6.153), letting $x = -k\lambda$, we are able to write (6.153) as

$$8\gamma^2 i\sqrt{x} + [\exp(if+g) + \exp(-if-g)](-x+\gamma^2+1+2i\sqrt{x})(1-i\sqrt{x})^2$$
$$- [\exp(if-g) + \exp(g-if)](-x+\gamma^2+1-2i\sqrt{x})(1+i\sqrt{x})^2 = 0$$
$$(6.185)$$

where

$$\begin{cases} f(x) = \dfrac{\sqrt{x}}{\sqrt{2}k}h(x), \quad g(x) = \dfrac{\sqrt{2}}{k}\dfrac{x}{h(x)}, \\[2mm] h(x) = \sqrt{-x+\gamma^2+1+\sqrt{(-x+\gamma^2+1)^2+4x}} > 0, \quad \text{for all } x \geq 0. \end{cases}$$
$$(6.186)$$

Comparing the real and imaginary parts, we see that (6.185) reduces to

$$4\gamma^2\sqrt{x} - 2\gamma^2\sqrt{x}[\exp(g)+\exp(-g)]\cos(f)$$
$$+ [(1+x)^2 + (1-x)\gamma^2][\exp(g)-\exp(-g)]\sin(f) = 0.$$
$$(6.187)$$

Theorem 6.36 *For any $0 < \gamma \leq 1$, there is at least one solution x to equation (6.187) which locates between 0 and $k^2\pi^2$ on the real axis.*

Proof. Denote by $F(x)$ the left-hand side of (6.187). By Taylor expansion, we find that $\lim_{x\to 0}\frac{F(x)}{\sqrt{x}x} = \frac{2(\gamma^2+1)^2}{k^2}$. This means that $F(x) > 0$ for $x > 0$ sufficiently small. The proof is complete if we can show that $F(k^2\pi^2) < 0$ because $F(x)$ is a continuous function on $[0, k^2\pi^2]$ with respect to x. To this end, let $a = k\pi$. Note that

$$(x+1-\gamma^2)^2 \leq (-x+\gamma^2+1)^2 + 4x \leq (x+\gamma^2+1)^2;$$

hence, when $0 < \gamma \leq 1$, $\sqrt{2} < h(x) \leq \sqrt{2}\sqrt{1+\gamma^2} < 2$ which gives, in particular,

$$\pi < f(a^2) < \sqrt{2}\pi. \qquad (6.188)$$

Taking $x = a^2$ in (6.186) and rearranging terms we have

$$\gamma^2 = -\pi^2 a^2/f^2 + f^2/\pi^2 + a^2 - 1, \quad g = \pi^2 a/f. \qquad (6.189)$$

Here, and in the following equation, f and g denote, respectively, $f(a^2)$ and $g(a^2)$ with an abuse of notation. Substituting (6.189) into $F(a^2)$ yields

$$f^2\pi^2 e^{\pi^2 a/f}F(a^2)$$
$$= 4a[f^4 - \pi^4 a^2 + \pi^2(a^2-1)f^2]e^{\pi^2 a/f}$$
$$- 2a[f^4 - \pi^4 a^2 + \pi^2(a^2-1)f^2][e^{2\pi^2 a/f} + 1]\cos(f)$$
$$+ [4\pi^2 a^2 f^2 + (1-a^2)(f^4 - \pi^4 a^2)][e^{2\pi^2 a/f} - 1]\sin(f). \quad (6.190)$$

It can be shown that the right-hand side of this equation is less than 0 [65]. Thus, $F(a^2) < 0$.

$\square$

We have actually proved the following important result.

Corollary 6.37 *Suppose $0 < \gamma \leq 1$.*

(i) *The spectrum-determined growth condition holds for system (6.125), i.e.,*

$$\omega(A) = S(A).$$

(ii) *For any $\epsilon > 0$ sufficiently small, there exists a positive number $M > 1$ such that*

$$E(t) \leq M e^{(\omega(A)+\epsilon)t} E(0), \tag{6.191}$$

where $E(t)$ is the energy function defined by (6.122). Since $\omega(A) < 0$, the energy decays exponentially.

6.6   Thermoelastic system with Dirichlet - Neumann boundary conditions

The methods used in the previous section can be applied easily to thermoelastic systems with Dirichlet-Neumann and Neumann-Dirichlet natural boundary conditions. We shall find that the problem with Neumann-Dirichlet or Dirichlet-Neumann boundary conditions is much simpler than that with Dirichlet-Dirichlet boundary conditions, because for the present problem the characteristic equation which an eigenvalue should satisfy is a third order polynomial. The roots of this polynomial can be directly solved, enabling us to prove the spectrum-determined growth condition.

Because the analyzes are very similar, we restrict ourselves to the following thermoelastic equation with Dirichlet-Neumann boundary condition:

$$\begin{cases} u_{tt}(x,t) - u_{xx}(x,t) + \gamma \vartheta_x(x,t) = 0, & 0 < x < 1, \\ \vartheta_t(x,t) + \gamma u_{xt}(x,t) - k\vartheta_{xx}(x,t) = 0, & 0 < x < 1, \\ u(i,t) = \vartheta_x(i,t) = 0, & i = 0,1. \end{cases} \tag{6.192}$$

To put this in the form of an abstract equation with state space $\tilde{H} = H_0^1(0,1) \times L^2(0,1) \times L^2(0,1)$, let us define the operator $A : D(A) \to \tilde{H}$ by

$$\begin{cases} A(u,v,\vartheta) = (v, u_{xx} - \gamma \vartheta_x, k\vartheta_{xx} - \gamma v_x), \\ D(A) = (H^2(0,1) \cap H_0^1(0,1)) \times H_0^1(0,1) \times H_e^2(0,1), \end{cases} \tag{6.193}$$

where $H_e^2(0,1) = \{\vartheta \in H^2(0,1) \mid \vartheta'(0) = \vartheta'(1) = 0\}$. Then (6.192) can be written as

$$\frac{dw(t)}{dt} = Aw(t), \tag{6.194}$$

with $w(t) = (u(\cdot,t), u_t(\cdot,t), \vartheta(\cdot,t))$. Since A is densely defined, we find, after a simple calculation, that

$$\begin{cases} A^*(f,g,h) = (-g_x, f_{xx} + \gamma h_x, \gamma g_x + k h_{xx}), \\ D(A^*) = (H^2(0,1) \cap H_0^1(0,1)) \times H_0^2(0,1) \times H_e^2(0,1), \end{cases} \tag{6.195}$$

and

$$\operatorname{Re}\langle A^*(f,g,h), (f,g,h)\rangle = -k \int_0^1 |h_x|^2 dx \leq 0, \quad \forall (f,g,h) \in D(A^*),$$

$$\operatorname{Re}\langle A(u,v,\vartheta), (u,v,\vartheta)\rangle = -k \int_0^1 |\vartheta_x|^2 dx \leq 0, \quad \forall (u,v,\vartheta) \in D(A).$$

Therefore, both A and A^* are dissipative. The closedness of the operator A is evident by definition. By Corollary 2.28, A generates a C_0-semigroup of contractions on $\tilde{H}$.

The operator A has a zero eigenvalue and

$$\Phi_0 = (0,0,1) \tag{6.196}$$

is the unique eigenvector associated with the zero eigenvalue. Physically, the zero eigenvalue has an easy interpretation in that it corresponds to the heated equilibrium. From (6.195), we see that Φ_0 is also the unique eigenvector of the adjoint operator A^* corresponding to the zero eigenvalue; hence, $\mathcal{R}(A)$ is orthogonal to $\mathcal{N}(A)$. Since we do not wish to consider nonzero equilibrium solutions, it is natural to define

$$H = \{\Phi \in \tilde{H} \mid \langle \Phi, \Phi_0\rangle = 0\} = H_0^1(0,1) \times L^2(0,1) \times L_e^2(0,1), \quad (6.197)$$

where $L_e^2(0,1) = \{f \in L^2(0,1) \mid \langle f, 1\rangle = 0\}$. For notational simplicity, we still use A to denote $A\mid_H$. More precisely, by operator A from now on, we mean the following unbounded operator defined in H:

$$\begin{cases} A(u,v,\vartheta) = (v, u_{xx} - \gamma \vartheta_x, k\vartheta_{xx} - \gamma v_x), \\ D(A) = (H^2(0,1) \cap H_0^1(0,1)) \times H_0^1(0,1) \times H_p^2(0,1), \end{cases} \tag{6.198}$$

where $H_p^2(0,1) = \{f \in H^2(0,1) \mid f'(0) = f'(1) = 0, \langle f, 1\rangle = 0\}$.

Lemma 6.38 *The operator A defined in (6.198) generates a C_0-semigroup of contractions on H. A^{-1} is compact and hence $\sigma(A)$ consists only of isolated eigenvalues with finite algebraic multiplicities.*

Proof. For any given $(f, g, h) \in H$, solving

$$A(u, v, \vartheta) = (v, u_{xx} - \gamma \vartheta_x, k \vartheta_{xx} - \gamma v_x) = (f, g, h),$$

we find that

$$u(x) = c_1 x + \int_0^x (x - \tau) g(\tau) d\tau + \gamma \int_0^x \vartheta(\tau) d\tau,$$

$$v(x) = f(x),$$

$$\vartheta(x) = \frac{\gamma}{k} \int_0^x f(\tau) d\tau + \frac{1}{k} \int_0^x (x - \tau) h(\tau) d\tau + c_2 x,$$

where

$$c_1 = -\int_0^1 (1 - \tau) g(\tau) d\tau - \gamma \int_0^1 \vartheta(\tau) d\tau,$$

$$c_2 = \frac{\gamma}{k} \int_0^1 (\tau - 1) f(\tau) d\tau + \frac{1}{k} \int_0^1 \frac{(1 - \tau)^2}{2} h(\tau) d\tau.$$

Therefore, A^{-1} is compact. Since A is dissipative, by the Lümer-Phillips theorem, A generates a C_0-semigroup of contractions on H. □

Again let $K = H_0^1(0, 1) \times L^2(0, 1)$ and let A_0 and $C(\lambda)$ be the operators defined in (6.133) and (6.134), respectively. Denote by D^2 the Laplacian with Neumann boundary conditions in $L_e^2(0, 1)$. Following exactly the proof procedure in Theorem 6.29 yields the following result.

Proposition 6.39 *The growth rate of A, defined in (6.198), satisfies the following relation:*

$$\omega_0(A) \leq \max\{S(A), -k\pi^2\}.$$

With the aid of this proposition, in order to show the exponential stability of the semigroup generated by A we need only to analyze the distribution of eigenvalues of A in the complex plane.

As before, $\lambda \in \sigma(A)$, which must satisfy $\mathrm{Re}\lambda \leq 0$ because A is dissipative, if and only if there exists $(\phi, \psi) \neq 0$ such that

$$\begin{cases} \lambda^2 \phi(x) - \phi''(x) + \gamma \psi'(x) = 0, \\ \lambda \psi(x) + \lambda \gamma \phi'(x) - k\psi''(x) = 0, \\ \phi(i) = \psi'(i) = 0, \quad i = 0, 1. \end{cases} \tag{6.199}$$

Eliminating ψ from this equation we obtain

$$\begin{cases} k\phi''''(x) - \lambda(k\lambda + \gamma^2 + 1)\phi''(x) + \lambda^3 \phi(x) = 0, \\ \phi(i) = \phi''(i) = 0, \quad i = 0, 1. \end{cases} \tag{6.200}$$

Once ϕ is obtained, ψ is determined from

$$\gamma\lambda\psi(x) = k\phi'''(x) - \lambda(\lambda k + \gamma^2)\phi'(x). \tag{6.201}$$

Solving the eigenvalue problem (6.199) is thus equivalent to finding a pair of $(\lambda, \phi) \in \mathbf{C} \times H^4(0,1)$ such that $\phi \neq 0$ and equation (6.200) is satisfied. The general solution of the first equation of (6.200) is given by

$$\phi(x) = c_1 e^{a_1 x} + c_2 e^{-a_1 x} + c_3 e^{a_2 x} + c_4 e^{-a_2 x}. \tag{6.202}$$

where $c_i, i = 1, 2, 3, 4$ are arbitrary constants to be determined by the boundary conditions and a_1, a_2 are the same as defined in (6.144).

We next derive the characteristic equation which an eigenvalue λ should satisfy. Since $(k\lambda + \gamma^2 + 1)^2 - 4k\lambda \neq 0$, we can then use the boundary condition at $x = 0$ and (6.202) to obtain

$$c_1 = -c_2, \; c_3 = -c_4; \tag{6.203}$$

hence, (6.202) becomes

$$\phi(x) = c_1(e^{a_1 x} - e^{-a_1 x}) + c_3(e^{a_2 x} - e^{-a_2 x}). \tag{6.204}$$

From the boundary conditions at $x = 1$, we get

$$(e^{2a_1} - 1)(e^{2a_2} - 1) = 0. \tag{6.205}$$

Proposition 6.40 $\lambda \in \sigma(A)$ *if and only if* $\lambda \neq 0$ *is a root of equation* (6.205).

Theorem 6.41 *Asymptotically, the nonzero solutions of (6.205) consist of a real sequence* $\{\lambda_n\}$ *and a sequence of conjugate pairs* $\{\sigma_n, \bar{\sigma}_n\}$ *with*

$$\sigma_n = n\pi i - \frac{\gamma^2}{2k} + i\frac{\gamma^2}{2k^2}\left(1 - \frac{\gamma^2}{4}\right)\frac{1}{n\pi} + O(n^{-2}),$$

$$\lambda_n = -k(n\pi)^2 + \frac{\gamma^2}{k} + O(n^{-2}),$$

where n *is a sufficiently large positive integer.*

Proof. Equation (6.205) is equivalent to

$$e^{2a_1} = 1 \quad \text{or} \quad e^{2a_2} = 1. \tag{6.206}$$

Things are now much simpler as compared to (6.164) because from $e^{2a_1} = 1$ we immediately obtain

$$\lambda = n\pi i - \frac{\gamma^2}{2k} - \frac{\gamma^2}{2k^2}\left(1 - \frac{\gamma^2}{4}\right)\frac{1}{\lambda} + O(|\lambda|^{-2})$$

for some integer n. This implies that $|\lambda| = O(n)$ which, when substituted into the right-hand side of the above equation, yields

$$\lambda = n\pi i - \frac{\gamma^2}{2k} - \frac{\gamma^2}{2k^2}\left(1 - \frac{\gamma^2}{4}\right)\frac{1}{n\pi i - \frac{\gamma^2}{2k} - \frac{\gamma^2}{2k^2}(1 - \frac{\gamma^2}{4})\frac{1}{\lambda} + O(|\lambda|^{-2})}$$

$$+ O(|\lambda|^{-2})$$

$$= n\pi i - \frac{\gamma^2}{2k} + i\frac{\gamma^2}{2k^2}\left(1 - \frac{\gamma^2}{4}\right)\frac{1}{n\pi} + O(n^{-2}).$$

This is the proof of the first part. From the definition of a_1 and a_2, it is easy to check that $a_2 = \frac{\lambda^{3/2}}{\sqrt{k}}a_1^{-1}$, so that we are able to show, in a similar manner as was used to solve $e^{2a_1} = 1$, that $e^{2a_2} = 1$ has solutions $\lambda_n = -k(n\pi)^2 + \gamma^2 + O(n^{-2})$. The proof is completed. □

Lemma 6.42 $\lambda \neq 0$ *is a solution of (6.206) if and only if there is a nonzero integer* n *such that*

$$f_n(\lambda) = \lambda^3 + k\mu_n\lambda^2 + (\gamma^2 + 1)\mu_n\lambda + k\mu_n^2 = 0 \tag{6.207}$$

where $\mu_n = n^2\pi^2$. *Moreover, for each* $n > 0$, *(6.207) admits one unique real solution* λ_n *and one conjugate pair solution* $\{\sigma_n, \bar{\sigma}_n\}$.

Proof. Obviously, the nonzero solutions λ of $e^{2a_1} = 1$ or $e^{2a_2} = 1$ must satisfy $a_1 = n\pi i$ or $a_2 = n\pi i$ for some nonzero integer n, i.e.,

$$\frac{\lambda}{2k}(k\lambda + \gamma^2 + 1 + \sqrt{(k\lambda + \gamma^2 + 1)^2 - 4k\lambda}) = -n^2\pi^2$$

or

$$\frac{\lambda}{2k}(k\lambda + \gamma^2 + 1 - \sqrt{(k\lambda + \gamma^2 + 1)^2 - 4k\lambda}) = -n^2\pi^2.$$

Rearranging terms we have

$$\pm\frac{\lambda}{2k}\sqrt{(k\lambda + \gamma^2 + 1)^2 - 4k\lambda} = -n^2\pi^2 - \frac{\lambda}{2k}(k\lambda + \gamma^2 + 1)$$

or

$$\lambda^3 + k(n^2\pi^2)^2 + \lambda(k\lambda + \gamma^2 + 1)n^2\pi^2 = 0.$$

This is exactly (6.207), completing the proof of the first part if we note that the above process is reversible.

When $k^2\mu_n - 3(\gamma^2 + 1) < 0$, $f_n'(\lambda) > 0$ for all real λ, so that there is a unique real solution λ_n of $f_n(\lambda) = 0$ since $f_n(-\infty) = -\infty$ and $f_n(\infty) = +\infty$. When $k^2\mu_n - 3(\gamma^2 + 1) \geq 0$, $f_n'(\lambda) = 3\lambda^2 + 2k\mu_n\lambda + (\gamma^2 + 1)\mu_n = 0$ has real roots

$$\eta_{1,2} = -\frac{1}{3}k\mu_n \pm \frac{1}{3}\sqrt{k^2\mu_n^2 - 3(\gamma^2 + 1)\mu_n} > -k\mu_n. \tag{6.208}$$

Since

$$9f_n(\eta_{1,2}) = -2[k^2\mu_n^2 - 3(\gamma^2 + 1)\mu_n]\eta_{1,2} + k(8 - \gamma^2)\mu_n^2 > 0,$$

and $f_n'(\lambda) = 3(\lambda - \eta_2)(\lambda - \eta_1)$, we see that $f_n'(\lambda) > 0$ for all $\lambda > \eta_1$ or $\lambda < \eta_2$. Therefore, there is a unique real solution λ_n to $f_n(\lambda) = 0$ with $\lambda_n < \eta_2$. $\square$

It is clear from the proof above that

$$f_n(\lambda) < 0 \quad \text{for all } \lambda < \lambda_n. \tag{6.209}$$

Theorem 6.43 *Let $f_n(\lambda)$ be defined by (6.207) and $\{\lambda_n, \sigma_n, \overline{\sigma}_n\}$ be the solution of $f_n(\lambda) = 0$. Then*

$$\lambda_n \leq -\frac{k\mu_1}{\gamma^2 + 1}, \quad Re\sigma_n \leq -\frac{\gamma^2}{2}\frac{k\mu_1}{k^2\mu_1 + \gamma^2 + 1} \tag{6.210}$$

for all $n > 0$.

Proof. Let $\epsilon = 1/(\gamma^2 + 1)$. It is seen that $f_n(\lambda) = (\lambda^2 + \mu_n)(\lambda + k\mu_n) + \gamma^2\mu_n\lambda$ and

$$f_n(-\epsilon k\mu_n) = (\epsilon^2 k^2\mu_n^2 + \mu_n)(1 - \epsilon)k\mu_n - \epsilon k\gamma^2\mu_n^2 = (1 - \epsilon)\epsilon^2 k^3\mu_n^3 > 0$$

which, together with (6.209), gives

$$\lambda_n \leq -\epsilon k\mu_n \leq -\frac{k\mu_1}{\gamma^2 + 1}.$$

Now, because

$$f_n(-\frac{k\mu_n}{3}) = \frac{2k^3\mu_n^3}{27} + \frac{k\mu_n^2}{3}(2 - \gamma^2) > 0$$

and

$$f_n(-k\mu_n) = -k\gamma^2\mu_n^2 < 0,$$

we see that $\lambda_n \in (-k\mu_n, -k\mu_n/3)$. In particular, we have

$$\lambda_1 > -k\pi^2. \tag{6.211}$$

Solving

$$\begin{aligned}
f_n(\lambda) &= \lambda^3 + k\mu_n\lambda^2 + (\gamma^2 + 1)\mu_n\lambda + k\mu_n^2 \\
&= (\lambda - \lambda_n)(\lambda - \sigma_n)(\lambda - \overline{\sigma}_n) \\
&= \lambda^3 - (\sigma_n + \overline{\sigma}_n + \lambda_n)\lambda^2 + [|\sigma_n|^2 + \lambda_n(\sigma_n + \overline{\sigma}_n)]\lambda - \lambda_n|\sigma_n|^2 \\
&= 0
\end{aligned}$$

we see that the real parts of the remaining two complex roots satisfy

$$2\mathrm{Re}\sigma_n = -k\mu_n - \lambda_n.$$

Since $f_n''(\lambda) = 6\lambda + 2k\mu_n < 0$ for all $\lambda < -k\mu_n/3$, we can apply Newton's method to get an approximation λ_n^1 of λ_n with an initial guess $-k\mu_n$, namely

$$\lambda_n^1 = -k\mu_n + \frac{k\gamma^2\mu_n}{k^2\mu_n + \gamma^2 + 1}.$$

Consequently,

$$
\begin{aligned}
2\mathrm{Re}\sigma_n &= -k\mu_n - \lambda_n < -k\mu_n - \lambda_n^1 = \frac{k\gamma^2\mu_n}{k^2\mu_n + \gamma^2 + 1} \\
&= -\frac{\gamma^2}{k}\left(1 - \frac{\gamma^2 + 1}{k^2\mu_n + \gamma^2 + 1}\right) \le -\frac{\gamma^2}{k}\left(1 - \frac{\gamma^2 + 1}{k^2\mu_1 + \gamma^2 + 1}\right).
\end{aligned}
$$

This is the desired result.   □

Combining (6.211) and Proposition 6.39, we have the following theorem.

Theorem 6.44 *The spectrum-determined growth condition holds for system (6.194) in the state space H, i.e.,*

$$\omega_0(A) = S(A)$$

and by Theorem 6.43,

$$\omega_0(A) \le \max\{\frac{-k\pi^2}{\gamma^2 + 1}, -\frac{\gamma^2}{2}\frac{k\pi^2}{k^2\pi^2 + \gamma^2 + 1}\}.$$

Consequently, the system (6.194) is exponentially stable.

6.7   Renardy's counter-example on spectrum-determined growth condition

In the previous sections of this chapter, we stated several typical physical systems in which the spectrum-determined growth condition hold. Unfortunately, it is not true that all physical systems are good enough to possess this property. We have already seen one counter-example in Chapter 3, but the example to be given below is more "physical", and the result is more surprising in the sense that even a lower order derivative perturbation to a two-dimensional wave equation can destroy the spectrum-determined growth condition. This counter-example is taken from Renardy [137], but we provide a detailed proof.

Consider the system

$$\begin{cases} u_{tt} = u_{xx} + u_{yy} + e^{iy}u_x, & -\infty < x, y < \infty, \\ u(x + 2\pi, y) = u(x, y), \ u(x, y + 2\pi) = u(x, y). \end{cases} \quad (6.212)$$

We proceed as usual by rewriting it as

$$\frac{d}{dt}(u, u_t) = A(u, u_t) \quad (6.213)$$

on the Hilbert space $H = H^1 \times L^2 = \{(f, g) \in H^1(\mathbf{C}^2) \times L^2(\mathbf{C}^2) | f, g \text{ are } 2\pi - \text{periodic in both directions}\}$ with the inner product

$$\langle (u_1, v_1), (u_2, v_2) \rangle = \int_{-\pi}^{\pi} \int_{-\pi}^{\pi} [u_1 \overline{u}_2 + u_{1x} \overline{u}_{2x} + u_{1y} \overline{u}_{2y} + v_1 \overline{v}_2] dx dy.$$

In (6.213), the operator A is defined by

$$\begin{cases} A(u, v) = (v, u_{xx} + u_{yy} + e^{iy}u_x), \\ D(A) = \{(u, v) \in H \mid (u, v) \in H^2(\mathbf{C}^2) \times H^1(\mathbf{C}^2)\}. \end{cases} \quad (6.214)$$

Lemma 6.45 *The operator A defined by (6.214) generates a C_0-semigroup on H and A has compact resolvent.*

Proof. First, since for any $(u, v) \in D(A)$,

$$\text{Re}\langle A(u, v), (u, v) \rangle$$

$$= \text{Re} \int_{-\pi}^{\pi} \int_{-\pi}^{\pi} [v\overline{u} + v_x \overline{u}_x + v_y \overline{u}_y + (u_{xx} + u_{yy} + e^{iy}u_x)\overline{v}] dx dy$$

$$= \text{Re} \int_{-\pi}^{\pi} \int_{-\pi}^{\pi} [v\overline{u} + e^{iy}u_x \overline{v}] dx dy$$

$$\leq \frac{1}{2} \int_{-\pi}^{\pi} \int_{-\pi}^{\pi} [2|v|^2 + |u|^2 + |u_x|^2] dx dy$$

$$\leq \|(u, v)\|^2$$

we see that $A - I$ is dissipative. We further show that $(A - I)^{-1}$ exists and is compact. To this end, take any $(f, g) \in H$ and solve $(A - I)(u, v) = (f, g)$ which is rewritten as

$$\begin{cases} v = u + f, \\ u_{xx} + u_{yy} + e^{iy}u_x - u = f + g. \end{cases} \quad (6.215)$$

Let $f(x, y) + g(x, y) = \sum_{n=-\infty}^{\infty} w_n(y)e^{inx}, u(x, y) = \sum_{n=-\infty}^{\infty} u_n(y)e^{inx}$. Then the second equation above leads to

$$(ine^{iy} - n^2 - 1)u_n(y) + u_n''(y) = w_n(y), \quad n = 0, \pm 1, \dots.$$

Let $u_n(y) = \sum_{m=-\infty}^{\infty} a_{nm} e^{imy}, w_n(y) = \sum_{m=-\infty}^{\infty} b_{nm} e^{imy}$. Then the above equation reduces to

$$(-n^2 - m^2 - 1)a_{nm} + ina_{n(m-1)} = b_{nm}, \quad m = 0, \pm 1, \ldots. \tag{6.216}$$

Therefore, for $m \geq 1$, we have

$$
\begin{aligned}
a_{nm} &= \frac{in}{n^2 + m^2 + 1} a_{n(m-1)} - \frac{1}{n^2 + m^2 + 1} b_{nm} \\
&= \frac{(in)^m}{[n^2 + m^2 + 1][n^2 + (m-1)^2 + 1] \cdots [n^2 + 1^2 + 1]} a_{n0} \\
&\quad - \frac{(in)^{m-1}}{[n^2 + m^2 + 1][n^2 + (m-1)^2 + 1] \cdots [n^2 + 1^2 + 1]} b_{n1} - \cdots \\
&\quad - \frac{1}{n^2 + m^2 + 1} b_{nm}
\end{aligned}
\tag{6.217}
$$

and

$$
\begin{aligned}
a_{n(-m)} &= \frac{n^2 + (m-1)^2 + 1}{in} a_{n(-m+1)} + \frac{1}{in} b_{n(-m+1)} \\
&= \frac{[n^2 + (m-1)^2 + 1][n^2 + (m-2)^2 + 1] \cdots [n^2 + 1]}{(in)^m} a_{n0} \\
&\quad + \frac{[n^2 + (m-1)^2 + 1][n^2 + (m-2)^2 + 1] \cdots [n^2 + 1^2 + 1]}{(in)^m} b_{n0} \\
&\quad + \cdots \\
&\quad + \frac{1}{in} b_{n(-m+1)}.
\end{aligned}
\tag{6.218}
$$

It follows that

$$
\begin{aligned}
&\frac{(in)^m}{[n^2 + (m-1)^2 + 1][n^2 + (m-2)^2 + 1] \cdots [n^2 + 1]} a_{n(-m)} \\
&= a_{n0} + \frac{1}{n^2 + 1} b_{n0} + \frac{in}{[n^2 + 1][n^2 + 1^2 + 1]} b_{n(-1)} + \cdots \\
&\quad + \frac{(in)^{m-1}}{[n^2 + (m-1)^2 + 1][n^2 + (m-2)^2 + 1] \cdots [n^2 + 1]} b_{n(-m+1)}.
\end{aligned}
$$

Let $m \to \infty$. The left-hand side above tends to zero for any fixed n because the requirement of $u \in H^1$ implies that $a_{n(-m)} \to 0$ as $m \to \infty$. Hence,

$$a_{n0} = -\sum_{m=1}^{\infty} \frac{(in)^{m-1}}{[n^2 + (m-1)^2 + 1][n^2 + (m-2)^2 + 1] \cdots [n^2 + 1]} b_{n(-m+1)}. \tag{6.219}$$

Substituting this into (6.218) yields

$$a_{n(-m)} = \sum_{k=m+1}^{\infty} \frac{(in)^{k-m-1}}{[n^2 + (k-1)^2 + 1] \cdots [n^2 + m^2 + 1]} b_{n(-k+1)}. \quad (6.220)$$

Thus, a_{nm} is uniquely determined from (6.217), (6.219) and (6.220). We now show that

$$\sum_{n=-\infty}^{\infty} \sum_{m=-\infty}^{\infty} n^2 |a_{nm}|^2 < \infty.$$

Indeed,

$$\sum_{n=-\infty}^{\infty} \sum_{m=1}^{\infty} n^2 |a_{n(-m)}|^2$$

$$\leq \sum_{n=-\infty}^{\infty} \sum_{m=1}^{\infty} n^2 \left(\sum_{k=m+1}^{\infty} \left[\frac{n^{k-m-1}}{[n^2 + (k-1)^2 + 1] \cdots [n^2 + m^2 + 1]} \right]^2 \right)$$

$$\times \left(\sum_{k=m+1}^{\infty} |b_{n(-k+1)}|^2 \right)$$

$$= \sum_{n=-\infty}^{\infty} \sum_{k=1}^{\infty} |b_{n(-k)}|^2 \sum_{m=1}^{k} \sum_{i=m+1}^{\infty} \left[\frac{n^{i-m}}{[n^2 + (i-1)^2 + 1] \cdots [n^2 + m^2 + 1]} \right]^2$$

$$\leq \sum_{n=-\infty}^{\infty} \sum_{k=1}^{\infty} |b_{n(-k)}|^2 \Delta_n$$

where

$$\Delta_n = \sum_{m=1}^{\infty} \sum_{i=m+1}^{\infty} \left[\frac{n^{i-m}}{[n^2 + (i-1)^2 + 1] \cdots [n^2 + m^2 + 1]} \right]^2$$

$$= \sum_{i=2}^{\infty} \sum_{m=1}^{i-1} \left[\frac{n^{i-m}}{[n^2 + (i-1)^2 + 1] \cdots [n^2 + m^2 + 1]} \right]^2$$

$$\leq 1 + \sum_{i=3}^{\infty} \left[\frac{1}{(i-1)^2} + \frac{1}{(i-1)^3} \right] < \infty.$$

Hence, $\sum_{n=-\infty}^{\infty} \sum_{m=1}^{\infty} n^2 |a_{n(-m)}|^2 < \infty$. Similarly, by (6.217) and (6.219),

$$\sum_{n=-\infty}^{\infty} n^2 |a_{n0}|^2 < \infty, \quad \sum_{n=-\infty}^{\infty} \sum_{m=1}^{\infty} n^2 |a_{nm}|^2 < \infty$$

which yields $\sum_{n=-\infty}^{\infty} \sum_{m=-\infty}^{\infty} n^2 |a_{nm}|^2 < \infty$. Using the relation $(n^2 + m^2 + 1)a_{nm} = -b_{nm} + ina_{n(m-1)}$, it is not difficult to verify that

$$\sum_{n=-\infty}^{\infty} \sum_{m=-\infty}^{\infty} (n^2 + m^2 + 1)^2 |a_{nm}|^2 < \infty$$

which shows that $u(x,y) = \sum_{n=-\infty}^{\infty} \sum_{m=-\infty}^{\infty} a_{nm} e^{inx} e^{imy} \in H^2(\mathbf{C^2})$ and so $v \in H^1(\mathbf{C^2})$ in view of (6.215). We have actually proved that $(A-I)^{-1}$ exists and is compact by the Sobolev imbedding theorem because $(A-I)^{-1}H \subset H^2(\mathbf{C^2}) \times H^1(\mathbf{C^2})$. These results, together with the dissipativity of $A - I$, guarantee the C_0-semigroup generation of A on H by the Lümer-Phillips Theorem.   □

Now, we state the main result of this section.

Theorem 6.46 *Let A be defined as in (6.214). Then $S(A) = 0, \omega_0(A) \geq 1/2$.*

Proof. Since the resolvent of A is compact, the spectrum of A consists only of its eigenvalues. First we prove that $S(A) = 0$ by showing that all eigenvalues of A locate on the imaginary axis. Let λ be an eigenvalue of A and $u(x,y)$ be the eigenvector associated with λ. Because $u(x,y)$ is 2π-periodic, $u(x,y)$ can be expanded as a Fourier series $u(x,y) = \sum_{n=-\infty}^{\infty} w_n(y)e^{inx}$ and each $w_n(y)e^{inx}$ is the eigenvector corresponding to λ by the orthogonality of e^{inx}. Therefore, we may set, without loss of generality, that $u(x,y) = w(y)e^{inx}$. The resulting spectral problem is

$$(\lambda^2 + n^2)w = w'' + ine^{iy}w, \tag{6.221}$$

where w is required to be 2π-periodic. The problem is trivial for $n = 0$ and we shall therefore assume $n \neq 0$. By Fourier series expansion

$$w(y) = \sum_{m=-\infty}^{\infty} w_m e^{imy},$$

equation (6.221) is transformed into

$$(\lambda^2 + n^2 + m^2)w_m = inw_{m-1}. \tag{6.222}$$

We can now start at some value $m = M$ such that $w_M \neq 0$ (not all w_m are identically zero), and then recursively compute $w_{M-1}, w_{M-2}, \cdots$ from (6.222). If $w_{M-k} \neq 0$ for all $k = 1, 2, \ldots$, then $|w_m| \to \infty$ as $m \to -\infty$, contradicting the hypothesis that w is square integrable. Hence, there is an $m \leq M$ such that $w_{m-1} = 0$ and $w_k \neq 0, m \leq k \leq M$. Applying this to

(6.222) yields $\lambda^2 + n^2 + m^2 = 0$, or $\lambda = \pm i\sqrt{n^2 + m^2}$, which shows that the spectrum is on the imaginary axis, that is, $S(A) = 0$.

Next, we demonstrate that $\omega_0(A)$ is at least $1/2$. If this is not true, then, by the Hille-Yosida theorem, the resolvent of A must be bounded on the line $\text{Re}\lambda = 1/2$. Consider now the sequence

$$\lambda_n = \sqrt{-n^2 + in}, \quad u_n(x, y) = e^{inx}\phi(y, \epsilon_n), \tag{6.223}$$

where $\epsilon_n > 0$ is such that $\epsilon_n \to 0$ and $\epsilon_n^2 n \to \infty$ as $n \to \infty$. We note that $\lambda_n = 1/2 + in + O(1)$ as $n \to \infty$, the real part of λ_n approaches $1/2$, and $|\lambda_n|$ is proportional to n as $n \to \infty$. The function $\phi(y, \epsilon_n)$ is constructed as follows. Let ϕ be a nonzero C^∞-function with compact support contained in $(-1, 1)$. Consider function $\tilde{\phi}(y, \epsilon_n) = \phi(y/\epsilon_n)$. Without loss of generality, we may assume $\epsilon_n < \pi$, so that the support of $\tilde{\phi}(y, \epsilon_n)$ is contained in $(-\pi, \pi)$. We can then take $\phi(y, \epsilon_n)$ to be the 2π-periodic continuation of $\tilde{\phi}(y, \epsilon_n)$. For such chosen $u_n(x, y)$, it is easy to verify that

$$\int_{-\pi}^{\pi} \int_{-\pi}^{\pi} [|u_n(x, y)|^2 + |u_{nx}(x, y)|^2 + |u_{ny}(x, y)|^2]dxdy$$

$$= 2\pi \int_{-\pi}^{\pi} [(1 + n^2)|\phi(y/\epsilon_n)|^2 + \epsilon_n^{-2}|\phi'(y/\epsilon_n)|^2]dy$$

$$= \int_{-1}^{1} \epsilon_n[(2 + n^2)|\phi(z)|^2 + \epsilon_n^{-2}|\phi'(z)|^2]dz,$$

which indicates that the H^1-norm of u_n is proportional to $n\sqrt{\epsilon_n}$ and L^2-norm of u_n is proportional to $\sqrt{\epsilon_n}$ as $n \to \infty$. On the other hand, the following equality holds

$$\lambda_n^2 u_n - \frac{\partial^2 u_n}{\partial x^2} - \frac{\partial^2 u_n}{\partial y^2} - e^{iy}\frac{\partial^2 u_n}{\partial x} = in(1 - e^{iy})u_n - e^{inx}\phi''(y, \epsilon_n).$$

$$\tag{6.224}$$

By using the fact that $|1 - e^z| \le |z|e^{|z|}$ which is true for any complex number z, we see that

$$n^2 \int_{-\pi}^{\pi} \int_{-\pi}^{\pi} |1 - e^{iy}|^2 |u_n(x, y)|^2 \, dxdy$$

$$= 2\pi n^2 \epsilon_n \int_{-1}^{1} |1 - e^{i\epsilon_n z}|^2 |\phi(z)|^2 dz$$

$$\le 2\pi n^2 \epsilon_n^3 e^{\epsilon_n} \int_{-1}^{1} |\phi(z)|^2 dz,$$

which shows that the L^2-norm of the first term on the right-hand side of (6.224) is proportional to $n\epsilon_n^{3/2}$. Similarly, it is shown that the L^2-norm of the second term on the right-hand side of (6.224) is proportional to $\epsilon_n^{-3/2}$. Let $v_n = \lambda_n u_n$. Then

$$(A - \lambda_n)(u_n, v_n) = (0, \ -in(1 - e^{iy})u_n + e^{inx}\phi''(y, \epsilon_n));$$

hence, $\|(A - \lambda_n)(u_n, v_n)\|_H$ is proportional to $n\epsilon_n^{3/2} + \epsilon_n^{-3/2}$. Since $\|(u_n, v_n)\|_H$ is proportional to $\sqrt{\epsilon_n}$, we have

$$\|(A - \lambda_n)(u_n, v_n)\|_H \ / \ \|(u_n, v_n)\|_H \to \infty, \quad \text{as } n \to \infty.$$

This means that the resolvent of A is not uniformly bounded on $\text{Re}\lambda = 1/2$, contradicting the assumption. Consequently, the growth rate of the semigroup generated by A is at least $1/2$, that is, $\omega_0(A) \geq 1/2$. $\qquad \square$

6.8 Notes and references

The study on the general linear hyperbolic system in Section 6.1 originates from Neves [124]. Section 6.2 is studied by Chen et al.[27]. Our interest here is to show that it can be recast into the form of the general hyperbolic system and thus the results obtained in the previous section can be applied to derive the exponential stability conditions [24]. The results in Section 6.3 are documented in Guo and Zhu [71]. The stabilization control problem for the vibration cable with a tip mass in Section 6.4 was first considered in Morgül et al. [118], but here we proved the exponential stability of the boundary controlled closed-loop system by verifying the spectrum-determined growth condition [69]. Theorems 6.28 and 6.29 appeared in Renardy [136] and the spectral analysis results in Section 6.5 are obtained by Guo and Yung [70] and [65]. The results in Section 6.6 have been given by Hansen [72] but our proof is new. The counter-example in Section 6.7 is taken from Renardy [137] to show that not every physically plausible system satisfies the spectrum-determined growth condition.

Bibliography

[1] N. U. Ahmed. *Semigroups Theory with Applications to Systems and Control*. Logman Scientific & Technical, 1991.

[2] B. D. O. Anderson and S. Vongpanitlerd. *Network Analysis and Synthesis, A Modern Systems Approach*. Prentice-Hall, Englewood Cliffs, 1973.

[3] W. Arendt. Resolvent positive operators. *Proc. London Math. Soc.*, 54(3):321–349, 1987.

[4] W. Arendt. Vector valued Laplace transforms and Cauchy problems. *Israel J. Math.*, 59(3):327–352, 1987.

[5] W. Arendt and C. J. K. Batty. Tauberian theorems and stability of one parameter semigroups. *Trans. Amer. Math. Soc.*, 306:837–852, 1988.

[6] A. V. Balakrishnan. *Applied Functional Analysis*. Springer-Verlag, New York, 1976.

[7] M. Balas. Direct velocity feedback control of large space structure. *J. Guidance and Control*, 2(3):252–253, 1979.

[8] H. T. Banks and K. Ito. A unified framework for approximation in inverse problems for distributed parameter systems. *Control-Theory and Advanced Technology*, 4:73–90, 1988.

[9] H. T. Banks and K. Kunisch. *Estimation Techniques for Distributed Parameter Systems*. Birkhäuser, Boston, 1989.

[10] H. T. Banks, R. S. Smith, and Y. Wang. *Smart Material Structures: Modeling, Estimation and Control*. John Wiley & Sons, 1996.

[11] H. T. Banks, Y. Wang, and D. J. Inman. Bending and shear damping in beams: frequency domain estimation techniques. *ASME J. Vibration and Acoustics*, 116(2):188–197, 1994.

[12] V. Barbu. *Nonlinear Semigroups and Differential Equations.* Noordhoff, 1976.

[13] J. F. Barman, F. M. Callier, and C. A. Desoer. l_2-stability and l_2-instability of linear time invariant distributed feedback systems perturbed by a small delay in the loop. *IEEE Trans. Autom. Control,* 18:479–484, 1973.

[14] C. J. K. Batty. Spectral conditions for stability of one parameter semigroups. *J. Diff. Eqns.,* 127:805–818, 1996.

[15] C. J. L. Batty and V. Q. Phong. Stability of individual elements under one-parameter semigroups. *Trans. Amer. Math. Soc.,* 322:805–818, 1990.

[16] C. D. Benchimol. Feedback stabilization in Hilbert spaces. *Appl. Math. Optim.,* 4:209–223, 1978.

[17] C. D. Benchimol. A note on weak stabilizability of contraction semigroups. *SIAM J. Control and Optim.,* 3:751–780, 1978.

[18] A. Bensoussan, G. D. Prato, M. C. Delfour, and S. K. Mitter. *Representation and Control of Infinite Dimensional Systems.* Birkhäuser, Boston, 1992.

[19] J. Bontsema and S. A. deVries. Robustness of flexible structures against small time delays. In *Proc. of 27th CDC,* pages 1647–1648, Austin, TX, 1988.

[20] W. J. Book, O. Maizza-Neto, and D. E. Whitney. Feedback control of two beam, two joint systems with distributed flexibility. *J. Dynamic Sys., Meas., and Control,* 97(4):424–431, 1975.

[21] C. I. Byrnes, A. Isidori, and J. C. Willems. Passivity, feedback equivalence, and global stabilization of minimum phase nonlinear systems. *IEEE Trans. Autom. Control,* 36:1228–1241, 1991.

[22] R. H. Jr. Cannon and E. Schmitz. Initial experiments on the end-point control of a flexible one-link robot. *Int. J. Robotics Res.,* 3(3):62–75, 1984.

[23] D. E. Carlson. Linear thermoelasticity. In C. Truesdell, editor, *Handbuck der Physik.* Springer-Verlag, 1972.

[24] W. L. Chan and B. Z. Guo. Pointwise stabilization for a chain of vibrating strings. *IMA J. Math. and Information,* 7:307–315, 1991.

[25] C. T. Chen. *Linear System Theory and Design.* Holt, Rinehart and Winston, New York, 1984.

[26] G. Chen. Energy decay estimates and exact boundary value controllability for the wave equation in a bounded domain. *J. Math. Pures. Appl.*, 58:249–273, 1979.

[27] G. Chen, M. Coleman, and H. H. West. Pointwise stabilization in the middle of the span for second order systems, nonuniform exponential decay of solutions. *SIAM J. Appl. Math.*, 47:751–780, 1987.

[28] G. Chen, M. C. Delfour, A. M. Krall, and G. Payre. Modeling, stabilization and control of serially connected beams. *SIAM J. Control and Optim.*, 25:526–546, 1987.

[29] G. Chen, S. A. Fulling, F. J. Narcowich, and S. Sun. Exponential decay of energy of evolution equations with locally distributed damping. *SIAM J. Appl. Math.*, 51:967–983, 1991.

[30] G. Chen, S. G. Krantz, D. W. Ma, C. E. Wayne, and H. H. West. The Euler-Bernoulli beam equation with boundary energy dissipation. In S. J. Lee, editor, *Operator Methods for Optimal Control Problems*. Marcel-Dekker, New York, 1987.

[31] G. Chen and D. L. Russel. A mathematical model for linear elastic systems with structural damping. *Quart. Appl. Math.*, 39:433–454, 1981-82.

[32] G. Chen and J. Zhou. *Vibration and Damping in Distributed Systems, Vol.I.* CRC Press, 1993.

[33] Y. Choquey-Bruhat, C. de Witt-Morette, and M. Dillard-Bleick. *Analysis, Manifolds and Physics*. Amsterdam, Noth-Holland, 1977.

[34] R. V. Churchill. *Operational Mathematics*. McGraw-Hill, 1972.

[35] Ph. Clément, H. J. A. M. Heijmans, S. A. Angenent, C. J. van Duijn, and B. de Pagter. *One-Parameter Semigroups*. North-Holland, 1987.

[36] F. Conrad and Ö. Morgül. On the stabilization of flexible beam with a tip mass. *SIAM J. Control and Optim.*, 1998. to appear.

[37] F. Conrad and M. Pierre. Stabilization of second order evolution equations by unbounded nonlinear feedback. Technical Report 14, Les prépublications de l'Institut Élie Cartan, Départment de Math., Université de Nancy 1, 1992.

[38] John B. Conway. *Functions of One Complex Variable*. Springer-Verlag, 1978.

[39] M. G. Crandall. A generalized domain for semigroup generators. *Proc. of Amer. Math. Soc.*, 37:434–439, 1973.

[40] M. G. Crandall and T. Liggett. A theorem and a counter example in the theory of semigroups of nonlinear transformations. *Trans. Amer. Math. Soc.*, 160:163–278, 1971.

[41] R. F. Curtain and H. J. Zwart. *An Introduction to Infinite Dimensional Linear Systems Theory.* Springer-Verlag, New York, 1995.

[42] C. M. Dafermos and M. Slemrod. Asymptotic behavior of nonlinear contraction semigroups. *J. Func. Anal.*, 13:97–106, 1973.

[43] R. Datko. Two questions concerning the boundary control of certain elastic systems. *J. Diff. Eqns.*, 92:27–44, 1991.

[44] R. Datko. Two examples of ill-posedness with respect to small time delays in stabilized elastic systems. *IEEE Trans. Autom. Control*, 38(1):163–166, 1993.

[45] R. Datko, J. Lagnese, and M. P. Polis. An example on the effect of time delays in boundary feedback stabilization of wave equations. *SIAM J. Control and Optim.*, 24:152–156, 1986.

[46] E. B. Davies. *One-parameter Semigroups.* Academic Press, London, 1980.

[47] W. A. Day. *Heat Conduction with Linear Thermoelasticity.* Springer-Verlag, 1985.

[48] K. Deleeuw and I. Glicksberg. Application of almost periodic compactifications. *Acta Math.*, 105:63–97, 1961.

[49] W. Desch and W. Schappacher. Spectral properties of finite dimensional perturbed linear semigroups. *J. Diff. Eqns.*, 59:80–102, 1985.

[50] C. A. Desoer and M. Vidyasagar. *Feedback Systems: Input-Output Properties.* Academic Press, New York, 1975.

[51] G. Doetsch. *Introduction to the Theory and Application of the Laplace Transforms.* Springer-Verlag, 1974.

[52] R. Dowson. *Spectral Theory of Linear Operators.* Academic Press, London, 1978.

[53] N. Dunford and J. T. Schwartz. *Linear Operators*, volume 3. Interscience, 1971.

[54] P. L. Duren. *Theory of H^p Spaces.* Academic Press, London, 1970.

[55] D. E. Evans. On the spectrum of a one-parameter strongly continuous representation. *Math. Scand.*, 39:80–82, 1976.

[56] S. R. Foguel. Powers of contraction in a Hilbert space. *Pacific J. Math.*, 13:551–562, 1963.

[57] H. Fujii, T. Ohtsuka, and S. Udou. Mission function control for a slew maneuver experiments. *J. Guidance, Control and Dynamics*, 14(5):986–992, 1991.

[58] T. Fukuda. Flexibility control of elastic robotic arm. *J. Robotic Sys.*, 2(1):73–88, 1985.

[59] L. Gearhart. Spectral theory for contraction semigroups on Hilbert space. *Trans. of Amer. Math. Soc.*, 236:385–394, 1978.

[60] J. S. Gibson. A note on stabilization of infinite dimensional linear oscillators by compact linear feedback. *SIAM J. Control and Optim.*, 18(3):311–316, 1980.

[61] E. A. Guilliemin. *Synthesis of Passive Networks*. Wiley, New York, 1957.

[62] B. Z. Guo. Asymptotic behavior of the spectrum of a direct strain feedback control system. *J. Australian Math. Soc. (Series B)*, 37:86–98, 1995.

[63] B. Z. Guo. A characterization of the exponential stability of a family of C_0–semigroups by infinitesimal generators. *Semigroup Forum*, 56:78–83, 1998.

[64] B. Z. Guo. On the exponential stability of C_0–semigroups on Banach spaces with compact perturbations. *Semigroup Forum*, 1998. to appear.

[65] B. Z. Guo and J. C. Chen. The real eigenvalues of a one-parameter linear thermoelastic system. *Appl. Math. Letter*, 1998. to appear.

[66] B. Z. Guo and Z. H. Luo. On the analytic solution of a controlled flexible arm system with structural damping. *Indian J. Pure and Appl. Math.*, 25(12):1223–1227, 1994.

[67] B. Z. Guo and Z. H. Luo. Initial-boundary value problem and exponential decay for a flexible beam vibration with gain adaptive direct stain feedback control. *Nonlinear Analysis, Theory, Methods and Appl.*, 27(3):353–365, 1996.

[68] B. Z. Guo and Z. H. Luo. Stability analysis of a hybrid system arising from feedback control of flexible robots. *Japan J. of Industrial and Appl. Math.*, 13(3):417–434, 1996.

[69] B. Z. Guo and C. Z. Xu. The spectrum-determined growth condition of the stabilization of a vibration cable with a tip mass. *submitted*, 1998.

[70] B. Z. Guo and S. P. Yung. Asymptotic behavior of the eigenfrequency of a one-dimensional linear thermoelastic system. *J. Math. Anal. Appl.*, 213:406–421, 1997.

[71] B. Z. Guo and W. D. Zhu. On the energy decay of two coupled strings through a joint damper. *J. Sound and Vibration*, 203:447–455, 1997.

[72] S. W. Hansen. Exponential energy decay in a linear thermoelastic rod. *J. Math. Anal. Appl.*, 167:429–442, 1992.

[73] G. H. Hardy and E. M. Wright. *An Introduction to the Theory of Numbers*. Oxford University Press, 5th edition, 1989.

[74] A. J. Helmicki, C. A. Jacobson, and C. N. Nett. Ill-posed distributed parameter systems: a control viewpoint. *IEEE Trans. Autom. Control*, 36(9):1053–1057, 1991.

[75] D. Henry. Linear autonomous neutral functional differential equations. *J. Diff. Eqns.*, 15:106–128, 1974.

[76] D. Hill and P. Moylan. The stability of nonlinear dissipative systems. *IEEE Trans. Autom. Control*, 21:708–711, 1976.

[77] E. Hille and R. S. Phillips. *Functional Analysis and Semigroups*. Amer. Math. Soc. Providence, R. I., 1957.

[78] F. L. Huang. Characteristic conditions for exponential stability of linear dynamical systems in Hilbert spaces. *Anal. of Diff. Eqns.*, 1:43–53, 1985.

[79] H. L. Huang. Strong asymptotic stability of linear dynamical systems in Banach spaces. *J. Diff. Eqns.*, 104:307–324, 1993.

[80] T. Kailath. *Linear Systems*. Prentice-Hall, Englewood Cliffs, 1980.

[81] R. E. Kalman. Lyapunov functions for the problem of Lur'e in automatic control. *Proc. of the National Academy of Sciences*, 49:201–205, 1963.

[82] H. Kanoh and H. G. Lee. Vibration control of one-link flexible arm. *Proc. 24th Conf. CDC*, 2:1172–1177, 1985.

[83] S. Kantorovitz. *Semigroups of Operators and Spectral Theory*. Pitman Research Notes in Math. Series 330. Longman, New York, 1995.

[84] T. Kato. *Perturbation Theory of Linear Operators*. Springer-Verlag, Berlin, 1966.

[85] J. U. Kim and Y. Renardy. Boundary control of the Timoshenko beam. *SIAM J. Control and Optim.*, 25:1417–1429, 1987.

[86] V. Komornik. *Exact Controllability and Stabilization – the Multiplier Method.* John Wiley & Sons, 1994.

[87] T. W. Körner. *Fourier Analysis.* Cambridge University Press, Cambridge, 1988.

[88] M. A. Krasnoselski, P. P. Zabreiko, E. L. Pustylik, and P. E. Sobolevskii. *Integral Operators in Space of Summable Functions.* Leyden, Netherlands, 1976.

[89] S. G. Krein. *Linear Differential Equations in Banach Space.* AMS Transl. 29. Providence, 1972.

[90] J. Lagnese. Decay of solutions of wave equations in a bounded domain with boundary dissipation. *J. Diff. Eqns.*, 50:163–182, 1983.

[91] B. J. Levin. *Distribution of Zeros of Entire Functions.* Translations of Math. Monographs, Vol.5, AMS, Providence, R.I., 1964.

[92] J. L. Lions. *Optimal Control of Systems Governed by Partial Differential Equations.* Springer-Verlag, New York, 1971.

[93] J. L. Lions. Exact controllability, stabilization and perturbations for distributed parameter systems. *SIAM Review*, 30:1–68, 1988.

[94] K. S. Liu, F. L. Huang, and G. Chen. Exponential stability analysis of a long chain of coupled vibrating strings with dissipative linkage. *SIAM J. Appl. Math.*, 49:1694–1707, 1989.

[95] Z. Y. Liu and S. M. Zheng. Exponential stability of semigroup associated with thermoelastic system. *Quart. of Appl. Math.*, 51:535–545, 1993.

[96] Z. Y. Liu and S. M. Zheng. Uniform exponential stability and approximation in control of a thermoelastic system. *SIAM J. Control and Optim.*, 32:1226–1246, 1994.

[97] H. Logemann, R. Rebarber, and G. Weiss. Conditions for robustness and nonrobustness of the stability of feedback systems with respect to small delays in the feedback loop. Technical Report 285, Institut für Dynamische Systeme, Universitat Bremen, June 1993.

[98] Z. H. Luo. On existence and uniqueness of solutions of a class of perturbed second order evolution equations. *Systems and Control Letters*, 17:401–408, 1991.

[99] Z. H. Luo. Direct strain feedback control of flexible robot arms: new theoretical and experimental results. *IEEE Trans. Autom. Control*, 38(11):1610–1622, 1993.

[100] Z. H. Luo and B. Z. Guo. Further theoretical results on direct strain feedback control of flexible robot arms. *IEEE Trans. Autom. Control*, 40(4):747–751, 1995.

[101] Z. H. Luo and B. Z. Guo. Shear force feedback control of a single link flexible robot with a revolute joint. *IEEE Trans. Autom. Control*, 42(1):53–65, 1997.

[102] Z. H. Luo, N. Kitamura, and B. Z. Guo. Shear force feedback control of flexible robot arms. *IEEE Trans. Robotics and Autom.*, 11(5):760–765, 1995.

[103] Z. H. Luo and Y. Sakawa. Dynamics and stabilization of a flexible orbiting spacecraft with rigid tip body. *Int. J. Systems Science*, 25(2):205–223, 1994.

[104] A. M. Lyapunov. The general problem of the stability of motion. *Int. J. Control*, 55(3), 1992.

[105] Y. I. Lyubich and V. Q. Phong. Asymptotic stability of linear differential equations in Banach spaces. *Sankia Math.*, 88:37–42, 1988.

[106] W. Malk. *Hilbert Spaces and Operator Theory*. Kluwer Academic Publishers, 1991.

[107] L. Meirovitch. *Analytical Methods in Vibrations*. MacMillan, New York, 1967.

[108] Ö. Morgül. Control and stabilization of a flexible beam attached to a rigid body. *Int. J. Control*, 51:11–33, 1990.

[109] Ö. Morgül. Boundary control of a Timoshenko beam attached to a rigid body: planar motion. *Int. J. Control*, 54(4):763–791, 1991.

[110] Ö. Morgül. Orientation and stabilization of a flexible beam attached to a rigid body: planar motion. *IEEE Trans. Autom. Control*, 36(8):953–963, 1991.

[111] Ö. Morgül. Dynamic boundary control of a Euler-Bernoulli beam. *IEEE Trans. Autom. Control*, 37(5):639–642, 1992.

[112] Ö. Morgül. Dynamic boundary control of a Timoshenko beam. *Automatica*, 28(6):1255–1260, 1992.

[113] Ö. Morgül. On the stabilization of the wave equation. In R. F. Curtain, editor, *Analysis and optimization of systems: state and frequency domain approaches for infinite dimensional systems*, volume 185 of *Lect. Notes in Control and Info. Sciences*, pages 531–542. Springer-Verlag, 1993.

[114] Ö. Morgül. A dynamic control law for the wave equation. *Automatica*, Nov. 1994.

[115] Ö. Morgül. On the stabilization and stability robustness against time delays of some damped wave equations. *IEEE Trans. Autom. Control*, 40:1626–1630, 1995.

[116] Ö. Morgül. On the boundary control of single link flexible robot arms. *Proc. IFAC'96*, pages 127–133, 1996.

[117] Ö. Morgül. Stabilization and disturbance rejection for the wave equation. *IEEE Trans. Autom. Control*, 43:89–95, 1998.

[118] Ö. Morgül, B. P. Rao, and F. Conrad. On the stabilization of a cable with a tip mass. *IEEE Trans. Autom. Control*, 39:2140–2145, 1994.

[119] R. Nagel, editor. *One-parameter Semigroups of Positive Operators*, volume 1184 of *Lect. Notes in Math.* Springer-Verlag, 1986.

[120] R. Nagel. Spectral and asymptotic properties of strongly continuous semigroups. In G. R. Goldstein and J. A. Goldstein, editors, *Semigroups of Linear and Nonlinear Operators and Applications*. Curracao Conference, 1992.

[121] K. S. Narendra and A. M. Annaswamy. *Stable Adaptive Systems*. Prentice-Hall, Englewood Cliffs, 1989.

[122] J. B. Neto. *An Introduction to the Theory of Distributions*. Marcel Dekker, New York, 1973.

[123] F. Neubrander. Integrated semigroups and their applications to the abstract Cauchy problem. *Pacific J. Math.*, 135:111–155, 1988.

[124] A. F. Neves, H. D. S. Ribeiro, and O. Lopes. On the spectrum of evolution operators generated by hyperbolic systems. *J. Func. Anal.*, 67:320–344, 1986.

[125] R. W. Newcomb. *Linear Multiport Synthesis*. McGraw-Hill, New York, 1966.

[126] R. D. Nussbaum. The radius of the essential spectrum. *Duke Math. J.*, 38:473–478, 1970.

[127] R. D. Nussbaum. A folk theorem in the spectrum theory of C_0-semigroups. *Pacific J. Math.*, 113:433–449, 1984.

[128] A. Pazy. *Semigroups of Linear Operators and Applications to Partial Differential Equations.* Springer-Verlag, New York, 1983.

[129] V. Q. Phong and Y. I. Lyubich. On the spectral mapping theorem for one-parameter groups of operators. *J. Soviet Math.*, 61(2):2035–2037, 1992.

[130] H. R. Pitt. A theorem on absolutely convergent trigonometrical series. *J. Math. Phys.*, 16:191–195, 1937.

[131] G. Da Prato and E. Sinestrari. Differential operators with nondense domain. *Ann. Scuola Norm. Sup. Pisa*, XIV:285–344, 1987.

[132] A. J. Pritchard and J. Zabczyk. Stability and stabilization of infinite dimensional systems. *SIAM Review*, 23:25–52, 1981.

[133] J. Prüss. On the spectrum of C_0-semigroups. *Trans. of Amer. Math. Soc.*, 284:847–857, 1984.

[134] B. Rao. Uniform stabilization of a hybrid system of elasticity. *SIAM J. Control and Optim.*, 33:440–445, 1995.

[135] M. Reed and B. Simon. *Methods of Modern Mathematical Physics, Vol.IV: Analysis of Operators.* Academic Press, San Diego, 1980.

[136] M. Renardy. On the type of certain C_0-semigroups. *Comm. in PDE.*, 18:1299–1307, 1993.

[137] M. Renardy. On the linear stability of hyperbolic PDE's and viscoelastic flows. *Z. Angew. Math. Phys.*, 45:854–865, 1994.

[138] M. Renardy. Spectrally determined growth is generic. *Proc. Amer. Math. Soc.*, 124:2451–2453, 1996.

[139] H. L. Royden. *Real Analysis.* MacMillan, New York, 2nd edition, 1968.

[140] D. L. Russell. Decay rates for weakly damped systems in Hilbert space obtained with control-theoretical methods. *J. Diff. Eqns.*, 19:344–370, 1975.

[141] D. L. Russell. Controllability and stabilizability theory for linear partial differential equations: recent progress and open problems. *SIAM Review*, 20:639–739, 1978.

[142] Y. Sakawa. Feedback control of second order evolution equations with damping. *SIAM J. Control and Optim.*, 22(3):343–361, 1984.

[143] Y. Sakawa and Z. H. Luo. Modeling and control of coupled bending and torsional vibrations of flexible beams. *IEEE Trans. Autom. Control,* 34(9):970–977, 1989.

[144] Y. Sakawa, F. Matsuno, and S. Fukushima. Modeling and feedback control of a flexible arm. *J. Robotic Systems,* 2(4):453–472, 1985.

[145] S. Saperstone. *Semidynamical Systems in Infinite Dimensional Spaces.* Springer-Verlag, New York, 1981.

[146] R. E. Showalter. *Hilbert Space Methods for Partial Differential Equations.* Pitman Publishing Ltd., London, 1977.

[147] M. Slemrod. Stabilization of boundary control systems. *J. Diff. Eqns.,* 22:402–415, 1976.

[148] J. J. E. Slotine and W. Pi. *Applied Nonlinear Control.* Prentice-Hall, New Jersey, 1991.

[149] S. M. Smith. Stability and dichotomy of positive semigroups on l^p. *Proc. Amer. Math. Soc.,* 124:2433–2437, 1996.

[150] H. Tanabe. *Equations of Evolution.* Pitman, 1979.

[151] A. E. Taylor. Theorems on ascent, descent, nullity and defect of linear operators. *Math. Annalen,* 163:18–49, 1996.

[152] A. E. Taylor and D. C. Lay. *Introduction to Functional Analysis.* John Wiley & Sons, 1980.

[153] S. Timoshenko. *Vibration Problems in Engineering.* Princeton, Van Nostrand, 3rd edition, 1955.

[154] E. Torricelli. *Opera Geometrica.* Florence:Masse, 1664.

[155] R. Triggiani. Lack of uniform stabilization for noncontractive semigroups under compact perturbation. *Proc. Amer. Math. Soc.,* 105:375–383, 1989.

[156] M. Vidyasagar. *Nonlinear Systems Analysis.* Prentice-Hall, 2nd edition, 1993.

[157] J. A. Walker. *Dynamic systems and Evolution Equations.* Plenum Press, New York, 1980.

[158] D. Wang and M. Vidyasagar. Control of a flexible beam for optimum step response. In *Proc. IEEE Conf. Robotics and Autom.,* pages 1567–1572, 1987.

[159] G. F. Webb. Compactness of bounded trajectories of dynamical systems in infinite dimensional spaces. *Proc. of the Royal Soc. of Edinburg*, 84A:19–33, 1979.

[160] L. Weis. Inversion of the vector-valued Laplace transform in $L^p(X)$ spaces. In G. Dore et al., editor, *Differential Equations in Banach Spaces*, volume 148 of *Lect. Notes in Pure and Appl. Math.*, pages 235–253. Marcel Dekker, 1993.

[161] L. Weis. The stability of positive semigroups on L^p spaces. *Proc. Amer. Math. Soc.*, 123:3089–3095, 1995.

[162] G. Weiss. The resolvent growth assumption for semigroups on Hilbert spaces. *J. Math. Anal. Appl.*, 145:154–171, 1990.

[163] D. V. Widder. *An Introduction to Transform Theory*. Academic Press, London and New York, 1971.

[164] J. C. Willems. *The Analysis of Feedback Systems*. MIT Press, Cambridge, MA, 1971.

[165] J. C. Willems. Dissipative dynamical systems, part 1: general theory. *Arch. Rational Mechanics and Analysis*, 45:321–351, 1972.

[166] C. Z. Xu and J. Baillieul. Stabilizability and stabilization of a rotating body-beam system with torque control. *IEEE Trans. Autom. Control*, 38(12):1754–1766, 1993.

[167] V. A. Yakubovich. The solution of certain matrix inequalities in automatic control theory. *Doklady Akademii Nauk, SSSR*, 143:1304–1307, 1962.

[168] P. F. Yao. *Stability and square root of damping of distributed parameter systems*. PhD thesis, Institute of Systems Science, Academia Sinica, 1994.

[169] P. F. Yao. On the inversion of the Laplace transform of C_0–semigroups and its applications. *SIAM J. Math. Anal.*, 26(5):1331–1341, 1995.

[170] P. F. Yao and D. X. Feng. Structure for nonnegative square roots of unbounded nonnegative selfadjoint operators. *Quart. Appl. Math.*, LIV(3):457–473, 1996.

[171] K. Yosida. *Functional Analysis*. Springer-Verlag, New York, 1978.

[172] J. Y. Yu, B. Z. Guo, and G. T. Zhu. $L(0, r_m)$ asymptotic expansion and controllability of the population dynamics. *J. Sys. Sci. and Math. Sci.*, 7:97–104, 1987. in Chinese.

[173] J. Zabczyk. A note on C_0−semigroups. *Bull. Acad. Pol. de Sc. Serie Math.*, 13:897–898, 1975.

[174] J. Zabczyk. *Mathematical Control Theory: An Introduction.* Birkhäuser, 1993.

[175] Q. Zheng. Perturbations and approximations of integrated semigroups. *Acta Math. Sinica (New Series)*, 9:252–260, 1993.

Index